Chirotechnology

Chirotechnology

INDUSTRIAL SYNTHESIS OF OPTICALLY ACTIVE COMPOUNDS

Roger A. Sheldon

**Delft University of Technology
Delft, The Netherlands**

MARCEL DEKKER, INC.　　　NEW YORK · BASEL

FIRST INDIAN REPRINT, 2012

Library of Congress Cataloging-in-Publication Data

Sheldon, Roger A.
   Chirotechnology : industrial synthesis of optically active
compounds / Roger A. Sheldon.
      p.   cm.
   Includes bibliographical references and index.
   ISBN 0-8247-9143-6
   1. Optical isomers--Biotechnology. 2. Chiral drugs-
-Biotechnology. I. Title.
TP248.65.065S44 1993
660'.63--dc20

93-18852
CIP

The publisher offers discounts on this book when ordered in bulk quantities. For more information, write to Special Sales/Professional Marketing at the address below.

MARCEL DEKKER, INC.
270 Madison Avenue, New York, New York 10016

Printed and bound in India by Bhavish Graphics.

FOR SALE IN SOUTH ASIA ONLY

To
Jetty,
Annemarie
and Frank

Foreword

Only a few people are able to present the topic of chirotechnology from both its academic and industrial sides. The academic side cannot be neglected because of the fundamental nature of chirality and its longstanding influence on the thinking of chemists. The other side, the industrial approach, has come into its own during the past ten years as purity and safety regulations have swamped the chemical industry. Roger Sheldon, formerly a Shell chemist and Andeno-DSM research director and presently Professor of Organic Chemistry at the University of Delft, is one of the few people who can handle this multifaceted task. He has done so in this excellent book, whose time has come.

The reader will come away with a new understanding not only of the importance of chirality as a dominating factor in the preparation of new drugs, but also of new reagents, conditions, catalysts, mechanisms, and syntheses. These are presented in a manner that allows even the inexperienced student to separate the useful from the merely exotic. The masters of our art, the role models in organic synthesis, are often those who have perfected the art of "total synthesis" of complex natural products containing upward of a dozen stereocenters. It is not surprising that many students complete their studies in organic chemistry with the notion that fame and fortune lie at the end of the "total synthesis" rainbow. Little do they realize that the successful preparation on an industrial scale of a totally synthetic (enantiomerically pure) new drug may require six months of laboratory scale synthesis and three years of process development work. Modesty demands us to admit that synthetic transformations that were perfected and scaled up at the end of the last century—such as classical resolutions, oxidations and reductions, esterifications and hydrolyses, and substitutions and eliminations—have lost none of their importance today, despite novel solvents, catalysts, crown ethers, and silicon-protecting groups.

The value of the contribution of Professor Sheldon's book to the literature of synthetic organic chemistry lies to a large extent in this aspect of the art. Professor Sheldon has been there and has seen what it takes to make enantiomerically pure compounds on a ton scale. He introduces the reader to the reality of organic preparations—of an important class of compounds—not as an idle curiosity published as a prestigious communication to the editor, but as a procedure of value to the plant manager.

In addition to the obvious value of this monograph to organic chemists everywhere, the educational value of this book should not be underestimated. Written in understandable prose and in an entertaining style, this book will prove of great value for advanced undergraduate and graduate students.

On a more personal note, I wish to thank Professor Sheldon for giving me the opportunity to speak my piece on the value of this book. I have known the author as research director and colleague for more than a dozen years and I have valued his friendship. The author's thorough knowledge of all phases of organic chemistry, his enthusiastic leadership, and his productive career have crystallized in this valuable monograph. I wish him all of the success he has surely earned.

Hans Wynberg
Emeritus Professor of Organic Chemistry
Groningen, The Netherlands

Preface

Optically active compounds are ubiquitous in our everyday lives. They are the active constituents of many medicines, vitamins, flavors and fragrances, and the herbicides and pesticides used in crop protection. Indeed, the essential components of life itself—proteins, carbohydrates, and DNA—are constructed from optically active building blocks.

The pioneering studies of Pasteur on optical activity date back to the middle of the last century. Since that time there has been a continuing interest in developing efficient methods for the synthesis of optically pure compounds. In more recent times the widespread recognition of the importance of using bioactive substances in optically pure form has provided an extra stimulus. More and more drugs, pesticides, etc., are now being sold as pure optical isomers. This has created a steadily increasing demand for efficient methods for their industrial production, on scales ranging from hundreds of kilos to thousands of tons per annum.

The major aim of this book is to provide an easy-to-read overview of the various methods that are available for the synthesis of optically pure compounds. Emphasis is placed on practical utility and relevance to industrial synthesis. The relative merits of the various approaches are discussed and compared using commercially relevant examples. Particular importance is attached to discussing the factors that play a role in determining the choice of a certain route for the industrial synthesis of the molecule in question. The book should be an invaluable aid to research chemists, in industry or academe, in planning their syntheses of optically pure products. It should also be eminently suitable for university courses on this important aspect of modern organic synthesis.

Although I was well acquainted with the subject at the outset, I was nevertheless surprised by the gargantuan task that the fitting together of this enormous subject proved to be. It embraces technologies (e.g., fermentation, crystallization, and catalysis by metal complexes) that are not usually dealt with in a single text. Books could be (and indeed have been) written on the separate topics. Rather than striving to be comprehensive I have attempted to cover the salient features of the various methodologies. Inevitably, somebody will search in vain for their favorite product or process. Indeed, I welcome suggestions from readers with regard to additions as well as errors and oversights in this book. Hopefully, the result of my endeavors is an unique monograph that presents a balanced view of all of the industrially relevant methods for the synthesis of optically active compounds.

The first chapter presents a general introduction to optical isomerism and the underlying aspects of stereochemistry that are essential for an understanding of the subject. Chapter 2 deals with the importance of chirality (optical isomerism) in conjunction with biological activity. Chapter 3 consists of a general introduction to the different synthetic methodologies that are covered in more detail in subsequent chapters. Fermentation, being the oldest production method, is treated first in Chapter 4. This is followed by the chirality pool (Chapter 5), crystallization techniques (Chapter 6), processes using free enzymes (Chapter 7), and catalytic asymmetric synthesis (Chapter 8). Finally, Chapter 9 compares the various approaches by considering different routes to commercially relevant products, for example, optically active beta-blockers, α-arylpropionic acids, antiinflammatory drugs, angiotensin-converting enzyme inhibitors, beta-lactam antibiotics, l-menthol, and α-phenoxypropionic acid herbicides.

An extensive bibliography of additional material is included at the end of each chapter. The index has been thoroughly cross-referenced and the reader is encouraged to seek the various topics under more than one entry. The principal literature has been covered through 1992.

I would like to express my sincere thanks to Joop Lie for taking my scrawled writings and sketches and transforming them into a beautiful manuscript. Also to Hans Wynberg for reading the manuscript and for helpful suggestions. And last but not least to my wife Jetty and children, Annemarie and Frank, without whose constant support and encouragement this book would not have been possible.

Roger A. Sheldon

Contents

Abbreviations

$[\alpha]$	specific optical rotation of an enantiomeric mixture
$[\alpha]_0$	specific optical rotation of a pure enantiomer, (i.e., its absolute rotation)
$[\alpha]_D^2$	specific optical rotation at 20°C and sodium D line (589 nm)
$[\alpha]_\lambda^t$	specific optical rotation at t°C and λ nm
α-AAD	α-aminoadipic acid
Ac	acetyl
7-ACA	7-aminocephalosporanic acid
ACE	angiotensin converting enzyme
ACL	α-aminocaprolactam
AcOH	acetic acid
7-ADCA	7-aminodesacetoxycefalosporanic acid
AEC	(S)-β-aminoethyl-L-cysteine
$[AgBF_4]$	catalytic amount of $AgBF_4$
AGP	α_1-acid glycoprotein
Am	n-amyl
3-AMA	3-aminomonobactamic acid
Amino acids	IUPAC abbreviations employed
AMP	adenosine monophosphate
AMS	anthraquinone-2-sulfonate

6-APA	6-aminopenicillanic acid
ATC	2-amino-2-thiazoline-4-carboxylic acid
ATP	adenosine triphosphate
BINAP or Binap	2,2′-bis(diphenylphosphino)-1,1′-binaphthalene
BOC = Boc	*tert*-butyloxycarbonyl; see also *t*-BOC
BSA	bovine serum albumin
Bu	*n*-butyl
BuLi	*n*-butyllithium
c	concentration in conjunction with $[\alpha]_D^{20}$
CAMP (ligand)	methylpropylcyclohexylphosphine
CAS	camphorsulfonic acid
CCL	*Candida cylindrica* lipase
CHT	chymotripsin
CMPA	chiral mobile phase additive in conjunction with HPLC
CoA	coenzyme A
COD	1,5-cyclooctadiene
CPO	chloroperoxidase
CSP	chiral stationary phase in conjunction with HPLC
d = (+)	dextrorotatory denotation
D	configurational symbol related to D-(+)-glyceraldehyde (Fischer convention)
de	diastereomeric excess
DEAE	diethylamino ethylcellulose
DHAP	dihydroxyacetone phosphate
DIOP	1,4-bis(diphenylphosphino)-1,4-dideoxy-2,3-O-isopropylidene-D-threitol
DIPAMP	bidentate analogue of PAMP
DIPT	di-isopropyl tartrate
DMF	dimethylformamide
DNA	deoxyribonucleic acid
E, (Z)	Entgegen; designation for *trans*-isomers
E	enantiomeric ratio
ee	enantiomeric excess; defined as $(R-S)/(R+S) \times 100\%$
EECE	electric eel acetylcholine esterase
EI	eudismic index (in Pfeiffer's rule)
ER	eudismic ratio (in Pfeiffer's rule)
Et	ethyl
Et_3N	triethylamine
FAD	flavine adenine dinucleotide
FADH	flavine adenine dinucleotide reduced

FDA	food drug administration
FDH	formate dehydrogenase
GLC	gas-liquid chromatography
HLADH	horse liver alcohol dehydrognase
HMPA	hexamethylphosphoramide
HPG	hydroxyphenylglycine
HPLC	high-performance liquid chromatography
IFO	Institute for Fermentation in Osaka
IPA	isopropanol
IPG	isopropylideneglycerol
^{i}Pr or i-Pr	*iso*-propyl
IR	infrared
$l = (-)$	*levo*rotatory denotation
L	configurational symbol related to L-(−)-glyceraldehyde (Fischer convention)
L^*	refers to a chiral ligand
$LiAlH_4$	lithium aluminum hydride
LSR	lanthanide shift reagent
Mal	maleyl
MCPA	4-chloro-2-methylphenoxyacetic acid
Me	methyl
Me_3N	trimethylamine
MEEC	membrane enclosed enzymatic catalysis
MS	mass spectrometry
NAD	β-nicotineamide adenine dinucleotide
NADH	β-nicotineamide adcnine dinucleotide reduced
NADP	β-nicotineamide adenine dinucleotide phosphate
NADPH	β-nicotineamide adenine dinucleotide phosphate reduced
NBA	*N*-bromoacetamide
NBS	*N*-bromosuccinimide
^{n}Bu or n-Bu	*n*-butyl
NMO	*N*-methylmorpholine *N*-oxide
NMR	nuclear magnetic resonance
NSAID	nonsteroidal antiinflammatory drug
[O] or "O"	any oxidant
PAC	phenylacetylcarbinol
PAL	phenyl ammonia lyase
PAMP (ligand)	methylanisylphenylphosphine
PEG	polyethylene glycol

PEG-NADH	polyethylene glycol attached to NADH
Pen-G	penicillin G = benzylpenicillin
Pen-V	penicilin V = phenoxymethylpenicillin
PES	phenylethanesulfonic acid
PFL	*Pseudomonas fluorescens* lipase
PG	phenylglycine
Ph	phenyl
Phβ-glup	chiral diphosphonite ligand derived from glucose
Ph_3P	triphenylphosphine
PLE	pig liver esterase
PN	productivity number of an enzyme
PNNP (ligand)	chiral diphosphonite ligand nitrogen based
PPI	inorganic phosphate
PPL	porcine pancreas lipase
PPMP	methylpropylphenylphosphine
Pr	*n*-propyl
pro-R	See *Re*
pro-S	See *Si*
PTC	phenylisothiocyanate
Pyr	pyridine
R, (*S*)	*Rectus* configurational symbol (Cahn-Ingold-Prelog convention)
RAMA	rabit muscle aldolase
r-DNA	recombinant-DNA
Re = *pro-R*	*Rectus* face specification; clockwise orientation of the substituents according to the Cahn-Ingold-Prelog priority rule
r.t.	room temperature
S	Fogassy parameter
S, (*R*)	*Sinister* configurational symbol (Cahn-Ingold-Prelog convention)
SDM	site-directed mutagenesis
SHMT	serine hydroxymethyl transferase
Si = *pro-S*	*Sinister* face specification; clockwise orientation of the substituents according to the Cahn-Ingold-Prelog priority rule
SIGNS	Stereochemically Informative Generic Name System
TAL	L-threonine acetaldehyde ammonia lyase
TBHP	*tert*-butyl hydroperoxide
t-BOC = Boc	*tert*-butyloxycarbonyl; see also BOC
^{t}Bu or *t*-Bu	*tert*-butyl
TFA	triflluoroacetic acid
$Ti(O\text{-}iPr)_4$	titanium(IV) isopropoxide

Tos	*p*-tolylsulfonyl
TLC	thin-layer chromatography
UV	ultraviolet
YAD	yeast alcohol dehydrogenase
Z, (**E**)	**Z**usammen; designation for *cis*-isomers
Z	benzyloxycarbonyl

Chirotechnology

1

Introduction to Optical Isomerism

The universe is dissymmetrical; for if the whole of the bodies which compose the solar system were placed before a glass moving with their individual movements, the image in the glass could not be superposed on reality . . . . Life is dominated by dissymmetrical actions. I can foresee that all living species are primordially, in their structure, in their external forms, functions of cosmic dissymmetry.

Louis Pasteur

I. EARLY HISTORY OF OPTICAL ISOMERISM

The history of our subject can be traced back to 1815, the year in which the French physicist Jean-Baptiste Biot [1] discovered the phenomenon of optical activity: the ability of a substance to rotate the plane of polarization of light. Subsequently, the legendary Louis Pasteur, a student of Biot, made a series of observations [2] that led him to a proposal of monumental significance: that the observed optical activity of certain organic substances is a consequence of their molecular asymmetry that produces nonsuperimposable mirror-image structures.

Louis Pasteur

A.   Pasteur and Molecular Asymmetry

Pasteur was intrigued by the crystal shapes and optical properties of two substances isolated from the tartar deposited in the barrels of maturing wines. A solution of the major component, *(+)*-tartaric acid, rotated the plane of polarized light clockwise, that is, in a positive, right-handed direction (hence the designation *(+)*). The minor component, paratartaric or racemic acid (Latin: *racemus*, a bunch of grapes), although chemically identical exhibited no optical activity. Pasteur observed that the sodium-ammonium double salt of racemic acid crystallized in two different forms and he succeeded in separating these crystals manually in a seminal experiment in 1848 [2]. The two sets of crystals differed only in the relative positions of their hemihedral facets (Figure 1-1), which bore a mirror-image relationship to one another. One of the sets proved to be truly isomorphous with crystals of *(+)*-sodium ammonium tartrate and exhibited the same positive (clockwise) optical rotation.

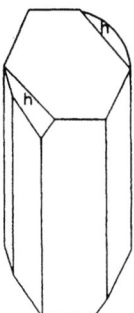 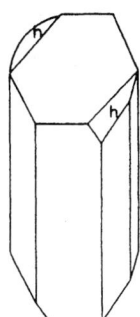

Figure 1-1. Hemihedral crystals of sodium ammonium tartrate showing hemihedral facets (h). *levo*-Tartrate is on the left and *dextro*-tartrate on the right.

The other set had a nonsuperimposable mirror-image crystal form and, in solution, exhibited a specific rotation of the same magnitude but negative sign (i.e., counter-clockwise). Consequently, Pasteur inferred that optically inactive racemic acid comprises equal amounts of *(+)*- and *(−)*-tartaric acid. Thereafter, the term racemate was used to denote an optically inactive equimolar mixture of optical isomers (enantiomers).

At this juncture, it is important to emphasize the fundamental difference between the observed optical activity of the tartaric acid isomers and that of, for example, hemihedral quartz crystals. The optical activity of the latter is only observed in the solid state and therefore it is related to **macroscopic** structure. On the other hand, the optical activity of tartaric acid is not only observed in the solid state but also persists when the crystals are dissolved. This led Pasteur to the conclusion that in this case it is a property of the constituent molecules and he coined the term 'dissymétrie' to describe this phenomenon. The term was translated as asymmetry and more recently has been replaced by chirality (Greek: *cheir*, hand), from the familiar analogy of the mirror-image relationship of the left and right hand.

In hindsight, it is worth noting that Pasteur was extremely fortunate in achieving this first resolution of a racemate. The tartrate salt that he used is one of the few racemates that undergo a spontaneous resolution into enantiomeric (hemi-hedral) crystals thereby allowing manual separation; most enantiomers do not form enantiomeric crystals and moreover, this separation only occurs below 27°C.

Pasteur went on to postulate that chiral molecules are a result of ubiquitous dissymmetric forces extant in the natural world. He was a fervent believer in the vital force (*vis vitalis*) theory which held that all substances derived from living matter possessed a mysterious, cosmic vital force and therefore could not be synthesized in the laboratory. Hence, it was logical for Pasteur to conclude that chiral forces are inherent in nature and form a demarcation criterion between the chemistry of the laboratory and that of living organisms. In his own words: "The essential products of life are asymmetric and possess such asymmetry that they are not superimposable on their images. . . . This establishes perhaps the only well-marked line of demarcation that can at present be drawn between the chemistry of dead matter and the chemistry of living matter."

B.   Resolution of Racemates

Although lesser mortals might have been tempted to rest on their laurels after such a monumental discovery, Pasteur went on to develop two important methods for the resolution of racemates: diastereomeric salt crystallization and biocatalytic kinetic resolution. Today, more than a hundred years later, these are still two of the most important techniques for industrial scale resolutions (see chapter 3).

The physical properties of enantiomers are identical in an achiral environment. However, reaction of the mixture of enantiomers with an optically active molecule (resolving agent) results in the formation of a pair of diastereomers with different physical properties (Reaction 1-1). This mixture of diastereomers can often be separated by crystallization.

$$(dl)\text{-A} \; + \; (l)\text{-B} \; \longrightarrow \; (d)\text{-A} \, (l)\text{-B} \; + \; (l)\text{-A} \, (l)\text{-B}$$

racemate   resolving agent   pair of diastereoisomers

(1-1)

Pasteur succeeded in 1853 in separating the two enantiomers of racemic tartaric acid by crystallization of the diastereomeric salts formed with the optically active alkaloid cinchonidine [3]. The diastereomeric salt which crystallized out contained $(-)$-tartaric acid while the $(+)$-tartrate salt remained in solution.

The third method developed by Pasteur (in 1858) involved fermentation, another subject in which he made far-reaching contributions [4]. This method depends on the ability of many microorganisms, because of their chiral nature, to discriminate between enantiomers and to selectively metabolize one of them. Thus, fermentation of the ammonium salt of racemic tartaric acid, mediated by the mold *Penicillium glaucum*, yielded $(-)$-tartrate. Pasteur concluded that the molecular asymmetry peculiar to organic substances also plays an important role in physiological phenomena. By doing so he made the important link between chirality and biological activity, a subject we shall discuss in more detail later (see chapter 2). Finally, it should be noted that this method suffers from the inherent disadvantage that one of the enantiomers is completely destroyed during the resolution process.

C.   van 't Hoff, Le Bel, and the Tetrahedral Carbon Atom

In speculating on the kind of structural arrangements that could produce molecular asymmetry, Pasteur had suggested that the atoms of a right-handed compound might be "arranged in the form of a right-handed spiral, or situated at the corners

Jacobus Henricus van 't Hoff

of an irregular tetrahedron." This idea was not developed any further. The problem was subsequently taken up and solved independently in 1874 by van 't Hoff and Le Bel. Following on from the ideas of Kekulé on the tetravalency of carbon, van 't Hoff published a paper entitled "Proposal for the extension of the structural formulae now in use in chemistry into space, together with a related note on the relation between the optical active power and the chemical constitution of organic compounds" [5]. This epochal paper, written at 22 years of age, laid the foundations of stereochemistry. In it he proposed that the four valences of a carbon atom are directed towards the corners of a tetrahedron with the carbon atom at the center. On this basis he could explain the molecular asymmetry inferred by Pasteur. Thus,

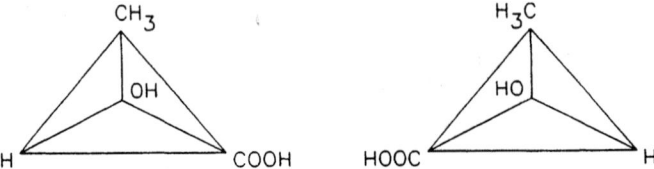

Figure 1-2. Tetrahedral model for lactic acid enantiomers as envisioned by van 't Hoff.

van 't Hoff introduced the concept of the asymmetric carbon atom as follows: "When the four affinities of the carbon atom are satisfied by four univalent groups differing among themselves, two and not more than two different tetrahedrons are obtained, one of which is the reflected image of the other, they cannot be superimposed; that is, we have here to deal with two structural formulas isomeric in space." van 't Hoff proposed that all carbon compounds that are optically active possess an asymmetric carbon atom. He illustrated his proposal with numerous examples, such as lactic acid (Figure 1-2), tartaric acid, malic acid, aspartic acid, camphor, and various sugars.

Two months later, Le Bel published his paper entitled "The relations that exist between the atomic formulas of organic compounds and the rotatory power of their solutions" [6]. He proceeded purely from symmetry arguments and spoke of the asymmetry of whole molecules rather than individual carbon atoms. Thus, Le Bel's proposal may be considered as the general theory of stereoisomerism and van 't Hoff's as the special theory (restricted to tetrahedral carbon).

II. PRINCIPLES OF STEREOCHEMISTRY

The science of organic chemistry is primarily concerned with the relationship between the molecular structure and properties of carbon compounds. That part of the science which deals with structure in three dimensions is called stereochemistry. One aspect of stereochemistry is stereoisomerism; isomers that differ from each other only in the way that the atoms are oriented in three dimensional space are called **stereoisomers**. Symmetry (or the lack of it) is one of the interesting features of geometric figures with two or more dimensions. For example, the alphabet contains both symmetrical and unsymmetrical two-dimensional letters. The latter have different appearances when they are reflected in a mirror. Thus, the mirror-images of the unsymmetrical letters **R** and **S** appear reversed (Figure 1-3). The property of nonidentical mirror-images also exists for objects with three dimensions. However, there is one important difference. The mirror-image of the letter **R** or **S** can be lifted out of the place of the paper, turned over, and placed exactly on top of the original letter. The mirror-image of a right hand, in contrast, is not superimposable on the real right hand because the two sides of the hand (palm and

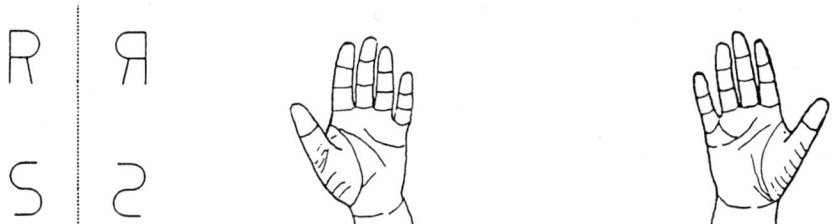

Figure 1-3. Mirror image relationships of asymmetrical figures in two and three dimensions.

back) are not identical. The mirror-image of a right hand is, however, identical with and superimposable on a left hand.

A. What Is Chirality?

Chirality is a geometrical attribute; an object that is not superimposable upon its mirror-image is said to be chiral while an achiral object is one that is super-imposable on its own mirror-image. Commonplace examples of chiral objects are an individual's right and left hands, clockwise- and counterclockwise-threaded screws, seashells, and dice. The two-dimensional letters are not chiral since the plane of the paper in which they lie is a plane of symmetry. An alternative definition of chirality is based on symmetry elements: a chiral object lacks reflectional symmetry.

The spatial arrangements that exist for macroscopic objects also exist at the molecular level. Thus, a molecule is chiral when it is not superimposable on its mirror-image (i.e., when it lacks reflectional symmetry). It is important to emphasize that chirality is a property of the molecule as a whole. It is incorrect, therefore, to say that a molecule contains a chiral center. It is preferable to refer to asymmetric centers or even better, stereogenic centers. Mislow and Siegel [7] have defined a stereogenic atom as: an atom bearing several groups of such nature that an interchange of two groups will produce a stereoisomer. Nevertheless, we shall sometimes use the more widely established term asymmetric center. Similarly, it is incorrect to describe a molecule containing more than one stereogenic center as multichiral; a molecule is either chiral or achiral.

B. Enantiomers

The most common type of chiral molecule contains a tetrahedral carbon atom attached to four different groups; the carbon atom is the stereogenic (asymmetric) center of the molecule. Such a molecule can exist as two different compounds that are stereoisomers and have identical chemical properties in an achiral (symmetrical) environment. The structures of these compounds are nonsuperimposable

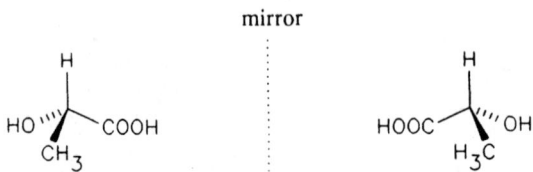

Figure 1-4. Enantiomers of lactic acid.

mirror-images of each other. Such stereoisomers are called **enantiomers** (Greek: *enantio*, opposite). Lactic acid, for example, is a chiral molecule and exists as a pair of enantiomers. The modern way of depicting such stereoisomers is shown in Figure 1-4.

It should be noted that a further condition that has to be fulfilled in order to observe optical activity is that the energy barrier for conversion of a chiral molecule into its enantiomer must be high enough (> 80 kJ/mol) to allow isolation or at least observation. The rapid inversion of trisubstituted nitrogen is a well-known phenomenon that precludes the isolation of stable optical isomers of molecules with stereogenic nitrogen atoms.

Although molecules containing an asymmetric tetravalent carbon atom comprise by far the largest class of chiral compounds, there are numerous examples of other chiral molecules. For example, the sedative methaqualone consists of a pair of enantiomers (Figure 1-5). In this molecule, asymmetry is the result of the hindered rotation around the axis between the nitrogen atom (N-1) and the phenyl group (C-2), which is due to steric hindrance between the methyl groups (M-3 and M-4). There are many other examples of molecules that are chiral as a consequence of axial asymmetry. Isomers that can be separated only because rotation about single bonds is prevented or greatly hindered are called **atropisomers**. Sulfur and phosphorus can also be stereogenic centers, such as in the organophosphate pesticide, cyanofenfos (Figure 1-6).

Figure 1-5. The enantiomers of methaqualone.

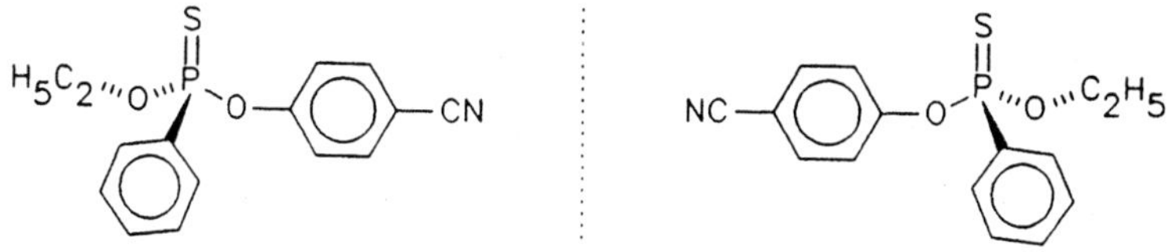

Figure 1-6. Enantiomers of cyanofenfos.

C. Optical Rotation

The specific distinguishing physical property of enantiomers is the rotation of plane-polarized light first observed by Biot. For this reason enantiomers were historically called optical isomers. The latter name is no longer preferred since it does not distinguish between enantiomers and diastereomers (see section II-E). Normal light consists of electric and magnetic fields oscillating in all directions perpendicular to each other and to the direction in which the light travels. In plane-polarized light the component electric and magnetic fields are contained within two perpendicular planes.

For symmetrical molecules every encounter of light with a molecule in a certain orientation is compensated by an encounter with the mirror-image molecular orientation, resulting in a net zero rotation of light. For a single enantiomer on the other hand, there is no canceling effect since no one molecule in solution can adopt the mirror-image of another. The result is a net rotation of the plane of the light. In other words, because enantiomers exist as mirror-images they rotate plane-polarized light to an equal but opposite extent (Figure 1-7). It is, moreover, obvious that a racemic mixture (i.e., a 1:1 mixture of the two enantiomers) does not rotate plane-polarized light.

The degree (angle) of rotation is easily measured using a polarimeter and has a specific value for each optically active substance. If the rotation is to the right

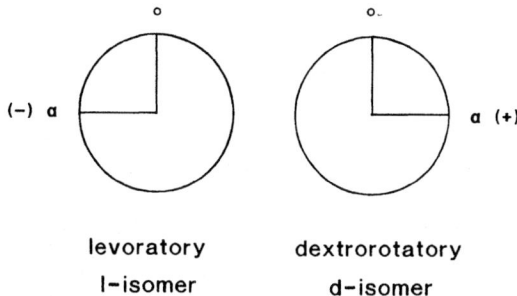

levoratory
l-isomer

dextrorotatory
d-isomer

Figure 1-7. The rotation of plane-polarized light by *l*- and *d*-enantiomers.

(clockwise) the substance is dextrorotatory (Latin: *dexter*, right); if the rotation is to the left (counterclockwise) it is levorotatory (Latin: *laevus*, left). The symbols *(+)* or *(d)* are used to denote dextrorotatory and *(−)* or *(l)* to denote levorotatory. For example, lactic acid extracted from muscle tissue is dextrorotatory and is referred to as *(d)-* or *(+)*-lactic acid.

The magnitude of the rotation of an enantiomer is reported as its specific rotation [α]. This rotation is dependent on the wavelength of light used, the length of the polarimeter tube, the temperature, the solvent, and the concentration. All of these variables should be given when a rotation is reported. The monochromatic light most often used is the light emitted by the sodium lamp at 589 nm (the sodium D line). Thus, the specific rotation of a substance at 20°C is given by:

$$[\alpha]_D^{20} = \frac{\text{observed rotation (degrees)}}{\text{length of sample tube (dm)} \times \text{conc. (g/ml)}} \tag{1-2}$$

The **optical purity** of a sample is the specific rotation of the particular sample divided by that of the pure enantiomer (the so-called absolute rotation) expressed as a percentage. The term optical purity has, however, been largely supplanted by **enantiomeric excess**, *ee* (see section V).

D.   Absolute Configuration and Its Notation

The earliest method of distinguishing between enantiomers was the sign of optical rotation, that is, the *d-* and *l-* or *(+)-* and *(−)*-forms, respectively. However, this does not say anything about the actual spatial arrangement of the substituents around the stereogenic center, which is the definition of **configuration**; thus, enantiomers have opposite configurations. An enantiomer can assume various **conformations** by rotation of groups around single bonds but a change in configuration requires the breaking of bonds at the stereogenic carbon.

Using the test of superimposability, we can conclude that there are two enantiomers of lactic acid. One form of lactic acid rotates the plane of polarized light to the right and the other to the left and are labeled *(+)-* and *(−)*-lactic acid, respectively. The two configurations of lactic acid are shown in Figure 1-4 and the question arises, which configuration corresponds to *(+)*-lactic acid and which to *(−)*-lactic acid? In other words, how do we assign configuration?

For almost a century after van 't Hoff's seminal paper the actual or **absolute configuration** of enantiomers remained largely unknown. This situation changed in 1951 when Bijvoet reported the first X-ray determination of the absolute configuration of an enantiomer [8]. The compound studied was also most appropriate: a salt of *(+)*-tartaric acid, the same acid that had been studied almost 100 years before by Pasteur.

Emil Fischer

Prior to the determination of absolute configuration by X-ray crystallography, the configuration of a stereogenic center was established by chemical transformation of the chiral molecule to an arbitrarily chosen standard, *(+)*-glyceraldehyde. This was the basis of the **Fischer convention** for the designation of configuration, first proposed by Emil Fischer [9] in 1919. The need for consistency in stereochemical designation prompted Fischer to use the C-5 of the *d*-enantiomer of glucose as a starting point. The molecule is first drawn as a Fischer projection (Figure 1-8) where all bonds are drawn as solid lines with the understanding that horizontal bonds point toward the observer and vertical bonds point away. Moreover, the molecule is drawn with the longest carbon chain vertical and the most highly oxidized end of the chain is at the top. *(+)*-Glucose was degraded by Fischer to

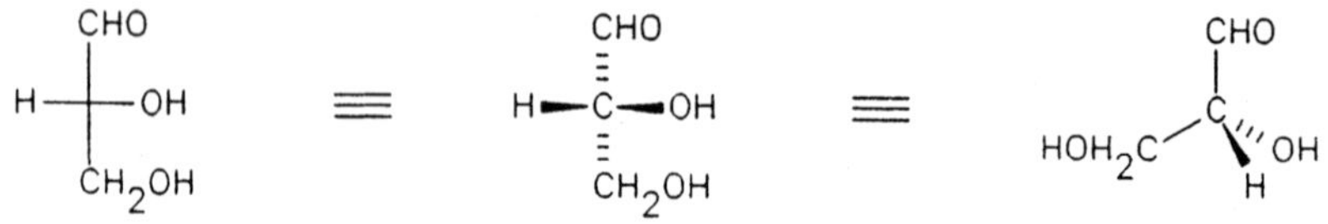

CHO

H——OH   ≡   CHO   ≡   CHO

CH₂OH   H—C—OH   HOH₂C—C—OH
                      CH₂OH   H

Fischer   flying-wedge   perspective
projection   representation   representation

Figure 1-8. Various representations of D-glyceraldehyde.

(+)-glyceraldehyde, in which the only remaining stereogenic center originates from the C-5 of glucose. Arbitrarily, Fischer assigned the configuration shown in Figure 1-8 to *(+)*-glyceraldehyde and called it D-*(+)*-glyceraldehyde, due to the position of the OH substituent on the right-hand side of the stereogenic center.

An alternative way of illustrating Fischer projections is the so-called flying-wedge representation in which bonds pointing towards the observer are symbolized by a wedge, bonds pointing away by a broken line (Figure 1-8). In the perspective representation, on the other hand, the bonds are arranged in a tetrahedral array, with the bonds pointing towards and away from the observer symbolized by a wedge and a broken line, respectively, and bonds in the plane of the paper by a continuous line (Figure 1-8).

All chiral molecules that could chemically be related to D-*(+)*-glyceraldehyde were assigned the configuration D (e.g., D-glucose), while molecules related to L-glyceraldehyde became the L-series. The Fischer convention is widely used in sugar chemistry and for α-amino acids. For sugars and other molecules that contain a number of stereogenic centers the Fischer convention defines a series as D or L according to whether the **configuration at the highest numbered stereogenic center** is equivalent to D- or L-glyceraldehyde. For example, the D-configuration is assigned to the erythrose and threose shown in Figure 1-9 because of the D-configuration at C-3.

^{1}CHO ^{1}CHO

H—^{2}C—OH   HO—^{2}C—H

H—^{3}C—OH   H—^{3}C—OH

^{4}CH₂OH ^{4}CH₂OH

D-Erythrose   D-Threose

Figure 1-9. Configuration of D-erythrose and D-threose according to the Fischer convention.

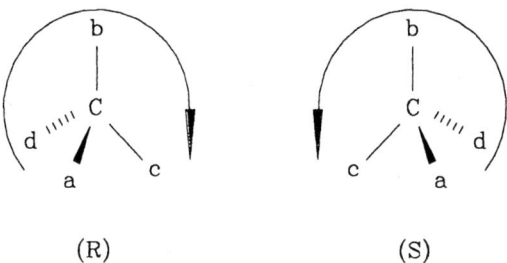

(R)   (S)

Figure 1-10. The Cahn-Ingold-Prelog convention.

There are, however, several drawbacks to the Fischer convention not the least of which is confusion of D and L with the sign of rotation d and l. As originally conceived, it indicated nothing more than relative configurations and it was not until 1951 that X-ray analysis [8] afforded absolute configurations and showed Fischer's inspired guess to have been correct. The ability to unequivocally determine the absolute configuration of an enantiomer brought the need for a simple, unambiguous nomenclature system. Consequently the Fischer convention has been almost entirely replaced by the **Cahn-Ingold-Prelog convention**, otherwise known as the sequence rule or the R and S nomenclature [10,11]. The Cahn-Ingold-Prelog convention can be used for the rapid and unambiguous designation of configurations. It is also extremely useful for describing diastereomers (see section II-D).

The Cahn-Ingold-Prelog convention is a set of arbitrary but consistent rules which allow a hierarchial assignment of the substituents at any stereogenic center (Figure 1-10). The four substituents are designated a > b > c > d in order of decreasing priority. The priority of substituents is assigned on the same basis that is used for the **E-Z** designation of *cis-trans* isomers. Thus, priority is established by the difference in atomic number of the atoms closest to the asymmetric center. The higher the atomic number the higher the priority of the substituent (e.g., Cl > S > F > O > N > C > H). For two atoms of the same atomic number the priority is determined by the next attached atoms until a difference is established (e.g., $CH_2Cl > CH_2OH > CH_2CH_3 > CH_3$); double bonds count as two single bonds (e.g., $CH=CH_2 > CH_2CH_3$).

For assignment of configuration, the stereogenic center is viewed with the group of lowest priority (d) pointing away from the observer. The configuration is then determined by whether the sequence of decreasing priority, a → b → c, proceeds in a clockwise or counterclockwise fashion. When the path a → b → c is clockwise, the configuration is denoted by R (Latin: *rectus*, right). When the path a → b → c is counterclockwise, the configuration is designated S (Latin: *sinister*, left). In order to help visualize the circular path of a → b → c, the comparison is

(i)

(ii)

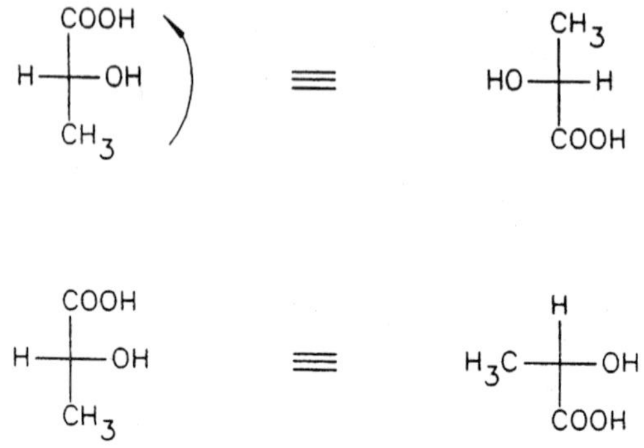

Figure 1-11. Allowed modes of rotation of Fischer projections.

often made with a steering wheel of a car (where the bond to the group d represents the steering column).

Sometimes it is desirable to rotate Fischer projections in order to render them more convenient for applying the Cahn-Ingold-Prelog rules. It is worth noting, however, that there are two and only two modes of rotation that are allowed: (i) rotation on the page by 180°, or (ii) one group is held steady and the other three are rotated clockwise or counterclockwise (Figure 1-11). All other modes of rotation result in a change of configuration.

Finally, the sort of confusion which abounds in the naming of stereoisomers can be illustrated by reference to optically active phenylglycine, an important antibiotic intermediate. The commercially relevant enantiomer is the levorotatory isomer (i.e., the (–)- or (l)-isomer). According to the Fischer convention, it has the D-configuration which causes confusion between the D- and l-assignment. This confusion is avoided if assignment follows the Cahn-Ingold-Prelog convention whereby the molecule is assigned the R-configuration. However, the use of the Fischer convention is so ingrained in amino acid chemistry that, despite the confusion that it causes, amino acids are usually still designated D or L (Figure 1-12).

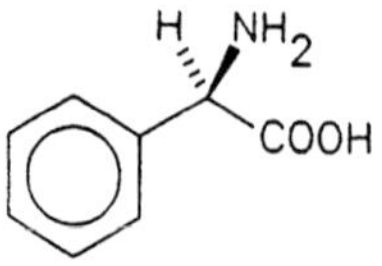

Figure 1-12. D-(–)-Phenylglycine.

E. Diastereomers

For compounds that contain more than one stereogenic carbon atom, there are more than two stereoisomers. If there are n stereogenic carbon atoms, the maximum number of stereoisomers is 2^n. When $n = 2$ the maximum number of stereoisomers is four. Since enantiomers exist only in pairs, some of the stereoisomers do not bear a mirror-image relationship to each other and therefore are not enantiomeric. Stereoisomers that are not mirror-images of each other are called **diastereo-isomers.** (Note that this definition includes not only chiral molecules but also achiral *cis-trans* geometric isomers). Unlike enantiomers, the physical and chemical properties of diastereomers differ, such as melting point, solubility, and density. Furthermore, their optical rotations can differ both in sign and magnitude.

The situation with two stereogenic carbon atoms is illustrated by reference to the drugs ephedrine and pseudoephedrine (Figure 1-13). The active drugs have the *(1R,2S)-* and *(1S,2S)-* configuration, respectively, as shown. The two compounds are stereoisomers but have a nonmirror-image relationship to each other (i.e., they are diastereomers). Both ephedrine and pseudoephedrine exist as a member of an enantiomeric pair as shown, *(1R,2S)-* and *(1S,2R)*-ephedrine and *(1S,2S)-* and *(1R,2R)*-pseudoephedrine. Alternatively, the two active drugs can be drawn in a perspective representation as shown in Figure 1-13. Diastereomers in which only

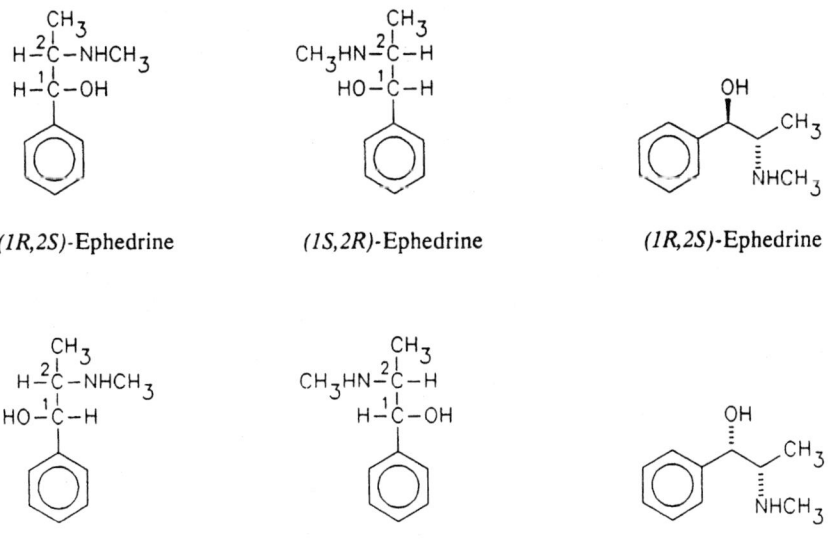

| *(1R,2S)*-Ephedrine | *(1S,2R)*-Ephedrine | *(1R,2S)*-Ephedrine |

| *(1S,2S)*-Pseudoephedrine | *(1R,2R)*-Pseudoephedrine | *(1S,2S)*-Pseudoephedrine |

Figure 1-13. Absolute configuration of ephedrine and pseudoephedrine shown as Fischer projections and as perspective representations.

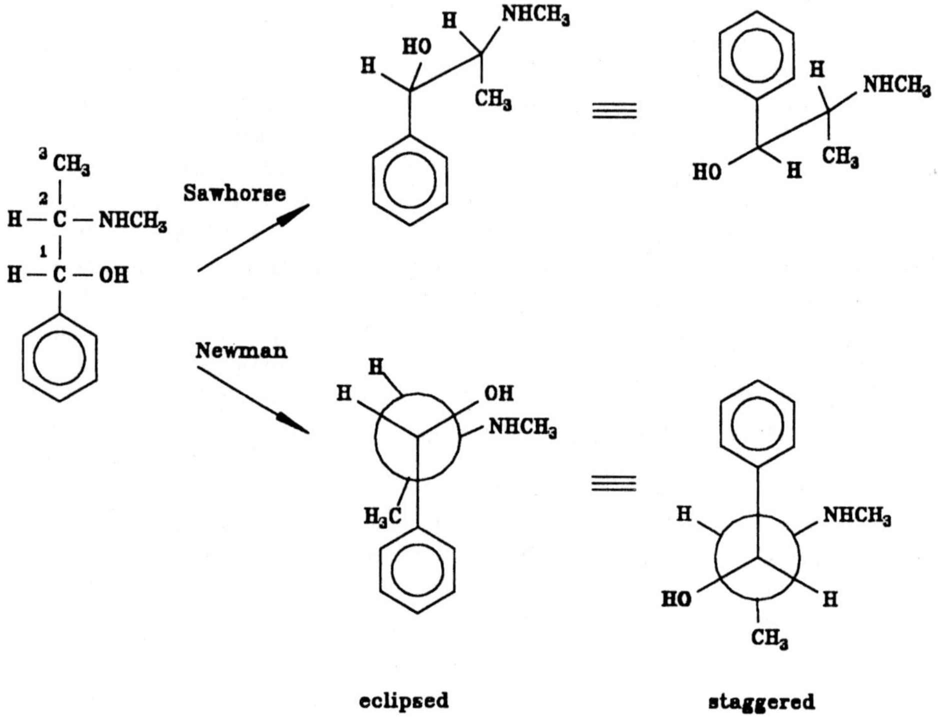

Figure 1-14. Transformation of Fischer projection of ephedrine to sawhorse or Newman formula.

one of two or more asymmetric centers is inverted are called **epimers**; ephedrine and pseudoephedrine are therefore epimers.

Molecules that contain two stereogenic atoms have special nomenclature and forms of notations that are derived from the names of the four-carbon sugars, erythrose and threose (Figure 1-9). If the two like groups a in systems of the type R-C_{ab}-C_{ac}-R^1 are on the same side in the Fischer projections, as the hydroxyl groups are in erythrose, the isomer is called the *erythro* form; if they are on the opposite sides, the isomer is called the *threo* form. Thus, ephedrine and pseudoephedrine are *erythro* and *threo* isomers, respectively.

Two convenient and widely used representations for molecules containing two (or more) stereogenic atoms are the so-called sawhorse representation and the Newman projection formula as illustrated for ephedrine in Figure 1-14. At first sight it might appear that Fischer projections are adequate for the projection of molecules with two stereogenic atoms. However, closer analysis reveals that Fischer projections represent the molecule in the so-called eclipsed conformation, the form in which C-2 and C-3 are so rotated with respect to each other that the groups attached to them approach each other as closely as possible. In reality, the molecule adopts the so-called staggered conformation whereby C-2 and C-3 are rotated with respect to each other by an angle of 60° so that their substituents are as far apart as possible. In considering stereochemical effects in chemical reactions it is convenient to depict molecules in their actual staggered form rather than in the hypothetical eclipsed form shown in the Fischer projection.

In the sawhorse representation, the molecule is simply shown in three dimensions with the bond between the stereogenic carbons oriented diagonally backward and somewhat exaggerated in length. In Newman projections, the molecule is viewed from front to back in the direction of the bond linking the stereogenic carbons. These two atoms thus exactly eclipse each other and are represented by two superimposed circles (i.e., one circle). The bonds and groups attached to the stereogenic carbon atoms are projected into a vertical plane and thus appear as the spokes of a wheel. Fischer projections can be rapidly translated into sawhorse or Newman projections by first translating into the eclipsed sawhorse or Newman projection and then rotating around the C2-C3 bond until one reaches the stable staggered conformation.

F. *Meso* Compounds

Let us now return to the molecule that has played such an important role in the development of stereoisomerism, tartaric acid. Although it has two stereogenic carbon atoms tartaric acid has only three stereoisomers. This is because one of the stereoisomers has a plane of symmetry while the other two do not (Figure 1-15).

The two isomers (R,R and S,S) that do not have a plane of symmetry are enantiomers. The R,S and S,R configurations, on the other hand, are superimposable since they have a plane of symmetry (i.e., they are achiral). A molecule that contains stereogenic carbon atoms but is optically inactive because it contains a plane of symmetry is said to possess a *meso* configuration. In other words, a *meso* compound is one with a

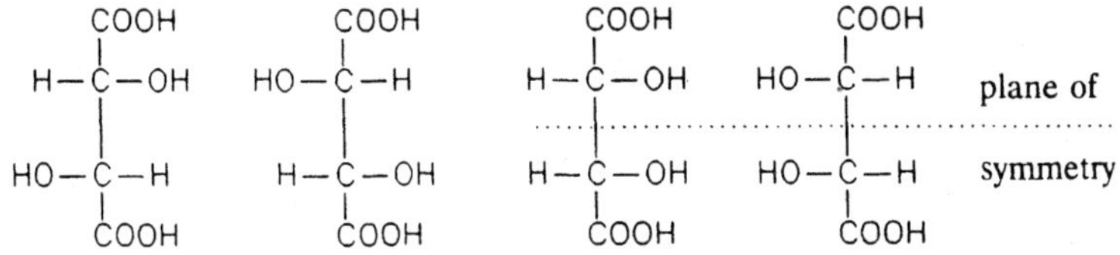

	COOH	COOH	COOH	COOH
	H—C—OH	HO—C—H	H—C—OH	HO—C—H	plane of
	HO—C—H	H—C—OH	H—C—OH	HO—C—H	symmetry
	COOH	COOH	COOH	COOH	
Configuration	(R, R)	(S, S)	(R, S)	(S, R)
Sign of rotation	(+)	(−)	none	none

Figure 1-15. Stereoisomers of tartaric acid.

structure that is superimposable on its mirror-image even though it contains stereo-genic centers. *Meso* structures are often easily recognized because, in at least one of its conformations, one half of the molecule is the mirror-image of the other half.

III. PRINCIPLES OF STEREODIFFERENTIATION: CHIRAL RECOGNITION

In the preceding section we were concerned with the stereochemistry of discrete molecules. In this section we shall deal with the stereochemical consequences of interactions between molecules. As noted earlier, the two enantiomers of a race-mate have identical chemical properties. When placed in a chiral environment, however, they can exhibit different chemical reactivities. This means that a chiral substance, such as an enzyme, is able to differentiate between the two enantiomers on the basis of their different stereochemical configuration. This phenomenon is referred to as chiral discrimination or chiral recognition, similar to the match between a right hand and a right-handed glove. Chiral recognition can also be observed in interactions between certain achiral molecules and a chiral reagent or catalyst. For both types of reactions there is one overriding principle: in order to make an optically active compound you need another one.

A. Asymmetric Synthesis Versus Kinetic Resolution

There are two possible approaches for the preparation of optically active products by chemical transformations of optically inactive starting materials: kinetic resol-ution and asymmetric synthesis. These two approaches are illustrated in Figure 1-16 for the synthesis of the antiinflammatory drug, *(S)*-naproxen. A kinetic resolution depends on the fact that the two enantiomers of a racemate react at different rates with a chiral reagent or catalyst, such as an enzyme. An asymmetric synthesis, on the other hand, involves the creation of an asymmetric (stereogenic) center which, in this case, is by chiral discrimination of equivalent groups in an achiral starting material. Such a starting material is said to be **prochiral** or to contain a center of prostereoisomerism.

Thus, a kinetic resolution involves substrate selectivity while an asymmetric synthesis involves product selectivity. In order to discriminate between these two types of enantioselectivity, some authors refer to the former as being enantio-specific (synonymous with substrate specific) and the latter as enantioselective. This is consistent with the widespread use of the term enantiospecific in enzymatic kinetic resolutions. Thus, enzymes that preferentially react with one enantiomer of a racemic substrate are usually referred to as being L- or D-specific.

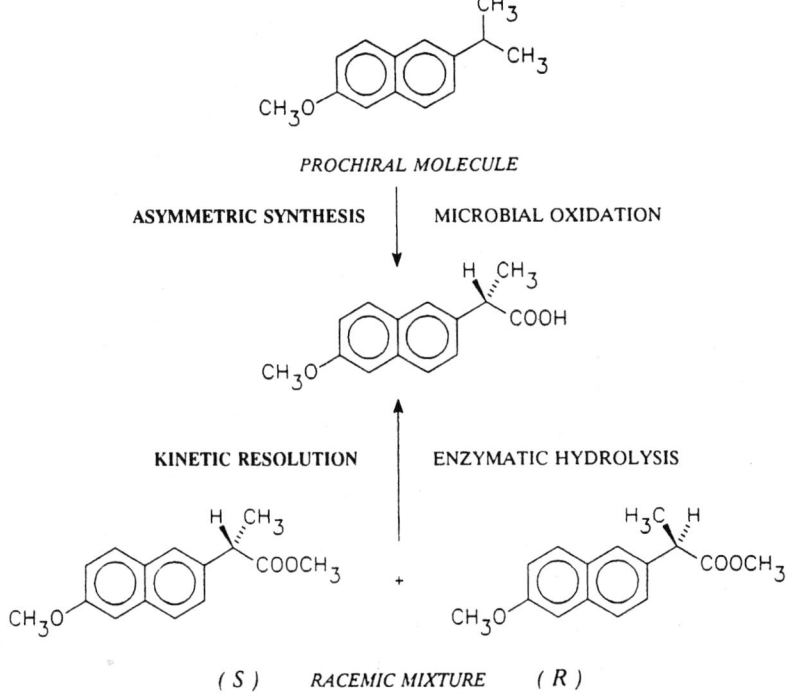

Figure 1-16. Asymmetric synthesis versus kinetic resolution.

B. Enantioselective Versus Diastereoselective Synthesis

We have seen that an enantioselective synthesis involves the reaction of a prochiral molecule with a chiral substance. A second type of asymmetric synthesis involves the preferential formation of a single diastereomer by the creation of a new asymmetric center in a chiral molecule. This is referred to as a **diastereoselective synthesis.** The two types of asymmetric synthesis are illustrated in Figure 1-17 for the epoxidation of an olefin.

In both cases we speak of asymmetric induction having taken place. In an enantioselective synthesis the asymmetry (i.e., stereoselectivity) is induced by the (external) chiral catalyst (Reaction 1-3). In contrast, a diastereoselective asymmetric synthesis does not require a chiral catalyst. The stereogenic center already present in the molecule is able to induce the observed stereoselectivity, assuming that we start with a single enantiomer. It is also worth noting that Reaction 1-4 is, according to the definitions outlined above, both enantiospecific and diastereoselective.

Enantioselective epoxidation:

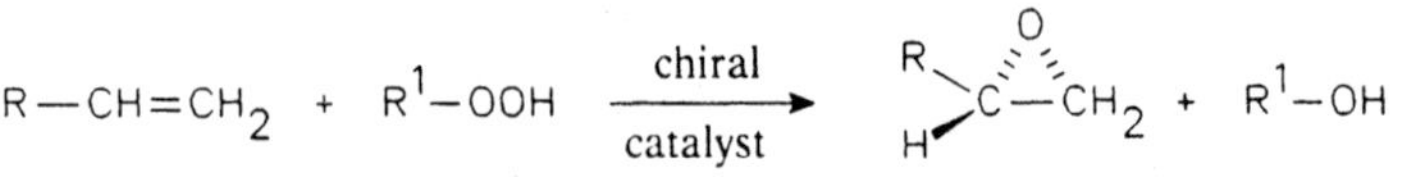

$$R-CH=CH_2 + R^1-OOH \xrightarrow[\text{catalyst}]{\text{chiral}} \underset{H}{\overset{R}{>}}C\overset{O}{-}CH_2 + R^1-OH \qquad (1\text{-}3)$$

prochiral olefin

Diastereoselective epoxidation:

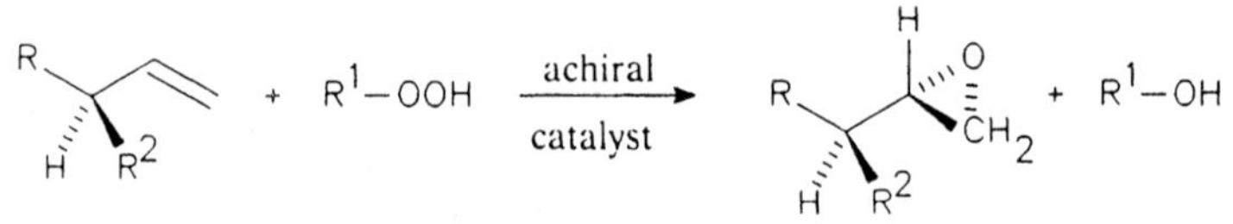

$$+ R^1-OOH \xrightarrow[\text{catalyst}]{\text{achiral}} + R^1-OH \qquad (1\text{-}4)$$

chiral olefin

Figure 1-17. Enantioselective versus diastereoselective epoxidation.

C. The Nomenclature of Asymmetric Synthesis: Prostereoisomerism

In order to understand fully the stereoselective reactions of prochiral molecules it is necessary to be able to specify the relationships of particular groups and faces within a molecule. The concept of prostereoisomerism or prochirality can be traced back to the work of Ogston [12] who pointed out that an enzyme could differentiate between the two terminal carboxyl groups of citric acid. As shown in Figure 1-18 citric acid possesses a mirror plane and is achiral. The central carbon atom can be

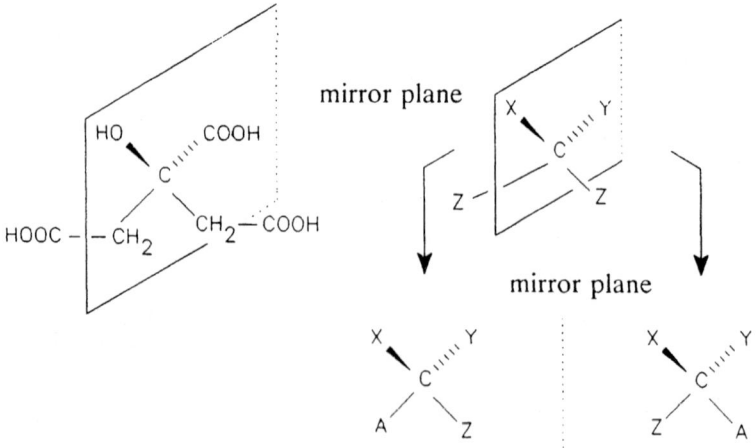

Figure 1-18. Enantiotopic groups in a prochiral molecule.

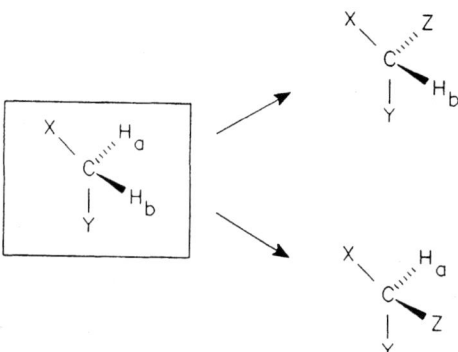

Figure 1-19. Application of the substitution rule to CH_2XY.

represented as C_{XYZZ}, that is, a tetrahedral carbon atom attached to two identical Z groups and two nonidentical groups X and Y. If either one or the other of the Z groups is substituted by A this leads to the formation of C_{XYZA} molecules that are enantiomers. The C_{XYZZ} center is, hence, described as a center of prostereo-isomerism [13] since replacement of a Z group creates a stereogenic center.

Topicity is the name given to relationships within molecules and it is usually studied by applying the substitution rule proposed by Mislow and Ralan [14]. The substitution rule operates by considering the consequences of replacing, sequentially, two (or more) constitutionally identical groups, by an achiral test group that is different from all the other groups. In any CH_3X molecule, for example, a stereogenic center cannot be produced by substitution of any single hydrogen atom. The hydrogen atoms are, therefore, indistinguishable in a chiral environment and are called **homotopic**; homotopic groups occupy stereochemically equivalent positions in space.

Stereoheterotopic groups, in contrast, occupy stereochemically nonequivalent positions in space. For example, if we apply the substitution rule to a CH_2XY molecule replacement of either one or the other (but not both) hydrogen atoms leads to a pair of enantiomers (Figure 1-19). Groups that produce enantiomers on application of the substitution rule are said to be **enantiotopic**.

If application of the substitution rule leads to a pair of diastereomers then the groups are said to be **diastereotopic**. In general, stereoheterotopic groups in an achiral molecule are enantiotopic and in a chiral molecule they are diastereotopic.

The analysis of prochirality can also be extended to the faces of molecules. Achiral molecules can be prochiral by virtue of the fact that they contain **enantiotopic or diastereotopic faces**. For example, the prochiral olefin shown in Figure 1-17 possesses enantiotopic faces since approach of the chiral epoxidizing agent from either one or the other face of the double bond affords a pair of enantiomers as illustrated in Figure 1-20. Similarly, if the approach of the reagent

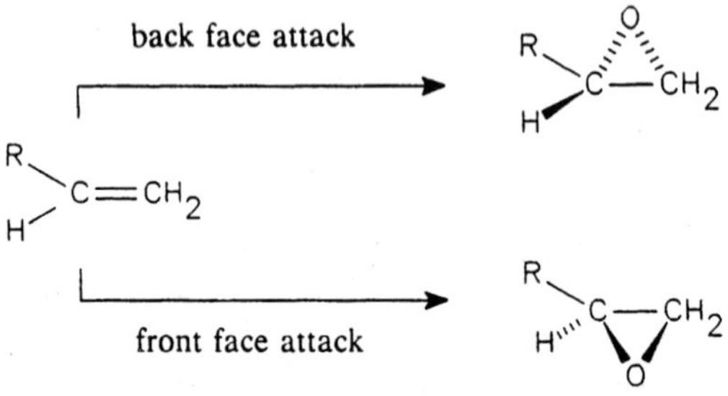

back face attack
R
H
C=CH2
O
C
CH2
front face attack

Figure 1-20. Enantiotopic faces of a prochiral olefin.

from either one or the other face of a molecule leads to a pair of diastereomers then the faces are diastereotopic (as in Reaction 1-4 in Figure 1-17). In this case, a chiral reagent is not required for stereoselectivity.

A convenient system for the specification of enantiotopic and diastereotopic groups or faces was developed by Hanson [15]. Thus, in a molecule with a trigonal center (e.g., an olefin), the three substituents are ranked according to the Cahn-Ingold-Prelog priority rule and, looking from a given side, their orientation is determined (Figure 1-21). If this is clockwise the face in question is specified as *Re* (Latin: *rectus*, right) and the opposite face is automatically *Si* (Latin: *sinister*, left). In the case of tetrahedral centers of the type C_{aabd} one of the enantiotopic *a* groups is arbitrarily assigned a higher priority and the center specified according to the priority rule. If this gives the *R*-configuration then this *a* group is designated *Re* and the other *a* group is accordingly designated *Si* (sometimes the notations *pro-R* and *pro-S* are used). This system can also be applied to diastereotopic faces and groups.

b
a
Re
C
Si
d
Si
a
C
a
Re
b
d

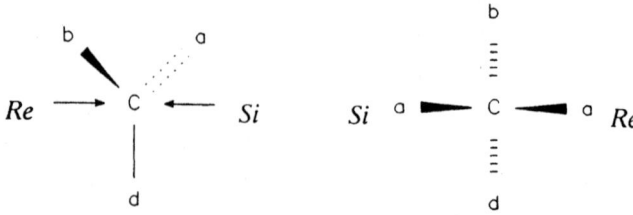

Figure 1-21. Illustration of the *Re*/*Si* convention for enantiotopic faces or groups.

D. Stereodifferentiation and Asymmetric Catalysis

The nomenclature of prostereoisomerism also forms the basis for describing different types of asymmetric transformations. The differentiation by reagents of stereoheterotopic units can be divided into two major types, **enantiodifferentiation** and **diastereodifferentiation**, each of which has three subcategories. In enantiodifferentiation the chiral agent is external to the substrate and the three categories of enantiodifferentiation are:

1. **Enantiotopos differentiation**, in which a chiral agent differentiates between enantiotopic units.

2. **Enantioface differentiation**, in which an enantiotopic face is discriminated by a chiral agent.

3. **Enantiomer differentiation** is another name for kinetic resolution and is not a method for creating chirality.

In diastereodifferentiation the chirality is present in the substrate and the reagent (catalyst) may be achiral.

4. **Diastereotopos differentiation**, in which diastereotopic units are differentiated by the reagent.

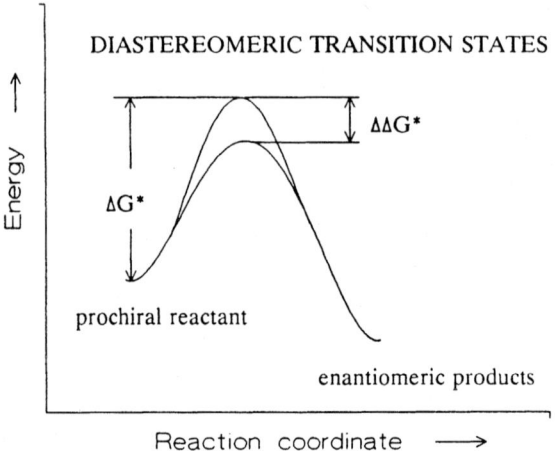

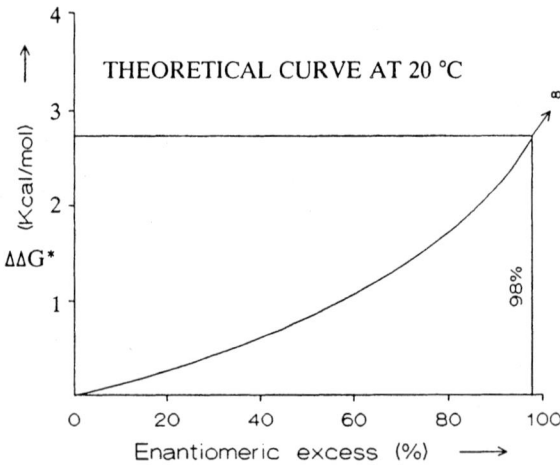

Figure 1-22. Reaction coordinate of an enantioselective synthesis.

5. **Diastereoface differentiation,** in which diastereotopic faces are differentiated by a reagent/catalyst.

6. **Diastereomer differentiation,** in which two diastereomers react at different rates with a reagent.

As with all catalysis, asymmetric catalysis is a kinetic phenomenon and all asymmetric inductions share the common principle that competing reactions via diastereomeric transition states proceed at different rates. In the absence of a chiral influence reactions producing enantiomers have transition states of identical energies and therefore have identical rates and produce equal amounts of the two enantiomers. A chiral reagent or catalyst, such as an enzyme or a chiral metal complex, differentiates between the enantiotropic features of a prochiral substrate, such as the enantiotropic faces of a prochiral olefin, to form energetically different diastereomeric transition states (Figure 1-22). The extent of asymmetric induction reflected in the *ee* of the product, is determined by the difference in free energy of these two diastereomeric transition states. As shown in Figure 1-22 a difference of 2.66 kcal/mole suffices to afford an *ee* of 98%. This is comparable to the energy of a fairly respectable hydrogen bond.

In contrast, the diastereotopic faces of a chiral olefin are distinguishable via the diastereomeric transition states that ensue even with an achiral catalyst. It should be noted, however, that with a chiral catalyst better facial discrimination may obtain. The energy profiles for an enantioselective and a diastereoselective synthesis are compared in Figure 1-23.

In the case of a diastereoselective synthesis the energies of the two diastereomeric products are also different. In other words, an enantioselective synthesis must be under kinetic control whereas a diastereoselective synthesis would produce unequal amounts of diastereomers even under thermodynamic control.

IV. ENANTIOMERIC PURITY DETERMINATION

The increasing awareness of the importance of optical purity in the context of biological activity (see chapter 2) has created a growing need for accurate, unequivocal methods for the determination of optical purity.

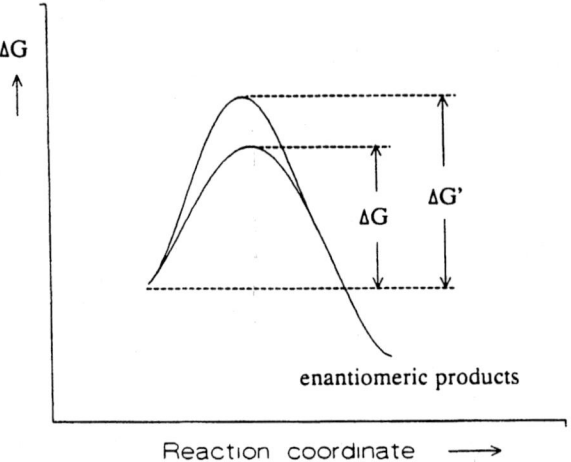

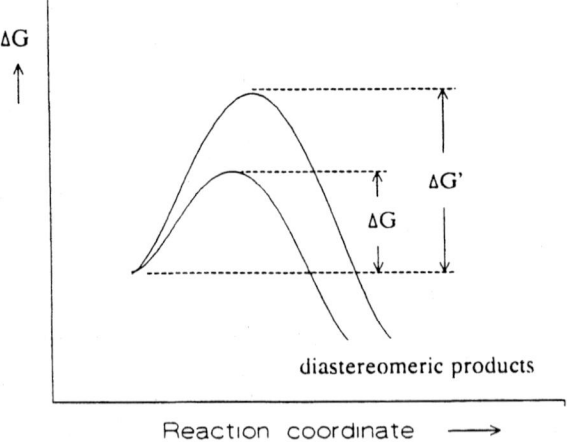

Figure 1-23.   Energy profiles for enantioselective and diastereoselective syntheses.

A. Optical Rotation

The classical method for the determination of enantiomeric purity is by measurement of the specific optical rotation, $[\alpha]^t_\lambda$, using a polarimeter. For a mixture of enantiomers, the **optical purity** is defined by the relation:

$$\text{Optical purity} = \frac{[\alpha]}{[\alpha]_0} \times 100\ \% \tag{1-5}$$

where $[\alpha]$ is the specific optical rotation of the enantiomeric mixture and $[\alpha]_0$ is the specific optical rotation of the pure enantiomer (i.e., its absolute rotation). However, this method of determining optical purity leaves a lot to be desired. In practice, it often leads to much confusion and animated discussions due to the comparison of samples with different chemical purities. Thus, a crude sample of a product with a lower specific rotation may nevertheless have a higher enantiomeric purity than a putative purified sample due, for example, to racemization taking place during distillation. Because of this and other shortcomings, optical purity as defined above has largely been supplanted by the unambiguous term enantiomeric excess (*ee*), which for $R > S$ is given by:

$$\text{Enantiomeric excess}\ \ (ee) = \frac{(R-S)}{(R+S)} \times 100\ \% \tag{1-6}$$

Thus, a sample containing the R and S enantiomers in a ratio of 95:5 has an *ee* of 90%. For the reason outlined above, optical rotation measurement is not an unequivocal method for *ee* determination and more reliable methods have been developed, many of which involve the chromatographic separation of enantiomers.

B. HPLC Methods

Because enantiomers have the same adsorption properties they are not amenable to direct chromatographic separation on regular (achiral) adsorbents. Separation can be achieved only via the formation of diastereomers that have different adsorption properties. This can be accomplished in two ways. One involves precolumn derivatization of the sample with a chiral reagent that yields diastereomeric molecules which can be separated on achiral columns [16]. The other employs the formation of transient diastereomers via the interaction of enantiomers with a chiral selector in the form of chiral mobile phase additive (CMPA) or a chiral stationary phase (CSP). A great deal of progress has been made with the latter technique and a wide variety of CSPs have been described, many of which are commercially available. The subject has been extensively reviewed [17–24] and the different types of CSPs have been classified as follows [17]: 1) chiral ligand exchange phases; 2) affinity phases; 3) helical polymers; 4) cavity phases; 5) Pirkle-type phases.

1.   Chiral ligand exchange phases

In chiral ligand exchange chromatography, introduced by Davankov and coworkers [25,26], an optically active ligand—usually an amino acid—is covalently bound to a solid support. Chloromethylated polystyrene was used originally; subsequently, Gübitz and coworkers [27] developed silica-bound CSPs (e.g., L-proline attached to silica by a 3-glycidylpropyl spacer). After loading the adsorbent with Cu^{2+} ions, the racemate is chromatographed. Each amino acid molecule displaces one of the polymer-bound proline ligands to form a mixed complex (Figure 1-24). The transient diastereomeric complexes formed with the R and S amino acid enantiomers, respectively, may differ in free energy by up to 8 kcal/mole.

2.   Affinity phases

Several serum proteins undergo enantioselective interactions with a wide variety of pharmacologically active compounds. This property has been used to develop CSPs that are based on bovine serum albumin (BSA) [23], or α_1-acid glucoprotein (AGP) [28] bound to HPLC-grade silica. Both are eminently suited for the separation of chiral drugs and often show high separation factors. As would be expected for a protein, the separation depends very much on parameters such as pH, ionic strength, and temperature. Hence, the optimum conditions have to be determined for each compound.

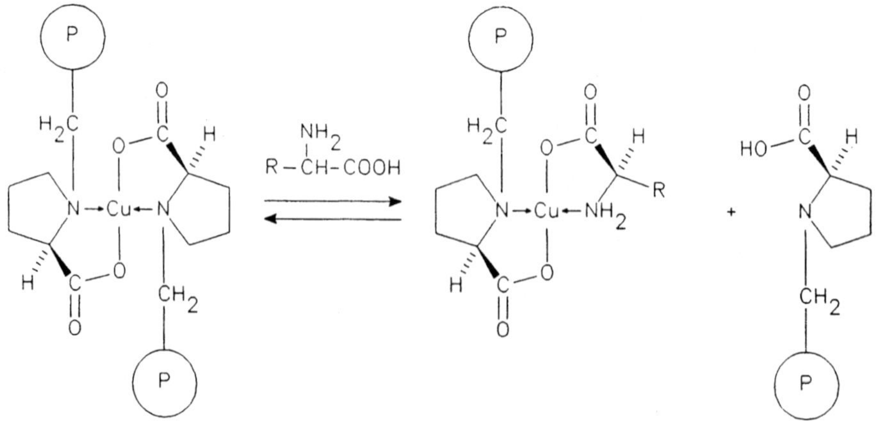

Figure 1-24. Chiral ligand exchange chromatography.

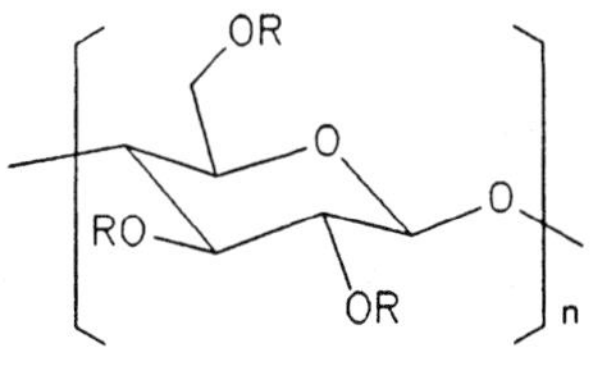

R = CH_3CO- ;

R = PhCO- ;

R = PhNHCO- ;

R = $PhCH_2-$;

R = PhCH=CHCO- ;

Figure 1-25. Cellulose derivatives used as CSPs.

3. Helical polymer phases

Polymers with a helical structure are able to separate enantiomers on the basis of steric effects. Helical polymer phases include several cellulose derivatives and synthetic polymers such as poly(triphenylmethyl)methacrylate. For example, a variety of cellulose derivatives (Figure 1-25) adsorbed on macroporous silica are commercially available and are widely used [29].

The polymerization of triphenylmethyl methacrylate in the presence of a chiral anionic initiator such as sparteine-butyllithium, yields an isotactic polymer that is chiral by virtue of its helical structure [29]. The bulky triphenylmethyl groups are assumed to direct the growing polymer chain to take a helical conformation. Good separations of a broad range of racemates are obtained using this polymer adsorbed on macroporous silica [29].

4. Cavity phases

Cyclodextrins are cyclic oligosaccharides composed of α-D-glucose units linked through the 1,4-position. The three most common forms are α-, β- and γ-cyclodextrin, containing six, seven, and eight glucose units, respectively (Figure 1-26). Cyclodextrins have a doughnut-shaped structure, the interior cavity of which is relatively hydrophobic. A variety of compounds fit into this cavity to form inclusion complexes. The β- and γ-forms have been successfully attached to silica to form CSPs that are finding wide applicability [17,24]. They are cheaper than many other CSPs and preparative scale columns are available.

5. Pirkle-type phases

Pirkle and coworkers [19] pioneered the development of a variety of CSPs consisting of amino acid derivatives, typically N-(3,5-dinitrobenzoyl)phenylglycine, immobilized on silica by either ionic or covalent attachment (Figure 1-27).

Many of these columns are commercially available and are widely used for analytical and preparative scale separations. The separation mechanism involves a combination of effects: charge-transfer, hydrogen-bonding, so-called 'dipole stacking' interactions and steric effects.

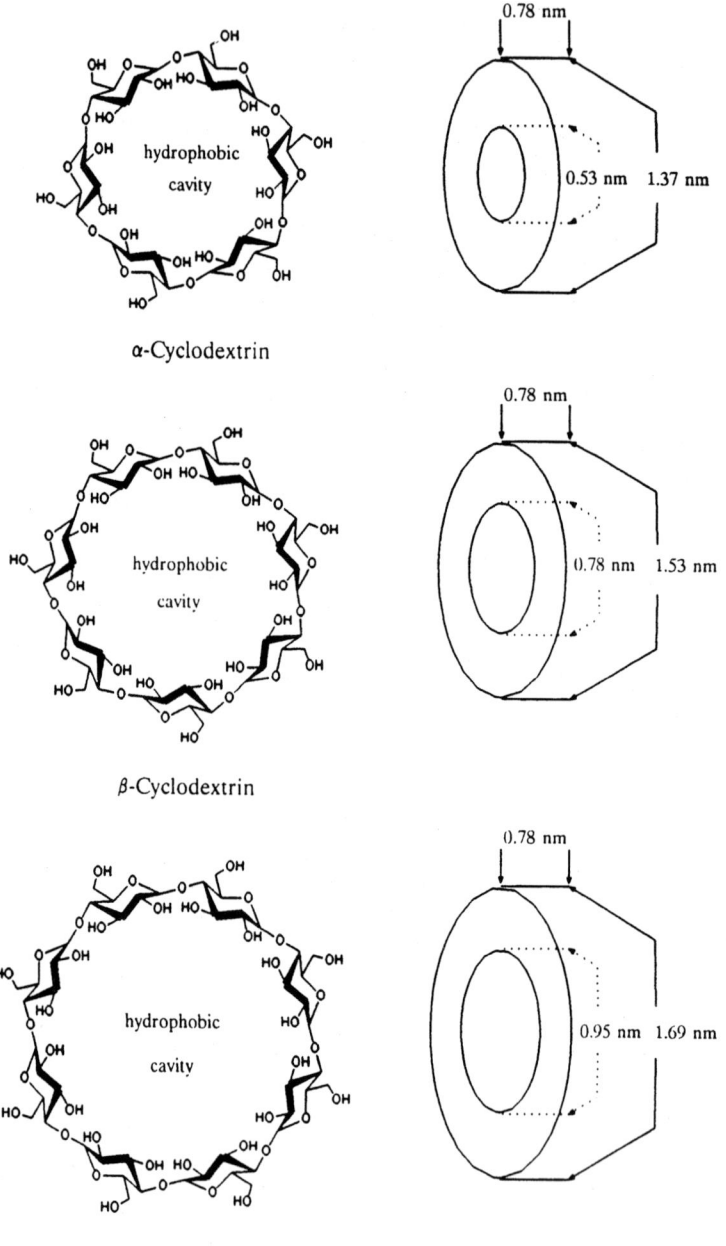

α-Cyclodextrin

β-Cyclodextrin

γ-Cyclodextrin

Figure 1-26. Structures of cyclodextrins.

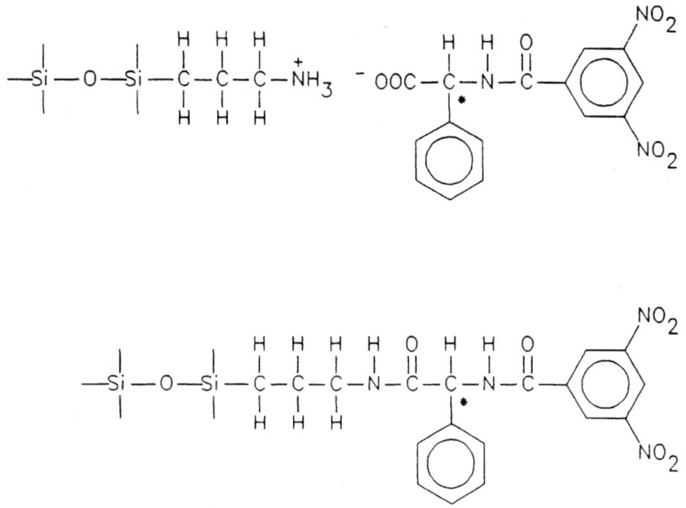

Figure 1-27. Pirkle-type CSPs.

In short, the range of commercially available CSPs for HPLC separation of enantiomers is comprehensive and the extensive literature [17–24] on the subject should be consulted before undertaking a particular separation. Some of the CSPs can be used for preparative scale separations of multikilo quantities.

An alternative approach is to use chiral mobile phase additives (CMPAs). They have certain advantages with respect to CSPs since, for example, less expensive nonchiral columns can be used and rapid screening of a range of chiral selectors is possible. Reversed phase ligand-exchange chromatography with Cu²⁺ and a chiral ligand (e.g., L-proline) in the mobile phase, for example, is a popular technique for amino acid separation. Cyclodextrins have also been widely used as CMPAs.

C. GLC Methods

For compounds that are readily vaporized without decomposition, gas chromatography on CSPs constitutes an accurate and reliable method for enantiomeric purity determination. The technique has developed rapidly and has the inherent advantages of simplicity, speed, reproducibility, and sensitivity [31]. CSPs that are used in gas-chromatographic separations of enantiomers can be divided into three classes:

1. Ligand exchange phases analogous to those used in HPLC determinations. This technique is referred to as complexation gas chromatography [32].
2. Amino acid derivatives [33].
3. Cyclodextrin derivatives coated on glass or fused silica surfaces [31].

In particular, the use of cyclodextrin derivatives has become increasingly important. They can be used over a wide temperature range (25–250°C) and exhibit high sensitivity for a variety of both polar and nonpolar compounds, such as alkenes and cyclic alkanes [31]. As in HPLC methods, GLC separations with cyclodextrin stationary phases involves the formation of inclusion complexes.

D. NMR Methods: Chiral Lanthanide Shift Reagents

NMR is also a widely used technique for enantiomeric purity determination, although it has suffered in recent years from the increasing popularity of chiral HPLC and GLC methods. One, well-tested method involves the conversion of a mixture of enantiomers to a mixture of diastereomers by reaction with an optically pure reagent. For example, Mosher's reagent (Figure 1-28; I) has been widely used for this purpose [34]:

If we examined the NMR spectrum of the mixture of enantiomers shown in Figure 1-28, we would find only one peak (split into a doublet by the C-H) for the methyl protons, since enantiomers have identical NMR spectra. In contrast, however, the two diastereomeric products exhibit methyl peaks at different chemical shifts. The molar ratio of diastereomers (and hence of the original enantiomers) can be determined from the relative intensities of the two doublets. Alternatively, the singlet peaks for the methoxy groups can be used.

A closely related method involves the use of chiral lanthanide shift reagents (LSRs). LSRs have the property of shifting the NMR signals of substrates via complex formation. The two diastereomeric complexes formed from a mixture of enantiomers exhibit different shifts. Whitesides and Lewis [35] reported the first chiral LSR (Figure 1-29), the europium(III) complex (**2**). Subsequently, the more widely used complexes (**1**) and (**3**) were introduced [36].

Summarizing, chiral HPLC and GLC techniques are the most accurate methods for the determination of enantiomeric purity. The method of choice is partly determined by the volitility and thermal stability of the product in question. NMR methods, although less accurate, are convenient for rapid analysis. Optical rotation is unreliable and should only be used as an indication.

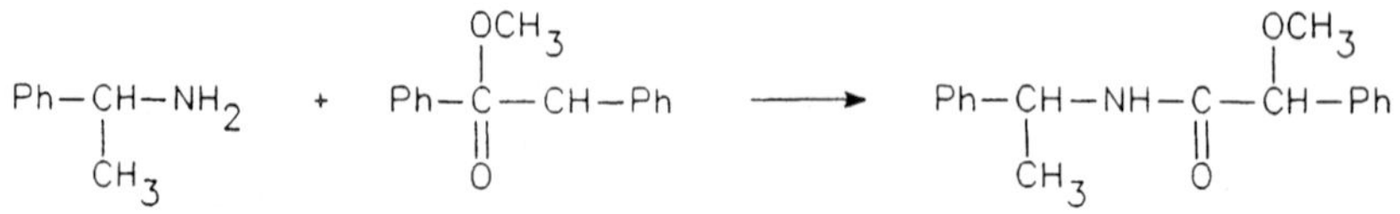

mixture of enantiomers   single enantiomer   mixture of diastereomers

Figure 1-28. Mosher's reagent.

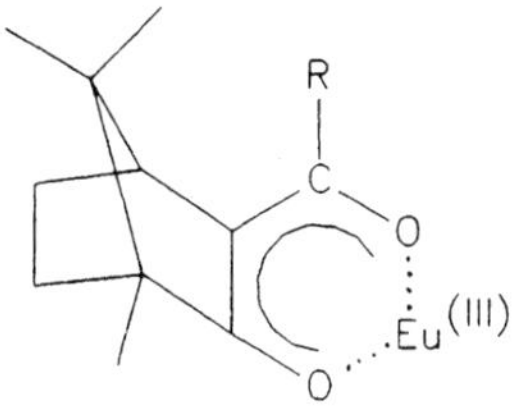

(1) R = *tert*-Trityl

(2) R = $CF_2CF_2CF_3$

(3) R = CF_3

Figure 1-29. Chiral lanthanide shift reagents.

V. THE ORIGIN OF BIOMOLECULAR CHIRALITY

A. Homochirality in Nature

One of the most intriguing aspects of living organisms is the homochirality of their molecular components. The term homochiral ('same hand') refers to objects or molecules of the same handedness or configuration. Thus, left hands are homochiral as are L-amino acids. The fact that proteins consist exclusively of L-amino acids and nucleic acids of D-sugars, respectively, raises a number of interesting questions. For example, why are proteins and nucleic acids not composed of mixtures of D- and L-amino acids? Moreover, why L- and not D-amino acids? Furthermore, how was the first optically pure molecule formed from the primordial soup of achiral molecules?

The first question is the easiest to answer. The main function of proteins that act as enzymes is to catalyze the synthesis of biomolecules, including other proteins. Their catalytic ability is crucially dependent on their three dimensional structure, which in turn depends on their L-amino acid sequence. Polypeptides comprising both D- and L-amino acids cannot form the regular winding structure—the alpha helix—that is essential for efficient enzymatic activity. In other words, it would lead to organisms that are not viable. Orgel [37] has likened this situation to that of a spiral staircase. Right-handed and left-handed staircases are equally useful, but a staircase that constantly changes its handedness is not a viable proposition, except perhaps to a rock climber.

B. Asymmetric Synthesis

Pasteur believed that external chiral forces are ubiquitous in nature and constitute a demarcation criterion between the chemistry of living organisms and that of the laboratory. This paradigm was tested and shown to be wrong by Emil Fischer in his monumental work on sugars [38]. Fischer showed that on ascending a particular sugar series (via the Kiliani cyanohydrin method), one of the two possible diastereomeric

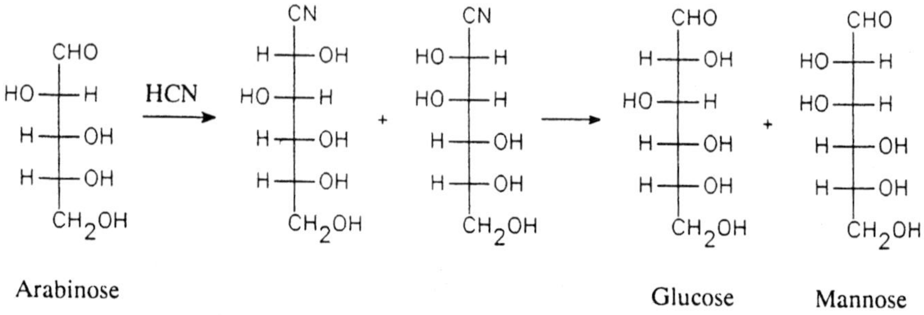

Figure 1-30. Homologation of arabinose.

products was formed predominantly (e.g., mannose rather than glucose from arabinose; Figure 1-30).

This was the first example of an asymmetric (diastereoselective) synthesis and demonstrated that Pasteur's external 'chiral force' is not necessary to account for the predominance of the D-series of sugars or the L-series of amino acids. Once a molecule is asymmetric, its extension proceeds also in an asymmetric manner. However, this does not explain the origin of the first primordial enantiomer from which stereoselective synthesis can begin.

C. The Chiral Force of Nature: Electroweak Interactions

A coherent theory of chemical evolution must, therefore, account for an initial production of an excess of molecules of a particular configuration. More recent theories tend to vindicate (at least partially) Pasteur's concept of universal dissymmetric forces playing a role in this initial stage. Thus, it is now generally accepted that universal electroweak interactions in nature are dissymmetric and can lead to minute but perhaps significant differences in the energy content of two enantiomers [39–41]. Moreover, calculations indicate that the particular enantiomers that have evolved, L-amino acids and D-sugars, are relatively more stable than their mirror-image isomers because of these interactions. The electroweak enantiomeric energy differences amount to no more than 10^{-14} J mol^{-1}, corresponding to an excess of about 10^6 molecules of the stabilized enantiomer per mole.

D. Chiral Symmetry Breaking and Chiral Amplification

Assuming that electroweak forces were able to produce a minute excess of one enantiomer, there must be a mechanism for amplifying this excess until one enantiomer completely predominates, somewhat akin to the Darwinian natural selection process. One mechanism for achieving this chiral amplification has been referred to as chiral symmetry breaking. It requires that the initially formed enantiomer catalyzes its own production (asymmetric autocatalysis) while inhibiting the production of its mirror-image.

An interesting example of chiral symmetry breaking has been reported [42]. Sodium chlorate crystallizes as enantiomorphic crystals and when it was crystallized from an unstirred aqueous solution equal numbers of d- and l-crystals obtained. In contrast, when the solution was stirred almost all of the crystals (99.7%) in a particular crystallization had the same structure, either dextro- or levorotatory. The explanation offered by the authors was that rapid production of secondary nuclei from primary nuclei in stirred systems is asymmetrically autocatalytic. Moreover, selective proliferation via secondary nucleation is accompanied by suppression of nucleation of crystals of the opposite handedness. In other words, depletion of the $NaClO_3$ from solution due to the rapid growth of secondary nuclei could reduce its concentration to a level at which the rate of primary nucleation is virtually zero. Consequently, the only nuclei that grow are the initial nucleus and the secondary nuclei derived from it, all of which have the same chirality. When the solution is not stirred there is no rapid autocatalytic production of nuclei. All of the nuclei are produced through primary nucleation and exhibit random handedness. As the authors noted, processes such as these may provide insights into the origins of biomolecular chirality.

E. Asymmetric Autocatalysis

One may ask: what does all this have to do with the industrial synthesis of optically active compounds? The answer is: perhaps very much. An understanding of the asymmetric autocatalytic processes involved in chiral symmetry breaking may lead to the development of new, efficient methods for asymmetric synthesis. Indeed, Wynberg has remarked that asymmetric autocatalysis may be the next generation of asymmetric synthesis [43].

The general principle is as follows. Two achiral substances A and B react to form a chiral product C which is a catalyst for its own production. Each of the enantiomers of C catalyzes its own production and inhibits the production of its isomer. If at any time slightly more of one enantiomer is formed, this immediately leads to rapid formation of that particular enantiomer and its predominance will rapidly increase. Hence, asymmetric autocatalysis must inevitably lead to molecular chirality without the need for an external chiral agent.

Obviously, if this principle could be put efficiently into practice, it could lead to remarkably effective methods for asymmetric synthesis. Some initial encouraging results have been reported [43] which provide a basis for optimism. We shall return to this topic later when we discuss catalytic asymmetric synthesis in more detail. (See chapter 8.)

REFERENCES

1. Biot, J. B., *Bull. Soc. Philomath., Paris*, 190 (1815).
2. Pasteur, L., *C. R. Acad. Sci*, **26**, 535-538 (1848).
3. Pasteur, L., *C. R. Hebd. Seances Acad. Sci*, **37**, 162 (1853).
4. Vallery-Radot, R., *The Life of Pasteur*, (Devonshire, R. L., transl.), Garden City Publishing Co., Inc., New York, 1926.
5. van 't Hoff, J. H., *Arch. Neerl. Sci. Exacts Nat.* **9**, 445-454 (1874).
6. Le Bel, J. A., *Bull. Soc. Chim. Fr.*, **22**, 337-347 (1874).
7. Mislow, K., and Siegel, J., *J. Am. Chem. Soc.*, **106**, 3319 (1984).
8. Bijvoet, J. M., Peerdeman, A. F., and Van Bommel, A. J., *Nature*, **168**, 271-272 (1951).
9. Fischer, E., *Ber. Dtsch. Chem. Ges.*, **524**, 129-138 (1919).
10. Cahn, R. S., Ingold, C. K., and Prelog, V., *Experientia*, **12**, 81 (1956).
11. Cahn, R. S., Ingold, C. K., and Prelog, V., *Angew. Chem. Int. Ed. Engl.*, **5**, 385-415 (1966).
12. Ogston, A. G., *Nature*, **162**, 963 (1948).
13. Hirschmann, H., and Hanson, K. R., *J. Org. Chem.*, **36**, 3293 (1971).
14. Mislow, K., and Raban, M., *Topics in Stereochemistry*, **1**, 1 (1967).
15. a) Hanson, K. R., *J. Am. Chem. Soc.*, **88**, 2731-2742 (1966). b) Prelog, V., and Helmchen, G., *Helv. Chim. Acta*, **55**, 2581-2589 (1972).
16. Buck, R. H., and Krummen, K., *J. Chromatogr.*, **315**, 279 (1984).
17. Däppen, R., Arm, H., and Meyer, V. R., *J. Chromatogr.*, **373**, 1-20 (1986).
18. Davankov, V. A., Kurganov, A. A., and Bochkov, A. S., *Advan. Chromatogr.*, **22**, 71 (1983).
19. Pirkle, W. H., and Pochapsky, T. C., *Advan. Chromatogr.*, **27**, 73 (1987).
20. Pirkle, W. H., and Finn, J., in *Asymmetric Synthesis, Vol. 1*, Morrison, J. D. (Ed.), Academic Press, New York, 1983, pp. 87-124.
21. Blaschke, G., *Angew. Chem. Int. Ed. Engl.*, **19**, 13-24 (1980).
22. Armstrong, D. W., *J. Liq. Chromatogr.*, **7**, (S2), 353 (1984).
23. Allenmark, S., *J. Biochem. Biophys. Methods*, **9**, 1 (1984).
24. Wainer, I. W., in *Drug Stereochemistry: Analytical Methods and Pharmacology*, Wainer, I. W., and Drayer D. E. (Eds.), Marcel Dekker, New York, 1988, pp. 147-173.
25. Davankov, V. A., Rogozin, S. V., Semechkin, A. V., and Sachkova, T. P., *J. Chromatogr.*, **82**, 359 (1973).
26. Davankov, V. A., *Advan. Chromatogr.*, **18**, 139 (1980).
27. Gübitz, G., Jellenz, W., and Santi, W., *J. Chromatogr.*, **203**, 377 (1981), and references cited therein.
28. Hermansson, J., *J. Chromatogr.*, **269**, 71 (1983).
29. Okamoto, Y., *Chemtech*, 176-181 (1987).
30. Wainer, I. W., and Doyle, T. D., *J. Chromatogr.*, **284**, 117 (1984).
31. a) Schürig, V., and Nowotny, H. P., *Angew. Chem. Int. Ed. Engl.*, **29**, 939-957 (1990). b) Schürig, V., in *Asymmetric Synthesis, Vol. 1*, Morrison, J. D. (Ed.), Academic Press, New York, 1983, pp. 59-86.
32. Schürig, V., *J. Chromatogr.*, **441**, 135 (1988).
33. Schürig, V., *Angew. Chem. Int. Ed. Engl.*, **23**, 747 (1984).
34. a) Dale, J. A., Dull, D. L., and Mosher, H. S., *J. Org. Chem.*, **34**, 2543 (1969). b) Dale, J. A., and Mosher, H. S., *J. Am. Chem. Soc.*, **90**, 3732 (1968). c) Dale, J. A., and Mosher, H. S., *J. Am. Chem. Soc.*, **95**, 512 (1973).

35. Whitesides, G. M., and Lewis, D. W., *J. Am. Chem. Soc.*, **92**, 6979 (1970).
36. For reviews see: a) Sullivan, G. R., *Topics in Stereochemistry*, **10**, 287-329 (1978). b) Fraser, R. R., in *Asymmetric Synthesis,Vol. 1*, Morrison, J. D. (Ed.),, Academic Press, New York, 1983, pp. 173-196.
37. Orgel, L., *The Origins of Life. Molecules and Natural Selection*, Chapman and Hall, London, 1973, p. 166.
38. a) Fischer, E., *Ber. Dtsch. Chem. Ges.*, **24**, 2683-2687 (1891). b) Freudenberg, K., *Advan. Carbohydr. Chem.*, **21**, 1-38 (1966).
39. a) Mason, S., *Chem. Soc. Rev.*, **17**, 347-359 (1988). b) Mason, S., *New Scientist*, **19** Jan, 1984, pp. 10-14. c) Mason, S., *Nouv. J. Chim.*, **10**, 739-747 (1986). d) Mason, S., *TIPS*, Jan 1986, pp. 20-23.
40. Mason, S., in *Chiral Separations by HPLC*, Krstulovic, A. M. (Ed.), Ellis Horwood, Chichester, 1989.
41. Hegstrom, R. A., and Kondepudi, D. K., *Sci. Am.*, Jan 1990, pp. 108-115.
42. Kondepudi, D. K., R. J. Kaufman, and N. Singh, *Science*, **250**, 975-976 (1990).
43. Wynberg, H., *Chimia*, **43**, 150-152 (1989).

ADDITIONAL READING

1. Rety, J., and Robinson, J. A., *Stereospecifity in Organic Chemistry and Enzymology*, Verlag Chemie, Basel, 1982.
2. Testa, B., *Principles of Organic Stereochemistry*, Marcel Dekker, New York, 1979.
3. Nogradi, M., *Stereochemistry. Basic Concepts and Applications*, Pergamon, Oxford, 1981.
4. Eliel, E. L., *Stereochemistry of Carbon Compounds*, McGraw Hill, New York, 1962.
5. Mislow, K., *Introduction to Stereochemistry*, Benjamin, New York, 1965.
6. Harmon, R. E. (Ed.), *Asymmetry of Carbohydrates*, Marcel Dekker, New York, 1979.
7. Simonyi, M. (Ed.), *Problems and Wonders of Chiral Molecules*, Akademiai Kiado, Budapest, 1990.
8. Bosnich, B., in *Asymmetric Catalysis*, Bosnich, B. (Ed.), NATO ASI Series, Martinus Nijhoff, Dordrecht, 1986.
9. Bonner, W. A., Origins of Chiral Homogeneity in Nature, in *Topics in Stereochemistry*, Eliel, E. L., and Wilen S. H. (Eds.), Vol. 18, Wiley, New York, 1988, p. 1096.
10. Krstulovic, A. M. (Ed.), in *Chiral Separations by HPLC*, Ellis Horwood, Chichester, 1989.
11. Bassindale, A., *The Third Dimension in Organic Chemistry*, Wiley, New York, 1984.
12. Buss, D. R., and Vermeulen, T., Optical Isomer Separation: Quest for a New Biological Technology, *Ind. Eng. Chem.*, **60(8)**, 12-28 (1968).

2

Chirality and Biological Activity

> The remarkable discrepancy between, on the one hand, the high degree
> of purity required for pharmaceuticals and, on the other hand, the
> acceptance of 50% impurity, as long as isomeric ballast is involved,
> should be a matter for serious concern.
>
> E. J. Ariëns, 1986

In the preceding chapter we addressed the question of what is chirality and what is its effect on chemical transformations? In order to do this we needed to understand the basic principles of stereochemistry and stereodifferentiation. In this chapter we shall discuss why chirality is important. Quite simply stated, chirality is important in the context of biological activity because, at a molecular level, asymmetry dominates biological processes. Chirality is not a prerequisite for bioactivity but in bioactive molecules where a stereogenic center is present, great differences are usually observed for the activities of the enantiomers. This is a general phenomenon and applies to all bioactive substances, such as drugs, insecticides, herbicides, flavors and fragrances, and food additives. Indeed, considering that the molecular components of living organisms are largely chiral, it should not be surprising to find that chirality plays a dominant role in their interactions with bioactive substances.

I. HISTORICAL DEVELOPMENT OF CONCEPTS

A. Biological Catalysis and Biological Activity

Pasteur was the first person to catch a glimpse of the close juxtaposition of biological catalysis and biological activity. In his epoch-making studies of the underlying causes of both fermentation and infectious diseases he became convinced that the omnipresent microbes, although capable of mediating useful biotransformations, could also bring death and destruction to mankind. Thus, at the molecular level biocatalysis and biological activity are two sides of the same coin; the former involving reaction of a substrate with a biopolymer (enzyme) and the latter a biopolymer (enzyme or receptor site) with a bioactive molecule. The common denominator is chirality.

It is not merely a coincidence, therefore, that many chiral natural products, such as the alkaloids quinine and quinidine (Figure 2-1) are both effective drugs and active catalysts for asymmetric synthesis [1].

B. Fischer's Lock-and-Key Concept

In his studies of sugars, Emil Fischer [2] observed that the enzyme emulsin catalyzes the hydrolysis of β-methyl-D-glucoside while maltase is active solely with the α-methyl-D-glucoside as substrate (Figure 2-2). Both enzymes were inactive towards the enantiomeric L-glucosides. These observations led Fischer to propose his famous 'lock-and-key' concept of enzyme specificity. In his own words, "to use a picture, I would say that the enzyme and the glucoside must fit each other like a lock and key, in order to effect a chemical reaction on each other." He also observed that "the difference frequently assumed in the past to exist between the chemical activity of living cells and of chemical reagents, in regard to molecular asymmetry, is nonexistent" [3].

The seemingly trivial difference between α- and β-glucoside units also accounts for why people eat potatoes and cows eat grass. Potatoes contain starch which is a

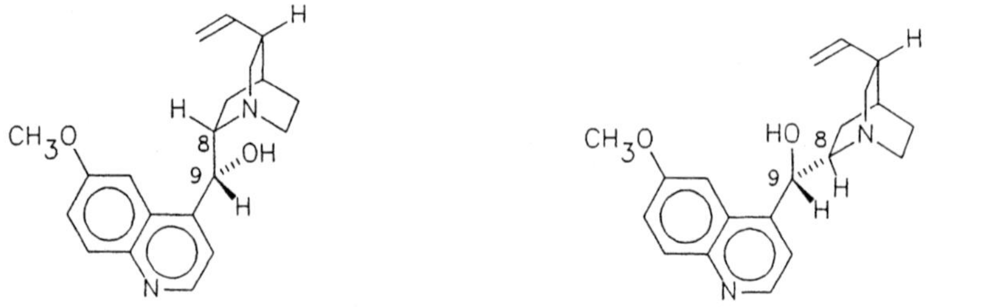

(8S,9R)-Quinine (antimalarial)   *(8R,9S)*-Quinidine (antiarrhythmic)

Figure 2-1. Structures of the cinchona alkaloids quinine and quinidine.

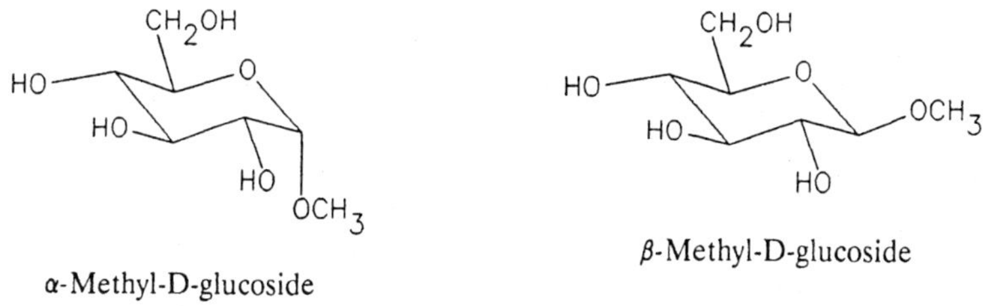

Figure 2-2. α- and β-methylglucosides.

polymer consisting of α-glucoside units (Figure 2-3). The mammalian enzyme α-amylase catalyzes the hydrolysis of starch into the energy source, glucose. Grass, on the other hand, contains cellulose which is a polymer consisting of β-glucoside units. Cellulose is not affected by α-amylase which means that we cannot digest cellulose. Cows possess β-amylase, an enzyme that is able to break cellulose down into glucose. Such is the power of stereochemistry!

The lock and key concept offered a simple explanation for the enigma of biological asymmetric synthesis and it has played an important role in the development of our understanding of enzyme mechanisms. It must be mentioned, however, that this has led to the popular 'one enzyme-one substrate' concept, a misconception that is prevalent among chemists. (See chapter 7.)

C.   Receptor Theory

Most drugs are specific and their action is usually explained on the basis of receptor theory. The basic idea of receptor sites can be traced back to Paul Ehrlich, the father of chemotherapy. Ehrlich envisaged a 'magic bullet' that could cure disease by attacking bacteria but not human tissues, just as certain dyes had an affinity for specific cells but left others unstained.

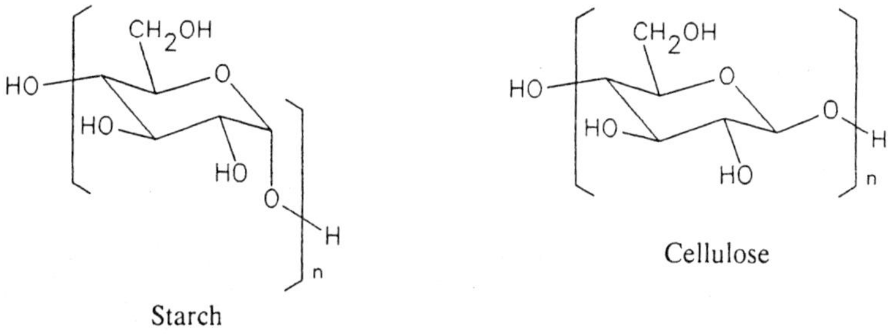

Figure 2-3. Structure of starch and cellulose.

The concept of a receptor site that binds selectively to certain bioactive agents (or chemical messengers) was introduced by Langley in 1906 in order to explain the observed effects of nicotine and curare on muscle tissue [4]. Ehrlich later coined the term **chemoreceptors** to describe these binding sites [5].

Receptor molecules in the body are proteins that exhibit high affinities for binding ligands of a certain molecular structure; this is completely analogous to enzyme substrate binding. The binding of a substrate to the receptor triggers a mechanism (e.g., modification of enzyme activity, transport of ions, etc.) that manifests itself in a biological response. Indeed, life processes are regulated by a sophisticated communication network involving specific interactions between chemical 'messengers' and receptor proteins. For example, endogenous messengers (e.g., neurotransmitters, hormones) regulate vital functions such as blood pressure, muscle contraction, and gastric acid secretion. On the other hand, drugs, pheromones, and pesticides can be regarded as exogenous messenger molecules. Similarly, specific interactions between olfactory receptors in the nose and flavor and fragrance substances are responsible for taste and odor perception. It should be noted, however, that taste is predominantly a nasal experience, as taste buds in the tongue can only distinguish salt, bitter, sweet, and sour tastes.

Whatever their physiological function, receptors have one thing in common: they are themselves chiral molecules and can, therefore, be expected to be enantioselective in their binding to messenger molecules.

D. The Three-Point Contact Model

The idea of enantioselectivity in drug-receptor interactions dates back to 1926 when Cushny [6] proposed that different bioactivities of two enantiomers arise from binding to sites of the same chirality. This was further elaborated by Easson and Stedman [7] in 1933 to become the widely-accepted **three-point contact** model (Figure 2-4). They postulated that the (more) active enantiomer binds more tightly because the sequence of the three groups around the asymmetric carbon atom, BCD, forms a triangular face of a tetrahedron that matches a complementary triad of the chiral binding site B'C'D' of the receptor protein. The less active enantiomer binds ineffectively since it has a mirror-image sequence of the three groups, DCB, which leads to a mismatch with the receptor binding sites.

Drug-receptor interactions can involve electrostatic effects, hydrophobic interactions, hydrogen bonding, and charge transfer processes, such as π–π interactions. The Easson-Stedman hypothesis provides a sound basis for understanding stereoselectivity in drug-receptor interactions, for example, those between adrenergic receptors and adrenergic blockers or stimulants [8]. When the body has a sudden need for energy (e.g., in a 'fight or flight' situation) the adrenergic hormone, noradrenaline, is released. This neurotransmitter, or chemical messenger,

A
B
D
C
A
D
B
C
B'
D'
C'
B'
D'
C'
MATCHING
MISMATCHING

Figure 2-4. The Easson-Stedman three-point contact model.

subsequently system that in turn gives rise to an increase in heart rate and force (i.e., a surge of energy).

Endogenous noradrenaline has the *(R)*-configuration and is produced in the brain by a series of enzyme-catalyzed reactions, starting from L-tyrosine (Figure 2-5). Only the *(R)*-isomer possesses the right stereochemical configuration for effective interaction with the adrenergic receptor binding sites.

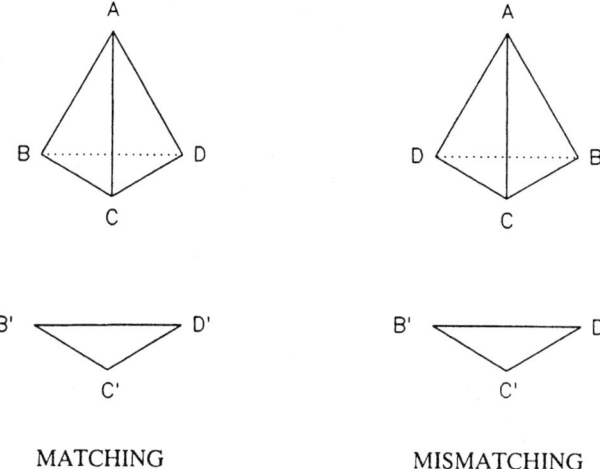

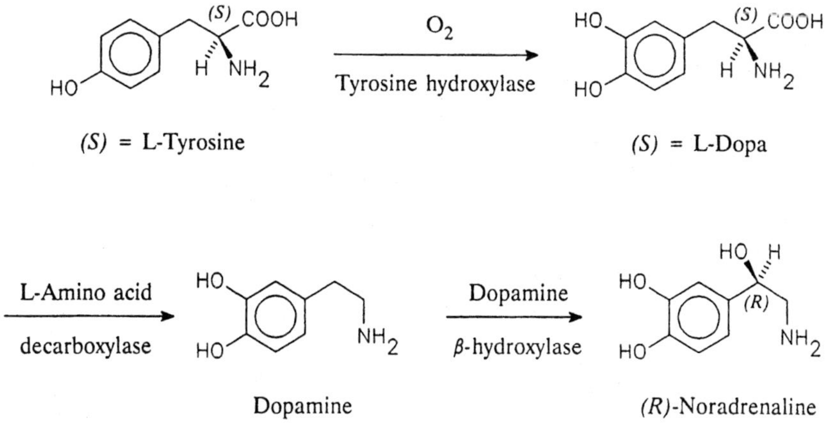

(S) COOH
H NH2
HO
O2
Tyrosine hydroxylase
HO
HO
(S) COOH
H NH2
(S) = L-Tyrosine
(S) = L-Dopa
L-Amino acid
decarboxylase
HO
HO
NH2
Dopamine
β-hydroxylase
HO H
HO
HO
(R)
NH2
Dopamine
(R)-Noradrenaline

Figure 2-5. Biosynthesis of *(R)*-noradrenaline.

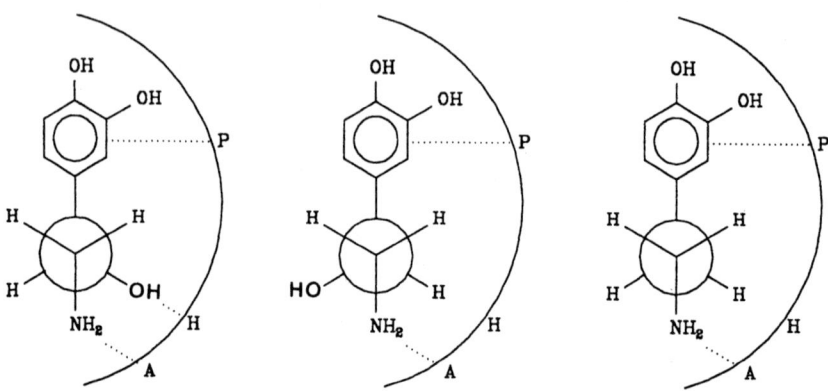

Figure 2-6. Schematic representation of the Easson-Stedman hypothesis applied to isomers of noradrenaline.

This is illustrated in Figure 2-6 for the interaction of *(R)*- and *(S)*-noradrenaline and the desoxy derivative, dopamine, with the receptor sites. The three binding sites of the messenger molecule are: i) the basic nitrogen atom, ii) the aromatic ring whose binding is enhanced by *m*- and/or *p*-hydroxyl groups, and iii) the benzylic hydroxyl group.

In the *(S)*-isomer and the desoxy derivative, dopamine, the benzylic hydroxyl group is incorrectly oriented or absent, respectively. This explains the significantly lower activity of the *(S)*- relative to the *(R)*- isomer and also accounts for the roughly equal activity of the *(S)*-isomer and the achiral analogue, dopamine. It should be noted that for a full appreciation of the three-point contact model, one has to bear in mind that we are dealing with an interaction between two three-dimensional molecules and that this is difficult to depict adequately in a two-dimensional drawing.

The three-point contact model (Figure 2-7) for pharmacological enantioselectivity became widely accepted and was later applied to understanding enzyme specificity following its rediscovery by Ogston [9] in 1948. Furthermore, the model has formed a useful basis for understanding chromatographic separations of enantiomers on chiral columns.

The current trend towards the rational design of drugs using computer graphics, based on X-ray diffraction and spectroscopic data, can be viewed as a further sophistication of the lock and key and three-point contact models. This has culminated in the design of a variety of enzyme inhibitors (e.g., the ACE-inhibitors) that block the action of particular enzymes in order to produce the desired therapeutic effect [10].

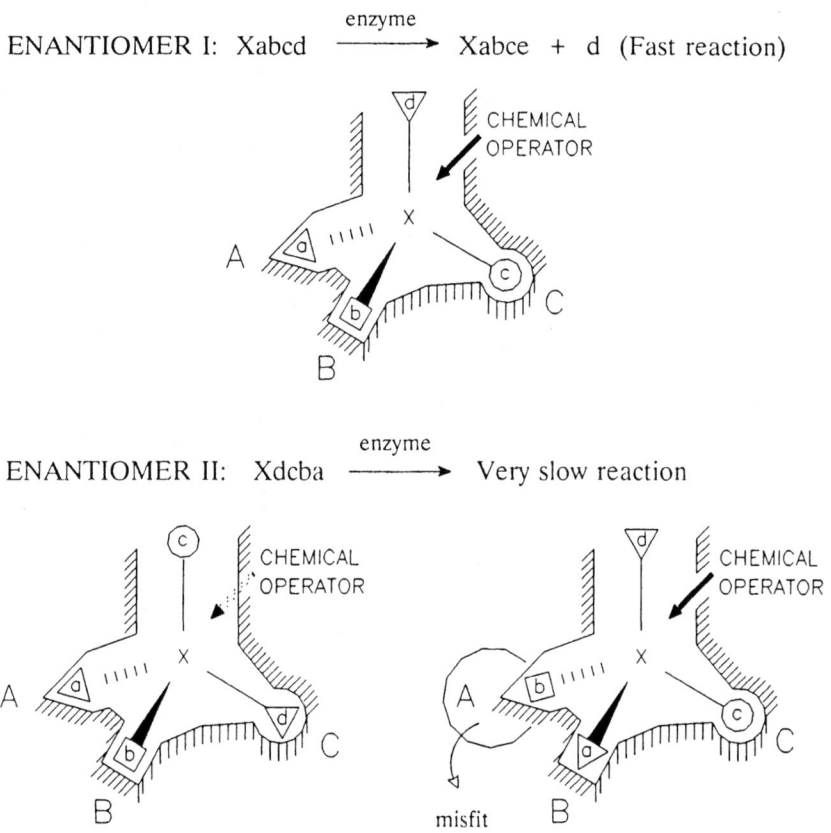

ENANTIOMER I: Xabcd $\xrightarrow{\text{enzyme}}$ Xabce + d (Fast reaction)

ENANTIOMER II: Xdcba $\xrightarrow{\text{enzyme}}$ Very slow reaction

Figure 2-7. Three-point contact model for enzyme-substrate binding.

E. Pfeiffer's Rule

A further consideration of the enantioselectivity of drug-receptor interactions led Pfeiffer [11] in 1956 to postulate that "the lower the effective dose of a drug, the greater the difference in the pharmacological effect of the optical isomers". This simple statement became known as Pfeiffer's rule. It is a logical corollary of the idea that a receptor-drug interaction involves a 'lock-and-key' fit of the molecule with the right configuration at the receptor site: the better the fit, the better the drug. If both enantiomers can fit into the active site, it is unlikely to be a good fit. This can be compared to a hand and a glove; if both a right and a left hand can fit equally well into a glove, it is unlikely to be a well-fitting glove (Figure 2-8).

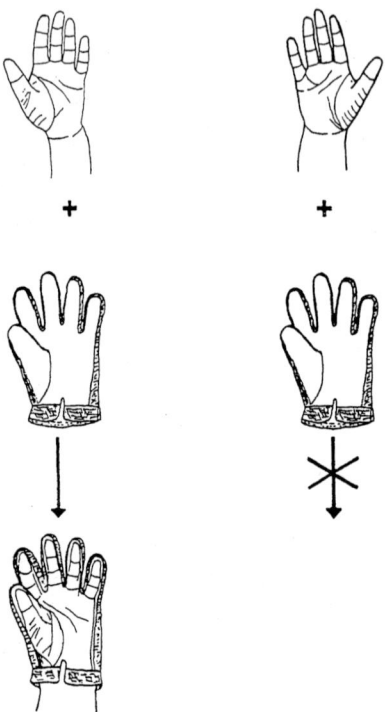

Figure 2-8. Enantioselectivity in glove-hand interactions.

Although Pfeiffer's rule was developed in the context of pharmacological activity, it is self-evident that it applies equally well to all bioactive substances. Bearing in mind that the concept of enantioselective drug action was already well-established in the 1930s, it is all the more remarkable that it was largely ignored in drug development until fairly recently.

F.   The Eudismic Ratio (ER)

Lehmann, Rodrigues de Miranda, and Ariëns [12] introduced the term **eudismic ratio** to describe Pfeiffer's rule (Table 2-1). The eudismic ratio is defined as the ratio of the activity of the active enantiomer (the eutomer) to that of the less active enantiomer (the distomer).

There is a direct relationship between the eudismic ratio and the effectiveness of the drug; that is, the higher the eudismic ratio, the more effective the drug or the lower its effective dosage. Therefore, it follows that if both enantiomers of a chiral substance show equal activity then it unlikely to be an effective bioactive agent.

Table 2-1 Nomenclature for Stereospecificity in Biological Activity

---

Eutomer: isomer with higher affinity (aff_{eu})
Distomer: isomer with lower affinity (aff_{dis})
Eudismic ratio (ER): aff_{eu}/aff_{dis}
Eudismic index (EI): $\log aff_{eu} - \log aff_{dis}$
ER and EI are measures of the stereospecificity of the substance

---

However, it should be noted that this principle strictly applies only to molecules that contain the stereogenic center in close proximity to the bioactive center, the 'business end' of the molecule as it were.

This historical development of concepts has hopefully shown that enantioselectivity in biological action is the rule and not the exception; we neglect it at our peril. We shall now consider the wide variety of effects that enantioselectivity in pharmacological action can lead to. Considering the enormous body of literature that has been generated on this subject, we shall only be able to treat a fraction of the known examples.

II. PHARMACEUTICALS

An overwhelming majority of naturally occurring medicinal agents are chiral molecules; moreover, they almost all exist in nature (and are marketed) as the single, active enantiomer. In contrast, for many years it was common practice to market synthetic chiral drugs as racemates (Figure 2-9). However, the situation is rapidly changing and a perusal of the top twenty drugs (based on revenues) in 1990 reveals a definite trend towards enantiopurity in chiral synthetic drugs (Table 2-2). Moreover, the trend towards rational design of drugs tends, almost by definition, to produce more complicated molecules and, hence, an increased probability that they are chiral.

A. The Main Phases in Biological Action

For an understanding of the biological effect of xenobiotics, such as drugs and pesticides, it is important to distinguish three main phases in their action [13]. The initial **exposure phase** is followed by the **pharmacokinetic phase** which involves the absorption, metabolic conversion, and excretion of the drug. The exposure and pharmacokinetic phase together determine the bioavailibility of the drug. The **pharmacodynamic phase** involves the interaction of the bioactive agent with the molecular site of action (receptors, enzymes, etc.) in the target tissue leading to the observed effect.

It is important to realize that enantioselectivity plays an important role not only in the pharmacodynamic phase but also in the pharmacokinetic. Metabolic conver-

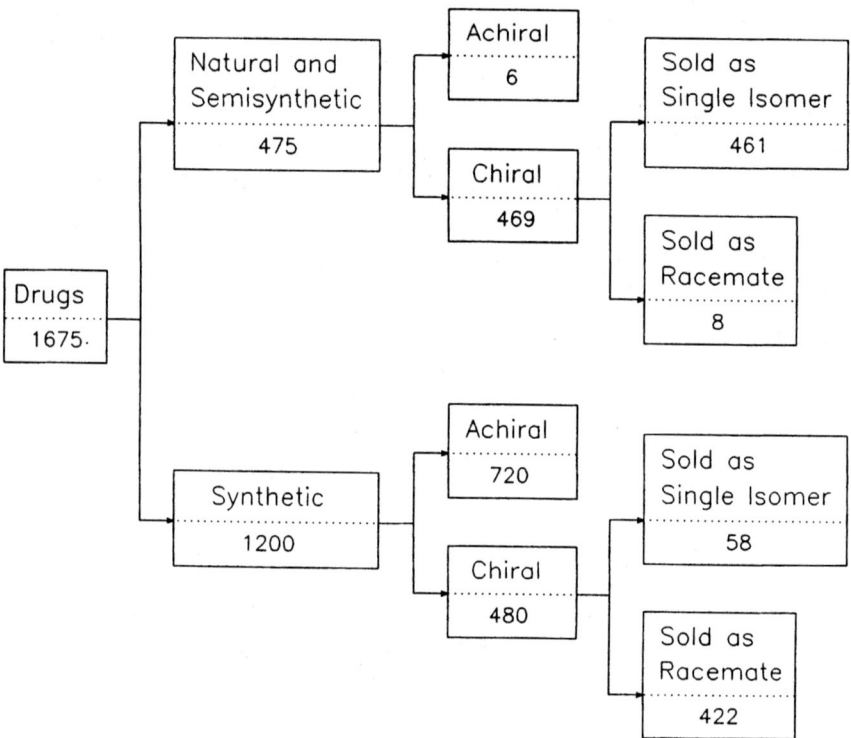

Figure 2-9. Chirality of drugs and their application as single isomers or racemates prior to 1982 (taken from ref. [30]).

Table 2-2 World-Wide Sales (Bulk Dosage Form) of the Top Twenty Drugs (Based on Sales) in 1990

	Drug	Therapeutic class	Sales ($ million)
‡	Ranitidine	Antiulcer	2370
*	Amoxicillin	Antibiotic	2000
*	Ampicillin	Antibiotic	1800
*	Captopril	ACE-inhibitor	1520
*	Enalapril	ACE-inhibitor	1500
†	Ibuprofen	NSAID	1400
‡	Nifedipine	Calcium antagonist	1290
*	Cefaclor	Antibiotic	1040
‡	Cimetidine	Antiulcer	1030
†	Atenolol	Beta-blocker	1020
‡	Diclofenac	NSAID	975
*	Diltiazem	Calcium antagonist	960
*	Naproxen	NSAID	950
*	Cefalexin	Antibiotic	900
*	Lovastatin	Antihypercholesteremic	750
‡	Famotidine	Antiulcer	630
†	Iohexol	Contrast medium	620
*	Cefatriaxone	Antibiotic	600
‡	Proxicam	NSAID	585
†	Albuterol	Bronchodilator	555

* optically pure (10 drugs)

† racemic (4 drugs)

‡ achiral (6 drugs)

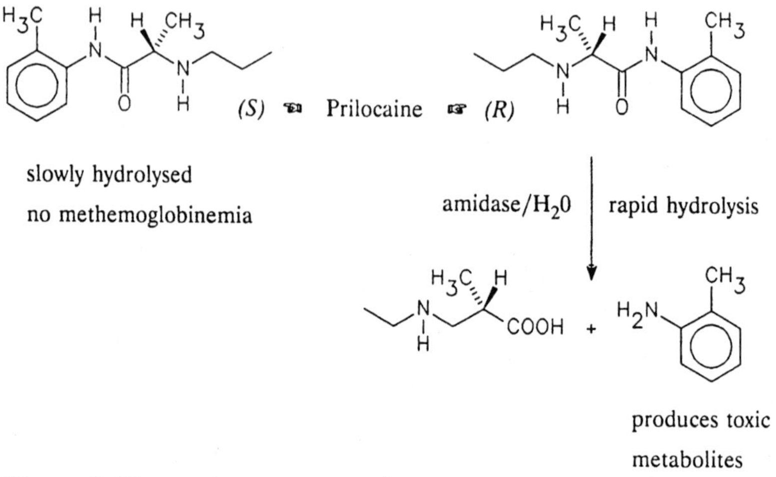

Figure 2-10. Enantiospecific biotoxification in prilocaine.

sion of xenobiotics involves enzyme-mediated processes that can be expected to be enantioselective with chiral and and enantioselective with prochiral substances. This can lead to the enantioselective or enantiospecific formation of metabolites exhibiting undesirable, possibly toxic, side-effects. An example of such an enantiospecific biotoxification is provided by the local anesthetic, prilocaine (Figure 2-10). The (R)-enantiomer of prilocaine is rapidly metabolized by enzymatic hydrolysis to give o-toluidine, a substance that causes methemoglobinemia. The (S)-enantiomer, in contrast, is only slowly hydrolyzed and shows no side-effects [14].

A good example of a pharmacologically important biotransformation involving prochiral stereospecificity is the in vivo conversion of dopamine to (R)-noradrenaline (Figure 2-5). However, in this case the product is not an undesirable metabolite but a hormone.

The overall pharmacological effect of a drug is thus influenced by both its pharmacodynamics and pharmacokinetics, and enantioselectivity plays an important role in both phases. The therapeutically inactive isomer (distomer) of a racemic drug should be regarded as an undesirable impurity that is a different (potentially toxic) pharmacological entity.

As has been pointed out by Ariëns, a neglect of stereoselectivity can lead to the generation of 'highly sophisticated scientific nonsense' in clinical pharmacology [15]. For example, nonclinical assays of plasma concentrations of drugs administered as racemates provide meaningless data since they only give information on the sum of the eutomer and distomer. This has been compared to drawing conclusions about individual spouses by measuring the age or body weight of a married couple [16].

In short, different pharmacodynamics and pharmacokinetics of the eutomer and the distomer in a racemic drug can lead to a variety of effects attributable to inactive isomers. We shall now consider the different situations that are encountered in practice.

B. The Distomer Shows No Serious Side-Effects: Isomeric Ballast

Beta-Blockers

In some cases chiral drugs are administered as racemates since the distomer displays no undesirable side-effects (or no more than the eutomer does). An example is provided by the beta-blockers that produce an antihypertensive effect (lowering of the blood pressure) by blocking adrenergic receptors. Beta-blockers are chiral aryloxypropanolamines which have the general structure shown in Figure 2-11. Their therapeutic effect is due almost entirely to the *(S)*-enantiomer that bears a strong structural resemblance to the adrenergic hormone, noradrenaline. *(S)*-propanolol is 130 times as active as its *(R)*-enantiomer (i.e., ER = 130) which effectively means that the latter is totally inactive (Note that low measured activities of distomers can be due to small amounts of the eutomer in the sample).

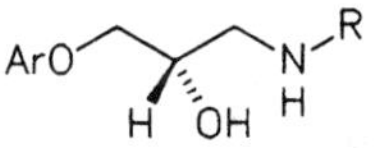

(S)- BETA-BLOCKER
(general structure)

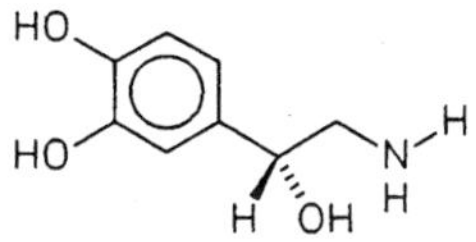

(R)- Noradrenaline (ER = 300)

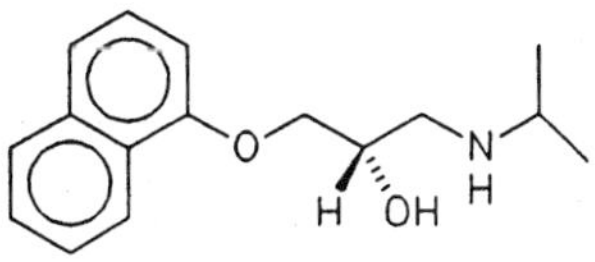

(S)- Propanolol (ER = 130)

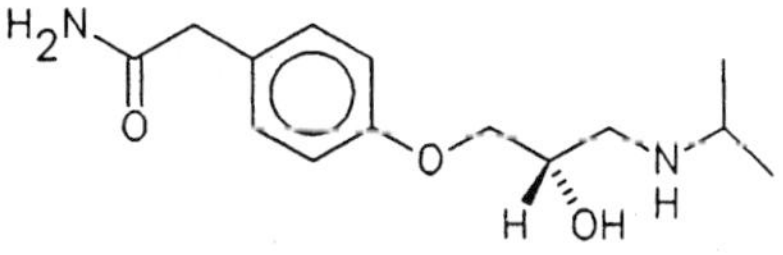

(S)- Atenolol (ER = 12)

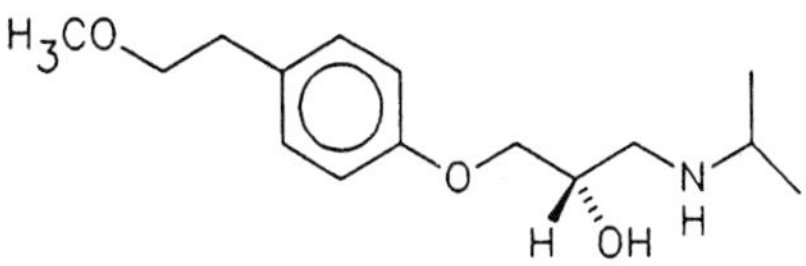

(S)- Metoprolol (ER = 270)

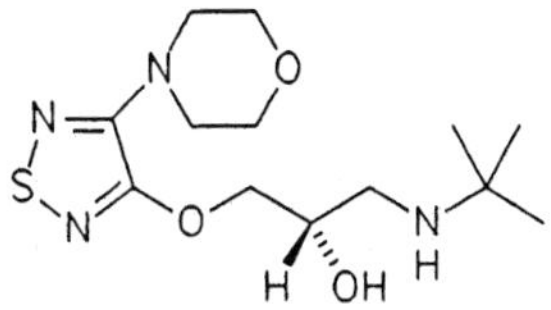

(S)- Timolol (ER = 50)

Figure 2-11. Structures of the eutomers of beta-blockers and the beta-adrenergic stimulant, noradrenaline.

The eudismic ratios of the various beta-blockers vary considerably (Figure 2-11) but in all cases the *(S)*-enantiomer is the eutomer [17]. The three most important beta-blockers—propanolol, atenolol, and metoprolol—are all marketed as the racemate. The decision to market these drugs as racemates was probably influenced by the fact that they are difficult to separate by classical methods and that the distomer exhibits no serious side-effects. In contrast, timolol is marketed as the pure *(S)*-enantiomer [18] and is mainly used for treatment of glaucoma. Interestingly, it has been pointed out [13,17] that the *(R)*-enantiomer would probably have been a better choice for this application where reduction of intraocular pressure is the desired effect and beta-blocking activity can be an undesirable side-effect.

If the beta-blockers were introduced today as new drugs they would almost certainly be marketed as single enantiomers. Thus, although the distomers generally exhibit no serious side-effects they still constitute unnecessary isomeric ballast [19] and produce the same side-effects as the eutomer without contributing to the desired therapeutic effect.

α-Methyldopa

The antihypertensive agent α-methyldopa stands out as one of the first and classic examples of a synthetic chiral drug that was developed as a single enantiomer [18]. α-Methyldopa suppresses the formation of noradrenaline by inhibiting the enzymatic decarboxylation of L-dopa. In doing so, α-methyldopa is itself decarboxylated, generating α-methyldopamine and, subsequently, α-methylnoradrenaline (Figure 2-12). The latter also contributes to the overall antihypertensive effect, presumably by blocking beta-adrenergic receptors.

It is only the L-*(S)*-enantiomer of α-methyldopa that acts as a substrate for the aromatic amino acid decarboxylase and hence produces an antihypertensive effect.

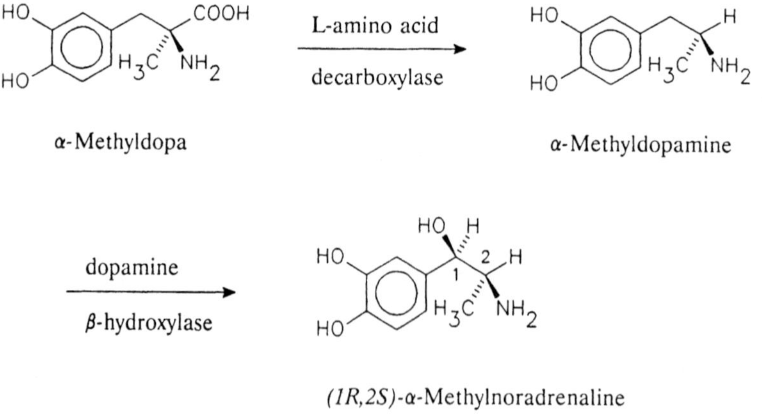

Figure 2-12. In vivo biotransformation of α-methyldopa.

The distomer only contributes to the overall toxicity of the drug. Consequently, α-methyldopa was marketed from the beginning as a single enantiomer. (See chapter 6 for its synthesis.)

C. The Distomer Exhibits Undesirable Side-Effects

In some cases the distomer may exhibit undesirable (toxic) side-effects that are not characteristic of the eutomer. This would seem to be a clear case for marketing the single eutomer. For example, ketamine is a widely used anesthetic and analgesic agent that is chemically similar to phencyclidine, a drug of abuse. Postanesthesia side-effects observed with ketamine include hallucination and agitation. Ketamine is a chiral molecule (Figure 2-13) that is administered as a racemate. However, a comparison of the two enantiomers [20,21] has shown that the (S)-(+)-ketamine is the active anesthetic but that the undesirable side-effects are overwhelmingly associated with the (R)-(−)-distomer.

Other examples where the distomer exhibits such toxic side-effects that the drug can only be administered as a pure enantiomer include the antiarthritic agent penicillamine and the antitubercular agent ethambutol (Figure 2-13). In the former the (S)-enantiomer is the eutomer while (R)-penicillamine is extremely toxic [13]. In the case of ethambutol it is the (S,S) isomer that is an active tuberculostatic with a eudismic ratio of about 200. The (R,R) enantiomer causes optical neuritis that can result in blindness. Similarly, L-dopa, which is used for treatment of Parkinson's disease, is marketed as the pure enantiomer because serious side-effects are attributable to the D-isomer, such as granulocytopenia [21].

Probably the most well-known and tragic example of a drug where the distomer causes serious side-effects is thalidomide, which was sold in the 1960s as a sedative and administered as a racemate. Unfortunately, it was not known at the time that although the (R)-enantiomer is an effective sedative, the (S)-enantiomer is highly teratogenic and causes fetal abnormalities [22]. Obviously in cases where the distomer alone is primarily responsible for side-effects, a drug should definitely be marketed as a single enantiomer (or not at all depending on the seriousness).

D. Both Isomers Have Independent Therapeutic Value

In some cases both isomers of a chiral drug can have a desirable, but different, therapeutic effect. This is indeed the case with many natural products (e.g., quinine and quinidine) and also with synthetic drugs. An elucidative example is provided by propoxyphene (Figure 2-14). Dextropropoxyphene is an analgesic agent while levopropoxyphene is devoid of analgesic properties but is an effective antitussive [13, 21,23]. Interestingly, their mirror-image relationship is reflected in the trade names under which the two drugs are marketed by Eli Lilly. Examples of separation of two different types of useful activity in two optical isomers is also observed with pesticides.

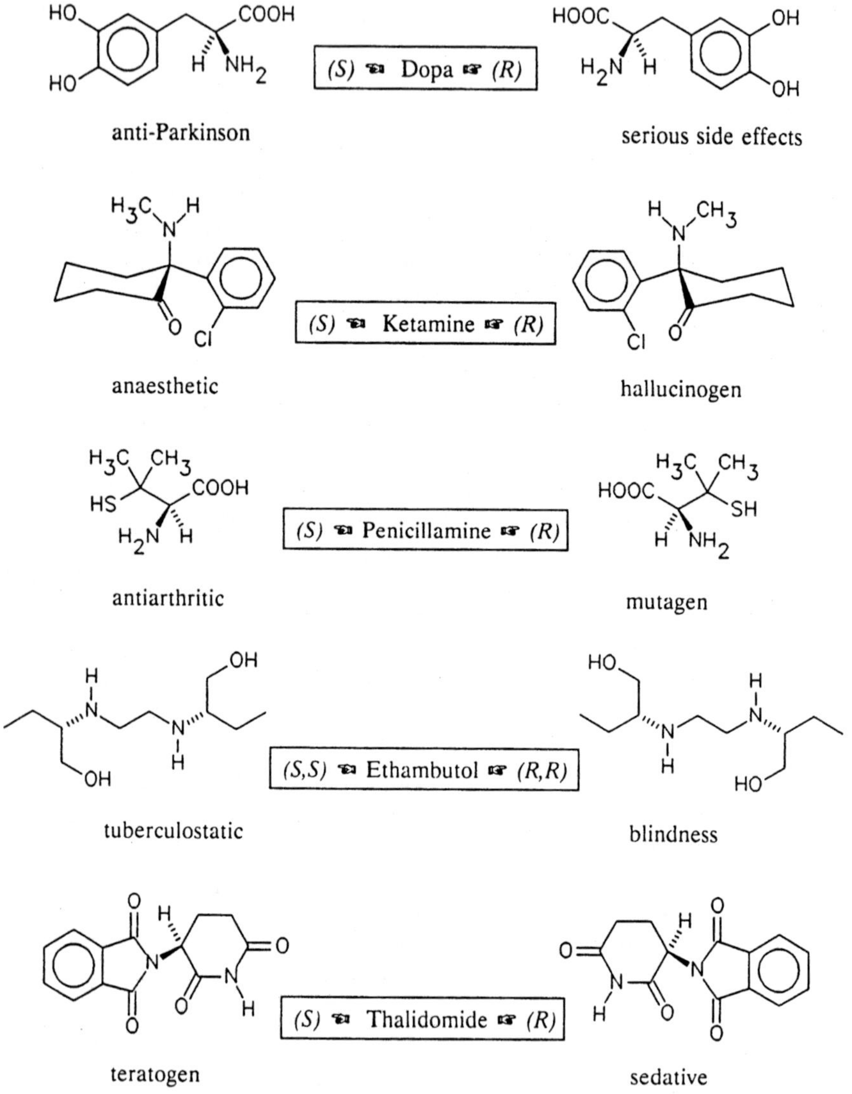

HO—⟨⟩—CH—COOH with H, NH₂
anti-Parkinson

(S) ← Dopa → (R)

HOOC—CH—CH₂—⟨⟩—OH, OH with H₂N, H
serious side effects

(S) ← Ketamine → (R)

anaesthetic

hallucinogen

(S) ← Penicillamine → (R)

antiarthritic

mutagen

(S,S) ← Ethambutol → (R,R)

tuberculostatic

blindness

(S) ← Thalidomide → (R)

teratogen

sedative

Figure 2-13. Examples of chiral drugs where the distomer exhibits undesirable or toxic side effects.

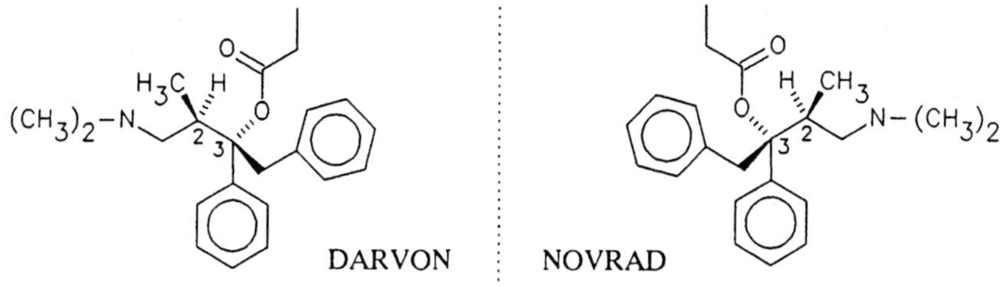

DARVON   NOVRAD

*(2R,3S)-(+)-*Dextropropoxyphene

(analgesic)

*(2S,3R)-(-)-*Levopropoxyphene

(antitussive)

Figure 2-14. Enantiomers exhibiting different types of therapeutic effect.

E. Combination Product Has Therapeutic Advantages

There are a few examples of chiral drugs where both enantiomers contribute, in different ways, to the overall desired effect. This can involve the distomer acting as an antagonist for undesirable side-effects of the eutomer. The investigational diuretic, indacrinone, is a good example of this phenomenon (Figure 2-15). The *(R)-(+)*-enantiomer is the active diuretic. In common with many other diuretics, the *(R)*-isomer also exhibits the undesired side-effect of uric acid retention. The *(S)*-enantiomer acts as an uricosuric, that is, it promotes uric acid secretion and, therefore, antagonizes the undesired side-effect of the *(R)*-isomer.

A cursory inspection of these facts may seem to provide a good argument for marketing this product as a racemate, since both enantiomers are required for the optimum effect. However, it would certainly be an amazing coincidence if both enantiomers showed the same quantitative activity, that is, that the 1:1 racemic mixture is the ideal combination. Indeed, studies have shown that a 9:1 mixture of the two enantiomers affords an optimal therapeutic profile [18,24].

(R)-Indacrinone   ☞ [*(R)* + *(S)*] ☞   *(S)*-Indacrinone

(diuretic)   (isouricemic diuretic)   (uricosuric)

Figure 2-15. Enantiomers of indacrinone.

The example of indacrinone indicates that, in cases where the distomer increases the efficacy of the therapeutic eutomer, marketing a tailor-made mixture of enantiomers can be beneficial. It is highly unlikely, however, that this will be the 1:1 racemic mixture.

F. Metabolic Chirality Inversion: The Arylpropionic Acid Antiinflammatory Drugs

The α-arylpropionic acids, typified by ibuprofen and naproxen (Figure 2-16), constitute the largest single group of drugs used in the treatment of rheumatoid arthritis and as general analgesics; ibuprofen, for example, is a widely used over-the-counter drug for headache and minor pains. These drugs are all chiral molecules and it is the *(S)*-enantiomer that is responsible for the desired therapeutic effect (e.g., the ER for naproxen is 30). Interestingly, the inactive *(R)*-enantiomers undergo an unidirectional metabolic inversion of configuration to afford the active *(S)*-enantiomer [25].

In other words, when the drug is administered as a racemate the distomer is converted in vivo into the eutomer while the latter is unaffected. Seemingly an open and shut case for marketing the drug as a racemate and, with the notable exception of naproxen, most of the α-arylpropionic acids are administered as racemates. However, the situation is more complex, as becomes evident on examining the mechanism of metabolic inversion (Figure 2-17).

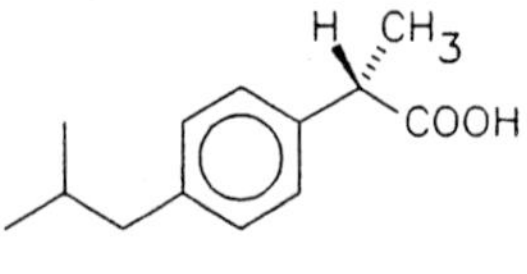

(S)-Ibuprofen

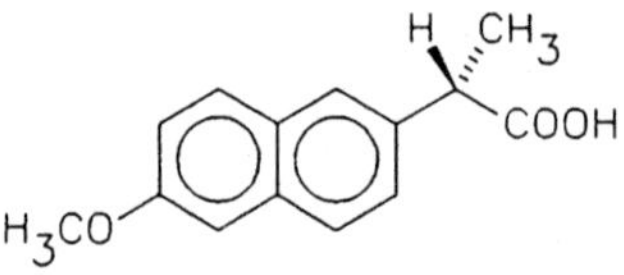

(S)-Naproxen *

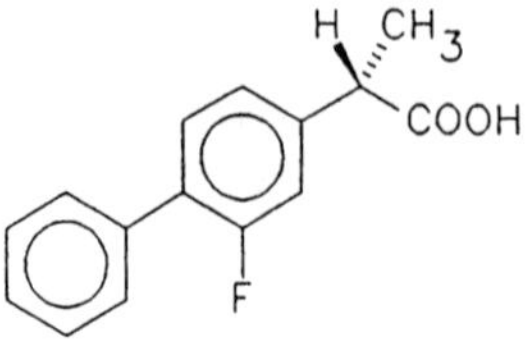

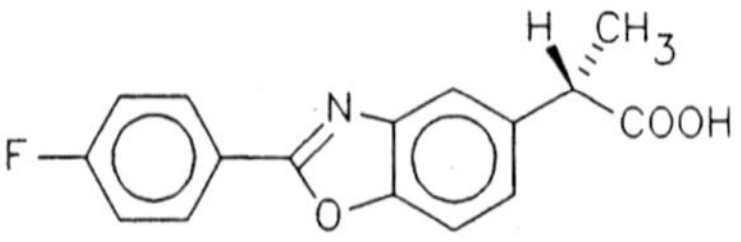

(S)-Flurbiprofen

(S)-Flunoxaprofen *

* marketed as a single enantiomer

Figure 2-16. Structures of α-arylpropionic acid nonsteroidal antiinflammatory drugs.

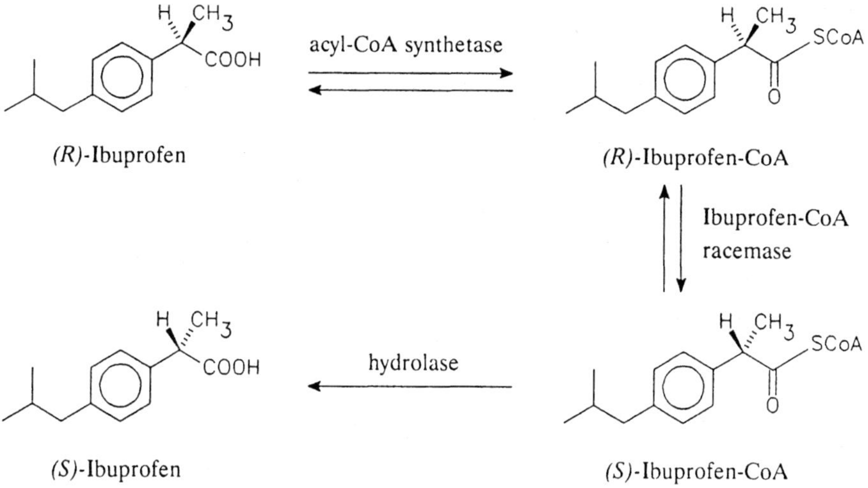

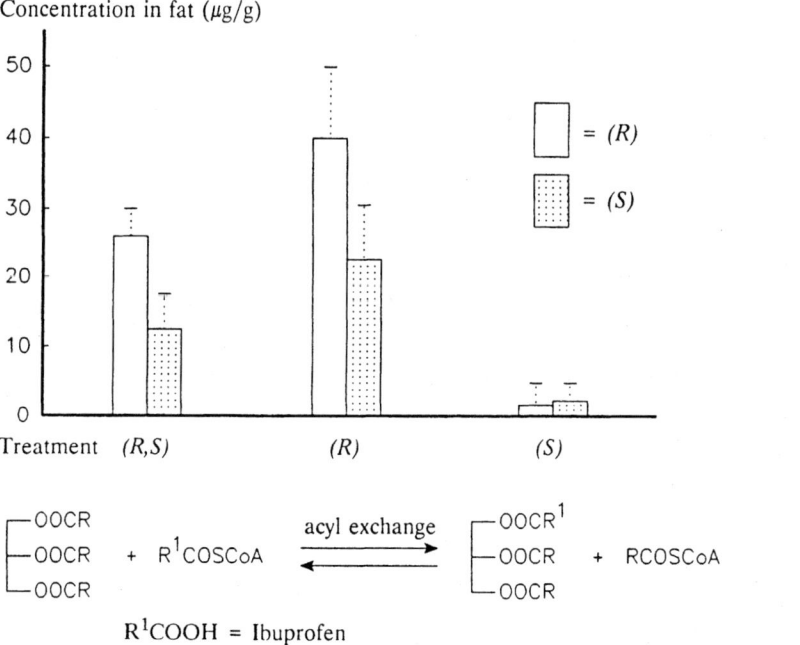

Figure 2-17. Mechanism of metabolic inversion of ibuprofen and related antiinflammatory drugs [25].

R^1COOH = Ibuprofen

Figure 2-18. Concentrations of the enantiomers of ibuprofen in fatty tissue following chronic treatment of rats with either racemic *(R,S)* drug or with the individual enantiomers (taken from reference [26]). (Note: Incorporation observed with the *(S)*-enantiomer is attributable to small amounts of the *(R)*-enantiomer present as an impurity.)

The enantiospecificity of the inversion is controlled by the enzyme, acyl-coenzyme-A synthetase, that converts *(R)*-ibuprofen to the corresponding coenzyme A thioester. Racemization of the latter and subsequent hydrolysis yields *(S)*-ibuprofen. In competition with the hydrolysis, the thioester intermediate can also undergo acyl exchange with the endogenous triacylglycerols [25,26]. This results in accumulation of ibuprofen residues in fatty tissue. Since the *(S)*-enantiomer does not form the coenzyme-A thioester, it cannot be incorporated into fatty tissue. This was shown by measuring the concentrations of ibuprofen in fatty tissue following chronic treatment of rats with either racemic ibuprofen or with the individual enantiomers (Figure 2-18).

The long term effects of accumulation of ibuprofen residues in fatty tissue are unknown but toxic side-effects cannot be completely ruled out. The words of Simonyi [27] are appropriate in this context: "if a substance remains in the body it is not a drug but a poison." The risk of side-effects can, in any case, be avoided by administering the pure *(S)*-enantiomer.

G. Modern Synthetic Drugs — The ACE-Inhibitors

A perfect illustration of the modern trend in the development of drugs is provided by the angiotensin converting enzyme (ACE) inhibitors [28,29], some examples of which are shown in Figure 2-19. What is remarkable about these antihypertensive drugs is that they are all synthetic, chiral substances, containing two or more asymmetric centers and that they have, without exception, been developed and marketed as single optical isomers. These drugs have revolutionized the treatment of hypertension and the first two to be launched, captopril (1981) and enalapril (1984) now have world-wide sales far exceeding $1000 million (Table 2-2). The prizes in the optical purity stakes must, however, go to perindopril and ramipril both of which have five asymmetric centers and are marketed as one of the 32 (2^5) possible optical isomers.

ACE-inhibitors, as their name suggests, suppress the formation of the powerful vasoconstrictor, angiotensin II by selectively binding to the active site of the enzyme responsible for its formation. This is illustrated in Figure 2-20 for captopril.

Bearing in mind the earlier discussion of the three-point contact model for stereoselective binding, it should be obvious that only one of the four possible optical isomers of captopril has the right stereochemistry for effective binding to the active site. Therefore, it is gratifying to conclude that these drugs have been developed with a full appreciation for the implications of chirality in the context of biological activity.

H. Semisynthetic Penicillins and Cephalosporins

Another important class of drugs that are overwhelmingly marketed as single optical isomers, comprises the semisynthetic antibiotics (penicillins and cephalo-sporins). The so-called nucleus part of these molecules, represented by 6-amino-

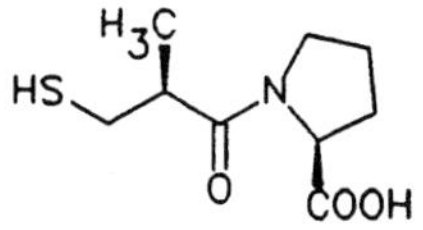

Captopril (Squibb)

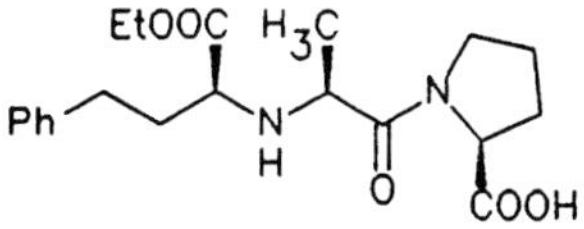

Enalapril (Merck)

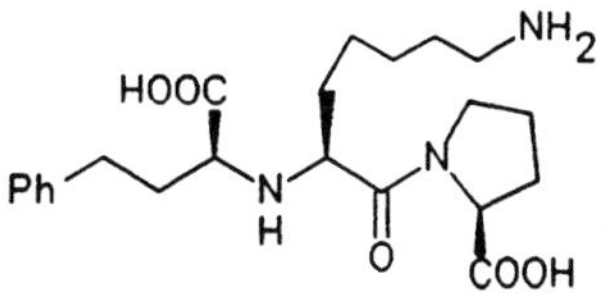

Lisinopril (Merck)

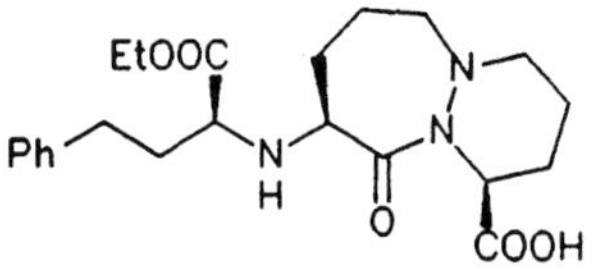

Cilazapril (Hoffmann-La Roche)

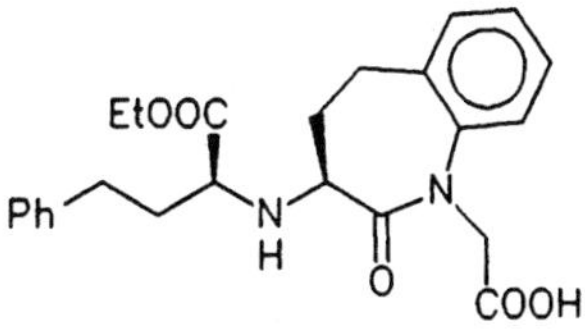

Benazapril (Ciba-Geigy)

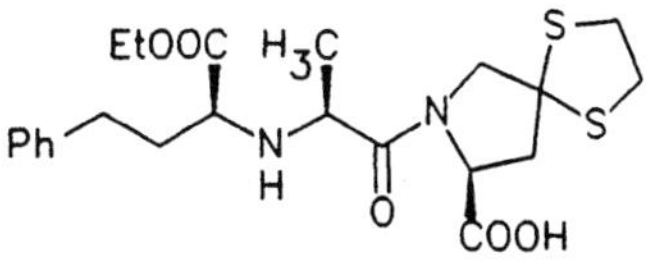

Spirapril (Schering-Plough Sandoz)

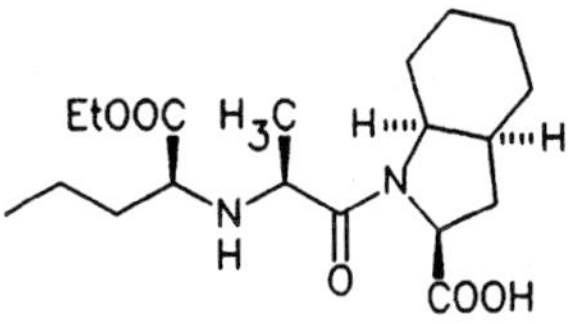

Perindopril (Servier)

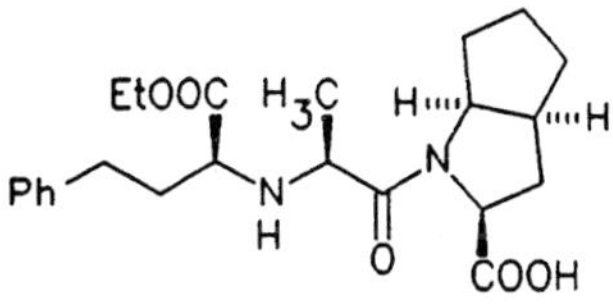

Ramipril (Hoechst)

Figure 2-19. Structure of various ACE-inhibitors.

penicillanic acid (6-APA) and 7-aminodesacetoxycephalosporinic acid (7-ADCA), respectively, are manufactured by fermentation. (See chapter 4.) The nucleus is attached by the amino group to a side-chain which is generally chiral and in many cases an unnatural amino acid. The most important members of this class of antibiotics, the penicillins ampicillin and amoxycillin, contain D-phenylglycine and D-p-hydroxyphenylglycine, respectively, as the side chain (Figure 2-21).

Finally, the importance of pharmaceuticals as an outlet for enantiomerically pure chemical intermediates is underscored by reference to the sales of the top ten optically active drugs in 1990 (Table 2-3). Moreover, based on the current trend

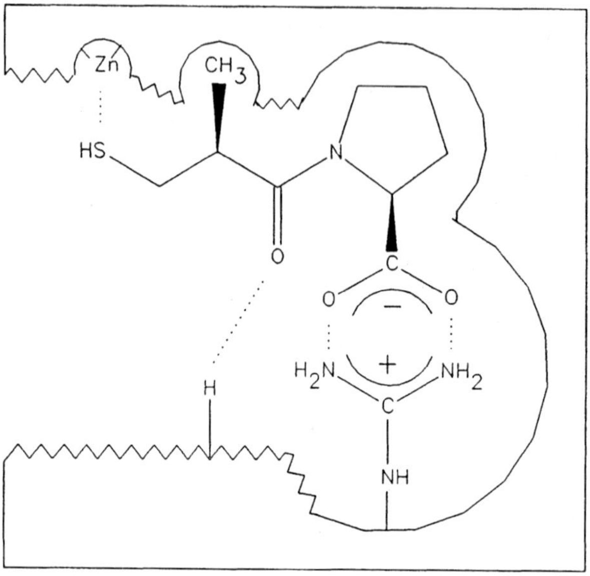

Figure 2-20. Schematic representation of the binding of captopril to the active site of angiotensin converting enzyme.

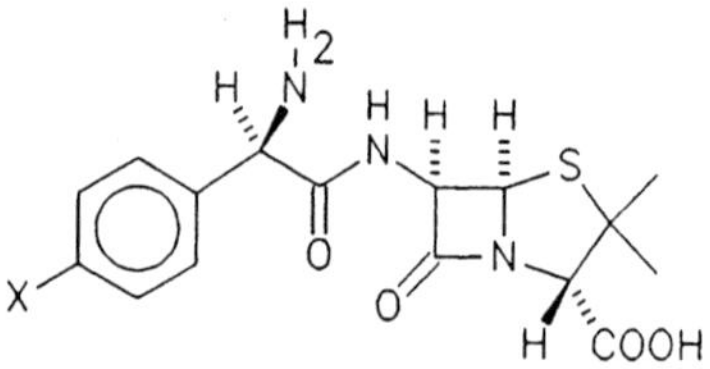

Ampicillin (X= H)

Amoxicillin (X= OH)

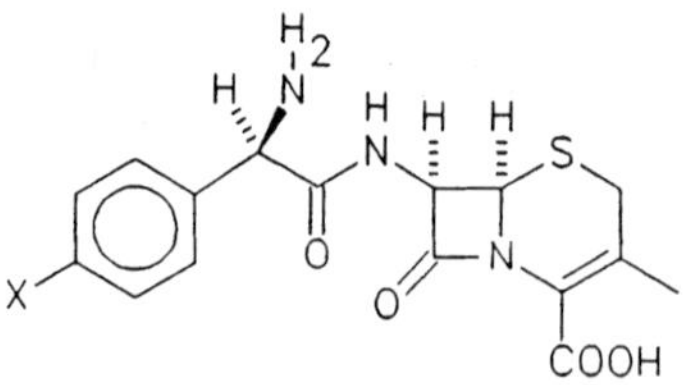

Cefalexin   (X= H)

Cefadroxil   (X= OH)

Figure 2-21. Structures of semisynthetic penicillins and cephalosporins.

Table 2-3 Optically Active Top Ten Drugs (1990)

Drug	Therapeutic Class	Sales ($ Million)
Amoxycillin	Antibiotic	2000
Ampicillin	Antibiotic	1800
Captopril	ACE-inhibitor	1520
Enalapril	ACE-inhibitor	1500
Cefaclor	Antibiotic	1040
Diltiazem	Calcium antagonist	960
Naproxen	NSAI	950
Cefalexin	Antibiotic	900
Lovastatin	Antihypercholesteremic	700
Ceftriaxone	Antibiotic	650

towards marketing drugs as pure enantiomers, the need for enantioselective production methods can only be expected to increase.

III. AGROCHEMICALS

The phenomenon of enantioselectivity in biological activity is universal. It is also exhibited by agrochemicals [30] that act on living organisms—plants, insects, and fungi. The molecular mechanisms of action are completely analogous to those encountered with drugs, that is, they involve interactions of the bioactive substance with receptor proteins or enzymes in the target organism. Here also, the desired effect will reside predominantly in one isomer, the eutomer, while the other isomer, the distomer, will either constitute unnecessary isomeric ballast or exhibit undesirable side-effects. One important difference between agrochemicals and drugs is that the former are targeted at the organism (i.e., they should be species selective) while drugs are targeted at tissues.

The overwhelming majority of chiral pesticides are marketed as racemates (Figure 2-22), even more so than with drugs (compare with Figure 2-9) but this situation is changing rapidly. Furthermore, it is worth emphasizing that with pesticides the quantities involved are of a different order of magnitude than with drugs, often several thousands of tons; therefore, the production of a chiral pesticide as a single optical isomer means a 50% (or more) reduction in the chemical burden on the environment. In the current climate of growing public awareness and concern about the environment, this would seem to be a laudable objective. Moreover, there is definite trend towards the rational design of pesticides that are more effective (i.e., lower dosage), more selective in their action, and more environmentally friendly. As with drugs, this is leading to the design of more

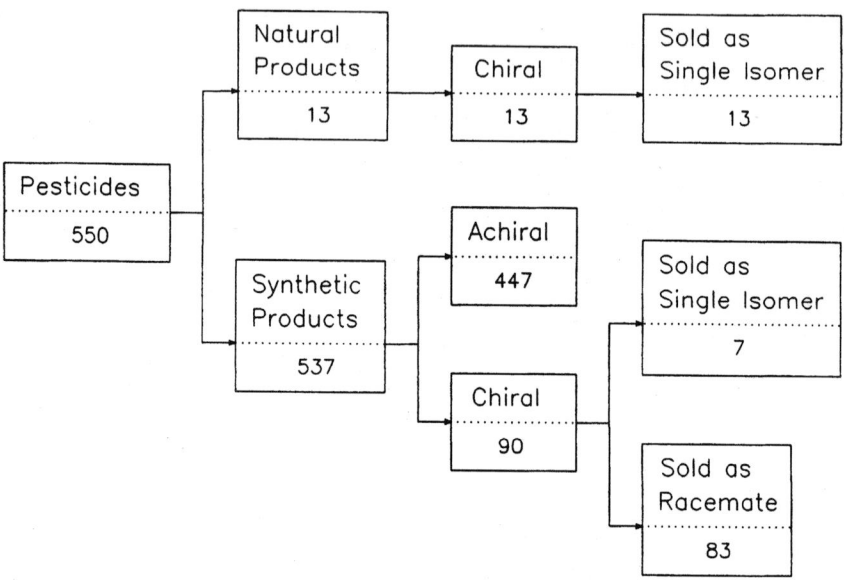

Figure 2-22. Chirality of pesticides and their application as single isomers or racemates prior to 1981 (taken from ref. [30]).

complicated molecules where the chance of them being chiral is increased. The pyrethroid insecticides (see section III-B) are a perfect illustration of this trend.

A. α-Aryloxypropionic Herbicides

The α-aryloxypropionic acids—such as mecoprop, diclofop, pyrenifop, and fluazifop—constitute a commercially important class of herbicides (Figure 2-23). They are chiral molecules and the herbicidal activity (inhibition of the activity of the plant growth hormone, indole-3-acetic acid) resides virtually entirely in the (R)-enantiomer [30,31].

The (R)-enantiomer of α-aryloxypropionic acid bears a striking structural resemblance to the active (S)-enantiomer of the α-arylpropionic acid antiinflammatory drugs. (Note: the configuration is the same but that the priority of the groups has changed.) Interestingly, both the α-arylprprionic acid drugs and the α-aryloxy-proprionic acid herbicides were developed following the same line of reasoning. In both cases, the activity was first discovered in the corresponding substituted acetic acid. Unfortunately, these initial compounds underwent rapid in vivo biodegradation to inactive metabolites. A commonly used approach in such cases is to introduce a suitable substituent, often a methyl group, at the metabolically vulnerable position in the molecule [30]. This tends to inhibit degradation of the molecule

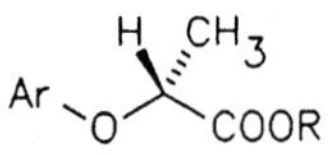

Alkyl *(R)*-α-aryloxypropionate
(herbicide)

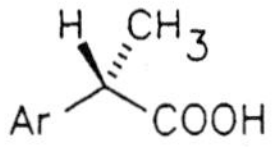

(S)-α-Arylpropionic acid
(anti-inflammatory)

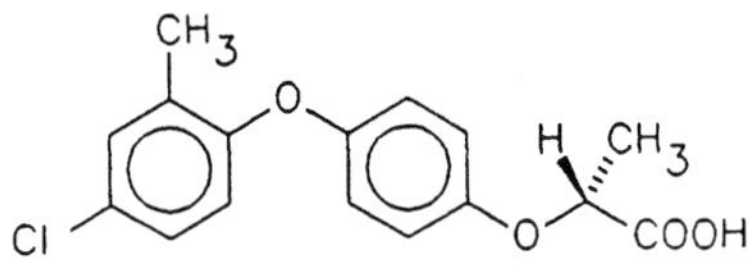

(R)-Mecoprop

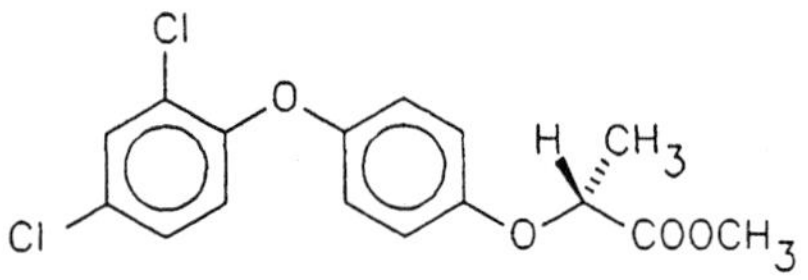

(R)-Diclofop-methyl

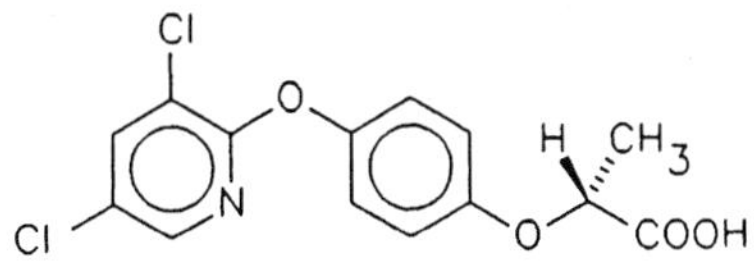

(R)-Pyrenifop

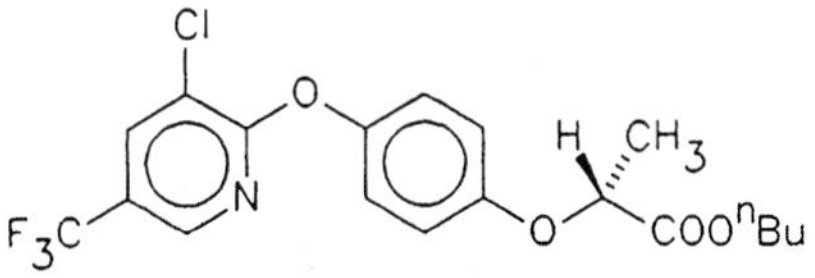

(R)-Fluazifop-butyl

Figure 2-23. Structures of the eutomers of α-aryloxypropionic acid herbicides.

by enzymes present in the target organism, thereby prolonging the duration of action and decreasing resistance. Hence, introduction of an α-methyl group led to the development of the active herbicides and antiinflammatory drugs, respectively. This is illustrated in Figure 2-24 for mecoprop.

The introduction of the α-methyl group also introduces chirality into the molecule and the steric blocking of enzymatic attack may only be significant for

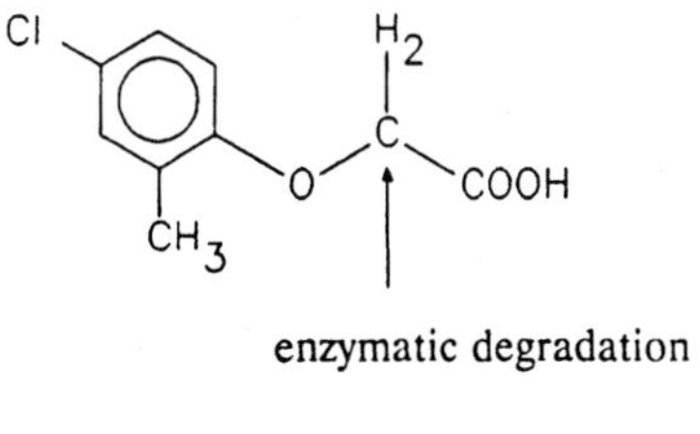

enzymatic degradation

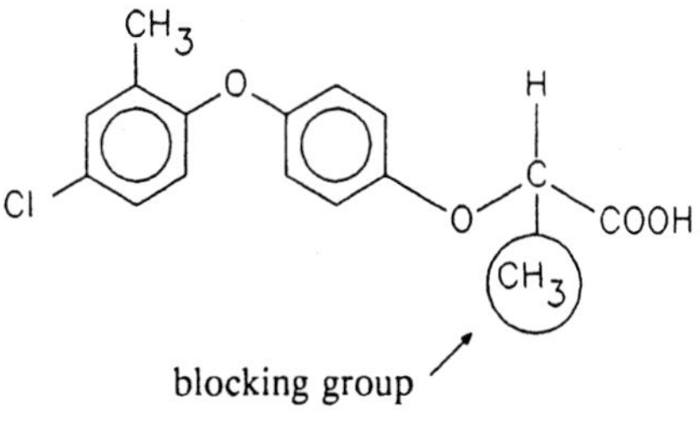

blocking group

MCPA = *4*-Chloro-2-methylphenoxyacetic acid

Mecoprop

rapid biodegradation (resistant)

slow biodegradation (not resistant)

Figure 2-24. Control of resistance in herbicides by stabilization to metabolic degradation.

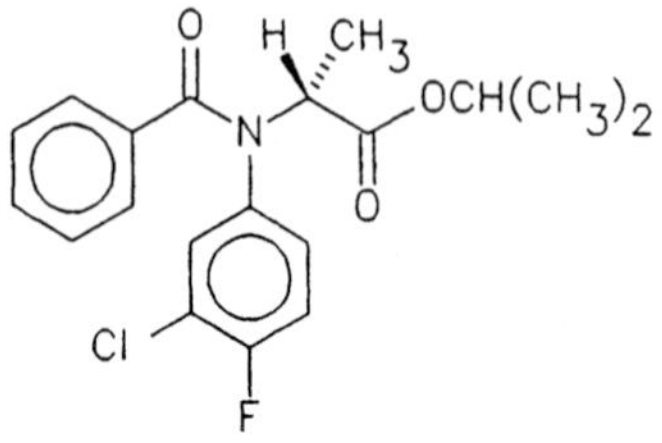

(R)-Flamprop-isopropyl

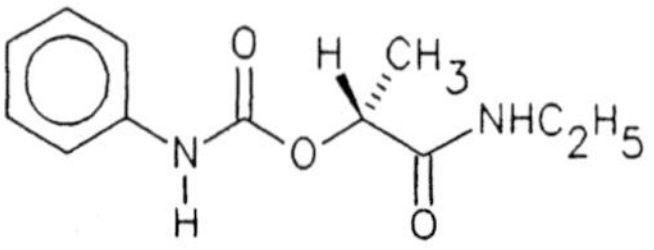

(R)-Carbetamide

Figure 2-25. *(R)*-Enantiomer of flamprop-isopropyl and of carbetamide.

one of the two enantiomers. In some cases, this isomer may still bind to the enzyme but not be converted, thus inhibiting degradation of the other isomer.

In common with the α-arylpropionic acid drugs, some of the α-aryloxypropionic acid herbicides also undergo metabolic inversion of configuration. For example, inversion of the inactive *(S)*-enantiomer of fluazifop-butyl to the active *(R)*-acid is observed in soil treated with this herbicide [30]; no inversion takes place when the soil is sterilized. The practical consequences of this bioinversion are dependent on the method of application of the herbicide and it has been suggested [30] that postemergence spraying of deep-rooted crops with the *(S)*-enantiomer as a 'propesticide' may be beneficial.

In the past, the α-phenoxypropionic acid herbicides have generally been sold as racemates; however, there is increasing environmental pressure to use single enantiomers. In The Netherlands, for example, the regulatory authorities have severely curtailed the use of racemates. It is interesting to note that the structurally related herbicides flamprop-isopropyl and carbetamide (Figure 2-25) were developed and produced as the pure *(R)*-enantiomers from the outset.

B. Pyrethroid Insecticides

The pyrethroids constitute a commercially important class of neuroactive insecticides [31,32] that have been structurally derived from the natural pyrethrin insecticides (e.g., pyrethrin I, Figure 2-26). Virtually all of the commercially important products are esters of *m*-phenoxybenzaldehyde cyanohydrin. They are all chiral molecules that contain up to three asymmetric centers.

Deltamethrin, for example, has three asymmetric centers and is sold as the single *(R,R,S)* isomer, one of the eight possible optical isomers. Cypermethrin and fenvalerate were initially sold as isomeric mixtures but were later replaced at least partially by optical isomers. The commercial manufacturing routes for these products, which are produced in multi-hundred ton quantities, constitute a synthetic 'tour de force'. (See chapter 9.)

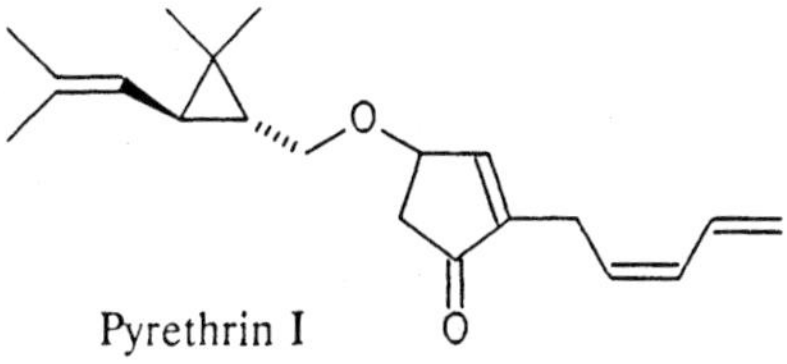

Pyrethrin I

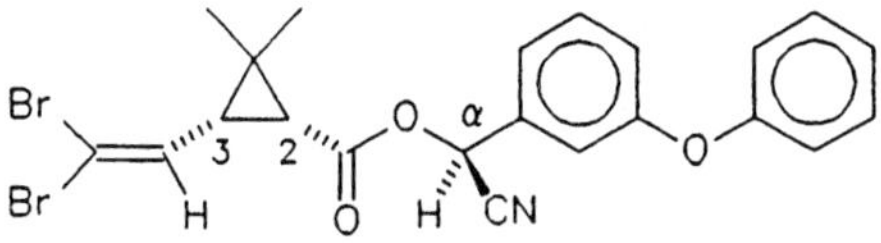

(αS,2R,3R)-Deltamethrin (Decis)

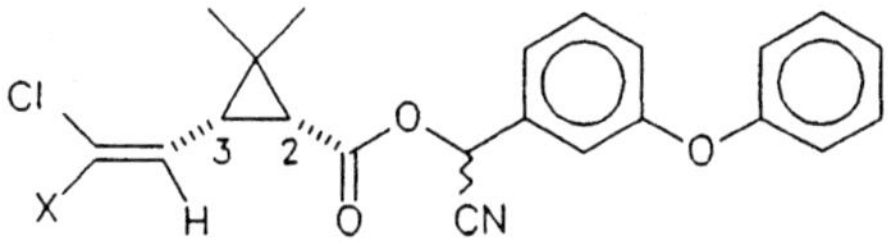

X = Cl : (2R,3R)-Alphametrin (Fastac)
X = CF₃ : (2R,3R)-Curare

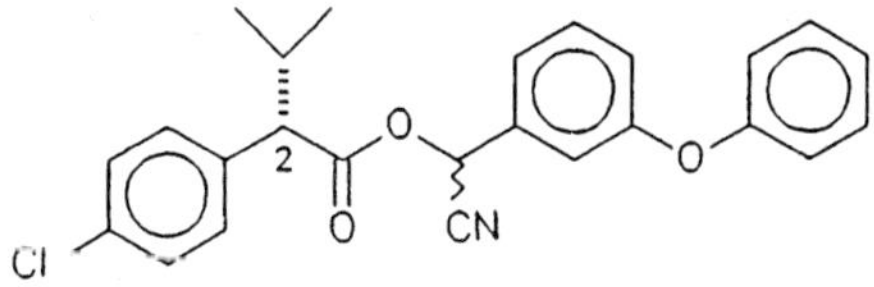

(2S)-Fenvalerate (Asana)

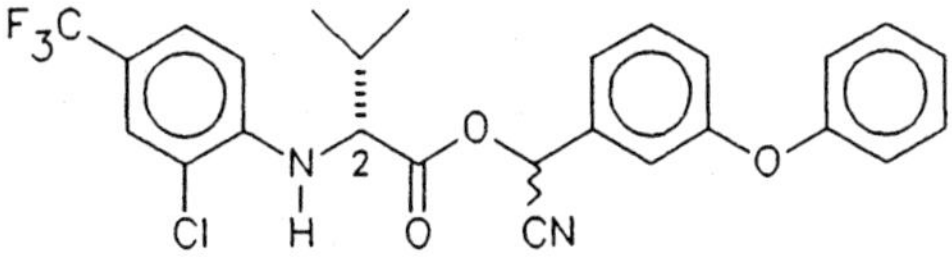

(2R)-Fluvalinate

Figure 2-26. Structures of pyrethroid insecticides.

The synthetic pyrethroids all exhibit the so-called 'rapid knockdown' effect without suffering from the instability to light which severely restricts the applications of their natural counterparts. Deltamethrin, for example, is the most potent insecticide to date, being 50 times as active as DDT, without being toxic to birds or mammals. Their low dosage combined with high species-selectivity make them a good example of the rational design of environmentally friendly bioactive agents. In short, we have come a long way since the days of *Silent Spring*.

C.   Chiral Fungicides

Many commercial fungicides are also chiral molecules [34], some examples of which are shown in Figure 2-27. Here also the respective enantiomers exhibit vastly different activities. The *(R)*-enantiomer of metaxalyl, for example, is about 1000 times more active in vitro than the *(S)*-isomer [34]. In vivo experiments, however, revealed much smaller differences. This is a commonly observed phenomenon with chiral pesticides and is probably due to racemization under the conditions of application. Interestingly the *(R)-(−)*-enantiomer of CGA 29212 is a potent fungicide with modest herbicidal activity while the *(S)-(+)*-isomer exhibits high herbicidal, but virtually no fungicidal activity.

A similar separation of two different types of biological activity is observed with paclobutrazol (Figure 2-28) [30]. The *(2R,3R)*-enantiomer is an excellent fungicide but the *(2S,3S)*-isomer shows plant growth regulatory activity.

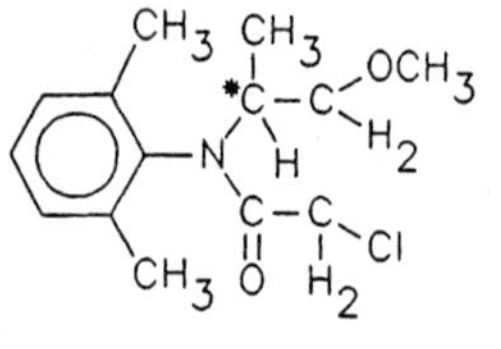

Metolachlor 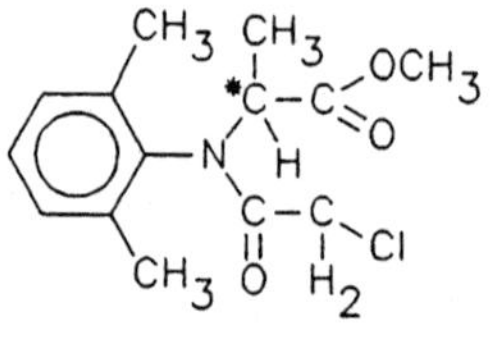   CGA 29212

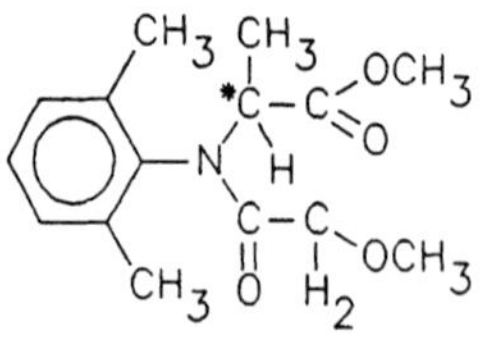

Metalaxyl 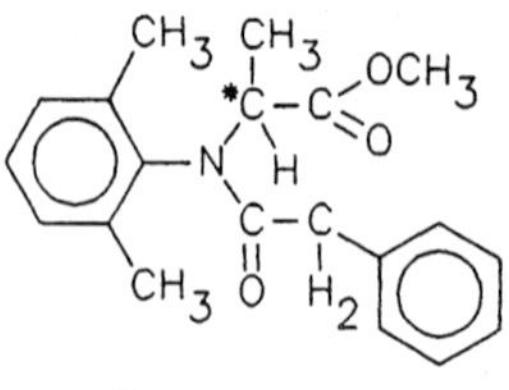   Venalaxyl

Figure 2-27. Structures of chiral fungicides.

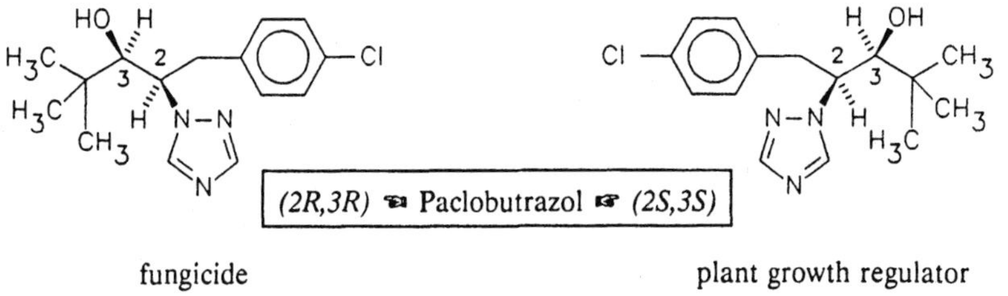

Figure 2-28. Separation of the biological activities of the paclobutrazol enantiomers.

These selected examples from the three major classes have hopefully shown that enantioselectivity is important in agrochemicals. Modern agriculture is almost unthinkable without the massive use of chemicals for crop protection. Nevertheless, a proper assessment of potential benefits and risks associated with agrochemicals requires a thorough knowledge of their interactions with living organisms and an adequate regard for stereochemical implications. The ultimate goal is to increase selectivity and reduce the chemical burden on the environment and removal of isomeric ballast may contribute to this end.

IV. FLAVORS AND FRAGRANCES

There is no accounting for taste, as they say, but one thing we can count on is that it is influenced by chirality. As noted earlier, taste and odor perception are determined by specific interactions between flavor and fragrance substances and the olfactory receptors in the nose. Such interactions are expected to be enantioselective, leading to vastly different odors and tastes for pairs of enantiomers (Figure 2-29). For example, the (R)-enantiomer of carvone tastes of spearmint while the (S)-enantiomer tastes of caraway. Gingergrass, in contrast, contains racemic carvone.

There is also a definite trend towards enantiopurity in flavors and fragrances, but here the underlying causes are different than in drugs and agrochemicals. In the food industry there is an increasing consumer demand for 'natural ingredients' as opposed to 'synthetic additives', whatever that may mean. Natural ingredients are products extracted from natural sources or produced by 'natural' processes, (e.g., fermentation) and involving a minimum number (preferably zero) of chemical steps. Since natural is often synonymous with 'enantiomerically pure', there is also an increasing demand for enantioselective syntheses in the flavor and fragrance industry. Even in cases where there is no significant difference between optical isomers, the natural isomer is preferred to a 'synthetic' racemate; for example, both d- and l-menthol have a minty aroma but the l-isomer is the naturally occurring form.

In short, there is a subtle difference between the issue of enantiomeric purity in flavor and fragrances and in the drug and pesticide industries. In flavors and

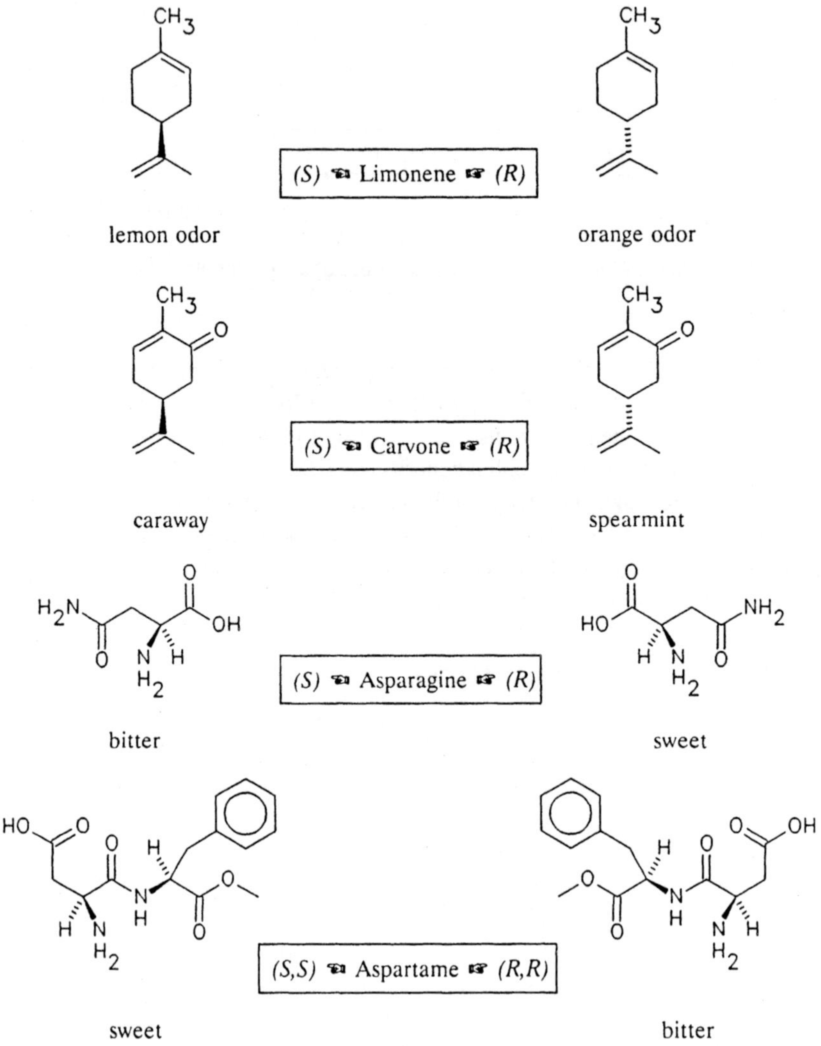

Figure 2-29. Enantioselectivity in taste and odor perception.

fragrances, the issue is related to whether the substance is the same as the naturally occurring one (and is made by natural methods) rather than to whether it is enantiomerically pure. A substance that is synthesized solely by natural methods is entitled to the accolade 'nature-identical' as opposed to the less favorable, 'synthetic'.

V. REGULATORY ASPECTS

The United States Food and Drug Administration (FDA) guidelines state that the development and marketing of chiral drugs as racemates should not be prohibited; however, the final approval (by the FDA) must be based on complete information with regard to the pharmacodynamics and pharmacokinetics of all the individual isomers and the racemic mixtures [35,36]. For example, regulatory requirements include a stipulation that the bioavailability of a drug be demonstrated. Since the two enantiomers of a chiral drug generally exhibit widely differing pharmacokinetics, it seems obvious that establishing the bioavailability of the drug from a racemate is a complex task which necessitates separation of the enantiomers and investigation of their pharmacokinetics as individual molecular entities.

The final approval must be based on a proper evaluation of clinical investigations that compare the safety and efficacy of a racemate and its individual enantiomers. It was concluded that "whenever a drug can be obtained in a variety of chemically equivalent forms (such as enantiomers), it is both good science and good sense to explore the potential for in vivo differences between these forms" [35].

The situation seems vaguely reminiscent of the question regarding unleaded gasoline. The regulatory authorities did not impose an outright ban on the use of leaded gasoline but chose the more subtle approach of making it increasingly unattractive (by tax penalties) compared to unleaded gasoline. The final outcome is the same but the method is more acceptable to those involved. Although regulatory attention has been initially focussed on chirality in food and drug applications, this is now being closely followed by agrochemicals. As noted earlier, the application of α-phenoxypropionic acid herbicides in the form of racemic mixtures has already been severely curtailed in The Netherlands. The writing is on the wall; the current climate of 'environmentality' is precipitating a dramatic move towards enantiomeric purity in bioactive agents.

Finally, it is worth mentioning that a frequently encountered problem is the fact that the present system of naming drugs and agrochemicals often gives no information with regard to whether or not the molecule is chiral and, if so, whether it is a pure stereoisomer or a mixture. The danger inherent in present usage is that it may be unwittingly assumed that each generic name refers to a single substance. In order to overcome this problem a Stereochemically Informative Generic Name System, with the acronym SIGNS has been proposed [37,38]. The recommended prefixes in the SIGNS nomenclature are summarized in Table 2-4. Although one

Table 2-4   SIGNS Prefixes for Stereoisomeric Drugs

Single Drugs		Mixtures	
Prefix	Meaning	Prefix	Meaning
dextro-	Dextrorotatory enantiomer	*rac-*	Racemate
levo-	Levorotatory enantiomer	*mep-*	Mixture of epimers
cis-	Geometrical isomer	*diam-*	Diastereomeric mixture
trans-	Geometrical isomer		(chiral or achiral)

can endlessly debate the appropriateness of individual prefixes, this is obviously a commendable step in the right direction.

REFERENCES

1. Wynberg, H., *Topics in Stereochemistry*, **16**, 87-129 (1986).
2. Fischer, E., *Ber. Dtsch. Chem. Ges.*, **27**, 2895-2993 (1894).
3. Fischer, E., *Ber. Dtsch. Chem. Ges.*, **27**, 3189-3932 (1894).
4. Langley, J.N., *Proc. Roy. Soc., B*, **78**, 170-194 (1906).
5. Ehrlich, P., *Brit. Med. J.*, *II*, 353-359 (1913).
6. Cushny, A.R., *Biological Relantionships of Optically Active Substances*, Bailliere, Tindall and Cox, London, 1926, pp. 53-67.
7. Easson, L.H., and Stedman, E., *Biochem. J.* **27**, 1257-1266 (1933).
8. Ruffolo, R.R., in *Stereochemistry and Biological Activity of Drugs*, Ariëns, E.J., Soudijn, W., and Timmermans, P.B.M.W.M. (Eds.), Blackwell, Oxford, 1983.
9. Ogston, A.G., *Nature*, **162**, 963 (1948).
10. Sandler, M., and Smith, H.J., *Design of Enzyme Inhibitors as Drugs*, Oxford University Press, Oxford, 1989.
11. Pfeiffer, C.C., *Science*, **124**, 29-30 (1956).
12. Lehmann, P.A., Rodrigues de Miranda, J.F., and Ariëns, E.J., *Progr. Drug Res.*, **20**, 101-142 (1976).
13. Ariëns, E.J., in *Chiral Separations by HPLC*, Krstulovic, A.M. (Ed.), Ellis Horwood, Chichester, 1989, pp. 31-68.
14. Ariëns, E.J., *TIPS*, **7**, 200-205 (1986), and references cited therein.
15. Ariëns, E.J., *Eur. J. Clin. Pharmacol.*, **26**, 663-668 (1984).
16. Ariëns, E.J., in *Chirality in Drug Design and Synthesis*, Brown, C. (Ed.), Academic Press, New York, 1990, pp. 29-43.
17. Main, B.G., in *Problems and Wonders of Chiral Molecules*, Simonyi, M. (Ed.), Akademiai, Kiado, Budapest, 1990, pp. 329-348.
18. Baldwin, J.J., and Abrams, W.B., in *Drug Stereochemistry. Analytical Methods and Pharmacology*, Wainer, I.W., and Drayer, D.E. (Eds.), Marcel Dekker, New York, 1988, pp. 311-356.

19. Ariëns, E.J., *Med. Res. Revs.*, **6**, 451-466 (1986).
20. Powell, J.R., Ambre, J.J., and Tsuen, I.R., in *Drug Stereochemistry. Analytical Methods and Pharmacology*, Wainer, I.W., and Drayer, D.E. (Eds.), Marcel Dekker, New York, 1988, pp. 245-270, and references cited therein.
21. Hyneck, M., Dent, J., and Hook, J.B., in *Chirality in Drug Design and Synthesis*, Brown, C. (Ed.), Academic Press, New York, 1990, pp. 1-28, and references cited therein.
22. For a comprehensive historical account see De Camp, W.H. *Chirality*, **1**, 2-6 (1989).
23. Drayer, D.E., *Clin. Pharmacol. Ther.*, **40**, 125-133 (1986).
24. Tobert, J.A., Cirillo, V.J., Hitzenberger, G., James, I., Pryor, J., Cook, T., Buntinx, A., Holmes, I.B., and Lutterbeck, P.M., *Clin. Pharmacol. Ther.*, **29**, 344-350 (1981).
25. Williams, K.M., in *Problems and Wonders of Chiral Molecules*, Simonyi, M. (Ed.), Akademiai, Kiado, Budapest, 1990, pp. 181-204, and references cited therein.
26. Williams, K., Day, R., Romualda, K., and Duffield, A., *Biochem. Pharmacol.*, **35**, 3403-3405 (1986).
27. Simonyi, M., *Med. Res. Revs.*, **4**, 359-413 (1984).
28. Sheldon, R.A., Zeegers, H.J.M., Houbiers, J.P.M., and Hulshof, L.A., *Chemistry Today*, **9** (5) 35-47 (1991), and references cited therein.
29. a) Cushman, D.W., and Ondetti, M.A., *CHEMTECH*, 620-624 (1982). b) Mackaness, G.B., *J. Cardiovascular Pharmacol.*, **7**(suppl.1), 30 (1985). c) Natoff, J.L., and Redshaw, S., *Drugs of the Future*, **12**, 475-483 (1987).
30. Ariëns, E.J., in *Stereoselectivity of Pesticides. Biological and Chemical Problems*, Ariëns, E.J., Van Rensen, J.J.S., and Welling, W. (Eds.), Elsevier, Amsterdam, 1988, pp. 39-108, and references cited therein.
31. Naber, J.D., and Van Rensen, J.J.S., in *Stereoselectivity of Pesticides. Biological and Chemical Problems*, Ariëns, E.J., Van Rensen, J.J.S., and Welling, W. (Eds.), Elsevier, Amsterdam, 1988, pp. 39-108.
32. Leahey, J.P., in *The Pyrethroid Insecticides*, Leahey, J.P. (Ed.), Taylor and Francis, London, 1985.
33. Vijverberg, H.P.M., and Oortgiesen, M., in *Stereoselectivity of Pesticides. Biological and Chemical Problems*, Ariëns, E.J., Van Rensen, J.J.S., and Welling, W. (Eds.), Elsevier, Amsterdam, 1988, pp. 151-182, and references cited therein.
34. Fuchs, A., in *Stereoselectivity of Pesticides. Biological and Chemical Problems*, Ariëns, E.J., Van Rensen, J.J.S., and Welling, W. (Eds.), Elsevier, Amsterdam, 1988, pp. 203-262.
35. De Camp, W.H., *Chirality*, **1**, 2-6 (1989).
36. Kumkumian, C.S., in *Drug Stereochemistry*, Wainer, I.W., and Drayer, D.E. (Eds.), Marcel Dekker, New York, 1988, pp. 299-310.
37. Simonyi, M., Gal, J., and Testa, B., *Trends Parmacol. Sci*, **10** (9), 349-354 (1989).
38. Simonyi, M., Gal, J., and Testa, B., in *Problems and Wonders of Chiral Molecules*, Simonyi, M. (Ed.), Akademiai, Kiado, Budapest, 1990, pp. 127-136.

ADDITIONAL READING

1. Wainer, I.W., and Drayer, D.E. (Eds.), *Drug Stereochemistry: Analytical Methods and Pharmacology*, Marcel Dekker, New York, 1988.
2. Brown, C. (Ed.), *Chirality in Drug Design and Synthesis*, Academic Press, New York, 1990.

3. Simonyi, M. (Ed.), *Problems and Wonders of Chiral Molecules*, Akademiai, Kiado, Budapest, 1990.

4. Ariëns, E.J., Van Rensen, J.J.S., and Welling, W. (Eds.), *Stereoselectivity of Pesticides. Biological and Chemical Problems*, Elsevier, Amsterdam, 1988.

5. E.J., Soudijn, W., and Timmermans, P.B.M.W.M. (Eds.), *Stereochemistry and Biological Activity of Drugs*, Ariëns, Blackwell, Oxford, 1983.

6. Simonyi, M., On Chiral Drug Action, *Med. Res. Revs.*, **4**, 359-413 (1984).

7. Ariëns, E.J., Racemates—an Impediment in the Use of Drugs and Agrochemicals, in *Chiral Separations by HPLC*, Krstulovic, A.M. (Ed.), Ellis Horwood, Chichester, 1989, pp. 31-68.

8. Testa, B., and Trager, W.F., Racemates versus Enatiomers in Drug Development: Dogmatism or Pragmatism?, *Chirality*, **2**, 129-133 (1990).

9. Testa, B., Mechanism of Chiral Recognition in Xenobiotic, Metabolism and Drug-Receptor Interactions, *Chirality*, **1**, 7-9 (1989).

10. Coutts, R.T., and Baker, G.B., Implications of Chirality and Geometric Isomerism in Some Psychoactive Drugs and Their Metabolites, *Chirality*, **1**, 99-120 (1989).

11. Borman, S., Chirality Emerges as Key Issue in Pharmaceutical Research, *Chem. and Eng. News*, 09 July, 1990, pp. 9-14.

3

Synthetic Methodology

> Synthesis must always be carried out by plan, and the synthetic frontiers
> can be defined only in terms of the degree to which the realistic planning
> is possible utilizing all of the intellectual and physical tools available.
>
> R. B. Woodward, 1956

Having dealt with the what and the why of enantiomeric purity, we shall now
concern ourselves with the question of how to approach the synthesis of pure
enantiomers. In this chapter we shall consider the general principles involved in
route selection for large-scale synthesis. In the following chapters the individual
methods will be treated in more detail.

I. THREE PRIMARY SOURCES OF PURE ENANTIOMERS

Basically, there are three primary sources of pure enantiomers for use in the
synthesis of enantiomerically pure drugs and agrochemicals (Figure 3-1). The first
source is the rich diversity of chiral molecules (e.g., carbohydrates, terpenes, and
alkaloids) that occur naturally as pure enantiomers. In many cases, the pure
substances can be easily recovered from plant or animal material by extraction
techniques.

The second source is de novo fermentation of an inexpensive, abundantly
available carbohydrate feedstock, such as sucrose or molasses. De novo fermentation is an important source of both relatively simple chiral molecules such as lactic,
tartaric and L-amino acids, and relatively complex substances such as antibiotics,
hormones, and vitamins. Fermentation is synonymous with microbial synthesis; it
is simply the reproduction of microorganisms which results from the consumption
of an energy source (e.g., sugar) in combination with various nutrients. The primary

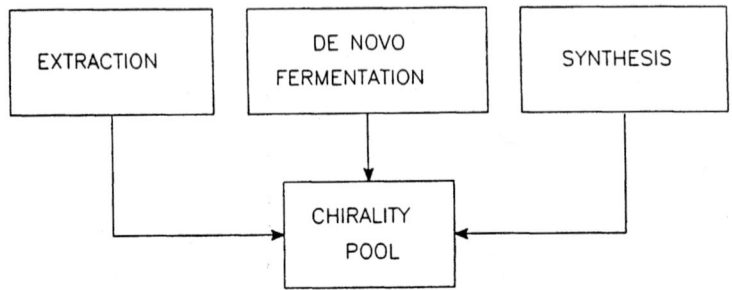

Figure 3-1. Primary sources of pure enantiomers.

product is more microorganisms, and chemicals (e.g., amino acids) may be considered as by-products of this process (Figure 3-2). The term chiral pool, which for purists should really be chirality pool as the pool itself is not chiral, usually denotes the vast array of enantiomerically pure products that are readily available from one or both of the above mentioned sources.

FERMENTATION = MICROBIAL SYNTHESIS

MICROORGANISM + SUBSTRATE + NUTRIENTS

$\Downarrow$

MORE MICROORGANISMS + PRODUCT(S)

Is a source of :

□   MICROBIAL CELLS (BIOMASS), *e.g.* YEAST

□   MICROBIAL ENZYMES

□   MICROBIAL METABOLITES, *e.g.* AMINO ACIDS

□   PRODUCTS OF MICROBIAL BIOTRANSFORMATION
       (PRECURSOR FERMENTATION)

Figure 3-2. What is fermentation?

The third source of pure enantiomers is by synthesis from either chiral or prochiral starting materials. Enantiomerically pure products that become readily available because they are synthesized on a large scale (e.g., D-phenylglycine) may also be considered to be part of the chirality pool.

II. SYNTHETIC METHODOLOGY

Synthetic routes to optically active compounds can be conveniently divided into three groups on the basis of the type of raw material used (Figure 3-3).

A. The Chirality Pool

As noted above, the chirality pool refers to inexpensive, readily available (natural) products (and their derivatives), such as carbohydrates, amino acids, and lactic acid. These substances can be transformed into synthetic products by chemical manipulation that may involve retention or inversion of configuration or chirality transfer. For example, the optically active α-phenoxypropionic acid herbicides discussed in chapter 2 can be prepared [1] from the appropriate enantiomer of lactic acid (Figure 3-4).

In the synthesis depicted in Figure 3-4, the chiral α-substituted propionic acid unit of the starting material remains as the key feature in the product; in this case, the chiral starting material is called a **chiral synthon.** Sometimes the asymmetric center in the product is created by **chirality transfer,** whereby the chiral starting material as such does not appear in the product; the chiral starting material is then referred to as a **chiral auxiliary.** A good example of the use of a chiral auxiliary, available from the chirality pool, is the Zambon synthesis of naproxen (Figure 3-5). In this elegant synthesis, the chiral auxiliary *(R,R)*-tartaric acid is recovered in the penultimate step and can be recycled.

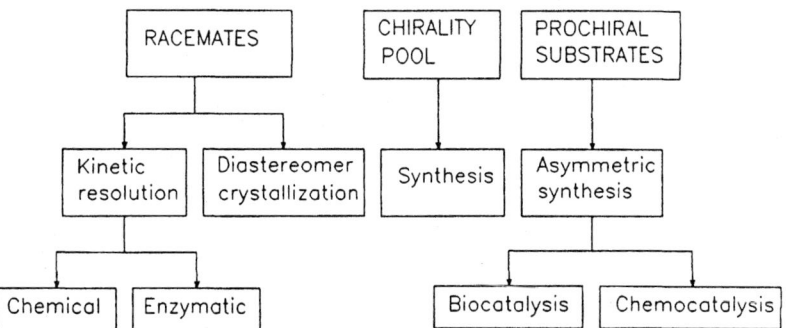

Figure 3-3. Synthetic methods for optical isomers.

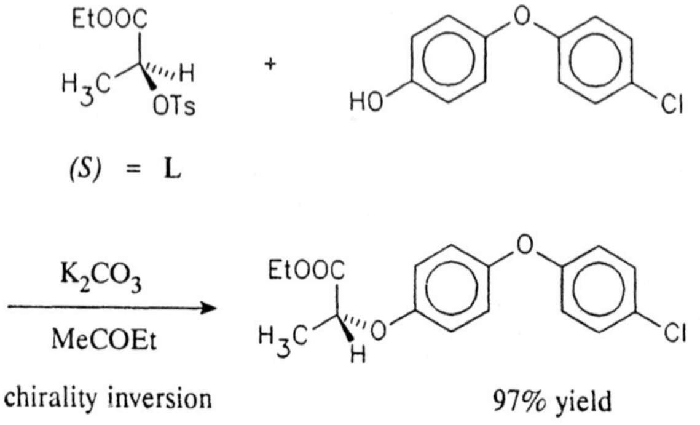

Figure 3-4. Synthesis of an optically active herbicide.

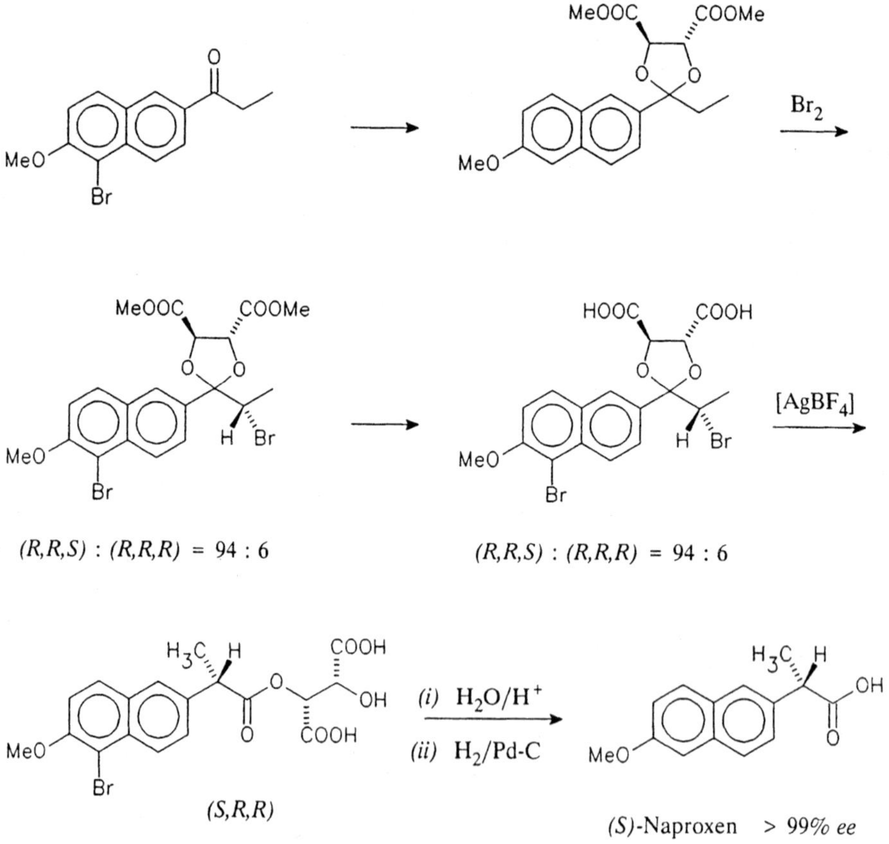

$(R,R,S) : (R,R,R) = 94 : 6$ $(R,R,S) : (R,R,R) = 94 : 6$

(S,R,R)

(S)-Naproxen $> 99\%$ ee

Figure 3-5. Zambon synthesis of (S)-naproxen.

B.   Racemate Resolution via Preferential Crystallization

Notwithstanding the revolutionary advances that are being made in catalytic asymmetric synthesis, the resolution of racemates still constitutes the main method for the industrial synthesis of pure enantiomers. Methods for their resolution can be divided into three categories (Figure 3-6): direct preferential crystallization, crystallization of diastereoisomeric salts and kinetic resolution.

Preferential crystallization, also referred to as resolution by entrainment, is widely used on an industrial scale, for example, in the manufacture of chloramphenicol [3] and α-methyl-L-dopa [4] (see also Chapter 6). It is technically feasible only with racemates that are so-called conglomerates, ones that consist of mechanical mixtures of the two enantiomers. Unfortunately, less than 20% of all racemates are conglomerates, the rest comprising true racemic compounds that cannot be separated by preferential crystallization (i.e., seeding with the crystals of one enantiomer). The success of preferential crystallization depends on the fact that for a conglomerate the racemic mixture is more soluble than either of the enantiomers. (For a more detailed discussion, see chapter 6.)

Preferential crystallization can be a particularly attractive method for industrial synthesis when it is accompanied by spontaneous in situ racemization which allows for a theoretical once-through yield of 100%. Such a process is referred to as a **crystallization-induced asymmetric transformation** of a racemate or simply a **deracemization**. An elegant example of deracemization has been reported by Okada and coworkers [5]. The two enantiomers of the 1,4-benzodiazepine derivative shown in Figure 3-7 undergo spontaneous racemization in solution at ambient temperature. When the solution is allowed to stand, one of the enantiomers crystallizes out in more than 50% yield.

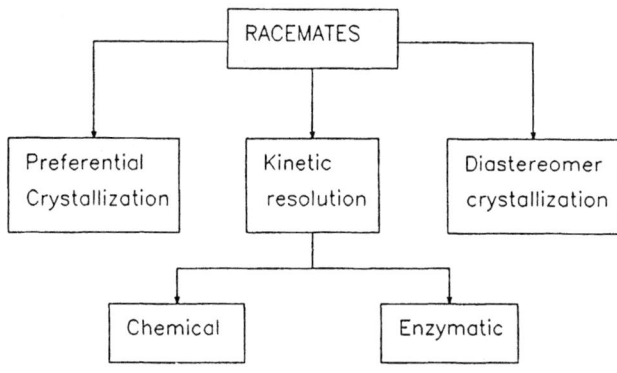

Figure 3-6. Methods for the resolution of racemates.

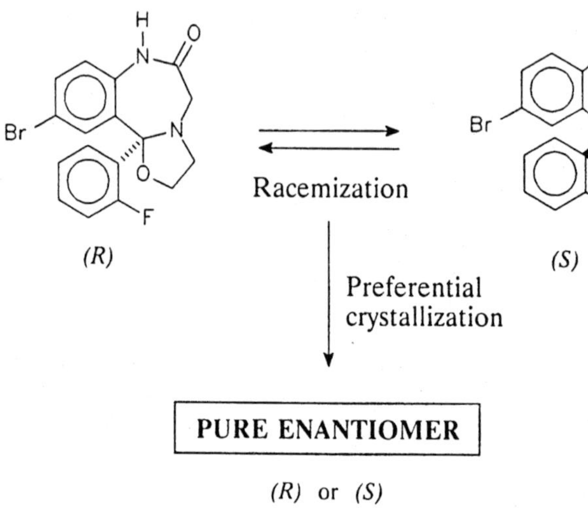

Figure 3-7. Preferential enantiomer (*(R)* or *(S)*) crystallization from spontaneous racemization.

C. Racemate Resolution via Diastereomer Crystallization

If a racemate is a true racemic compound—a homogeneous solid phase of the two enantiomers that coexist in the same unit cell—it cannot be separated by preferential crystallization. In this case, diastereomer crystallization can be used, the method pioneered by Louis Pasteur in 1854. A solution of the racemic mixture in water or ethanol, for example, is allowed to interact with a pure enantiomer (the resolving agent), thereby forming a mixture of diastereomers that can be separated by crystallization. This usually involves the formation of diastereomeric salts by the reaction of a racemic acid with an optically active base (Figure 3-8).

Diastereomer crystallization is widely used for the synthesis of pure enantiomers [6,7]. The most commonly used resolving agents are based on (natural) products available from the chirality pool, e.g., L-(+)tartaric acid, D-(−)-camphorsulfonic acid and various alkaloid bases. (See chapter 5.) A pertinent example is

$$(DL)\text{-RCOOH} \ + \ (L)\text{-R}^1\text{NH}_2 \longrightarrow$$

racemate   resolving agent

$$(D)\text{-RCOOH (L)-R}^1\text{NH}_2 \ + \ (L)\text{-RCOOH (L)-R}^1\text{NH}_2$$

mixture of diastereomeric salts

Figure 3-8. Resolution via diastereomer crystallization.

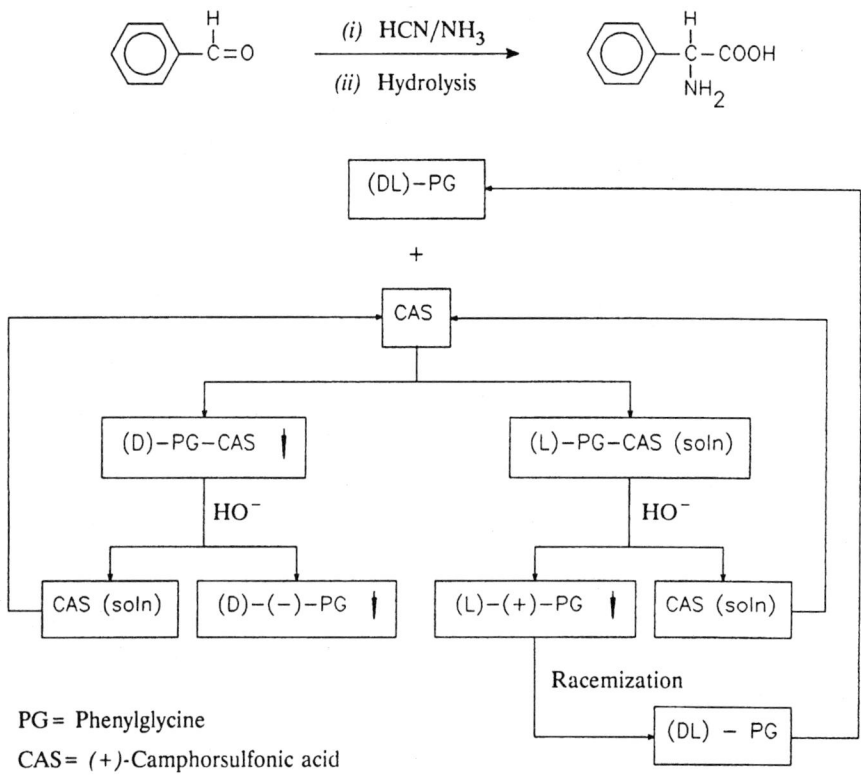

Figure 3-9. Manufacture of D-phenylglycine by diastereomer crystallization.

the process used by Andeno/DSM for the manufacture of D-(–)-phenylglycine, a key intermediate for semisynthetic penicillins and cefalosporins (see chapter 2), that uses optically pure camphorsulfonic acid as the resolving agent (Figure 3-9).

The theoretical once-through yield of a resolution by diastereomer crystallization is 50%. In practice, chemical yields of > 40% of material of high optical purity (> 95%) in a single crystallization are considered to be good resolutions. Facile racemization and recycling of the unwanted isomer is often a prerequisite for an economically viable process.

Once-through yields of > 50% are feasible when the diastereomeric salt remaining in solution undergoes spontaneous epimerization, often referred to as **diastereomer interconversion**. The overall process is called a crystallization-induced asymmetric transformation of a diastereomeric mixture (or simply an asymmetric transformation), and allows for a theoretical yield of 100%. An example of such a process is the resolution of a series of arylglycine esters with tartaric acid in the presence of a catalytic amount of a carbonyl compound such as benzaldehyde or acetone (Reaction 3-1) [8]; the latter catalyzes the racemization of the substrate in situ via the formation of a Schiff base. (See chapter 5.)

$$\text{(DL)-ArCHCOOMe} \atop \underset{\text{NH}_2}{|} \quad \xrightarrow[\text{MeOH, R}^1\text{COR}^2]{\text{L-}(+)\text{-tartaric acid}} \quad \text{(D)-ArCHCOOMe} \atop \underset{\text{NH}_2}{|} \qquad (3\text{-}1)$$

D. Catalytic Kinetic Resolution

A third method for the resolution of racemates is kinetic resolution [9], the success of which depends on the fact that the two enantiomers react at different rates with a chiral entity. From an economic point of view, the latter should preferably be used in catalytic amounts; it may be a biocatalyst (enzyme or a microorganism) or a chemocatalyst (chiral acid or base or a chiral metal complex). The first example of a kinetic resolution is Pasteur's fermentation of an aqueous solution of racemic ammonium tartrate by a *Penicillium glaucum* mold. (See chapter 1.)

Kinetic resolution can be defined as a process in which one of the enantiomers of a racemic mixture is more readily transformed than is the other (Reactions 3-2 and 3-3):

$$R \xrightarrow{k_R} P \qquad\qquad\qquad (3\text{-}2)$$

$$S \xrightarrow{k_S} Q \qquad\qquad\qquad (3\text{-}3)$$

Kinetic resolution occurs when $k_R \neq k_S$ and the reaction is stopped somewhere between 0 and 100% conversion. Ideally one enantiomer reacts much faster than the other,; for example, the R isomer ($k_R \gg k_S$). In this case, a 50% conversion of the initial 50/50 mixture leads to a final mixture of 50% S starting material and 50% product P. The nature of the products is irrelevant to the kinetic resolution process itself; the products P and Q may be chiral or achiral. In the Pasteur experiment, for example, microorganisms are the chiral catalyst, the products are metabolites of *(R,R)*-tartrate, and the recovered starting material is enantiomerically pure *(S,S)*-tartrate. A more modern example of an enzymatic kinetic resolution is the L-specific acylase-catalyzed hydrolysis of racemic N-acetylamino acids (Reaction 3-4), which was commercialized by Tanabe [10].

$$\text{(DL)-RCHCOOH} \atop \underset{\text{NHAc}}{|} \quad \xrightarrow[\text{H}_2\text{O}]{\text{L-acylase}} \quad \text{(L)-RCHCOOH} \atop \underset{\text{NH}_2}{|} \quad + \quad \text{(D)-RCHCOOH} \atop \underset{\text{NHAc}}{|} \qquad (3\text{-}4)$$

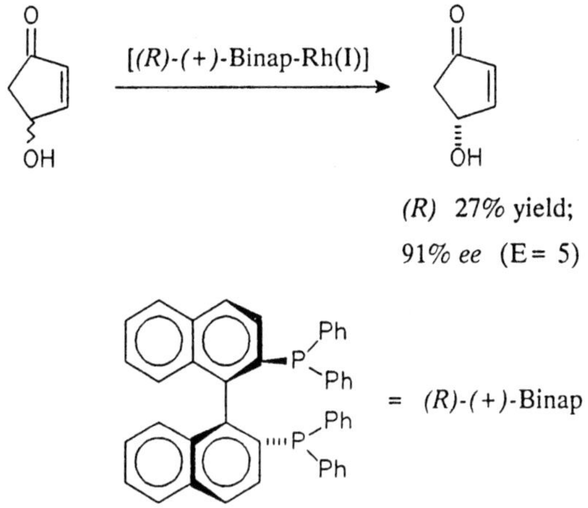

Figure 3-10. Catalytic kinetic resolution.

An example of a chemocatalytic resolution is the rhodium(I)/Binap-catalyzed isomerization (Figure 3-10) [11]. In this reaction, the *(S)*-enantiomer is selectively isomerized to the achiral 1,3-diketone, leaving the optically enriched *(R)*-enantiomer as unreacted substrate.

Similarly, the asymmetric epoxidation method developed by Sharpless and coworkers [12] has also been used [13] for the catalytic kinetic resolution of secondary allylic alcohols (Reaction 3-5) and amino alcohols (Reaction 3-6).

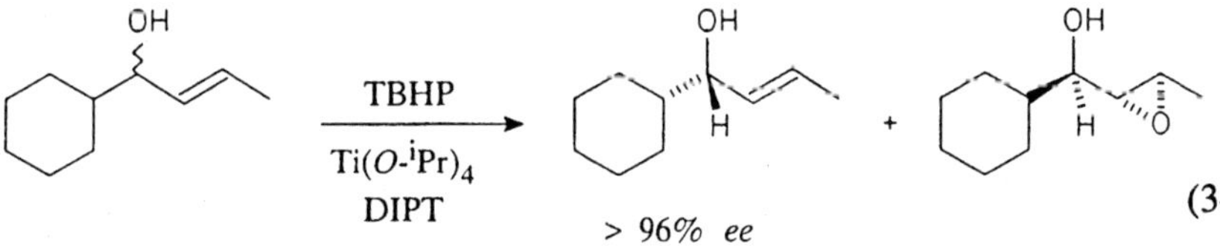

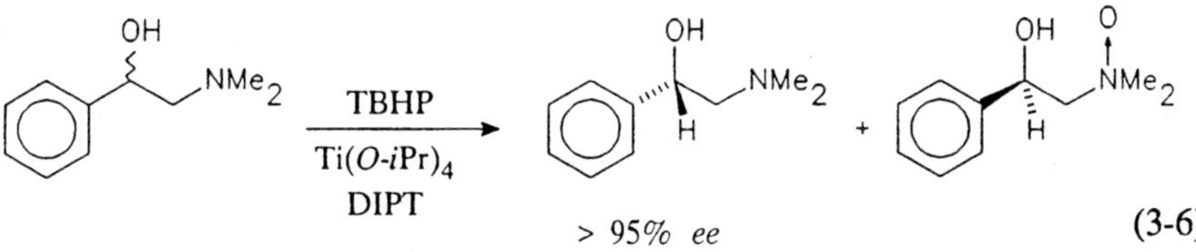

TBHP = *tert*-butyl hydroperoxide
Ti(*O*-^{i}Pr)$_4$ = titanium(IV) isopropoxide
DIPT = diisopropyl tartrate

E. The Enantiomeric Ratio

The enantiomeric ratio, E, is a measure of the efficacy of a kinetic resolution. It is simply the ratio of the pseudo first order rate constants for the two enantiomers, that is, for a reaction in which the S-enantiomer selectively reacts $E_S = k_S / k_R$.

Plots of the enantiomeric excess of remaining substrate and product against conversion for various E values are illustrated in Figures 3-11 and 3-12, respectively. An examination of these plots reveals that the less reactive enantiomer can be easily obtained in high optical purity by carrying out the reaction to the appropriate conversion. The same is not true of the product. In this case, a product of high optical purity (> 95%) is obtained, at attractive conversions, only in reactions with high E values (> 100). For this reason, kinetic resolutions are generally employed for synthesizing the less reactive enantiomer of the substrate.

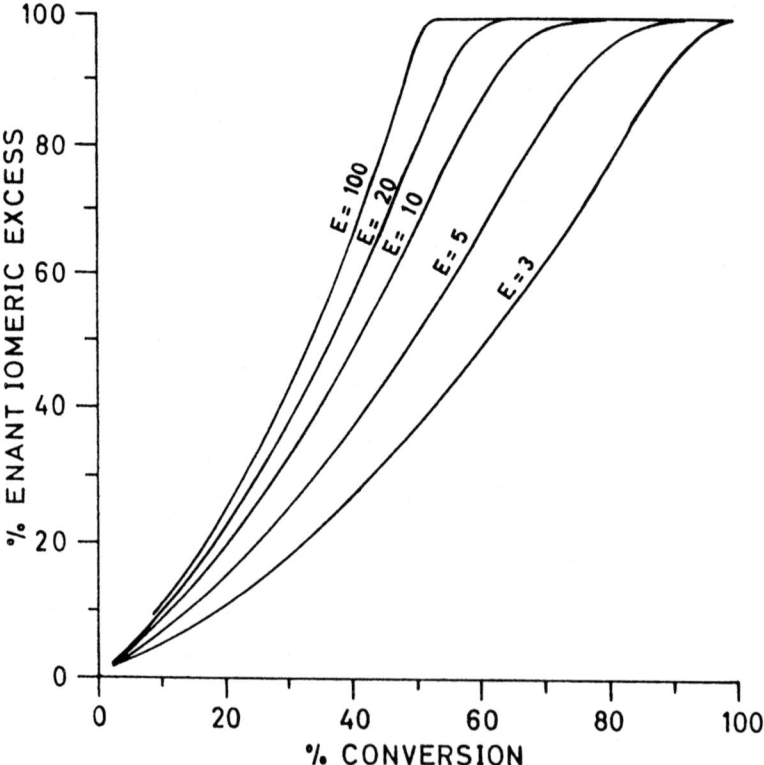

Figure 3-11. Dependence of the *ee* of remaining substrate on the conversion for various E values.

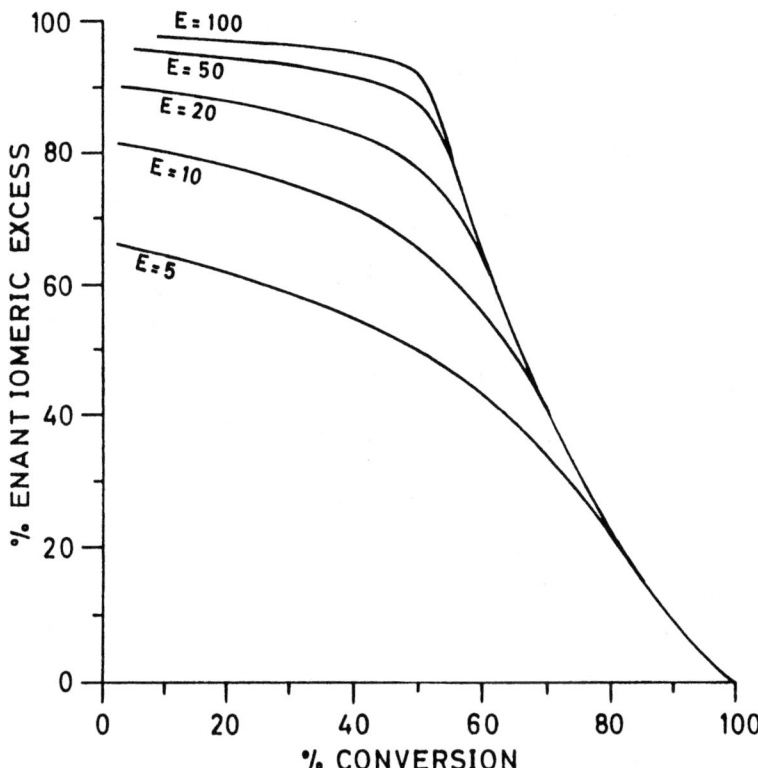

Figure 3-12. Dependence of the *ee* of product on the conversion for various *E* values.

When the *E* value for a particular reaction is high (> 100) then one obtains remaining substrate of high enantiomeric purity by carrying the reaction to about 50% conversion. With low *E* values, much higher conversions are needed (Figure 3-11). However, carrying out the reaction to higher conversions means a lower yield of remaining substrate (maximum yield = 100% – % conversion). Generally speaking, in order to obtain economically viable yields of product of sufficient enantiomeric purity, *E* should be at least 20 and preferably > 100. Interestingly, the enantiomeric ratio of the kinetic resolution illustrated in Figure 3-10 is low (*E* = 5), which manifests itself in the observed low yield (27%) of material of moderate optical purity (*ee* = 91%).

The conversions required to attain *ee*'s of 99% for remaining substrate for various *E* values are compiled in Table 3-1. It is readily seen that for *E* values greater than 200 the required conversion is close to 50%. Many enzymatic kinetic resolutions have *E* values > 1000 and are effectively completely enantiospecific.

Table 3-1 Conversion Required for 99% *ee* of Remaining Substrate for Various *E* Factors

E	Conversion (%)	*E*	Conversion (%)
5	86.6	100	52.3
10	72.1	200	51.1
20	61.9	300	50.6
50	54.9	500	50.3
		1000	50.01

Source: From ref. [9]

F. Prochiral Substrates Catalytic Asymmetric Synthesis

The enantioselective conversion of a prochiral substrate to an optically active product (by reaction with a chiral entity) is referred to as an asymmetric synthesis. Here again from an economic viewpoint the chiral entity should function in catalytic quantities. This may involve a simple chemocatalyst (chiral acid or base or a chiral metal complex) or a biocatalyst. A process that uses a chemocatalyst is the well-known Monsanto process for the manufacture of L-dopa by catalytic asymmetric hydrogenation (Figure 3-13) [15]. The Genex process [17] uses a biocatalyst for the manufacture of L-phenylalanine by the asymmetric addition of ammonia to *trans*-cinnamic acid in the presence of phenylalanine ammonia lyase (PAL) (Reaction 3-7).

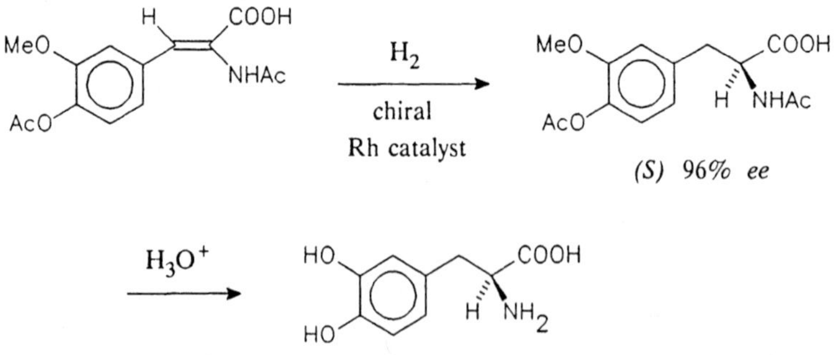

Figure 3-13. Monsanto process for L-dopa via asymmetric hydrogenation.

$$(3\text{-}7)$$

PAL = phenylalanine ammonia lyase

Chemocatalysis is not limited to chiral metal complexes but can also be achieved with chiral acids and bases. A highly enantioselective example [17] employing quinidine as a chiral basic catalyst is shown in Reaction 3-8.

$$(3\text{-}8)$$

98% *ee*

G.   Enantioproductivity

With kinetic resolutions the enantiomeric ratio (E) can be used to quantitatively compare the efficiencies of different processes. In order to compare the efficiencies of different catalytic asymmetric syntheses, Selke [18] introduced the term enantioproductivity (Q) which is simply the ratio of the enantiomers formed, S/R. Thus, whereas in kinetic resolutions E is the ratio of the pseudo first-order rate constants of the two enantiomers, Q is directly proportional to the ratio of the rates of formation of the two enantiomeric products. Obviously Q, in contrast to E, does not vary with conversion since an asymmetric synthesis is a reaction of a single compound whereas a kinetic resolution is a reaction of a mixture.

Q is obviously a more useful parameter than *ee* for comparing the effect of different variables (e.g., catalyst, substrate, solvent, etc.,) in asymmetric syntheses. Equivalent Q and *ee* values are shown in Table 3-2.

H.   *Meso* **Compounds and the** *Meso* **Trick**

Meso compounds constitute a special case of prochiral substrates. Theoretical yields of 100% can be obtained when a prochiral *meso* diester is subjected to enzymatic hydrolysis, as shown in the synthesis of the prostaglandin intermediate (Figure 3-14). It is interesting to compare this synthesis with the analogous synthesis of the prostaglandin intermediate via kinetic resolution (Figure 3-10). With a *meso* compound as the substrate the theoretical yield is 100% while with a chiral (racemic) substrate it is 50%; this is known as the '*meso*' trick.

Table 3-2 Comparison of % *ee* with Enantioproductivity

Enantiomeric excess	Enantioproductivity
% *ee* = (*S* − *R*) / (*S* + *R*) × 100%	Q = *S* / *R*
20	1.5
50	3.0
90	19.0
99	199.0
99.5	399.0
99.99	19999.0

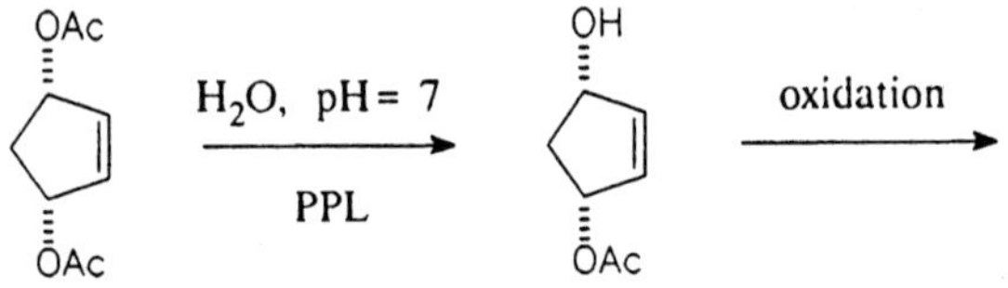

prochiral (*1R,4S*) 86% yield; 95% *ee*

PPL= porcine pancreas lipase

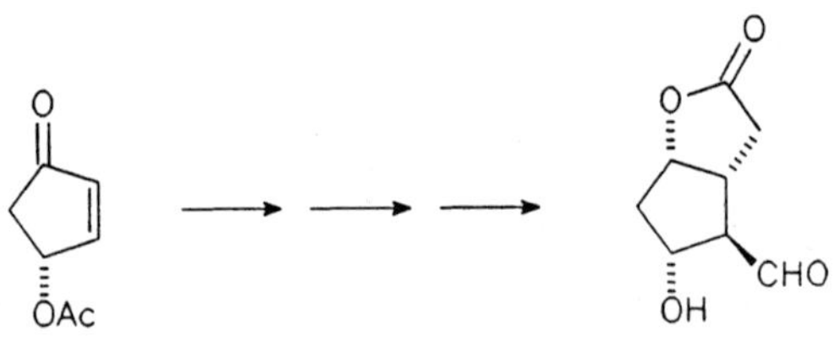

Corey lactone

Figure 3-14. Enantiotopic group differentiation.

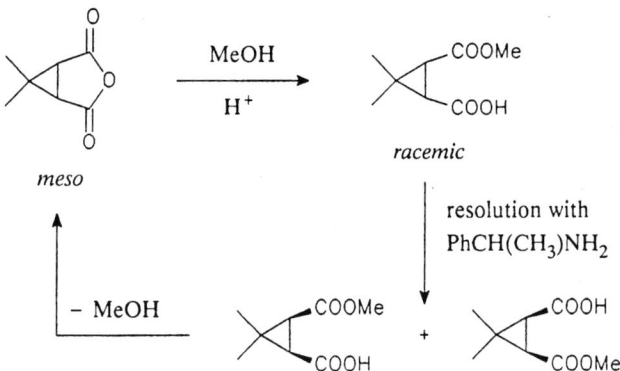

Figure 3-15. Resolution via the *meso* trick [20].

The *meso* trick can also be used in classical resolutions that use diastereomer crystallization. The *meso* compound is first converted to a racemate which is subsequently resolved by diastereomeric salt crystallization. Subsequent recycling of the unwanted enantiomer via the *meso* compound allows for a theoretical yield of 100% of the desired enantiomer (Figure 3-15).

III. CATALYTIC ASYMMETRIC SYNTHESIS VERSUS KINETIC RESOLUTION

A cursory appraisal of the relative economics of asymmetric synthesis versus kinetic resolution would seem to indicate a clear preference for the former since it has a theoretical yield of 100%, as compared to 50% for kinetic resolution. This is illustrated with a hypothetical example in Figure 3-16.

Unfortunately, it is not as simple as this, as kinetic resolutions have some advantages over asymmetric syntheses. First, they tend to be simpler processes, often giving much better productivities (volume yields) for equivalent optical purities. Second, as noted earlier, kinetic resolutions have the advantage that the *ee* of the remaining substrate can be tuned to any required value simply by adjusting the degree of conversion. In practice, it is often worthwhile sacrificing a few percent of chemical yield in order to obtain material of higher optical purity. This is not possible with asymmetric syntheses, since these reactions pertain to a single (prochiral) substrate and enantioselectivity is (by definition) independent of conversion.

The major disadvantage of kinetic resolutions is that they require at least one extra step—racemization of the unwanted isomer. This can be circumvented if conditions can be found whereby the unwanted enantiomer undergoes spontaneous in situ racemization, leading to a kinetic resolution that is formally equivalent to an asymmetric synthesis (i.e., with a maximum yield of 100%). Such a process is

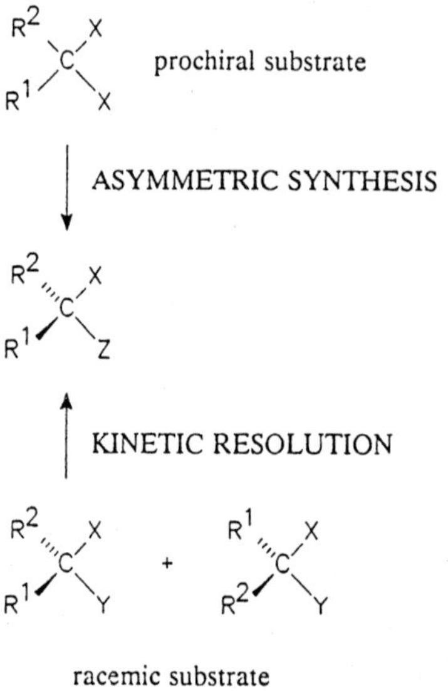

Figure 3-16. Asymmetric synthesis versus kinetic resolution.

referred to as a **dynamic kinetic resolution.** The asymmetric coupling of Grignard reagents with vinyl halides, catalyzed by chiral nickel or palladium complexes [21], is an example of such a process (Figure 3-17).

All of the Grignard reagent is consumed and the optical purity of the product is independent of the conversion, indicating that equilibration of the

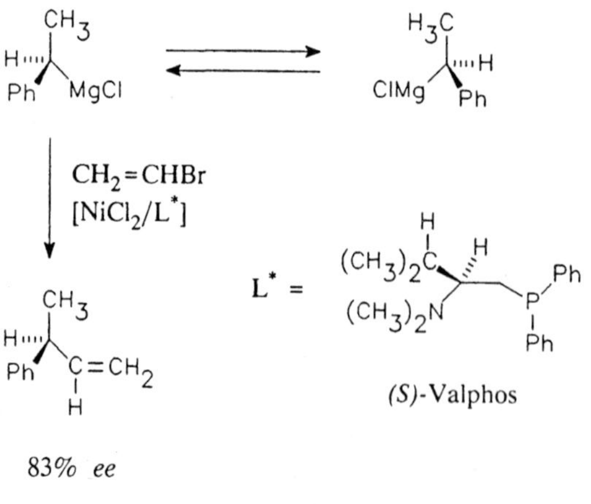

Figure 3-17. Asymmetric Grignard cross-coupling via dynamic kinetic resolution.

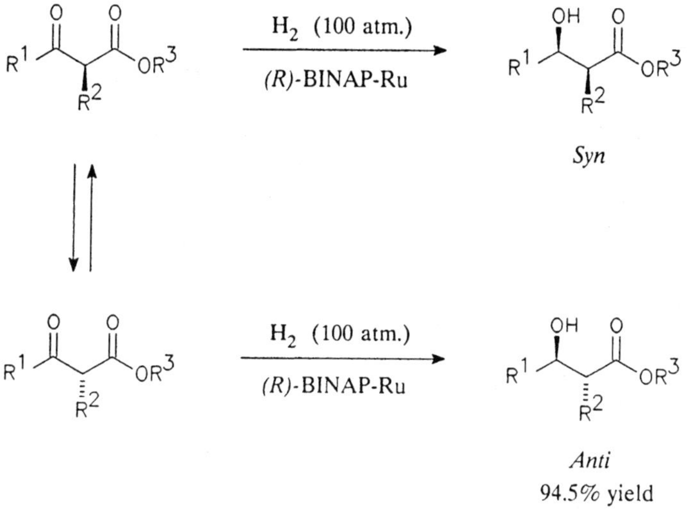

Figure 3-18. Dynamic kinetic resolution in Ru-Binap-catalyzed hydrogenation of β-keto esters.

Grignard reagent is fast compared to the transmetallation step. The latter must be highly stereoselective and may involve retention (as shown) or inversion of configuration.

A particularly elegant example is provided by the Ru-Binap catalyzed asymmetric hydrogenation of β-keto esters (Figure 3-18) [22]. This example constitutes a dynamic kinetic resolution and an asymmetric synthesis, all rolled into one. One of the four possible diastereomers is obtained in high yield (94.5%). In order to obtain such a remarkable selectivity, three requirements have to be simultaneously met: i) racemization of the substrate is faster than the hydrogenation, ii) efficient stereochemical control by the chiral catalyst, and iii) efficient differentiation between the *syn* and *anti* transition states.

IV. FACTORS AFFECTING THE COST-PRICE OF PRODUCT

The major factors having a bearing on the economics of processes for the synthesis of optically active compounds are listed in Table 3-3. Obviously, the economics are influenced by the substrate costs and this can have a bearing on the choice of method (resolution versus asymmetric synthesis). Similarly, the price and ease of recycling of the resolving agent or (bio)catalyst and the chemical and optical yields are obviously important. Another factor that is often overlooked in small-scale experiments, is the volume yield or productivity (kilo product per unit reactor volume per unit time). For an economically viable process, it is generally essential

Table 3-3 Factors Affecting Cost-Price of Product

1. Substrate costs: asymmetric synthesis versus resolution
2. Cost of resolving agent or (bio)catalyst
3. Chemical
4. Ease of racemization of unwanted isomer
5. Total number of steps
6. Position of resolution step in overall synthesis

to obtain high chemical and optical yields at high substrate concentrations and it is sometimes necessary to sacrifice percentage points on the former to the benefit of the latter.

A. Racemization of the Unwanted Isomer

The ease of racemization of an unwanted isomer is obviously of paramount importance to the economics of resolution processes and it is not generally recognized that the racemization is often the most difficult step in the overall process. Racemization generally entails treating the molecule in question with a strong acid or base at elevated temperatures. For example, carboxylic acids containing an asymmetric, hydrogen-bearing carbon atom adjacent to the carboxyl group are often amenable to racemization in strongly basic media (Figure 3-18). However, the forcing conditions required for acid- and base-promoted racemizations often lead to substantial decomposition of the substrate. There is a definite need, therefore, for milder racemization methods.

A good method for racemizing primary amines involves the formation of Schiff bases in the presence of catalytic amounts of carbonyl compounds. Subsequent reversible isomerization (Figure 3-19), leads to racemization at an asymmetric center adjacent to the amino group (example 2 in Figure 3-19).

One area that is largely virgin territory is the use of homogeneous or heterogeneous transition metal catalysts for promoting racemization. For example, typical (de)hydrogenation catalysts (e.g., Ni, Pd, Pt, etc.) would be expected to racemize chiral amines and alcohols as shown in Figure 3-19. Indeed, racemization is often observed as a side-reaction in hydrogenation of chiral substances. This is an area that is worthy of further systematic investigation. Finally, there are examples of enzymes, appropriately called racemases, that are able to mediate racemizations. The reactions generally involve the formation of Schiff base intermediates or redox pathways.

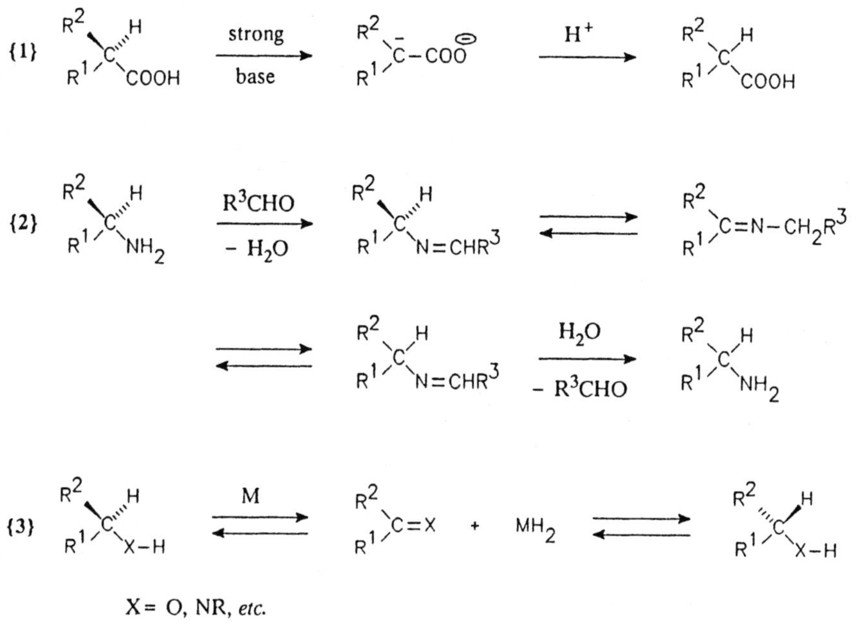

Figure 3-19. Pathways for racemization.

B. Attractive Versus Subtractive Resolutions

The total number of steps is important for the economics of any process. More steps means longer total reaction times, more labor costs and lower productivities. In resolution processes the total number of steps is determined by whether or not the required product is formed directly; a hypothetical example is shown in Figure 3-20. If the *(R)*-enantiomer of ROH is the required product we speak of an **attractive** process. In contrast, if the *(S)*-enantiomer of ROH is required the process involves two extra steps and we speak of a **subtractive** process. The difference between an attractive and a subtractive process is always two extra chemical steps, and the subtractive process is less economical, all other things being equal.

C. Precursor Versus Derivative Resolution

Another factor that influences the total number of steps of resolution processes is the nature of the substrate; is the substrate a precursor or a derivative of the racemic product? As illustrated in Figure 3-21, resolution of a precursor requires two steps less than an equivalent process that involves resolution of a derivative of the substance to be resolved.

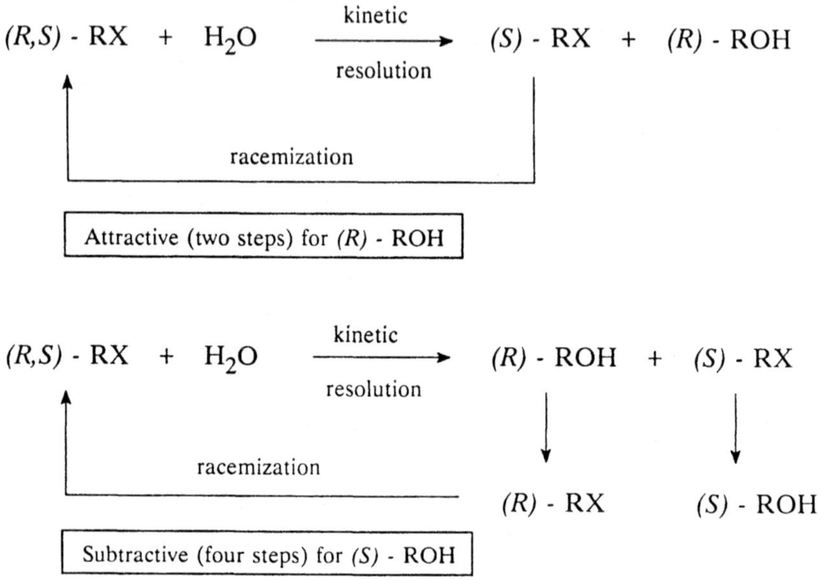

(R,S) - RX   +   H$_2$O $\xrightarrow[\text{resolution}]{\text{kinetic}}$ (S) - RX   + (R) - ROH

racemization

Attractive (two steps) for (R) - ROH

(R,S) - RX   +   H$_2$O $\xrightarrow[\text{resolution}]{\text{kinetic}}$ (R) - ROH   + (S) - RX

racemization

(R) - RX (S) - ROH

Subtractive (four steps) for (S) - ROH

Figure 3-20.   Attractive versus subtractive resolution.

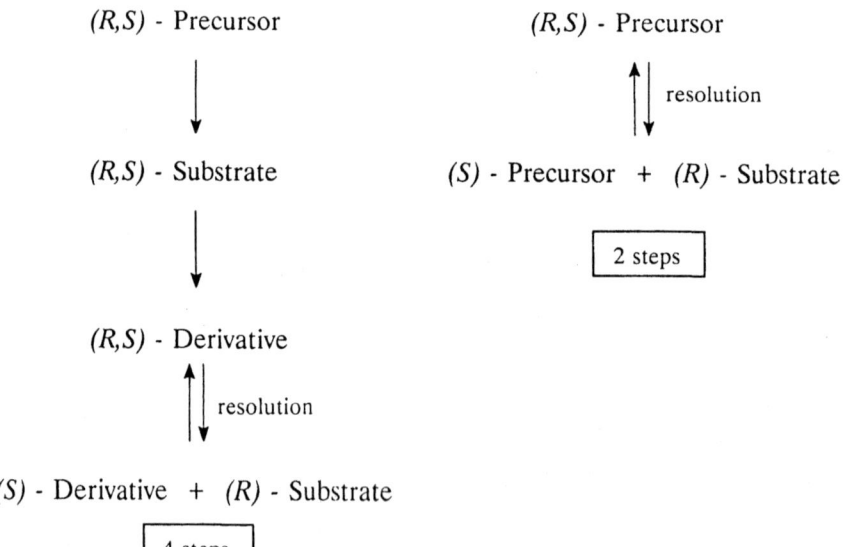

(R,S) - Precursor (R,S) - Precursor

                                                         resolution

(R,S) - Substrate (S) - Precursor   + (R) - Substrate

                                                         2 steps

(R,S) - Derivative

          resolution

(S) - Derivative   + (R) - Substrate

          4 steps

Figure 3-21.   Precursor versus derivative resolution [(R)-substrate is the desired product].

D. Amino Acid Resolutions

The general principles are best illustrated by references to practical examples such as the various processes for the enzymatic kinetic resolution of amino acids (Reactions 3-9 to 3-11). The first observation which can be made is that they are all attractive processes for L-amino acids but subtractive for D-amino acids. It is significant that the acylase process (Reaction 3-9), commercialized by Tanabe [23,24], is used only for the production of L-amino acids. From the viewpoint of precursor versus derivative resolution, the esterase process (Reaction 3-10) would appear to be the least attractive as it involves a double derivative [25]. It is not surprising, therefore, that this process is not used on an industrial scale. The potentially most attractive of the three routes would appear to be the amino-peptidase process [26] developed by DSM (Reaction 3-11), since the substrate is readily obtained from the appropriate aldehyde via the Strecker synthesis without going through the racemic amino acids (Figure 3-22). It may be considered, therefore, as a precursor process. An added advantage of the aminopeptidase process is its broad substrate specificity, allowing for the resolution of a broad range of amino acids. (See chapter 7 for a further discussion of this topic.)

$$\underset{\overset{|}{NHAc}}{(DL)\text{-}ArCHCOOH} \xrightarrow[H_2O]{acylase} \underset{\overset{|}{NH_2}}{(L)\text{-}ArCHCOOH} + \underset{\overset{|}{NHAc}}{(D)\text{-}ArCHCOOH}$$

$$(3\text{-}9)$$

$$\underset{\overset{|}{NHAc}}{(DL)\text{-}ArCHCOOMe} \xrightarrow[H_2O]{esterase} \underset{\overset{|}{NHAc}}{(L)\text{-}ArCHCOOH} + \underset{\overset{|}{NHAc}}{(D)\text{-}ArCHCOOMe}$$

$$(3\text{-}10)$$

$$\underset{\overset{|}{NH_2}}{(DL)\text{-}ArCHCONH_2} \xrightarrow[H_2O]{\substack{amino\text{-}\\peptidase}} \underset{\overset{|}{NH_2}}{(L)\text{-}ArCHCOOH} + \underset{\overset{|}{NH_2}}{(D)\text{-}ArCHCONH_2}$$

$$(3\text{-}11)$$

An example of an enzymatic resolution that is attractive for D-amino acids and uses a precursor as substrate is the hydantoinase-mediated, D-specific hydrolysis of hydantoins. The racemic hydantoins, readily available from the corresponding aldehydes via the Bucherer-Berg reaction, are converted by microbial cells of *Bacillus brevis*, which contain a D-specific hydantoinase, to a mixture of the D-N-carbamoyl amino acid and the L-hydantoin [26–28].

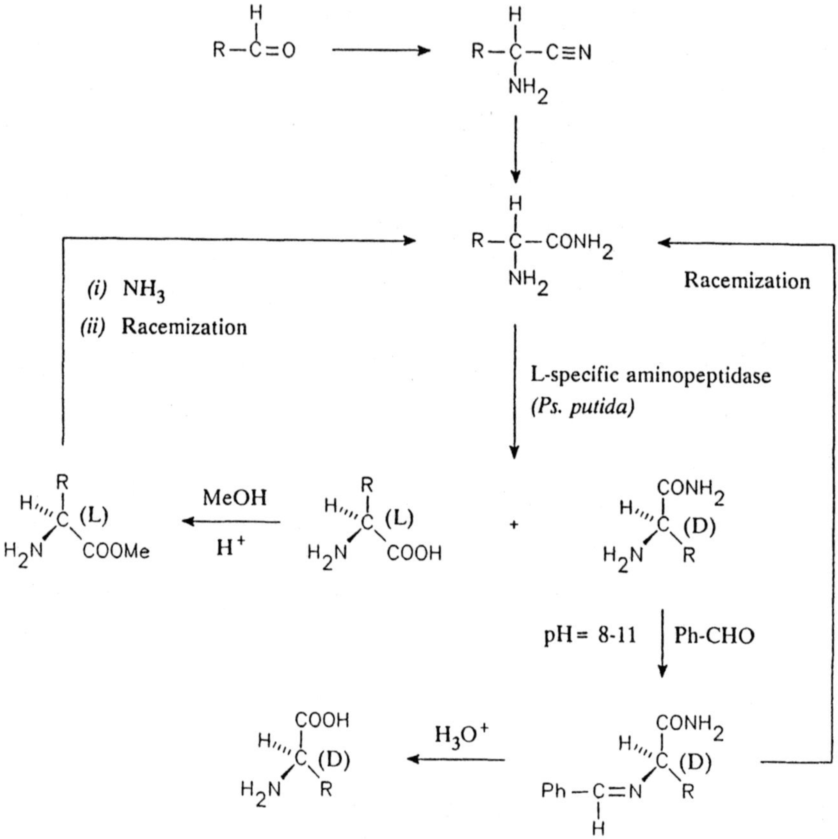

Figure 3-22. DSM aminopeptidase process.

A further advantage derives from the fact that the hydantoin contains an acidic C-H bond which results in spontaneous racemization of the unwanted enantiomer under the alkaline reaction conditions, thus allowing for a theoretical yield of 100%. The D-amino acid is obtained by subsequent treatment of the D-carbamoyl derivative with nitrous acid. This process is used on an industrial scale by Kanegafuchi [27,28] for the production of D-p-hydroxyphenylglycine (Figure 3-23), an important intermediate in the synthesis of amoxycillin and other antibiotics. In this case the substrate is readily prepared by condensation of phenol with a mixture of glyoxylic acid and urea [29].

Recordati (Italy) uses an even more elegant variation on this theme. The microorganism *Agrobacterium radiobacter* contains both a D-specific hydantoinase and a second enzyme that catalyzes the hydrolysis of the N-carbamoyl-D-amino acid,

(D)-HPG
Snamprogetti (Recordati)

(D)-*N*-Carbamoyl HPG
(Kanegafuchi)

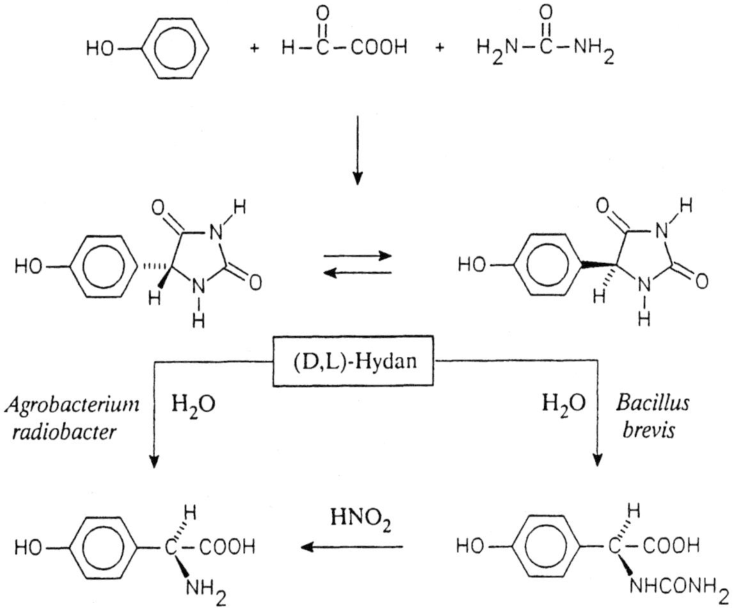

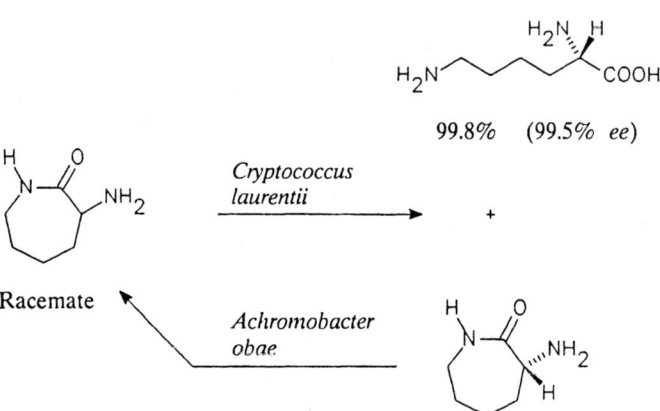

Figure 3-23. D-*p*-Hydroxyphenylglycine via enzymatic resolution of the DL-hydantoin.

Figure 3-24. Enzymatic kinetic resolution and enzymatic racemization.

allowing for a one-step, two-enzyme process with a theoretical yield of 100% [30]. In short, a beautiful example of an attractive, one-step, dynamic kinetic resolution process that utilizes a precursor as starting material.

In the above example, spontaneous racemization occurs in alkaline media due to the acidic character of the C-H bond in the hydantoin ring. In cases where such a spontaneous racemization is not feasible, one may be able to employ a racemase to effect in situ racemization. The racemase may be contained in the same microorganism as the hydrolytic enzyme or in a different organism. For example, the microorganism *Cryptococcus laurentii* contains an enzyme (Figure 3-24) that mediates the enantioselective hydrolysis of α-amino-ε-caprolactam to L-lysine. In the presence of a second microorganism, *Achromobacter obae*, the remaining D-α-amino-ε-caprolactam undergoes a racemase-mediated racemization [31].

E. Position of the Resolution Step

The position of the resolution step in the total synthesis also has an important bearing on the economics of the reaction. Here the golden rule is simple: carry out the resolution (or asymmetric synthesis) as early as possible in the synthesis. This can be readily understood if one reflects on the fact that every step performed on a racemate is carrying 50% ballast through the process. Removal of this ballast automatically halves the amount of reagents, solvents, reactor volume, etc., required in subsequent steps (Figure 3-25).

A further illustration of the golden rule is provided by a comparison of two alternative routes to dextromethorphan (Figure 3-26). Route A involves resolution via a diastereomeric salt crystallization as the final step in the synthesis [32]; in contrast,route B involves resolution of the first chiral intermediate in the process. Although both routes comprise the same number of steps (and assuming the same yield), route B is economically more attractive and is, in fact, the industrially applied route.

It should be noted that much of the extra costs involved in a late resolution can be recovered if an efficient racemization is possible. As a general rule, however, the structures of intermediates increase in complexity as the synthesis advances, making efficient racemization more of a problem. The dextromethorphan synthesis is a case in point: racemization of the early intermediate is relatively simple while racemization of the final product is not feasible.

F. Inversion of the Unwanted Isomer

An alternative to racemization of the unwanted isomer in resolution processes is to devise a means of inverting the configuration of this enantiomer. A particularly elegant example of this has been reported by Giordano and coworkers of Zambon [33] concerning the synthesis of the antiinfective agent thiamphenicol. The key step in the manufacture of thiamphenicol involves resolution of a precursor,

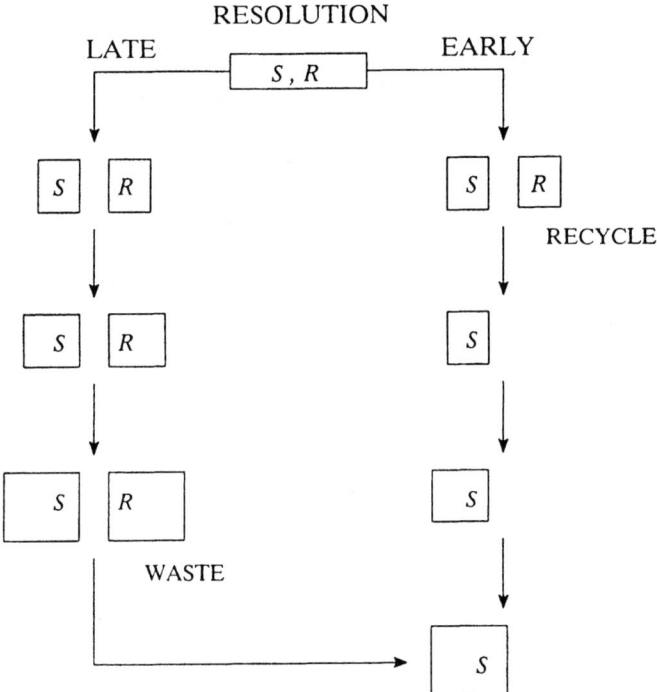

Figure 3-25. Early versus late resolution processes.

threo-2-amino-(4-methylthiophenyl)-1,3-propanediol, by preferential crystallization of the racemate, which yields the required *(1R,2R)* isomer. The method is completely analogous to that used for the industrial synthesis of the related drug, chloramphenicol. (See chapter 6.) The *(1S,2S)* isomer of the key intermediate is subsequently converted to its *(1R,2R)*-enantiomer by a sequential inversion of configuration at C_2 and C_1 in which each stereogenic center controls the other and maintains its stereochemical integrity (Figure 3-27).

Epimerization at C_2 is achieved by heating the labile aldehyde intermediate in the presence of a catalytic amount of base (DABCO) in toluene at 35°C. The equilibrium is continually displaced by crystallization of the required epimer, providing for stereoconvergence into a single diastereomer by a crystallization induced asymmetric transformation. Epimerization at C_1 is achieved by conversion to an oxazoline methanesulfonate which is formed as a 95:5 diastereomeric mixture, probably via a ring-opened carbonium ion intermediate [33]; subsequent hydrolysis gives a 95:5 mixture of the *(1R,2R)* and *(1S,2R)* diols. Recrystallization produces optically pure *(1R,2R)*-diol, the required precursor of thiamphenicol. In short, this example is a perfect illustration of the heights of ingenuity to which

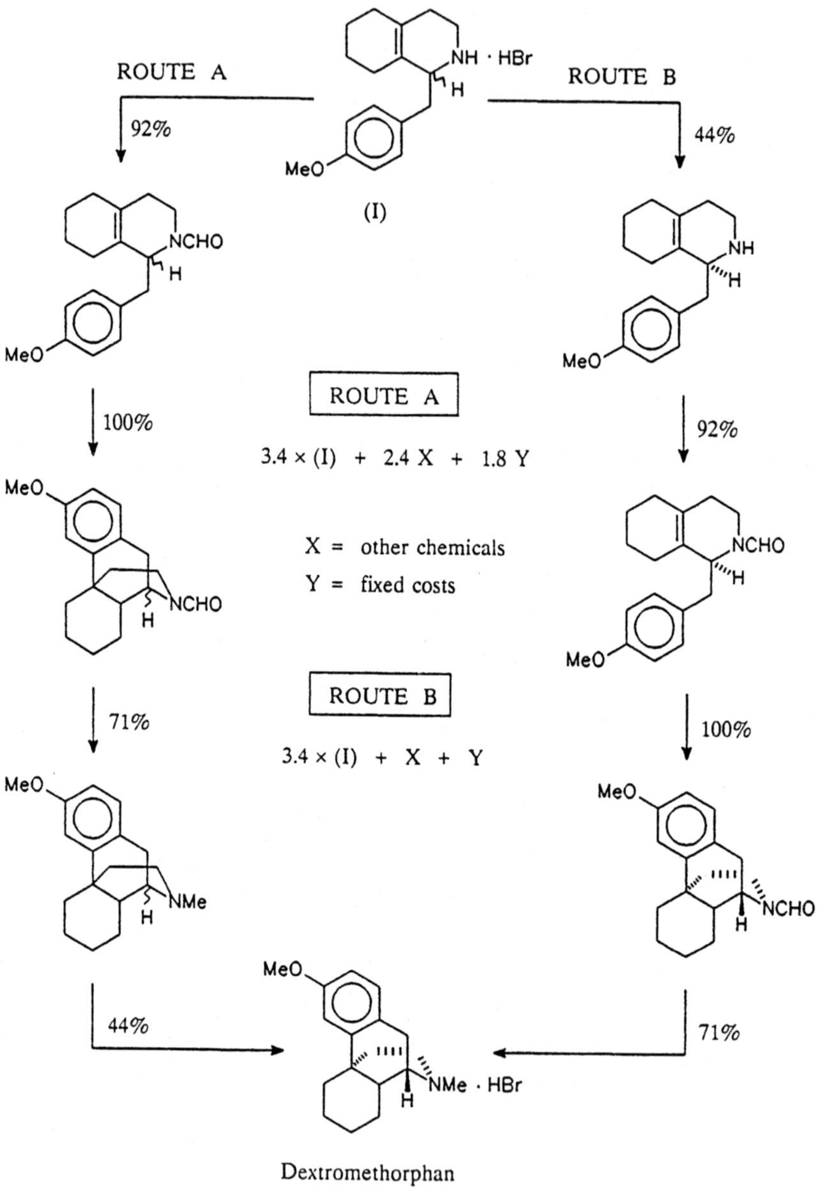

Figure 3-26. Two routes to dextromethorphan.

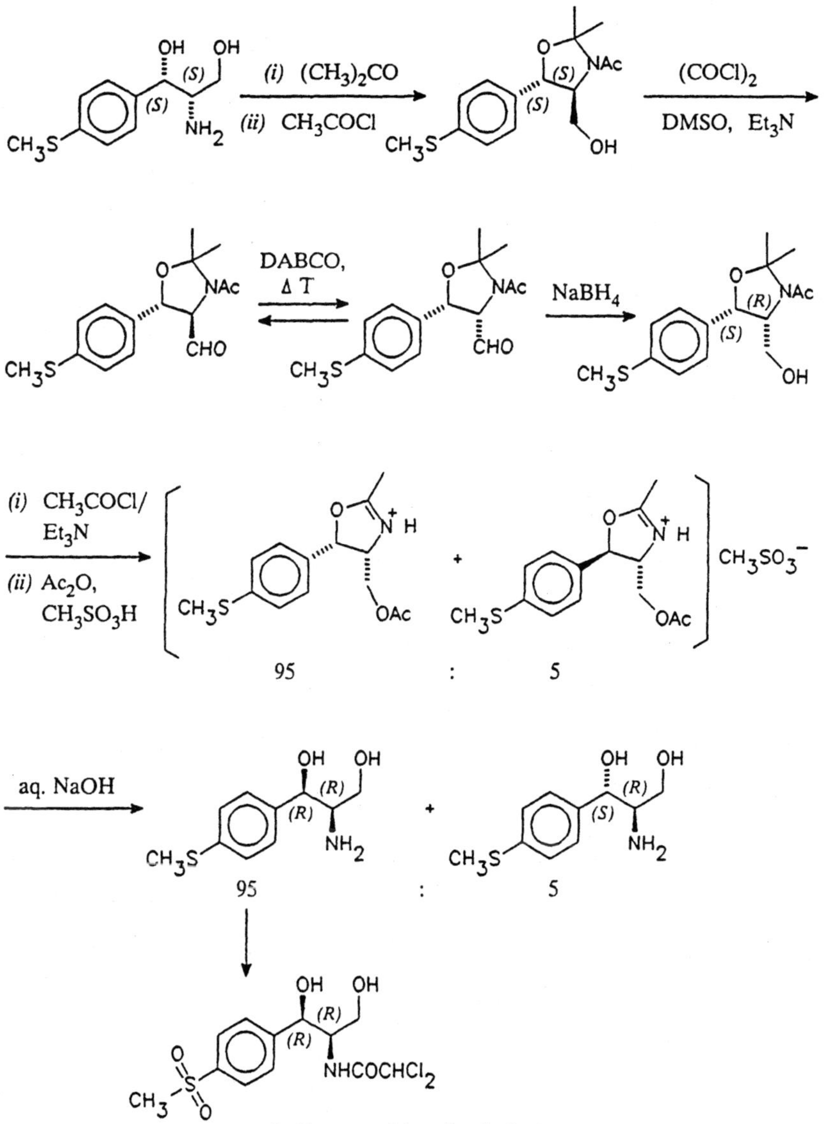

optically pure thiamphenicol

Figure 3-27. Synthesis of a thiamphenicol intermediate by sequential inversion of its enantiomer.

industrial organic chemists are aspiring in their quest for selective syntheses of optically pure compounds.

To summarize, a variety of technologies are now available for the efficient synthesis of optically active compounds on an industrial scale. With regard to the method of choice (e.g., enzymatic kinetic resolution, asymmetric synthesis, or crystallization techniques), there is no simple all-encompassing answer. Numerous factors are involved in determining the economics of alternative routes to a particular product and the method of choice obviously will vary from one product to another. There is just one general rule: plan your resolution early.

Finally, another general conclusion can be drawn: with the rapidly growing arsenal of cost-effective methods at our disposal, a lack of economically viable technology is no longer a tenable excuse for marketing biologically active products as racemic mixtures.

REFERENCES

1. Nestler, H. J., Hoerlein, G., Handte, R., Bieringer, H., Schwerdtle, F., Langellueddeke, P., and Frisch, P., British Patent Appl. 2,042,503 (1980) to Hoechst.
2. a) Giordano, C., Castaldi, G., Cavicchioli, S., and Villa, M., *Tetrahedron*, **45**, 4243-4252 (1989). b) Cavicchioli, S., Giordano, C., and Uggeri, F., *J. Org. Chem.*, **52**, 3018-3027 (1987).
3. Amiard, G., *Experienita*, **15**, 1-7 (1959).
4. Reinhold, D. F., Firestone, R. A., Gaines, W. A., Chemerda, J. M., and Sletzinger, M., *J. Org. Chem.*, **33**, 1209-1213 (1968).
5. a) Okada, Y., Takebayashi, T., Hashimoto, M., Kasuaga, S., Sato, S., and Tamura, C., *J. Chem. Soc. Chem. Commun.*, 784-785 (1983). b) Okada, Y., and Takebayashi, T., *Chem. Pharm. Bull.*, **36**, 3787-3792 (1988).
6. Bruggink, A., Hulshof, L. A., and Sheldon, R. A., *Pharmaceutical Manufacturing International*, 139-146 (1990).
7. Sheldon, R. A., Hulshof, L. A., Bruggink, A., Leusen, F. J. J., Van der Haest, A. D., and Wynberg, H., *Chemistry Today* 9: 23-29 July/August, (1991)
8. Clark, J. C., Phillips, G. H., and Steer, M. R., *J. Chem. Soc. Perkin I.*, 475 (1986).
9. Kagan, H. B., and Fiaud, J. C., in *Topics in Stereochemistry*, Eliel, E. L., and Wilen, S. H. (Eds.), Vol. 18, Wiley, New York, 1988, pp. 249-330.
10. Chibata, I., *Pure Appl. Chem.*, **50**, 667 (1978)
11. Kitamura, M., Manabe, K., and Noyori, R., *Tetrahedron Lett.*, **28**, 4719 (1987).
12. Katsuki, T., and Sharpless, K. B., *J. Am. Chem. Soc.*, **102**, 5974-5976 (1980).
13. Martin, V.S., Woodard, S. S. Katuski, T., Yamada, Y., Ikeda, M., and Sharpless, K. B., *J. Am. Chem. Soc.*, **103**, 6237 (1981).
14. Miyano, S., Lu, L. D. L., Viti, S. M., and Sharpless, K. B., *J. Org. Chem.*, **48**, 3611 (1983).
15. Knowles, W. S., Sabacky, M. J., Vineyard, B. D., and Weinkauf, D. J., *J. Am. Chem. Soc.*, **97**, 2567 (1975).
16. Hamilton, B. K., Hsiao, H., Swann, W. E., Anderson, D. M., and Delente, J. J., *Trends in Biotechnol.*, **3**, 64-68 (1985).

17. Wynberg, H., and Staring, E. G. J., *J. Am. Chem. Soc.*, **104**, 166 (1982).
18. Selke, R., Foken, H., and Facklam, C., *Tetrahedron Asymm.*, in press.
19. Laumen, K., and Schneider, M. P., *J. Chem. Soc. Chem. Commun.*, 1298 (1986).
20. De Vos, M. J., and Krief, A., *J. Am. Chem. Soc.*, **104**, 4282 (1982).
21. Hayashi, T., in *Asymmetric Reactions and Processes in Chemistry*, Eliel, E. L., and Otsuka, S. (Eds.), American Chemicsl Society, Washington, 1982, 177-186. ACS Symp. Ser. 185.
22. Kitamura, M., Ohkuma, T., Tokunaga, M., and Noyori, R., *Tetrahedron Asymmetry*, **1**, 1-4 (1990).
23. Chibata, I., Tosa, T., Sato, T., and Mori, T., in *Methods in Enzymology, Vol. 44*, Mosbach, E., (Ed.), Academic Press, New York, 1976, pp. 746-759.
24. a) Chibata, I., and Tosa, T., *Ann. Rev. Biophys. Bioengin.*, **10**, 197-216 (1981). b) Chibata, I., *Pure Appl. Chem.*, **50**, 667-675 (1978).
25. Schutt, H., German Patent, 2,807,286 (1979) and 2,927,535 (1981) to Bayer.
26. Sheldon, R. A., Schoemaker, H. E., Kamphuis, J., Boesten, W. H. J., and Meijer, E. M., in *Stereoselectivity of Pesticides: Biological and Chemical Problems*, Ariëns, E. J., Van Rensen, J. J. S., and Welling, W. (Eds.), Elsevier, Amsterdam, 1988, pp. 409-451, and references cited therein.
27. Takahashi, S., Ohashi, T., Kii, Y., Kumagai, H., and Yamada, H., *J. Ferment. Technol.*, **57**, 328-332 (1979).
28. Takahashi, S. in *Progress in Industrial Microbiology, Vol. 24*, Aida, K., Chibata, I, Nakayama, K., Takanami, T., and Yamada, H. (Eds.),, Elsevier, Amsterdam, 1986, pp. 269-279.
29. Ohashi, T. Takahashi, S., Nagamachi, T., Yoneda, K., and Yamada, H., *Agric. Biol. Chem.*, **45**, 831 (1981).
30. Olivieri, R., Fascetti, E., Angelini, L., and Degen, L., *Biotechnol. Bioeng.*, **23**, 2173 (1985).
31. Fukumura, T., *Agr. Biol. Chem.*, **41**, 1321-1325 (1977).
32. Schnider, O., and Grüssner, A., *Helv. Chim. Acta*, **34**, 2211 (1951).
33. a) Giordano, C., Cavicchioli, S., Levi, S., and Villa, M., *J. Org. Chem.*, **56**, 6114-6118 (1991). b) See also: Giordano, C., Cavicchioli, S., Levi, S., and Villa, M., *Tetrahedron Lett.*, **29**, 5561-5564 (1988).

ADDITIONAL READING

1. Sheldon, R. A., Porskamp, P., and Ten Hoeve, W., Advantages and Limitations of Chemical Optical Resolution, in *Biocatalysis in Organic Synthesis*, Tramper, J., Van der Plas, H. C., and Linko, P. (Eds.), Elsevier, Amsterdam, 1985.
2. Sheldon, R. A., The Industrial Synthesis of Pure Enantiomers, *Drug Information Journal*, **24**, 129-139 (1990).
3. Sheldon, R. A., Industrial Synthesis of Chiral Compounds, *Speciality Chemicals*, **Feb.**, 1990, pp. 30-45.
4. Sheldon, R. A., Industrial Synthesis of Optically Active Compounds, in *Problems and Wonders of Chiral Molecules*, Simonyi, M., (Ed.), Akademiai Kiado, Budapest, 1990.
5. Sheldon, R. A., Industrial Synthesis of Optically Active Compounds, *Chem. Ind., (London)* 212-219 (1990).
6. Crosby, J., Synthesis of Optically Active Compounds. A Large Scale Perspective, *Tetrahedron*, **47**, 4789-4846 (1991).

4

Fermentation Processes

Messieurs, c'est les microbes qui auront le dernier mot.

Louis Pasteur

In this chapter we shall discuss in more detail one of the oldest industrial methods for the synthesis of optically active compounds, namely fermentation, which is broadly defined as transformations mediated by growing microorganisms [1–3]. The harnessing of microorganisms in the service of mankind must surely be one of the most important and far-reaching of all scientific achievements.

I. DEFINITION OF FERMENTATION

As noted in the previous chapter, fermentation refers to the reproduction of microorganisms in the presence of a source of carbon and energy (e.g., sugar) and various nutrients. The commercially important products of fermentation fall into five major categories:

1. The microbial cells themselves (e.g., yeast and single-cell protein).
2. The large molecules that they synthesize (e.g., enzymes and polysaccharides).
3. The primary metabolic products (e.g., amino acids, nucleotides, vitamins) and organic acids (e.g., citric acid) that are essential for cellular growth.
4. The secondary metabolic products (e.g., antibiotics, dyes, pigments, perfumes, and poisons) that are not essential for cellular growth but serve the survival of

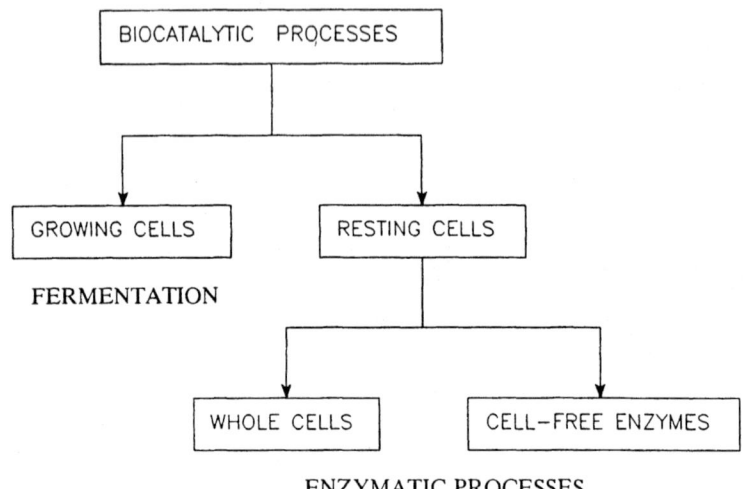

Figure 4-1. Fermentation versus enzymatic processes.

the species. Secondary metabolites constitute the chemical interface between microbes and the outside world. They are the downstream products of microbial synthesis (i.e., they are biosynthesized from one or more primary metabolites).

5. Microbial transformations of foreign substrates, often referred to as precursor fermentation (e.g., the microbial hydroxylation of steroids).

All fermentations involve fundamental reactions that are catalyzed by enzymes but it is nevertheless convenient to distinguish between fermentation processes and enzymatic transformations. As is shown in Figure 4-1, the distinction lies in whether or not living (i.e., growing) cells are involved.

Many fermentations are complex multistep reactions, involving several different enzymes, that are not possible outside the living cell. Enzymatic transformations, on the other hand, generally involve a single step mediated by one enzyme and the whole machinery of the cell is not necessary. It is also important to emphasize that fermentation refers to the method and not to the type of reaction. Thus, a fermentation may involve an asymmetric synthesis or the kinetic resolution of a racemate.

II. HISTORY OF INDUSTRIAL FERMENTATION

The art of fermentation has its roots deep in antiquity. Indeed, without knowing of their existence our early ancestors put microorganisms to work in the production of food and beverages. Beer brewing was practiced by the Sumerians and Baby-

Ionians before 6000 B.C. and references to winemaking can be found in the Book of Genesis. The use of yeast for the leavening of bread was known to the ancient Egyptians. Other long-standing fermentation processes include the production of vinegar, yoghurt, cheese, sauerkraut, and soy sauce. Thus, microbes provided food and drink for more than 8000 years before Anton van Leeuwenhoek, in the 17th century, saw these tiny 'animalcules' through his primitive microscope.

Although it had previously been suggested by others that these 'animalcules' were the cause of fermentation, it was Pasteur who, in 1857, showed unequivocally that alcoholic fermentation is caused by yeasts and that yeasts are living microorganisms. He subsequently postulated that microorganisms are the root cause of infectious diseases, thus launching the germ theory of disease, one of the landmarks in the history of civilization.

The first optically active compound to be produced industrially, in 1880 in the USA, was lactic acid. In the late nineteenth century, the scientific basis was laid for understanding fermentation that proved invaluable with the outbreak of World War I. In 1914 wartime shortages of acetone, the raw material for the explosive, cordite, stimulated the development of the acetone-butanol fermentation employing the anaerobic bacterium *Clostridium acetobutylicum*. The process was developed in the UK by the Russian-born chemist, Chaim Weizmann. In so doing he may have altered the course of history and in return for this remarkable achievement the British government enacted the Balfour declaration of 1917. This eventually led to the establishment of the state of Israel with Weizmann as its first president.

Parallel to the acetone-butanol development German chemists developed a fermentation process for glycerol and chemists at Pfizer in the USA investigated the production of citric acid by fermentation. In 1923, this culminated in the first commercial process for the production of citric acid by fermentation of sugar with the mold *Aspergillus niger*.

In the late nineteenth century, following Pasteur's observations that microbes present in soil were able to kill germs present in diseased cadavers, many microbial preparations were tested as antibiotics (antiseptic compounds of microbial origin). Unfortunately, they were either too toxic or were inactive in vivo. Finally, in 1928, Alexander Fleming made a chance observation of profound significance. He observed that spores of the mold *Penicillium notatum* inhibited the growth of cultures of *Staphylococcus aureus*, a particularly nasty germ responsible for boils. He went on to show that the cell-free liquid derived from the growing mold was active against a variety of bacteria and gave the name penicillin to this active ingredient. Initial attempts to isolate penicillin were thwarted by the compounds limited stability. Eventually, in 1940, Florey and Chain successfully prepared a stable form of penicillin and confirmed its remarkable antibacterial activity. This research eventually led to the commercial production of penicillin by a related mold *Penicillium chrysogenum*.

The advent of penicillin signaled the beginning of the antibiotic era and was closely followed by the discovery of a plethora of antibiotics, such as streptomycin, erythromycin, the cephalosporins and tetracyclins, and many more. The postwar period saw a further proliferation of fermentation technology for the production of amino acids, vitamins, enzymes, steroid hormones, and many other pharmaceuticals. Indeed the success story of industrial fermentation still continues unabated, having been given a new boost of energy with the advent of recombinant DNA technology.

III. THE BASICS OF INDUSTRIAL FERMENTATION

A. The Catalyst

Underlying the plethora of microbial processes are certain characteristics common to all microorganisms [4]. The most fundamental is the small size and high surface-to-volume ratio of the microbial cell that allows for rapid transport of nutrients into the cell which thereby supports its high rate of metabolism. Thus, the rate of protein production in bacteria is several orders of magnitude higher than in plants, which in turn is an order of magnitude higher than in animals. This extremely high rate of microbial synthesis allows some microorganisms to reproduce in only 15 minutes. In other words, compared to higher organisms, they are very active catalysts with high turnover frequencies.

The broad range of environments that can support microbial life reflects their tremendous versatility. Microorganisms have been found quite happily reproducing at temperatures ranging from the freezing point of water almost to its boiling point, in salt or fresh water and in the presence or absence of air.

From the viewpoint of industrial fermentation, there are four important classes of microorganisms: yeasts, molds, bacteria, and the soil-inhabiting actinomycetes (filamentous bacteria). The yeasts and molds together constitute the fungi. Organisms of this type are eukaryotic—their cells, like those of animals and plants, have a membrane-enclosed nucleus and more than one chromosome. The single-cell bacteria and actinomycetes, on the other hand, are prokaryotic—they have no nuclear membrane and only one chromosome.

Microorganisms are also classified on the basis of their environmental requirements. The strict aerobes and anaerobes grow only in the presence or absence of oxygen, respectively. The so-called facultative organisms, in contrast, can switch from an aerobic (respiratory) to an anaerobic (fermentative) mode, depending on the environment. For example, industrial yeasts are facultative organisms that can either respire or ferment certain substrates.

Only one subgroup of the bacteria, the Eubacteria, has industrial utility. These include the acetic acid bacteria (*Gluconobacter* and *Acetobacter*) and the lactic acid bacteria (*Streptococcus, Leuconostoc,* and *Lactobacillus*). The genera *Coryne-bacteria* and *Brevibacteria* are major industrial sources of amino acids. The most important microorganisms for the manufacturing of antibiotics are molds of the genera *Penicillium* (penicillins) and *Cephalosporium* (cephalosporins), and actinomycetes of the genus *Streptomyces* (tetracyclins and streptomycins).

The yeasts, typified by *Saccharomyces cerevisiae* (baker's yeast) and various *Candida* species have tremendous industrial utility, as sources of industrial enzymes and for a variety of microbial transformations. Indeed, more than 1.8 million tons of *Saccharomyces cerevisiae* are produced each year worldwide, making it one of the most efficient cell processes in the entire fermentation industry.

B.   Catalyst Screening, Selection, and Development

A microorganism is a finely tuned machine that has evolved to optimize its own survival and not for the sole purpose of manufacturing chemicals with commercially interesting applications. Modern biotechnology calls for the manipulation and enhancement of natural traits for the benefit of mankind; this 'artificial evolution' has lead to the production of metabolic products in amounts that would be a disastrous drain on a 'wild' organisms' resources of energy and nutrients, or even to produce substances that are not part of its normal repertoire. In other words, the catalyst has to be modified and developed for the specific task at hand.

The first step involves the screening and selection of wild strains of bacteria and fungi for the given task. A suitable microorganism is able to maintain its activity from generation to generation and is reasonably resistant to infections. Typical sources are soil, the mud of lakes and rivers, or the sludge from effluent treatment facilities of chemical plants. Over the years, stock cultures of bacteria, yeasts, molds, and actinomycetes with industrial utility have been accumulated and are maintained in permanent culture collections as, for example, at the Institute for Fermentation in Osaka (IFO).

In the primary microbial screening, culture plates or small-scale fermentations in shaker vessels are used to select microbes with desirable properties. Growth inhibitors (e.g., antibiotics) may be added to prevent the growth of undesirable microorganisms. Interesting activities from the primary screen are retested using more specialized media and conditions. When a useful organism emerges from a screening program, the next step is to develop its productivity (i.e., catalytic activity) for the required metabolite. The initial phase may involve optimizing the medium and fermentation conditions. The greatest improvements, however, are generally achieved in the performance of the microorganism itself. This involves the development of tailor-made industrial microorganisms from wild bacteria or by genetic manipulation such that undesirable properties are eliminated, desirable ones are accentuated, or entirely new ones are introduced [5].

The classical method for reprogramming the genetic information of micro-organisms is induced mutation. This involves subjecting the organism to ultraviolet light, ionizing radiation (e.g., X-rays) or one of a variety of chemical mutagens that react with DNA bases and thus interfere with the organism's genetic code. A typical mutation procedure may destroy 99% of the microbes. The surviving population is then screened to determine the suitability of new mutants for the required task.

For example, to select mutants resistant to inhibition by a particular metabolite the cells are spread on a culture plate containing the inhibitor. Only the resistant mutant will proliferate and form colonies. Mutants that are defective in some metabolic step are called **auxotrophic mutants**. For example, auxotrophic mutants of the bacteria *Corynebacterium glutamicum* and *Brevibacterium flavum* form the basis for the large scale production of a range of proteinogenic amino acids in Japan. The skillful exploitation of an understanding of the biosynthetic pathway for L-lysine production in these bacteria led to the development of highly productive mutant strains capable of converting more than one third of the sugar in a fermentation broth into this single amino acid [6]. L-Lysine concentrations as high as 75 g l^{-1} have been obtained in this way.

In these bacteria, L-lysine is produced, together with L-methionine and L-threonine, through the intermediacy of L-aspartic acid, as outlined in Figure 4-2. The main control mechanism for ensuring that the bacterium does not produce more amino acids than it needs is feedback inhibition of the first enzyme in this pathway, aspartate kinase, by threonine and lysine (Figure 4-2). In other words, the accumulation of excess amino acids blocks the enzyme responsible for their synthesis.

One type of auxotrophic mutant capable of lysine overproduction lacks the gene encoding for the enzyme homoserine dehydrogenase. This precludes the formation of threonine and methionine. When this auxotroph is cultured in a medium with just enough threonine and methionine to support growth, threonine inhibition is circumvented and lysine production proceeds at full speed. In a second type of auxotroph the gene encoding for aspartate kinase is mutated. The modified enzyme still functions but is no longer inhibited by lysine, even at high concentrations. Auxotrophic mutants containing aspartic kinase insensitive to feedback inhibition were found among those resistant to *(S)*-β-aminoethyl-L-cysteine (AEC), an analog of L-lysine.

One of the most notable success stories of strain improvement is the development of the mold *Penicillium chrysogenum* for the commercial manufactures of penicillin [7,8]. Over a period of several decades the yield of penicillin has been increased from a few milligrams per liter up to the 20 grams or more now being produced by highly developed industrial organisms. These truly unusual strains are the result of many successive rounds of mutation and selection.

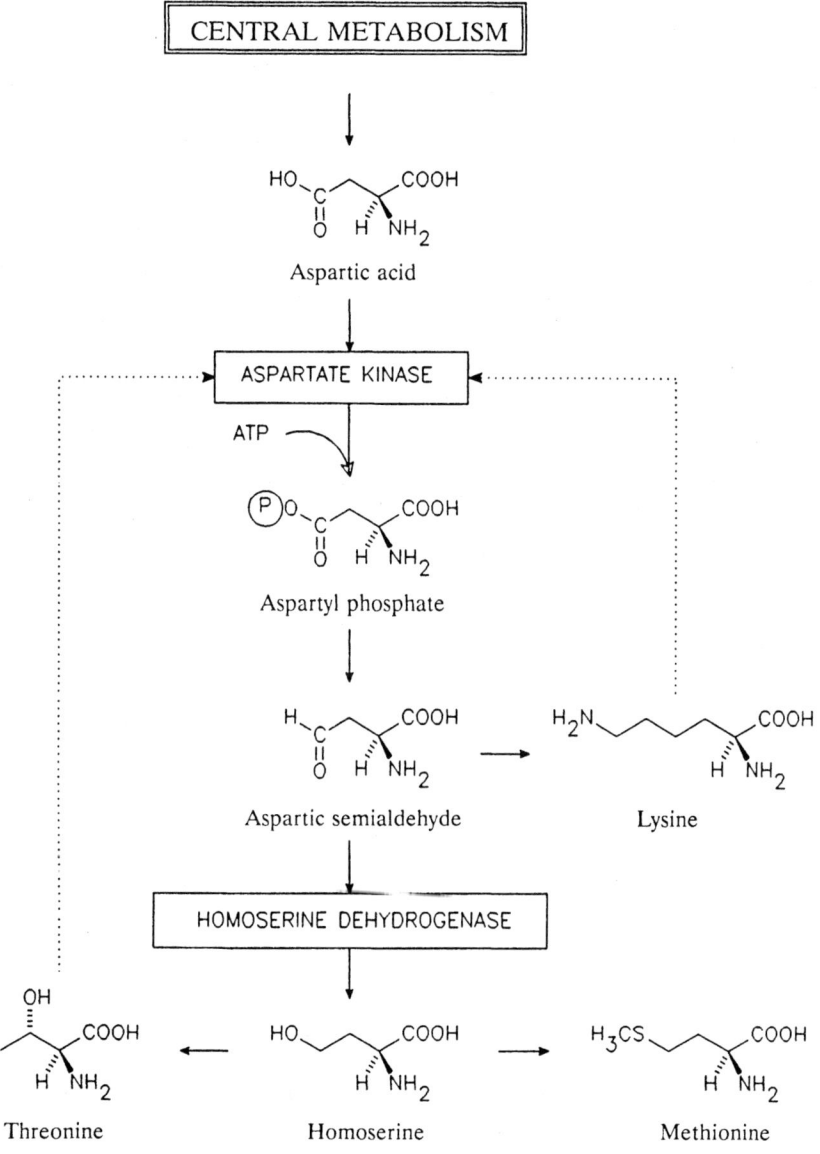

Figure 4-2. Biosynthetic pathway for L-lysine production (dotted lines denote feedback inhibition).

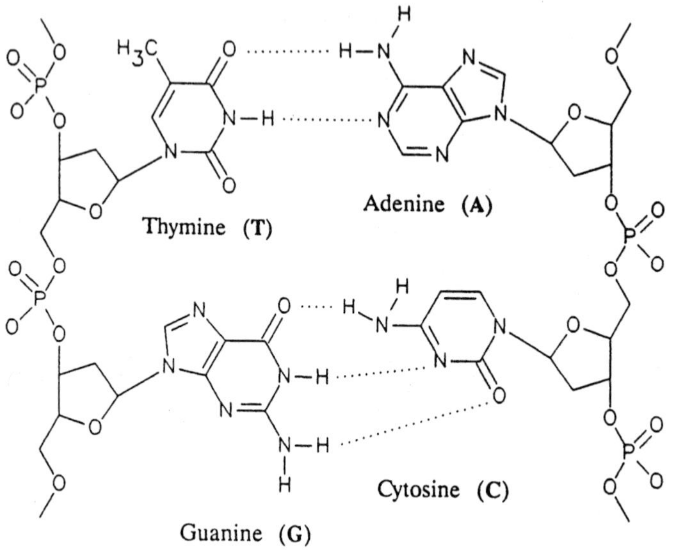

Figure 4-3. Section of a DNA double helix showing the base pairing. Each base pair has one purine base (**A** or **G**) and one pyrimidine base (**C** or **T**).

C. Genetic Engineering: Recombinant DNA Technology

The classical method of genetic manipulation via induced mutation with chemical mutagens or ionizing radiation is a shotgun, trial and error approach. Since the whole organism is bombarded with mutagens, the selective mutation of a particular gene is unlikely. This situation changed dramatically in 1973 [9] with the advent of recombinant DNA (r-DNA) technology that revolutionized microbiology and made specific mutagenesis feasible. With the help of r-DNA technology, it is now possible to rationally design highly productive microbial strains (so-called super-bugs) with desirable catalytic properties [10].

Genetic information is stored in living cells in DNA. In a typical bacterium most of the information is encoded in a single molecule of DNA: the bacterial chromosome. The molecule consists of a double helix, each strand of which contains several billion nucleotide bases. The bases in each strand are complementary; adenine (**A**) and guanine (**G**) pairing with thymine (**T**) and cytosine (**C**), respectively (Figure 4-3).

The total information content of the bacterial chromosome comprises several thousand genes, each one perhaps 1000 bases long, that code for the same number of proteins, mostly enzymes. The information is encoded in the sequence of bases on one strand of the DNA, each three-base 'codon' specifying a particular amino acid (e.g., **AAG** = lysine, **TGT** = cysteine, etc.).

In addition to the bacterial chromosome many bacteria contain smaller extra chromosomal DNA fragments called plasmids, which contain only a few genes, up to a dozen at most. Many bacteria owe their antibiotic resistance to genes carried on plasmids; these genes are capable of autonomous replication within a cell and are inherited by daughter cells. Moreover, they can also be carried from one bacterium to another by a bacterial virus (bacteriophage). The basis of r-DNA technology is the insertion of a gene or genes from an unrelated organism, or even an artificially synthesized gene, into a plasmid that is subsequently introduced into a new host microbe.

r-DNA techniques are used, for example, to induce a microorganism to produce a protein (e.g., an enzyme or a hormone) that is not part of its normal repertoire. The individual gene encoding for the desired product is inserted into a plasmid that is introduced into the host microbe and the latter is cloned in large quantities. The bacterium *Escherichia coli*, the workhorse of molecular biology, is often used as the host microbe since it contains effective plasmids (e.g., pBr322) with known DNA sequences.

The mechanics of r-DNA involve selective cleavage of a chromosome by a restriction endonuclease, an enzyme that recognizes particular short sequences in a DNA molecule, cleaving it at a specific site. Many endonucleases generate DNA fragments with single strand protrusions, the so-called 'sticky ends' (Figure 4-4).

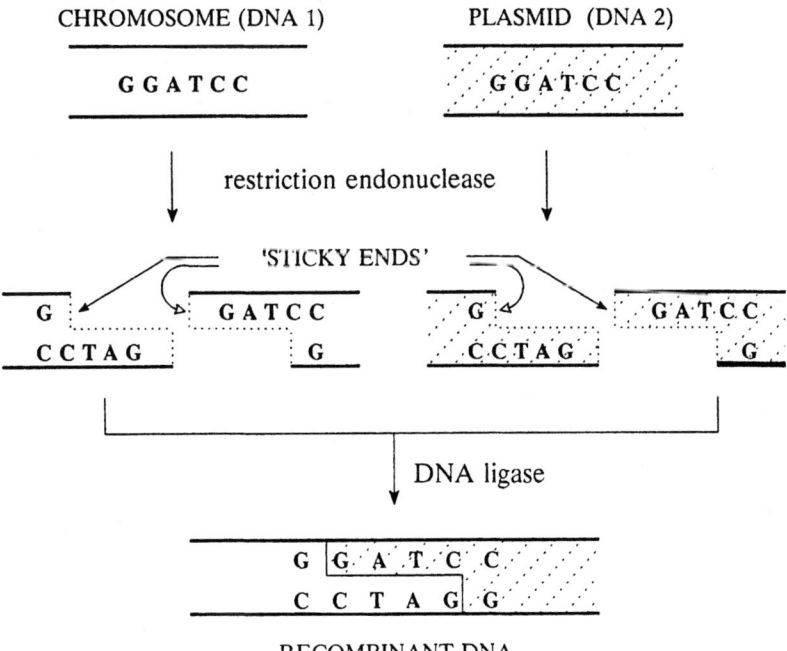

Figure 4-4. The basic method of recombinant DNA technology.

Fragments containing the gene to be transformed are joined together (through phosphodiester bonds) with matching sticky ends of the plasmid vector which has been cut open with the same endonuclease. This is achieved with a DNA ligase enzyme. The resulting recombinant plasmid is then introduced into the host cell which is cloned into billions of cells, each of which carries a copy of the foreign gene.

As a result of r-DNA technology bacteria, notably *Escherichia coli* and *Bacillus subtilisin* are now synthesizing a variety of proteins (e.g., human insulin, human growth hormone, interferon, and somatostatin) that they never made in nature [8]. Another elegant example of the use of r-DNA technology is the genetic engineering of a microbial strain for the single stage production of a vitamin C precursor.

D. Site-Directed Mutagenesis

A particularly exciting recent development in r-DNA technology is **Site-Directed Mutagenesis** (SDM). SDM represents a further fine-tuning of genetic manipulation, allowing for the specific replacement of one (or more) amino acid residue(s) in a protein (enzyme) [14–20]. Thus SDM, one variant of which is **oligonucleotide directed mutagenesis**, makes the rational design of proteins and enzymes with a predetermined amino acid sequence a viable proposition. The dream of the enzymologist—to be able to alter the amino acid sequence of proteins at will—is now reality. In other words, enzymes with novel catalytic properties are now amenable to rational design.

Typically, SDM involves the procedure outlined in Figure 4-5. The gene of interest is inserted into a vector (plasmid or phage) and expressed in a host bacterium such as *E. coli*. One strand of the DNA is isolated and an oligonucleotide is synthesized that is complementary to the region of the gene to be mutated except for a single (or double) base. This allows the replacement of a single amino acid in the protein. For example, by changing the codon from TTG to TTC, the corresponding amino acid changes from glycine to glutamic acid. The oligonucleotide is then annealed to the DNA strand. The resulting molecule is treated with a polymerase whereby the attached oligonucleotide is selectively polymerized to give a new strand of DNA. Finally, joining the ends of the new strand of DNA, by treatment with DNA ligase, affords a heteroduplex DNA that is expressed in a host bacterium and cloned. Clones containing the desired mutant DNA can be selected from the wild-type DNA on the basis of their enhanced ability to from hydrogen bonds with the original oligonucleotide. The required enzyme can then be produced in large quantities from the transformed cells.

SDM has tremendous potential for the rational design of enzymes with desirable properties, such as compatibility with organic solvents and enhanced enantioselectivity. It has been used [21] to engineer oxidative resistance into subtilisin, an important laundry detergent additive, by selective substitution of a methionine residue containing an easily oxidizable methylthio (CH_3S) group.

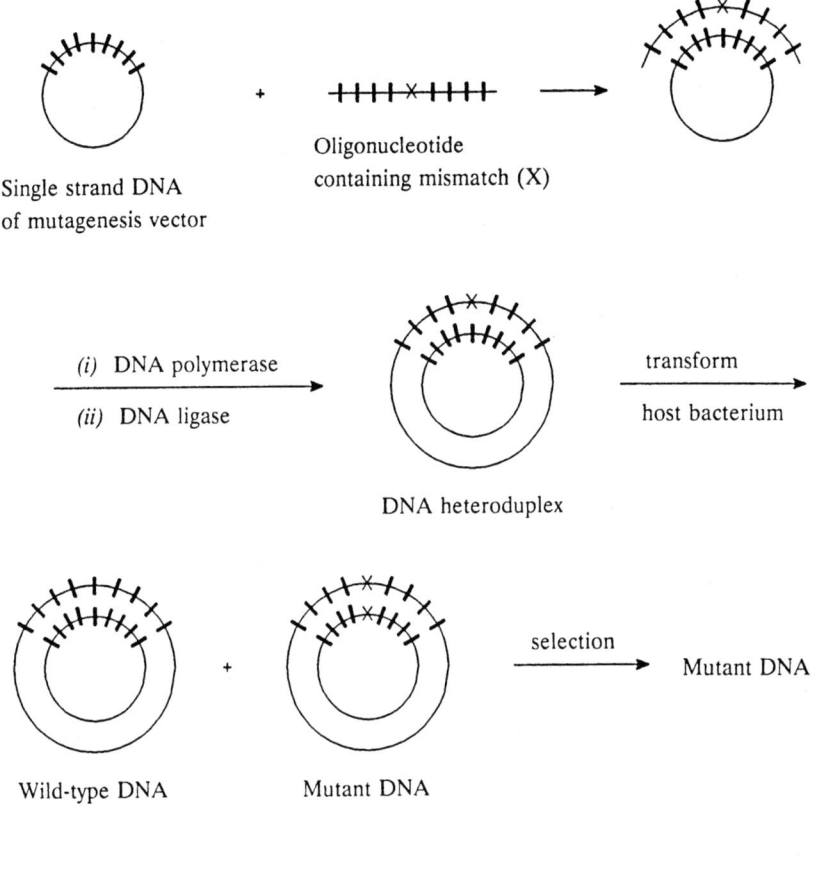

Figure 4-5. Typical procedure for site directed mutagenesis (SDM).

E. The Process

Genetic engineering of microbes, is only one factor, albeit an important one, contributing to the success of modern fermentations. Process engineering has also played an important role [22].

A microbe is merely a catalyst of exceptional complexity and sensitivity, and for optimum performance, the environment must be carefully controlled. Temperature and pH, for example, must be kept within a narrow range and the reaction medium must supply all of the nutrients in precisely the right amounts necessary to sustain growth. Furthermore, metabolic byproducts must be removed.

A primary requirement is a source of carbon which supplies the energy needed for metabolism and may also be the substrate. Carbohydrates, such as starch or sugar, are commonly used but many industrial microbes can utilize hydrocarbons, methanol, or natural fats. In addition to carbon, substantial quantities of nitrogen and phosphorus are needed as nutrients, together with smaller amounts of vitamins and metal ions as cofactors for enzymes. The availability of oxygen must also be taken into consideration; O_2 is essential for aerobic microbes but lethal for anaerobic ones.

A common feature of all fermentations is the susceptibility to contamination by foreign organisms. This necessitates sterilization of all equipment and raw materials including the large volumes of air required for aerobic fermentations.

Because fermentations generally involve rather dilute aqueous solutions, reactors can be very large, up to several hundred cubic meters. Both batch and continuous processes are used, the latter being better suited to large volumes but less amenable to maintaining aseptic conditions. Continuous fermentations employ an immobilized catalyst (e.g., cells adsorbed on alumina, charcoal, cellulose, etc.) or entrapped in a polymer matrix (e.g., starch). More recently, the technique of the fluid bed reactor has been borrowed from the petrochemical industry.

Because of the complex nature of the reaction mixtures, downstream processing is an important feature of all fermentations. The product is often thermally labile and is dissolved in a large volume of water. Distillation is energy intensive and may be accompanied by product decomposition, so solvent extraction is often used. Membrane separations would seem to offer many advantages and are finding increasing use in downstream processing [23].

F. Advantages and Limitations of Fermentation

When is fermentation the method of choice for the industrial synthesis of optically active compounds? In order to answer this question we need to assess the advantages and limitations of fermentation compared with alternative technologies. These are summarized in Table 4-1.

Table 4-1   Advantages and Limitations of Fermentation

Advantages	Limitations
1. Diversity of cheap raw materials	1. Low productivities and product concentrations
2. Complex molecules in one-step processing (de novo fermentation)	2. Complex, (i.e. expensive downstream processing)
3. Regio- and stereoselectivities sometimes unattainable by alternative means	3. High fixed costs (volume-factor)

Table 4-2 Productivities of Fermentations Compared to Other Processes [24]

Process/Product	Productivity $(g\ l^{-1}\ hr^{-1})$	Price $(\$\ kg^{-1})$
Fermentations		
Citric acid	7.0	1.5
L-Lactic acid	4.0	2.5
Glutamic acid	3.0	2.0
Ethanol	1.5	0.70
Lysine	0.6	13
Penicillin G	0.17	20
Vitamin B$_{12}$	0.001	8000
Enzymatic		
Glucose —> fructose	200	2.0
Fumaric acid —> aspartic acid	20	3.0
Cinnamic acid —> phenylalanine	15	10
Chemical		
Acetic acid (MeOH + CO)	500	0.75
Ethanol from ethylene	80	0.70
DL-Methionine	15	3.0

A major advantage is the diversity of inexpensive raw materials, ranging from petroleum hydrocarbons to a variety of agricultural (waste) products such as starch, molasses, corn-steep liquor, and sulfite liquor from paper manufacture. Major limitations are the low product concentrations and the accompanying large amounts of biomass that are produced. The desired product often represents only 1–10% of the total weight of organic material, thus necessitating complex and expensive downstream processing. It should be noted, however, that efficiencies are constantly being improved by genetic manipulation.

A comparison of the productivities of fermentation, enzymatic, and chemical processes (Table 4-2) shows that even relatively efficient fermentations (e.g., glutamic acid) have a significantly lower productivity than many enzymatic and chemical processes. Nevertheless, fermentation is attractive even for simple chiral molecules when they are produced efficiently. A case in point is L-lactic acid, produced in 90% yield on glucose in concentrations of more than 100 g l^{-1}. Fermentation is also the method of choice for complex natural products that can be made in a single stage (de novo fermentation), even though product concentrations may be of the order of mg l^{-1}. For example, it is difficult (or impossible) to imagine a chemical synthesis that can compete with de novo fermentation for the production of such complex molecules as vitamin B$_{12}$, lovastatin, or cyclosporin (Figure 4-6).

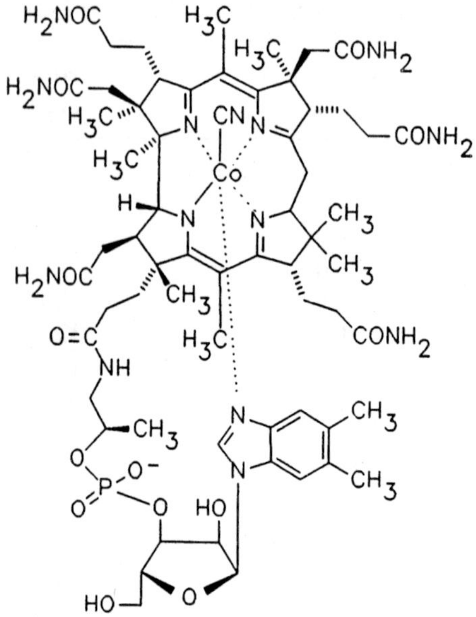

Vitamin B$_{12}$

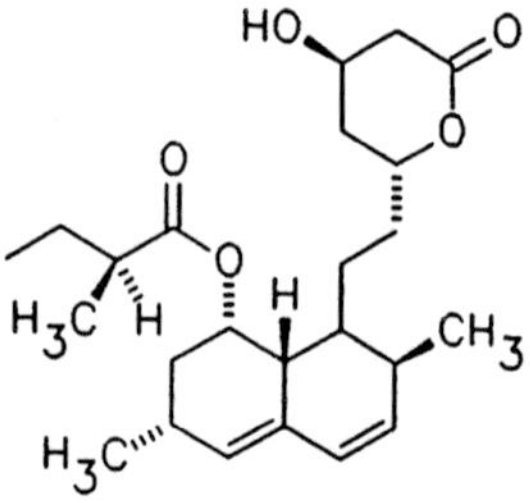

Lovastatin
(antihypercholesteremic)

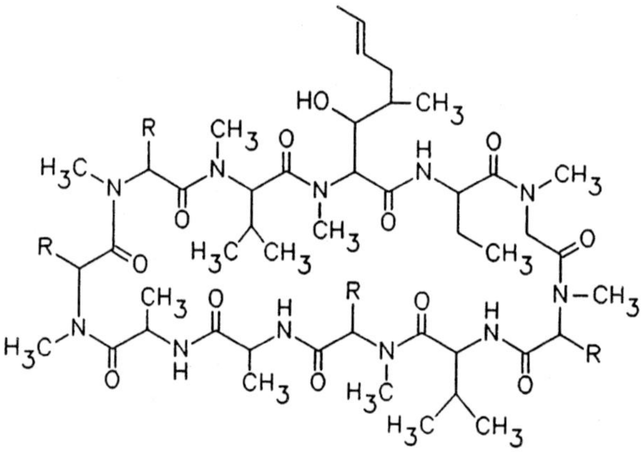

Cyclosporin A,   (R = CH$_2$–CH(Me)$_2$)
(immunosuppressant)

Figure 4-6. Complex molecules manufactured by fermentation.

Similarly, precursor fermentation can be the method of choice when highly regio- and/or stereoselective conversions are achieved in complex molecules; commercially relevant examples include the steroid hydroxylations. Indeed, the discovery [25] of the selective microbial hydroxylations of progesterone to 11α-hydroxyprogesterone (Figure 4-7) led to a commercially viable synthesis of cortisone that replaced a 31 step chemical synthesis from a bile acid.

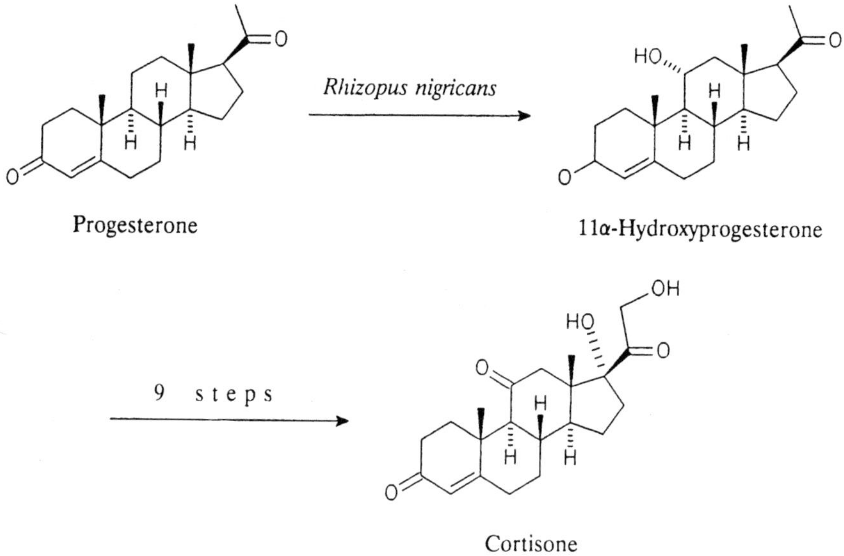

Figure 4-7. Regioselective microbial hydroxylation of progesterone.

In these examples, the choice in favor of fermentation is clear cut; there are no economically viable syntheses. In many other cases, however, the choice is not so clear cut. In amino acid manufacture, for example, fermentation is the method of choice for some (e.g., glutamic acid) but chemical or enzymatic methods are favored for others and a few amino acids are even produced by both fermentation and chemical synthesis. Because of the complexity of a fermentation process, fixed costs tend to be high and dedicated units the rule. This tends to make the economics more sensitive to the scale of production and fermentation can conceivably become the method of choice for a particular product above a certain production volume.

IV. FERMENTATION PRODUCTS

A. Chiral Hydroxy Acids from Carbohydrate Metabolism

More than 5 million tons of starch are produced per year in the European Economic Community (EEC) from agricultural products such as maize, wheat, barley and potatoes [26]. Some of this is used as a fermentation feed stock. D-Glucose, the basic building unit of starch, can be formed in situ by the action of amylase enzymes. The glucose can then be converted to a variety of commercially interesting hydroxy acids which are primary products of carbohydrate metabolism. The commercially most important is citric acid, which is achiral, but a variety of chiral hydroxy acids are also produced by carbohydrate metabolism (Figure 4-8).

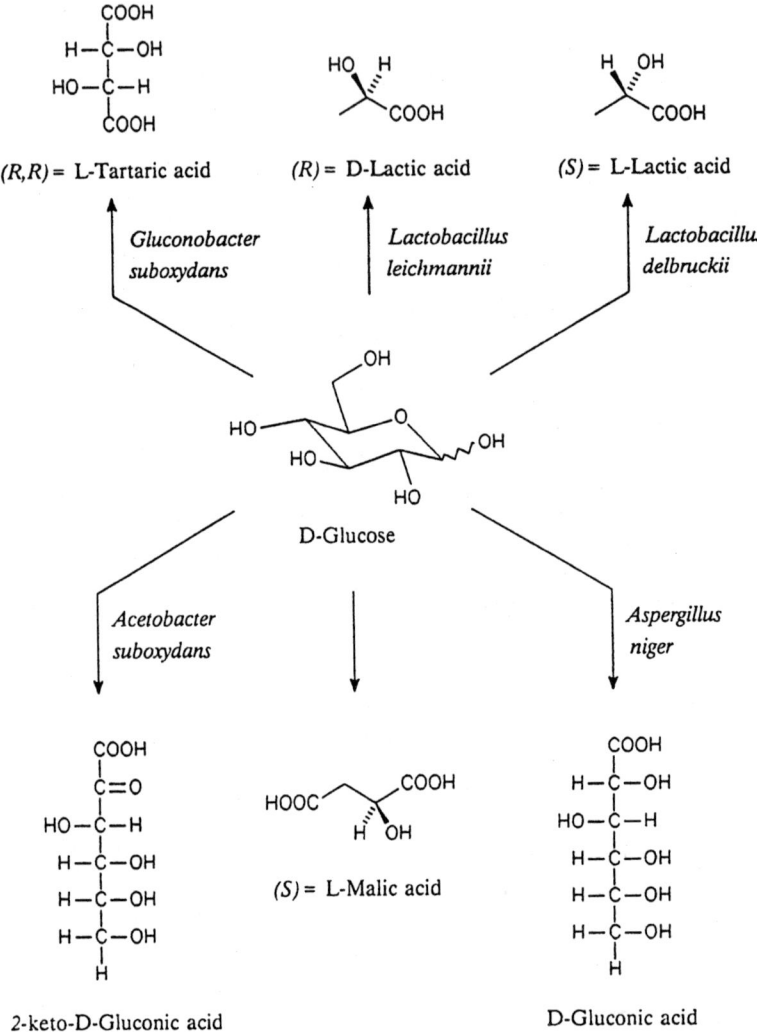

Figure 4-8. Chiral hydroxy acids available from industrial fermentations.

Both enantiomers of lactic acid are produced commercially, the more important one being the *(S)*=L-isomer. About 30,000 tons per annum are produced by fermentation of D-glucose with *Lactobacillus delbruckii* or lactose (whey from cheese manufacture) with *Lactobacillus bulgaricus* [27]. The *(R)*=D-enantiomer is produced on about a 1000 ton scale as an intermediate for optically active herbicides.

L-Tartaric acid can be produced by fermentation but the most economical source is still the cream of tartar byproduct of winemaking. Similarly, L-malic acid is available by fermentation of glucose but is more economically produced by the enzymatic hydration of fumaric acid (Reaction 4-1). The latter can also be carried out as a fermentation with the fumarase-containing *Brevibacterium ammoniagenes*.

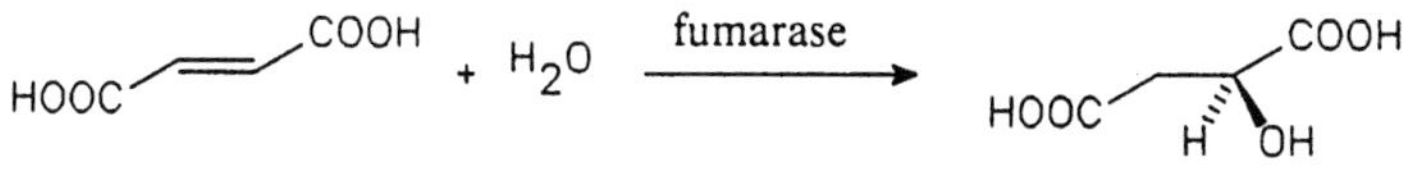

$$\text{HOOC}\diagup\!\!\!=\!\!\!\diagup^{\text{COOH}} \; + \; H_2O \xrightarrow{\text{fumarase}} \text{HOOC}\diagup\!\!\diagdown_{\substack{|\\ H \;\; OH}}^{\text{COOH}} \qquad (4\text{-}1)$$

About 45,000 tons per annum of D-gluconic acid are produced worldwide using a variety of microbes (e.g., *Aspergillus niger*). 2-Keto-D-gluconic acid is produced by fermentation of glucose with *Acetobacter suboxydans* and is an intermediate in the production of isoascorbic acid (isovitamin C) (Reaction 4-2).

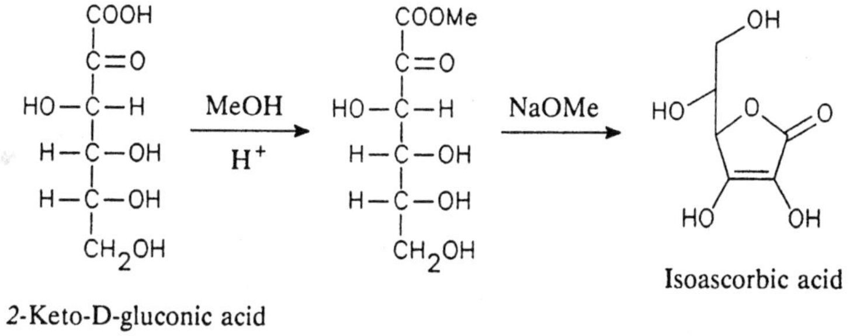

2-Keto-D-gluconic acid   Isoascorbic acid

$$(4\text{-}2)$$

About 30,000 tons per annum of ascorbic acid (vitamin C) are produced by the Reichstein-Grussner process which dates from 1934 [28]. The key step in this process is the microbial oxidation of D-sorbitol to L-sorbose mediated by *Acetobacter suboxydans* (Figure 4-9 and see chapter 5).

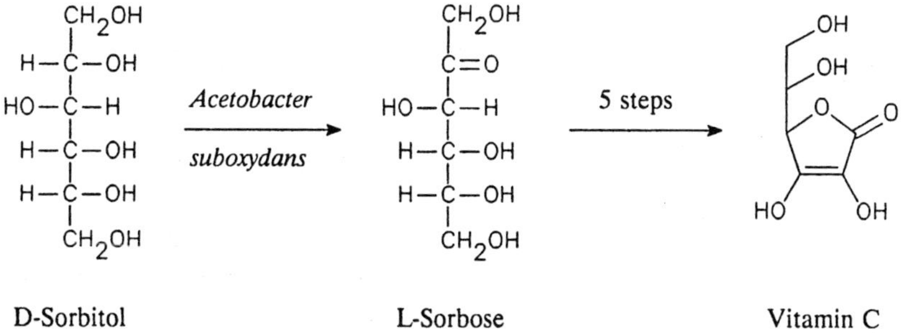

D-Sorbitol   L-Sorbose   Vitamin C

Figure 4-9. L-Sorbose production by fermentation of D-sorbitol.

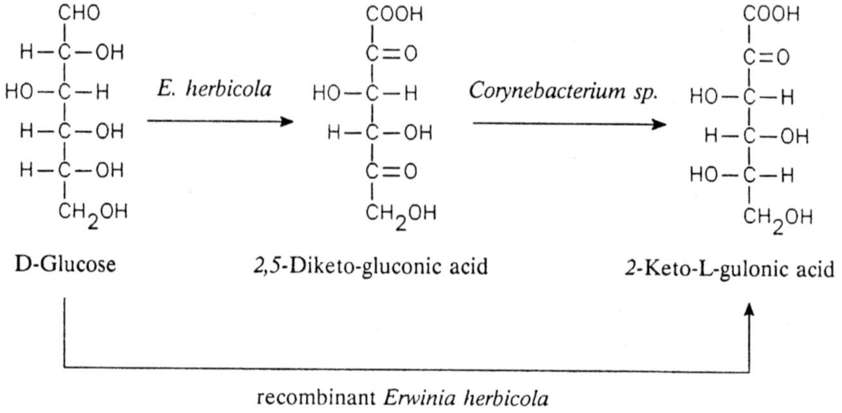

Figure 4-10. Direct fermentation of glucose to 2-keto-L-gulonic acid by a recombinant strain of *Erwinia herbicola*.

r-DNA technology has been used to engineer a microorganism capable of producing 2-keto-L-gulonic acid, the precursor of ascorbic acid, by direct fermentation of glucose [29,30]. Thus, several species of *Acetobacter*, *Gluconobacter* and *Erwinia* are able to efficiently oxidize glucose to 2,5-diketogluconic acid (Figure 4-10). The conversion of the latter to 2-keto-L-gulonic acid is mediated by *Corynebacterium sp.* In order to produce a recombinant strain for the one step conversion of glucose, the 2,5-diketo gluconic acid reductase gene was isolated from *Corynebacterium sp.* DNA and inserted into a plasmid, which in turn was inserted into *Erwinia herbicola*. However, this elegant process cannot as yet compete with the more circuitous Reichstein-Grussner process due to the low productivities obtained.

B. Amino Acids

All 19 of the proteinogenic amino acids are (in principle) available as primary metabolites in fermentation processes. However, most wild strains isolated from nature are not able to produce industrially significant amounts because over-production of amino acids is prevented by metabolic regulation mechanisms (e.g., feedback inhibition). Moreover, the cell membrane permeability barrier prevents the life-sustaining amino acids from escaping into the environment.

In 1957, two groups in Japan [31,32] discovered wild bacteria capable of producing high yields of glutamic acid. Monosodium glutamate had been used in Japan as a flavor enhancer from the beginning of this century and the discovery of a commercially viable microbial synthesis laid the foundation for an amino acid fermentation industry that is still dominated by Japanese companies.

Subsequently many strains of *Brevibacterium* and *Corynebacterium* were found to overproduce glutamic acid. Although glucose and molasses are the major carbon sources, strains have also been developed that utilize ethanol, acetic acid, or hydrocarbons. Further research led to the development of mutant strains with enhanced cell membrane permeability. One strategy adopted for this purpose is to grow *C. glutamicum* in a medium containing less than the optimum amount of the vitamin biotin; the cell membrane then becomes deficient in phospholipids and develops leaks, which allows more glutamic acid to be excreted.

Following the success achieved with glutamic acid fermentation, methods were developed for the microbial synthesis of other amino acids (e.g., lysine, threonine, proline, phenylalanine, and tryptophan; Table 4-3) [33,34]. The two most important examples, glutamic acid and lysine, are produced on scales of 350,000 and 70,000 tons, respectively, which rank with bulk chemicals.

The extensive application of r-DNA technology to amino acid fermentations [33,34] has resulted in further strain improvements. The only limiting factor appears to be the volume of product necessary to warrant the large development costs. L-Phenylalanine is a good example; once a large volume application emerged, in the form of the artificial sweetener aspartame, the fermentation process was significantly improved. Similarly, demand for L-proline as an intermediate for ACE-inhibitors stimulated the further development of the fermentation method for its production and a substantial reduction in its price.

Fermentation is not the most economical production method for all the natural amino acids. For example, aspartic acid, the raw material for aspartame, is more economically made [35] by enzymatic addition of ammonia to fumaric acid (Reaction 4-3, analogous to the formation of L-malic acid by enzymatic hydration of the same substrate (Reaction 4-1).

Table 4-3 Fermentation Processes of Interest for Industrial Amino Acid Production [34]

Amino acid	Organism	Concentration (g l^{-1})
L-Arginine	*Brevibacterium flavum*	35
L-Glutamic acid	*Corynebacterium* sp.	100
L-Glutamine	*Corynebacterium glutamicum*	37
L-Histidine	*Corynebacterium glutamicum*	15
L-Isoleucine	*Brevibacterium flavum*	30
L-Leucine	*Brevibacterium lactofermentum*	30
L-Lysine	*Brevibacterium lactofermentum*	70
L-Phenylalanine	*Brevibacterium lactofermentum*	25
L-Proline	*Brevibacterium flavum*	35
L-Threonine	*Serratia narcescens*	25
L-Tryptophan	*Brevibacterium flavum*	20
L-Valine	*Corynebacterium glutamicum*	30

$$\text{HOOC} \diagup\!\!=\!\!\diagup^{\text{COOH}} + NH_3 \xrightarrow{\text{aspartase}} \text{HOOC} \diagup\overset{\text{COOH}}{\underset{H\ NH_2}{\diagdown}}$$

L-Aspartic acid   (4-3)

Commercial production of aspartic acid, by Tanabe, involves the use of a packed column of immobilized cells of *E. coli* in a continuous process.

C.   Beta-Lactam Antibiotics

An enormous number of antibiotics are produced by fermentation. The most important class is the beta-lactam antibiotics, comprising the penicillins and cephalosporins. Their commercial importance is underscored by the fact that five of the top ten pharmaceuticals are beta-lactam antibiotics (see Table 2-2).

The fermentation process for the manufacture of penicillin has a long and colorful history and has earned the accolade 'queen of the fermentations'. The worldwide production of penicillin G and penicillin V (mostly the former) is about 25,000 tons. The current commercial penicillin-producing strains of *Penicillium chrysogenum* are the result of many rounds of strain improvement programs [7]. The fermentation is carried out on a very large scale, with fermentors having volumes up to 200 cubic meters. Productivities have also been significantly improved by developments in process engineering (e.g., the introduction of fed-batch process) [7]. Penicillin G concentrations of more than 20 g l^{-1} can be obtained, which represents a 1,000 fold increase over the original yields obtained by Florey and Chain.

The biosynthetic pathway for penicillin production involves the formation of a tripeptide from valine, cysteine, and L-α-aminoadipic acid (L-α-AAD). The tripeptide is converted to isopenicillin N by isopenicillin N synthase. Penicillin G and penicillin V are formed by enzymatic transacylation with phenylacetic or phenoxyacetic acid, respectively, that are added to the fermentation broth (Figure 4-11). Penicillin G (Pen-G) is converted to a wide range of semisynthetic penicillins and cephalosporins via the intermediacy of 6-aminopenicillanic acid (6-APA) and 7-aminodeacetoxycephalosporanic acid (7-ADCA) (Figure 4-12).

Originally 7-ADCA was produced from cephalosporin C, the latter being obtained by fermentation of *Cephalosporium acremonium*. However, due to the low productivity of this fermentation, production of 7-ADCA by chemical conversion of 6-APA proved more economical. An enzymatic process has been developed by Gist Brocades [7] for the deacylation of Pen-G to 6-APA and the analogous synthesis of 7-ADCA from its phenylacetyl derivative (Figure 4-12). The enzymatic process for 6-APA uses an amidase (penicillin acylase) that was originally developed in the 1960s [36] but could not compete with the chemical method. However, dramatic improvements in the enzyme production using genetic engin-

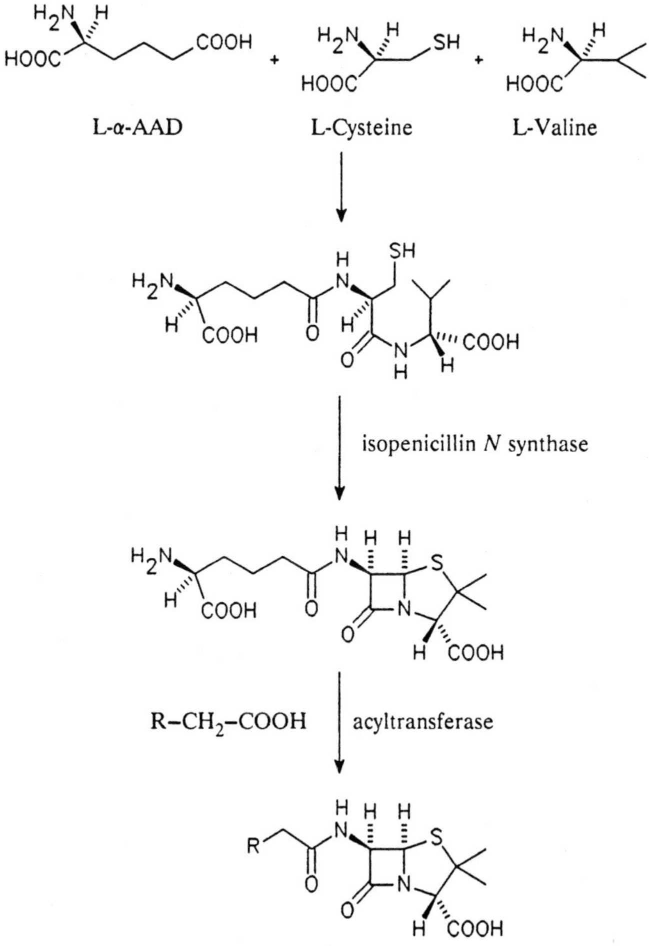

Pen-G (R= Ph)
Pen-V (R= PhO)

Figure 4-11. Biosynthetic pathway for penicillin.

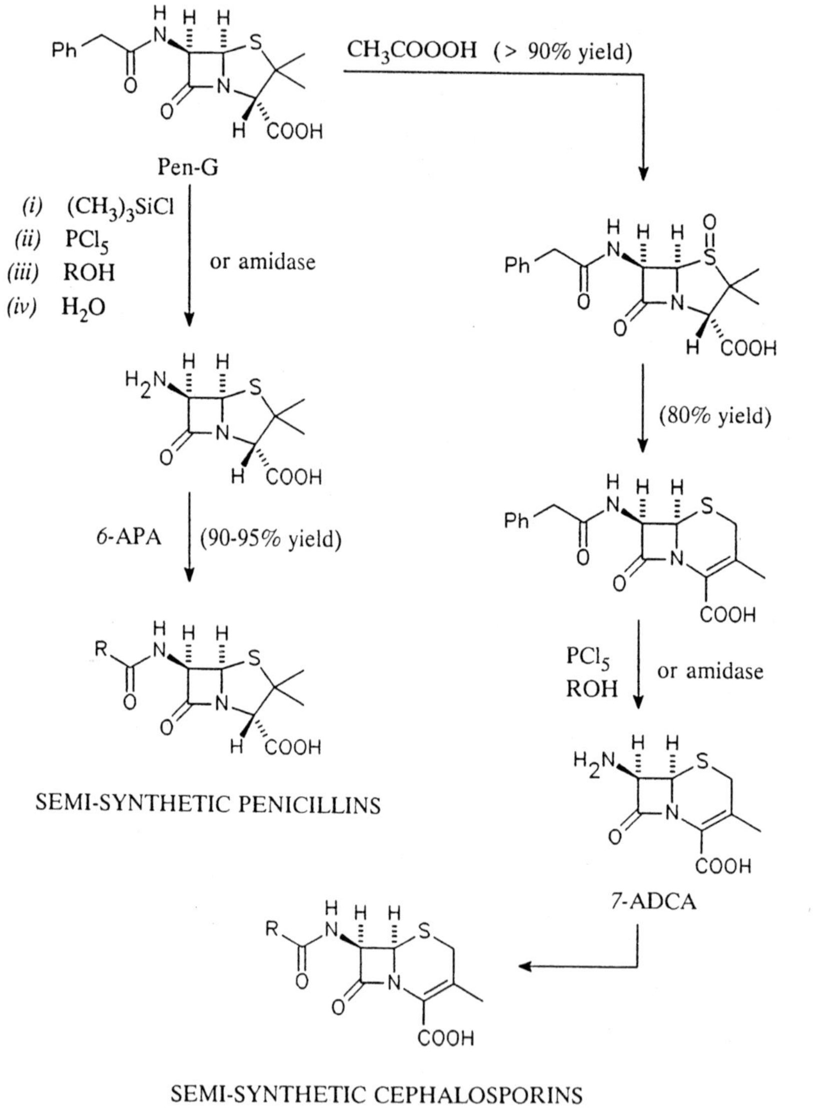

Figure 4-12. Conversion of Pen-G to semisynthetic penicillins and cephalosporins.

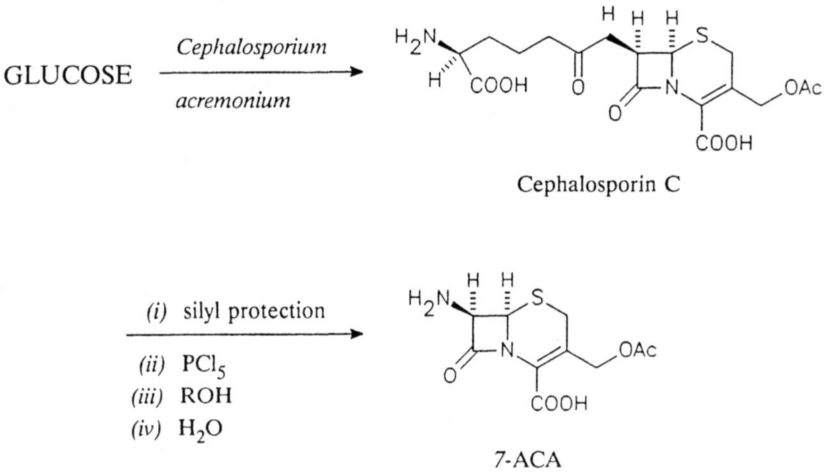

Figure 4-13. Production of 7-ACA from cephalosporin C.

eering techniques and more efficient utilization via immobilization have led to substantial cost reductions in the enzymatic process. This, coupled with increasing environmental pressure on the chemical process (e.g., the use of chlorinated hydrocarbon solvents, etc.), appears to have tipped the balance in favor of the enzymatic process which is more 'environment friendly'.

A second group of semisynthetic cephalosporins is derived from 7-amino-cephalosporanic acid (7-ACA). The latter is produced by chemical deacylation of cephalosporin C, the product of fermentation (Figure 4-13).

In the penicillins, only modifications at position 6 have led to therapeutically useful products whereas useful cephalosporins arise from modifications at both the 7 and 3 positions. A few examples of the many commercially important semisynthetic penicillins and semisynthetic cephalosporins are illustrated in Figure 4-14.

The discovery and development of new beta-lactam antibiotics still continues unabated and has led to the introduction of new classes such as the carbapenems [37] (e.g., thienamycin) and the monobactams [38] (e.g., azthreonam):

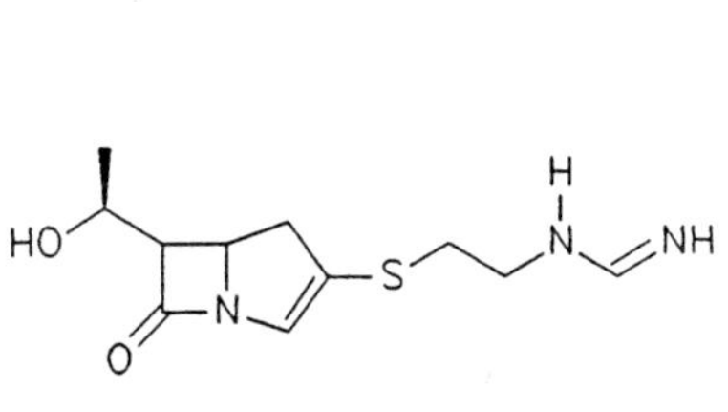

Thienamycin

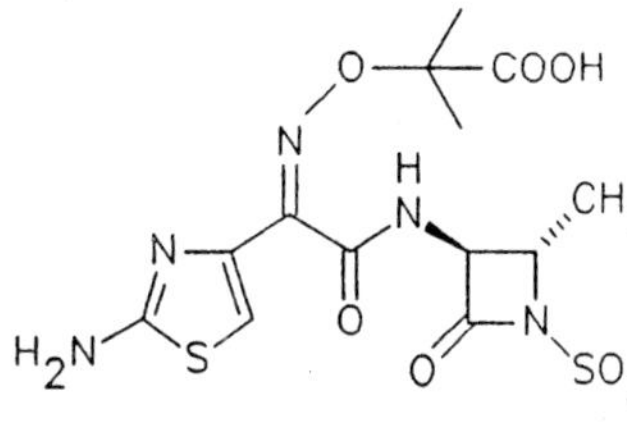

Azthreonam

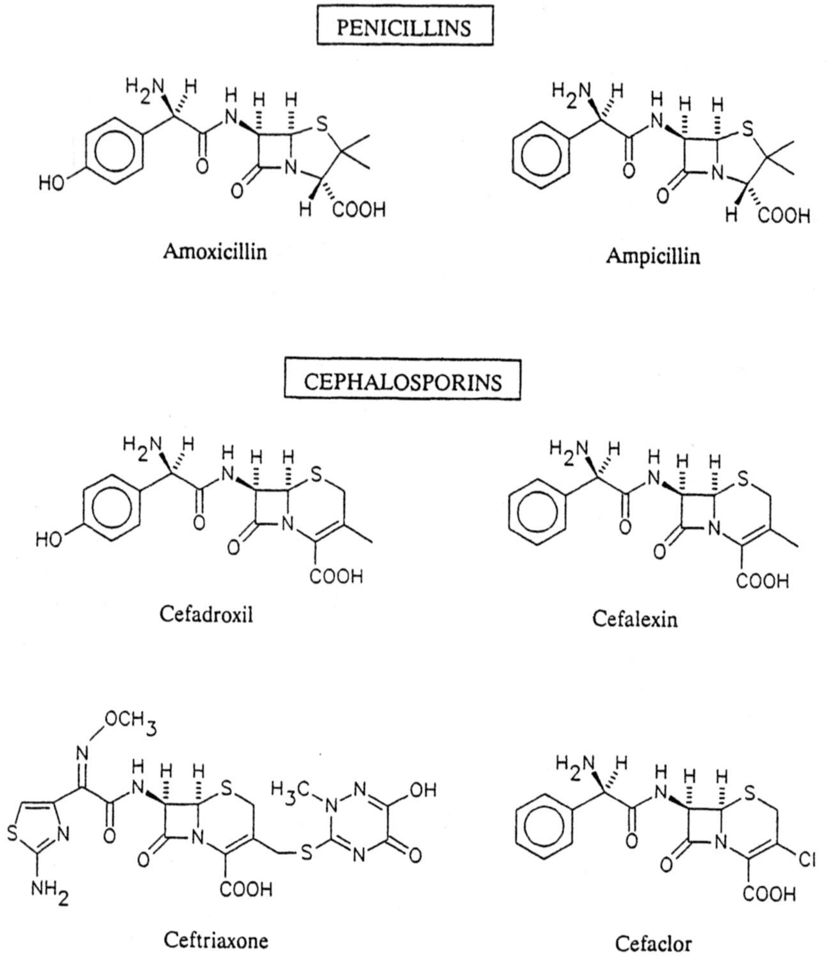

Figure 4-14. Examples of semisynthetic penicillins and cephalosporins.

Interestingly, although they are available by fermentation, thienamycin and azthreonam are both made by chemical synthesis, presumably due to very low productivities in fermentation.

D. Vitamins

Vitamins are organic substances that are needed, generally in trace quantities, for the functioning of most life forms, mostly as coenzymes in biosynthetic pathways [39]. Most of the 13 mammalian vitamins (Table 4-4) are produced by chemical synthesis

Table 4-4   Vitamins; Function and Production Method

Vitamin		Function	Method of production
Lipid soluble			
A	11-*cis*-Retinal	Vision	Chemical
* D	Ergosterol (provitamin D)	Calcium regulation	Byproduct of yeast fermentation
* E	α-Tocopherol	Intracellular antioxidant	Chemical
* K	Phylloquinone	Prothombin biosynthesis	Chemical/resolution
Water soluble			
B_1	Thiamine	Coenzyme	Chemical
* B_2	Riboflavin	Coenzyme	Chemical/fermentatiom
B_6	Pyridoxal phosphate	Coenzyme	Chemical
* B_{12}	Coenzyme B12	Coenzyme	Fermentation
* B_5	Pantothenic acid (coenzyme A)	Coenzyme	Chemical
* B_c	Tetrahydrofolic acid	Coenzyme	Chemical [†]
* C	Ascorbic acid	Cosubstrate of monooxygenases	Chemical
* H	Biotin	Coenzyme	Chemical
* PP	Niacin (nicotinic acid)	Coenzyme	Chemical

[*]   Chiral compounds; note, however, that vitamins E and K are produced synthetically as mixtures of optical isomers
[†]   Tetrahydrofolic acid is manufactured from L-glutamic acid which is available from fermentation

but a notable exception is the complex molecule vitamin B_{12}. It is produced by fermentation with *Pseudomonas denitrificans*. Strain improvement programs have resulted in productivity increases from 0.6 mg l^{-1} to about 60 mg l^{-1} at reaction times of 90 hrs (about 0.001 g l^{-1} h^{-1} !).

Some vitamins are produced by a combination of fermentation with chemical steps (e.g., vitamin C). Similarly, riboflavin is available by de novo fermentation but is more economically produced by the high yield fermentation (70 g l^{-1}) of glucose to D-ribose, using mutant strains of *B. subtilis* or *B. punilus*, followed by chemical conversion (Figure 4-15).

V.   PRECURSOR FERMENTATION—MICROBIAL TRANSFORMATION

The products discussed in preceding sections were mainly primary or secondary metabolites produced by de novo fermentation from glucose or a glucose equivalent. A second important category of fermentation processes is the so-called precursor fermentations or microbial transformations, whereby the conversion of a foreign substrate is mediated by a growing microbe. It usually involves a single

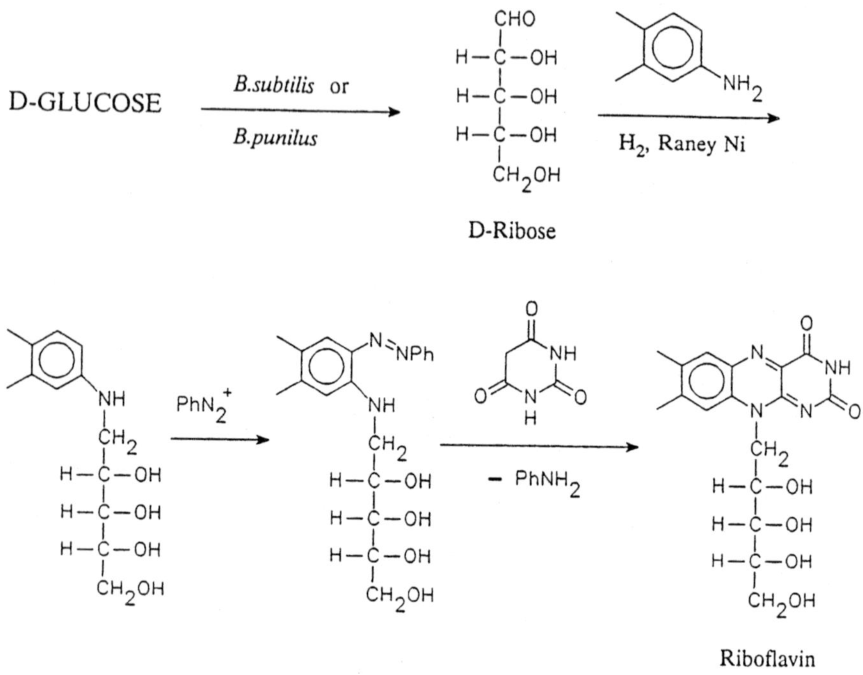

Figure 4-15. Production of riboflavin (vitamin B2) by fermentation/chemical synthesis.

step and a cofactor-dependent enzyme [40]. Carrying out the reaction as a fermentation obviates the need for expensive cofactor regeneration. Precursor fermentations generally involve redox processes or condensation reactions.

A. Microbial Oxidations

The commercially most important type of precursor fermentation is undoubtedly microbial oxidation. The archetype of such a process is the regioselective microbial oxidation of D-sorbitol to L-sorbose mentioned earlier. The most important examples of microbial oxidation are probably the regio- and stereoselective hydroxylations of the steroid nucleus which constitute key steps in the manufacture of steroid drugs with a value of more than a billion dollars [41,42]. The selective conversion of progesterone to 11α-progesterone has already been mentioned. By a suitable choice of microorganism (fungi, actinomycetes and bacteria are used) the same substrate can be converted to other important derivatives (Figure 4-16).

Similarly, microbial oxidation can also be utilized for the regioselective hydroxylation of aromatic compounds. A commercially relevant example is the Lonza process [43] for the regioselective hydroxylation of nicotinic acid (niacin)

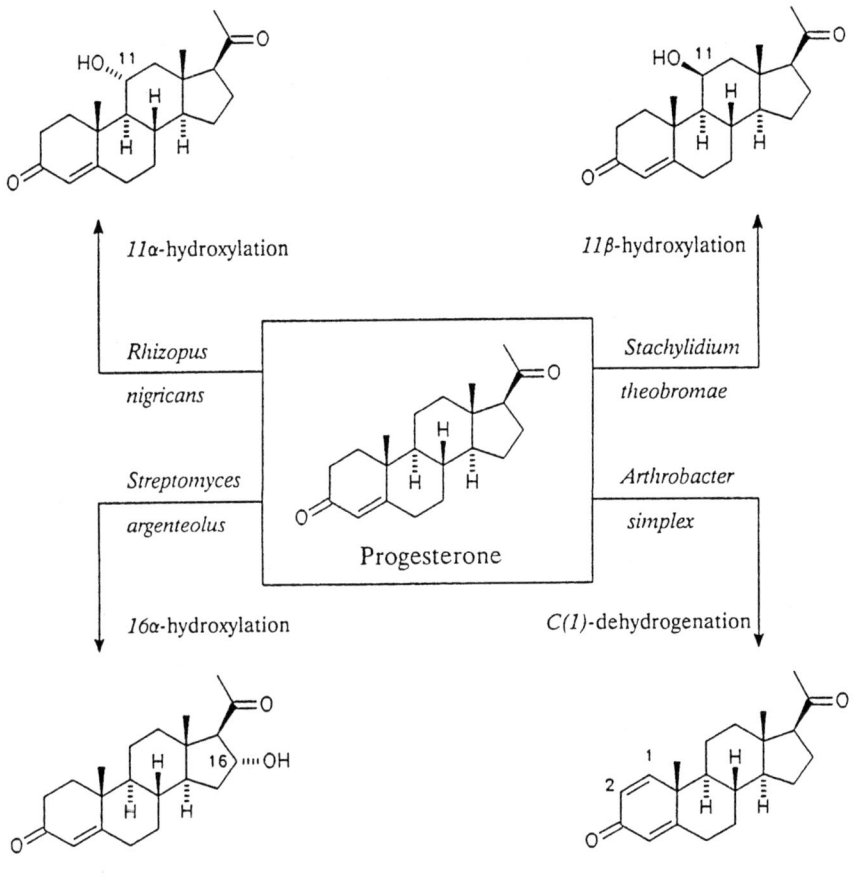

Figure 4-16. Microbial oxidations of progesterone.

to afford 6-hydroxynicotinic acid, an insecticide intermediate (Reaction 4-4). The commercial process takes advantage of the much lower water solubility of the magnesium salt of the product compared to the starting material. The magnesium salt of 6-hydroxynicotinic acid precipitates during the fermentation and is collected in a settler.

$$\text{(4-4)}$$

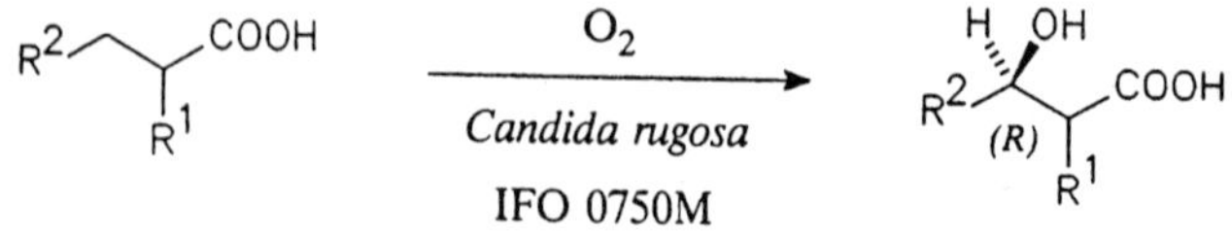

Figure 4-17. Selective microbial hydroxylation of carboxylic acids.

R^1	R^2	Optical purity
H	CH$_3$	95% ee
H	CH$_3$CH$_2$	93% ee
H	CH$_3$CH$_2$CH$_2$	96% ee
H	CH$_3$CH$_2$CH$_2$CH2	95% ee
CH$_3$	H	97% ee
CH$_3$CH$_2$	H	99% ee

Another industrially useful microbial transformation is the regio- and enantioselective beta-hydroxylation of aliphatic carboxylic acids mediated by yeasts such as *Candida rugosa* and *C. parapsitosis* [44–47]. For example. mutant strains of *C. rugosa* gave concentrations of the *(R)*-β-hydroxy carboxylic acid of the order of 5–10 g l^{-1} in 24 hours (Figure 4-17). The reaction proceeds via initial enzymatic dehydrogenation to the α,β unsaturated acid followed by enantio-selective enzymatic hydration, as illustrated in Figure 4-18 for isobutyric acid hydroxylation. Kanegafuchi has commercialized processes for the production of *(R)*-β-hydroxyisobutyric acid and *(R)*-β-hydroxy-*n*-butyric acid using this technology [46]. The former is a chiral synthon in the manufacture of the ACE-inhibitor, captopril [47], and the latter in a route [48] to carbapenem intermediates (see also chapter 5).

An alternative route to *(R)*-β-hydroxybutyric acid (esters) involves acid-catalyzed depolymerization of poly-*(R)*-3-hydroxybutyrate [49]. The latter is produced commercially by ICI by fermentation of glucose with *Alcaligenes eutrophus* bacteria. In practice, the fermentation can be controlled to give a copolymer containing varying proportions of *(R)*-3-hydroxybutyrate and *(R)*-3-hydroxy-valerate, with yields up to 80% of the dry weight of the biomass [50]. Depolymerization thus provides access to both optically pure acids or their esters.

Analogous to the above microbial hydroxylations, Lonza has commercialized a process for the production of *(R)*-carnitine by fermentation of butyrobetaine or crotonobetaine (Figure 4-19). A soil microorganism (taxonomically situated

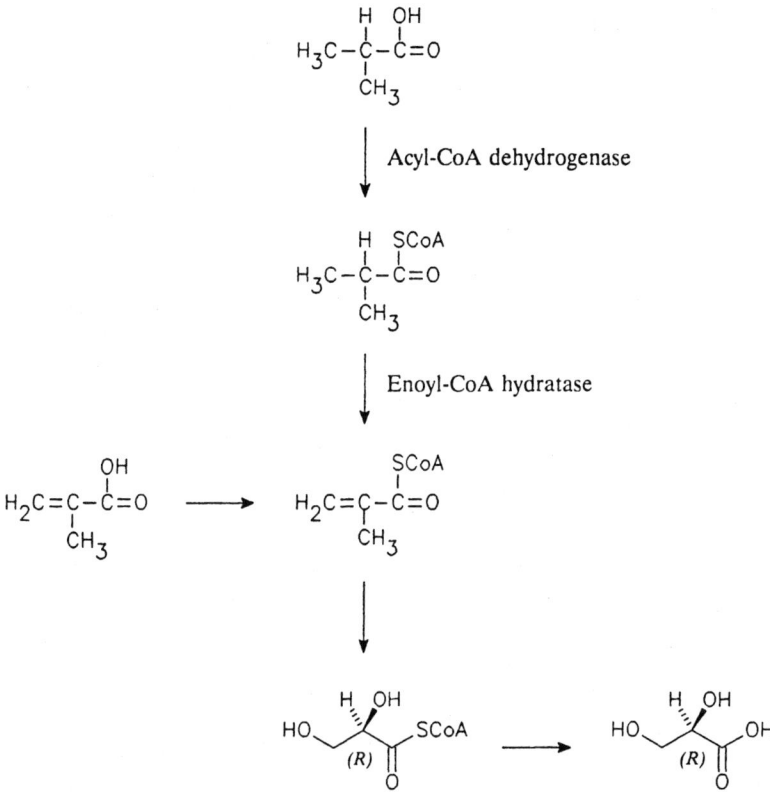

Figure 4-18. Mechanism of *Candida rugosa*-mediated hydroxylation of isobutyric acid.

between *Agrobacterium* and *Rhizobium*) was isolated that grows on crotonobetaine or butyrobetaine as the sole source of carbon, nitrogen and energy under aerobic conditions. A mutant strain of this microbe produced *(R)*-carnitine in concentrations up to 100 g l^{-1} and productivities of about 5 g l^{-1} h^{-1} [43].

Microbial oxidation can also be used for the kinetic resolution of racemates, the classic example being Pasteur's resolution of tartaric acid with *Penicillium glaucum*. (See chapter 1.) An example is the conversion of racemic 3-chloro-1,2-propanediol to the *(R)*-isomer by fermentation with *Serratia marescens* [46]. The *(R)*-isomer is a versatile chiral C_3-synthon that can, in principle, be converted to a variety of commercially interesting products (Figure 4-20).

Unfortunately, this method involves sacrificing half of the starting material; however, the methodology has been refined to provide an elegant method for complete conversion of racemic diols to the *(S)*-enantiomer by microbial stereoinversion (Figure 4-21) [46]. For example, fermentation of racemic 1,2-pentanediol

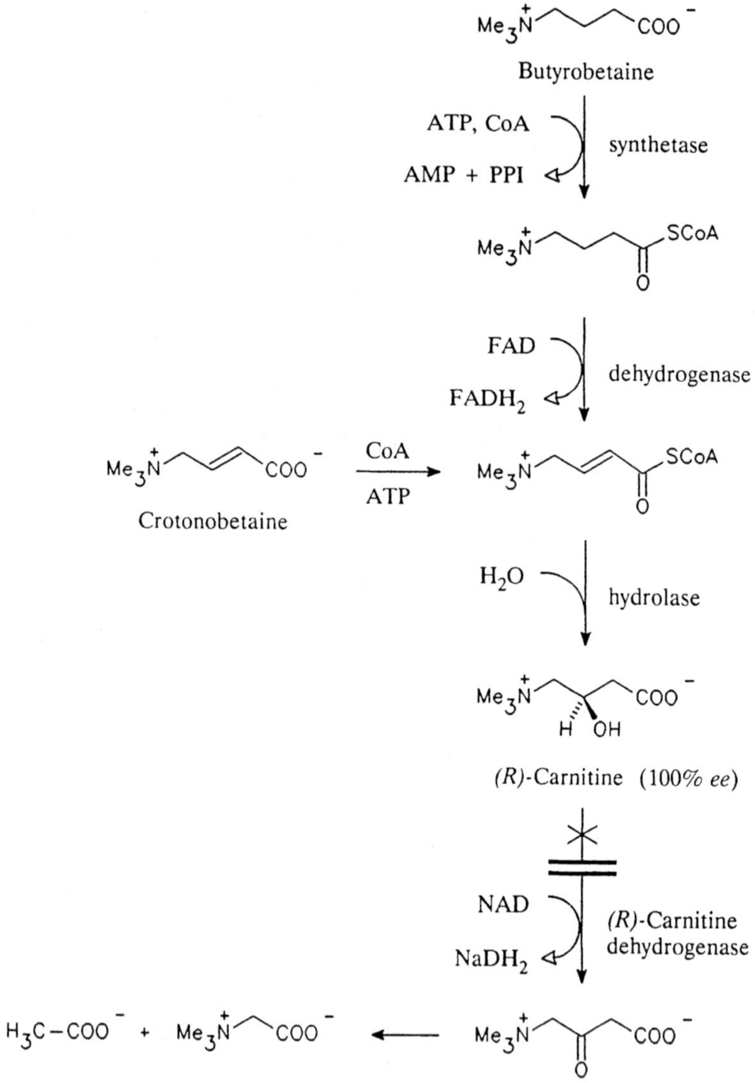

Figure 4-19. Pathway for the formation of *(R)*-carnitine from butyrobetaine. *(R)*-Carnitine dehydrogenase is blocked in the mutant strain [43]

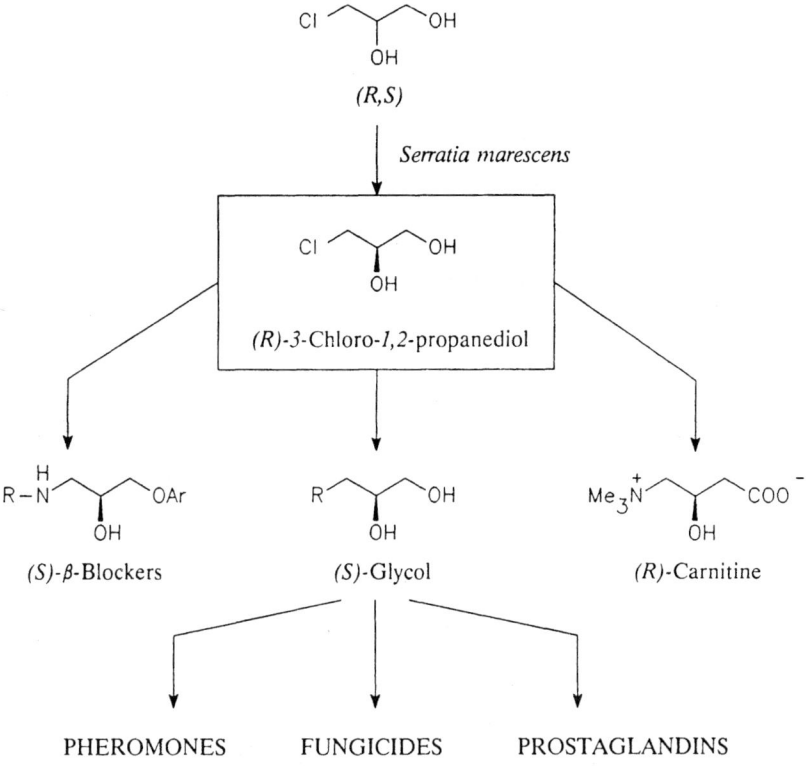

Figure 4-20. *(R)*-3-Chloro-1,2-propanediol by microbial kinetic resolution.

with *Candida parapsilosis* produced 28 g l⁻¹ of the *(S)*-enantiomer in 95% yield and 100% *ee* [46]. Stereoinversion involves the coupling of NAD-dependent, *(R)*-specific alcohol dehydrogenase with a NADPH-dependent *(S)*-specific keto-reductase (Figure 4-21).

Another potentially useful precursor fermentation is the enantioselective microbial epoxidation of prochiral olefins. For example, *Nocardia corallina* converts a variety of olefins to the *(R)*-epoxides (e.g., Reaction 4-5) with *ee*'s in the range of 60–94% [51]. However, productivities are generally low due to severe product inhibition, which raises doubts about the commercial potential of the method.

$$R-CH=CH_2 \xrightarrow[N.corallina]{O_2} R-CH-CH_2 \overset{O}{\triangle}$$

60-94% *ee*

(4-5)

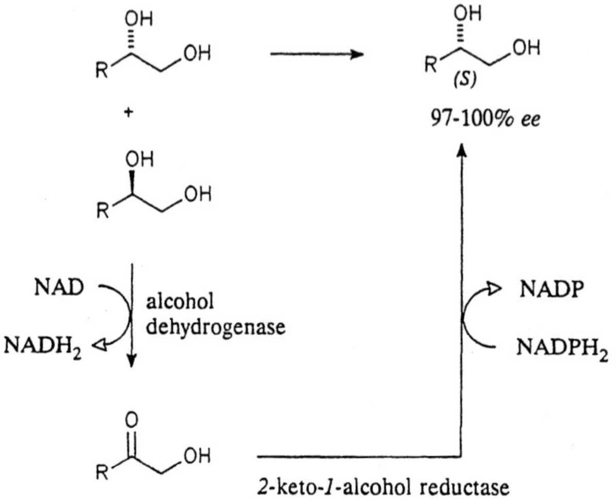

R = Et, n-Pr, n-Bu, n-Am, i-Bu, Ph, PhCH$_2$CH$_2$, etc.

Figure 4-21. Mechanism of *Candida parapsilosis*-mediated stereoinversion of 1,2-diols.

B. Microbial Reductions

Enantioselective reductions with fermenting Baker's yeast (*Saccharomyces cerevisiae*) is a widely used technique in organic synthesis [52]. The observation, by Mamoli and Vercellone [53] that fermenting yeast reduces 17-keto steroids to 17β-hydroxy steroids eventually led to commercial processes for the manufacture of steroid hormones (Reaction 4-6).

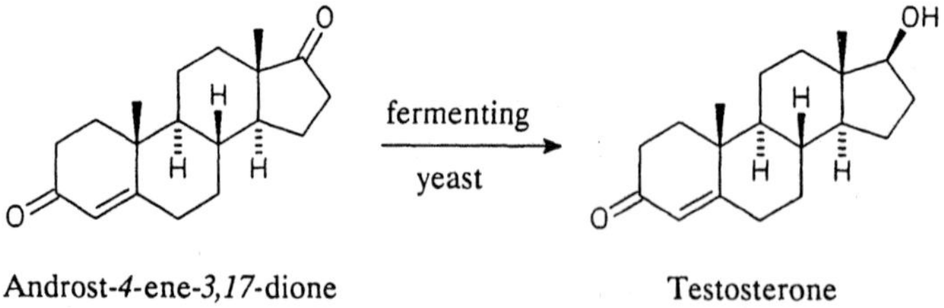

Androst-*4*-ene-*3,17*-dione   Testosterone

(4-6)

Reduction of β-keto esters to the corresponding *(S)*-alcohols is probably the most studied reaction [52]. In common with most yeast reductions, high *ee*'s are obtained only when substrate concentrations are kept at 1 g l^{-1} or less [53]. As

Table 4-5 A Comparison of Yeast Reduction with Catalytic Asymmetric Hydrogenation

Baker's yeast reduction   Asymmetric hydrogenation

Baker's yeast reduction	Asymmetric hydrogenation
Methyl ester, 40 g	Ethyl ester, 40 g
Water, 2600 ml	Methanol, 40 ml
Yeast, 200 g	Ru-BINAP, 0.14 g
Sucrose, 500 g	H_2, 20-100 bar
Room temperature, 80 h	Room temperature, 40 h
Filtration + extraction + distillation	Isolation by distillation
59–76% yield; S in 85% ee^*	96% yield; R or S in > 99% ee
Productivity: ≈ 0.01 g l^{-1} h^{-1}	Productivity: ≈ 12 g l^{-1} h^{-1}
(< 0.001 for > 97% ee)	

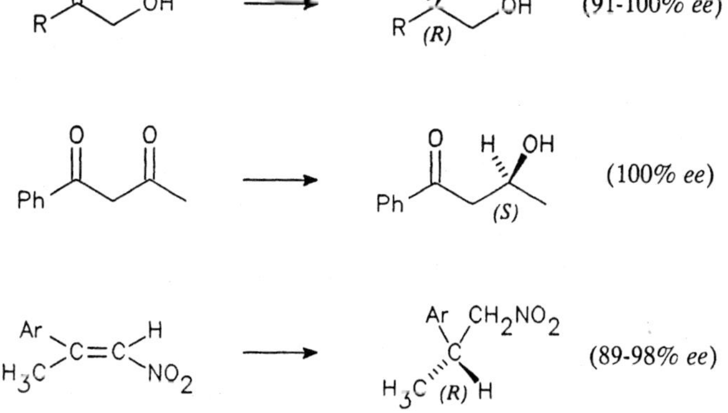

Figure 4-22. Variety of transformations by yeast reduction.

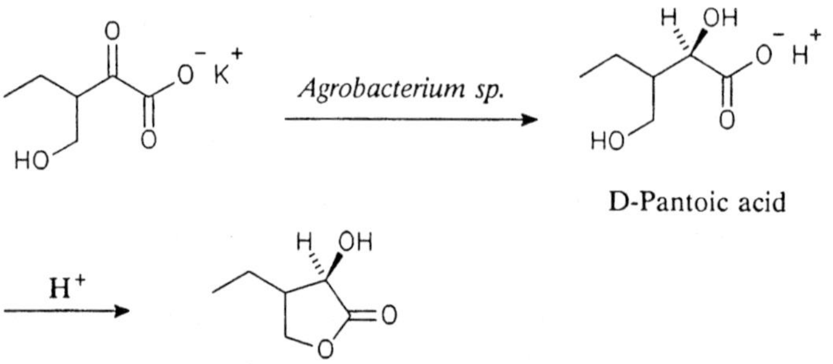

D-Pantoic acid

D-Pantoyl lactone

Figure 4-23. Enantioselective reduction of potassium ketopantoate to D-pantoic acid.

typical reaction times are 24–28 hrs this translates into meager productivities (0.02–0.04 g l^{-1} h^{-1}). Interestingly, Seebach [51] has compared yeast reduction with catalytic asymmetric hydrogenation (see chapter 8) of a β-keto ester (Table 4-5). Obviously, the asymmetric hydrogenation wins hands down on productivity and ease of work-up. Indeed, this is a serious handicap for yeast reductions on an industrial scale and in most cases alternative methods (e.g., catalytic asymmetric hydrogenation or enzymatic kinetic resolution) will be more economically viable. Nevertheless, yeast reductions are interesting for a variety of transformations as shown in the following examples (Figure 4-22) and perhaps the productivities are amenable to future improvement by genetic engineering.

Microbial reductions are not limited to baker's yeast, as is illustrated by the enantioselective reduction of potassium ketopantoate to D-pantoic acid (Figure 4-23), the precursor of panthothenic acid (vitamin B_5), by species of *Agrobacterium* [56]. In order to find a suitable microbe, 188 strains of bacteria, 84 actinomycetes, 231 yeasts, 203 molds, and 353 microorganisms isolated from soil samples were screened! Optimum activities were found with *Agrobacterium* spp. that yielded D-pantoic acid in > 98% *ee*. Under optimized conditions, concentrations of about 120 g l^{-1} were obtained at 90% conversion in 5 days (i.e., a productivity of 1 g l^{-1} h^{-1}). This compares favorably with the reduction of ketopantoyl lactone by *Candida* cells (81 g l^{-1} and 80% *ee* at substrate concentrations maintained below 3%) [57] and is a perfect example of perseverance pays.

C. Condensation Reactions

An example of a microbial condensation reaction of commercial significance is the baker's yeast-mediated acyloin condensation of aromatic and α,β-unsaturated aldehydes with acetaldehyde [58–62], the latter being formed in situ by glucose fermentation (Reactions 4-7 and 4-8). When benzaldehyde is the substrate, this constitutes the key step in the Knoll process for the manufacture [59] of ephedrine and pseudoephedrine (Figure 4-24). Furthermore, Fuganti and coworkers [60–62] have used the general reactions 4-7 and 4-8 as a key step in the synthesis of a variety of natural products.

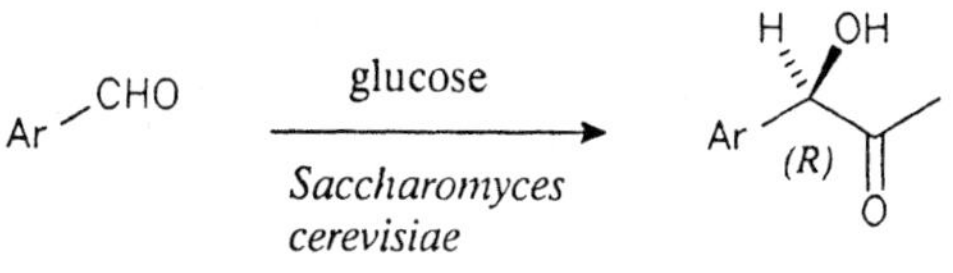

$$(4\text{-}7)$$

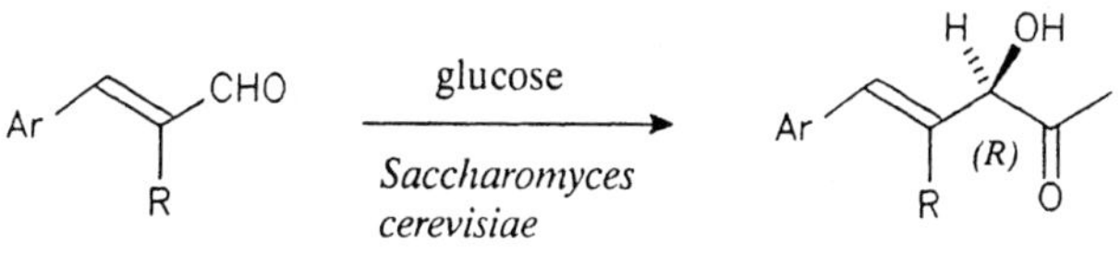

$$(4\text{-}8)$$

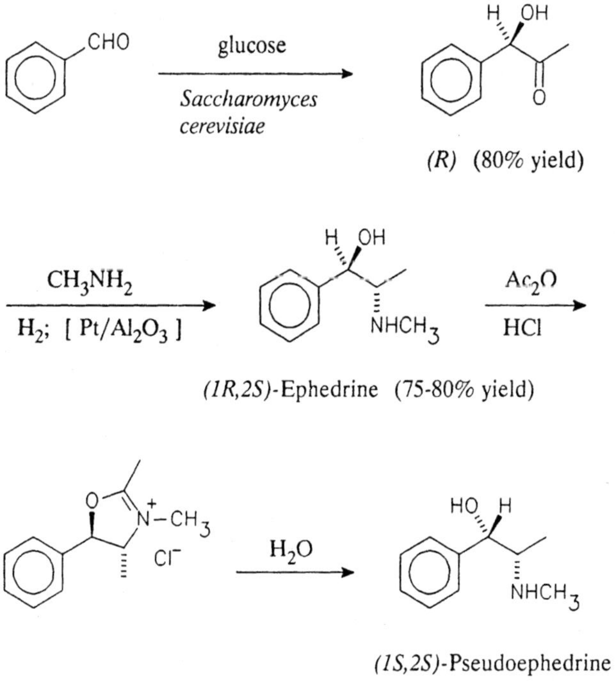

(R) (80% yield)

(1R,2S)-Ephedrine (75-80% yield)

(1S,2S)-Pseudoephedrine

Figure 4-24. Manufacture of ephedrine and pseudoephedrine.

VI. THE FUTURE

Fermentation is a technology with a long history and a bright future. Improvements resulting from advances in genetic and process engineering do not appear to have reached a plateau and only a small fraction of the potentially available microorganisms have been developed for industrial utilization. Moreover, as far as the utilization of microorganisms that flourish under extreme conditions of temperature, pH, etc., is concerned, we have hardly scratched the surface. In short, recent developments in fermentation technology seem to justify the optimism of J. B. S. Haldane who remarked in 1926: "why make a chemical when a bug can do it for you?"

REFERENCES

1. Demain, A. L., and Solomon, N. A., *Sci. Am.*, **245** (3), 67-76 (1981).
2. Wise, D. L., *Organic Chemicals from Biomass*, Benjamin/Cumnings, Menlo Park, California, 1983.
3. Bushell, M. E., and Gräfe, U. (Eds.), *Bioactive Metabolites from Microorganisms*, Elsevier, Amsterdam, Progr. Ind. Microbiol., Vol. 27, 1989.
4. Phaff, H. J., *Sci. Am.*, **245** (3), 77-90 (1981).
5. Hopwood, D. A., *Sci. Am.*, **245** (3), 91-102 (1981).
6. Tosaka, O., and Takinami, K., in *Biotechnology of Amino Acid Production*, Aida, K., Chibata, I., Nakayama, K., Takinami, K. and Yamada, H. (Eds.), Elsevier, Amsterdam, 1986, pp. 152-172. Progr. Ind. Microbiol., Vol. 24.
7. Hersbach, G. J. M., Van der Beek, C. P., and Van Dijk, P. W. M., in *Biotechnology of Industrial Antibiotics*, Vandamme, E. J. (Ed.), Marcel Dekker, New York, 1984.
8. Aharonowitz, Y., and Cohen, G., *Sci. Am.*, **245** (3), 141-152 (1981).
9. Cohen, S. N., Boyer, H. W., and Helling, R. B., *Proc. Natl. Acad. Sci. USA*, **70**, 3240-3244 (1973); see also Cohen, S. N., Cabello, F., Chang, A. C. Y., and Timmis, K., in *Recombinant Molecules: Impact on Science and Society*, Beers, R. F., and Bassett, E. G. (Eds.), Raven Press, New York, 1977.
10. Oxender, D. L., and Fox, C. F. (Eds.), *Protein Engineering*, A. R.Liss, New York, 1987.
11. Old, R. W., and Primrose, S. B., *Principles of Gene Manipulation*, Blackwell, Oxford, 1985.
12. Ryser, S., and Weber, M., *Genetic Engineering - What's Happening at Roche*, F. Hoffman-LaRoche, Basel, Switzerland, 1991.
13. Szekely, M., *From DNA to Protein. The Transfer of Genetic Information*, MacMillan, London, 1980.
14. Ward, W. H. J., and Fersht, A. R., in *Redesigning the Molecules of Life*, Benner, A. (Ed.), Springer-Verlag, Heidelberg, 1988, pp. 59-86.
15. Leatherbarrow, R. J., and Fersht, A. R., *Protein Eng.*, **1**, 7-16 (1986).
16. Shaw, W. V., *Biochem. J.*, **246**, 1-17 (1987).
17. Grandi, G., Toma, S., Margarit, I., Campagnoli, S., Gianna, R., Bellini, A. V., Zamai, M. and Carrera, P., *Chimicaoggi (Chemistry Today)*, July-August, 1990, pp. 9-13.
18. Botstein, D., and Shortle, D., *Science*, **229**, 1193 (1985).
19. Zoller, M. J., and Smith, M., *Nucl. Acids Res.*, **10**, 6487 (1982).

20. Rossi, J. R., and Zoller, M., in *Protein Engineering*, Oxender, D. L., and Fox, C. F. (Eds.), A. R. Liss, New York, 1987, pp. 51-64.
21. Wells, J. A., Powers, D. B., Bott, R. R., Katz, B. A., Ultsch, M. H., Kossiakoff, A. A., Power, S. D., Adams, R. M., Heynecker, H. H., Cunningham, B. C., Miller, J. V., Graycar, T. P., and Estell, D. A., in *Protein Engineering*, Oxender, D. L., and Fox, C. F. (Eds.), A.R.Liss, New York, 1987, pp. 279-287.
22. Schmidt-Kastner, G., and Gölker, C. F., in *Basic Biotechnology*, Bu'lock, J., and Kristiansen, B. (Eds.), Academic Press, New York, 1987, pp. 173-196.
23. Prenosil, J. E., Dunn, I. J., and Heinzle, E., in *Biotechnology*, Rehm, H. J., and Reed, G. (Eds.), Vol. 7a, Kennedy, J. F. (Ed.), VCH, Weinheim, 1987, pp. 489-545.
24. Personnal communication, A. Bruggink.
25. a) Peterson, D. H., and Murray, H. C., *J. Am. Chem. Soc.*, **73**, 5513 (1951). b) See also: Peterson, D. H., *Steroids*, **45**, 1 (1985).
26. a) Röper, H., in *Carbohydrates as Organic Raw Materials*, Lichtenthaler, F. W. (Ed.), Verlag Chemie, Weinheim, 1991, pp. 267-285; b) Röper, H., *Starch*, **42** (9), 342-349 (1990).
27. Buchta, K., in *Biotechnology*, Vol. 3, Verlag Chemie, Weinheim, 1983, pp. 410-417.
28. a) Reichstein, T., and Grussner, A., *Helv. Chim. Acta*, **17**, 311-328 (1934). b) See also: Crawford, T. C., and Crawford, S .A., *Adv. Carbohydrate Chem. Biochem.*, **37**, 79-155 (1980).
29. Lazarus, R. A., Seymour, J. L., Stafford, R. K., Dennis, M. S., Lazarus, M. G., Marks, C. B., and Anderson, S., in *Biocatalysis*, Abramowicz, D. A. (Ed.), van Nostrand Reinhold, New York, 1990, pp. 135-155.
30. Anderson, S., Marks, C. B., Lazarus, R., Miller, J., Stafford, K., Seymour, J., Light, D., Rastetter, W., and Estell, D. A., *Science*, **230**, 144-149 (1985).
31. Kinoshita, S., Udaka, S., and Shimono, M., *J. Gen. Appl. Microbiol.*, **3**, 193 (1957).
32. Asai, T., Aida, K., and Oishi, K., *Bull. Agric. Chem. Soc. Japan*, **21**, 134 (1957).
33. Aida, K., Chibata, I., Nakayama, K., Takinani, K., and Yamada, H. (Eds.), *Biotechnology of Amino Acid Production*, Elsevier, Amsterdam, Prog. in Industr. Microbiol., Vol. 24, 1986.
34. Bigelis, R., in *Application of Gene Technology*, Rehm, H. J., and Reed, G. (Eds.), Biotechnology, Vol 7b., 1989, pp. 229-259.
35. Chibata, I., Tosa, T., and Sato, T., in *Biotechnology of Amino Acid Production*, Aida, K., Chibata, I., Nakayama, K., Takinani, K., and Yamada, H. (Eds.), Elsevier, Amsterdam, 1986, Prog. in Industr. Microbiol., Vol. 24, pp. 144-151.
36. a) Vandamme, E. J. and Voets, J. P., *Advan. Appl. Microbiol.*, **17**, 311-369 (1974). b) Shimizu, M., Masuike, T., Fujita, H., Kimura, K., Okachi, R., and Nara, T., *Agr. Biol. Chem.*, **39**, 1225-1232 (1975).
37. Salzman, T. N., Ratcliffe, R. W., Christensen, B. G., and Boufford, F. A., *J. Am. Chem. Soc.*, **102**, 6161 (1980); Hagahara, T., and Kametami, T. *Heterocycles*, **25**, 729 (1987)
38. Parker, W. L., O'Sullivan, J., and Sykes, R. B., *Advan. Appl. Microbiol.*, **31**, 181-205 (1986).
39. Florent, J., in *Biotechnology, Vol. 4*, Rehm, G. J., and Reed, G. (Eds.), Verlag Chemie, Weinheim, 1989, pp. 115-158.
40. Vezina, C., in *Basic Biotechnology*, Bu'lock, J., and Kristiansen, B. (Eds.), Academic Press, New York, 1987, pp. 463-482.

41. Sebek, O. K., and Perlmann, D., in *Microbial Technology Vol. 2, 2nd. Ed.,* Peppler, H. J., and Perlman, D. (Eds.), Academic Press, New York, 1979, pp. 483-496.
42. Smith, L. L., in *Biotechnology, Vol. 6a,* Rehm, G. J., and Reed, G. (Eds.), Kieslich, K. (Ed.), Verlag Chemie, Weinheim, 1984, pp. 31-78.
43. Kulla, H. G., *Chimia,* **45,** 81-85 (1991).
44. Hasegawa, J., Ogura, M., Hamaguchi, S., Shimazaki, M., Kawaharada, H., and Watanabe, K., *J. Ferment. Technol.,* **59,** 203-208 (1981).
45. Hasegawa, J., Ogura, M., Kanema, H., Noda, N., Kawaharada, H., and Watanabe, K., *J. Ferment. Technol.,* **60,** 501-508 (1982).
46. Ohashi, T., *Proc Chiral 90 Symp.,* Spring Innovations, Stockport, UK, 1990, pp. 65-71.
47. Shimazaki, M., Hasegawa, J., Kan, K., Nomura, K., Nose, Y., Kondo, H., Ohashi, T., and Watanabe, K., *Chem. Pharm Bull.,* **30,** 3139-3146 (1982).
48. Ohashi, T., and Hasegawa, J., *J. Synth. Org. Chem., Japan,* **45,** 331-345 (1987).
49. a) Seebach, D., and Zuger, M., *Helv. Chim. Acta,* **65,** 495-503 (1982). b) Seebach, D., Imwinkelried, R., and Weber, T., in *Modern Synthetic Methods,* Scheffold, R. (Ed.), Vol. 4, Springer Verlag, Heidelberg, 1986, pp. 125-259.
50. a) *Chem. Ind.* (London), 1990, p. 274. See also: Lafferty, R. M., Korsatko, B., and Korsatko, W., in *Biotechnology, Vol. 6b,* Rehm, G.J., and Reed, G. (Eds.), Kieslich, K. (Ed.), Verlag Chemie, Weinheim, 1987, pp. 135-176.
51. Furuhashi, K., *Chem. Econ. Eng. Rev.,* **18** (7), 21-26 (1986).
52. For a recent review see Servi, S., *Synthesis,* 1-25 (1990).
53. Wipf, R., Kupfer, E., Bertazzi, R., and Leuenberger, H. G. W., *Helv. Chim. Acta,* **66,** 485 (1983).
54. Seebach, D., *Org. Synth.,* **63,** 1 (1985).
55. Ohta, H., Ozaki, K., and Tsuchihashi, G., *Chem. Lett.,* 191-192 (1987).
56. Kataoka, M., Shimizu, S., and Yamada, H., *Agric. Biol. Chem.,* **54,** 177-182 (1990).
57. a) Hata, H., Shimizu, S., and Yamada, H., *Agric. Biol. Chem.,* **51,** 3011 (1987). b) Hata, H., Morishita, T., Akutsu, S., and Kawanura, M., *Agric. Biol. Chem.,* **51,** 289 (1987).
58. a) Neuberg, C., and Hirsch, J., *Biochem. Z.,* **115,** 282 (1921). b) Groger, D., Schmader, H.P., and Frommel, H., German Patent, 1.543.691 (1966).
59. Rose, A. H., *Industrial Microbiology,* Butterworths, Washington, 1961, p. 264.
60. Fuganti, C., and Graselli, P., in *Biocatalysis in Agricultural Biotechnology,* Whitaker, J. R., and Sonnet, P. E. (Eds.), ACS Symp. Ser., **389,** 1989, pp 359-370.
61. Fuganti, C., and Graselli, P., in *Enzymes in Organic Synthesis,* Porter, R., and Clark, S. (Eds.), Pitman, London, 1985, pp. 112-127.
62. Fuganti, C., in *Enzymes as Catalyst in Organic Synthesis,* Schneider, M. P. (Ed.), Reidel, Dordrecht, NATO ASI Series, Vol. 178, 1986, pp. 3-17.

ADDITIONAL READING

1. Malik, V. S., Biotechnology: The Golden Age, in *Advan. Appl. Microbiol.,* **34,** 263-306 (1989).
2. Hacking, A. J., *Economic Aspects of Biotechnology,* Cambridge University Press, Cambridge, 1986.

3. Demain, A. L., and Soloman, N. A., *Biology of Industrial Microorganisms*, Benjamin Cummings, Menlo Park, CA, 1985.

4. Vanek, Z., and Hostalek, Z., (Eds.), *Overproduction of Microbial Metabolites*, Butterworths, London.

5. Cheremisinoff, P. N., and Ouellette, R. P., (Eds.), *Biotechnology: Applications and Research*, Technomic Publ. Co., Lancaster, PA, 1985.

6. Rehm, H., J., and Reed, G., (Eds.), *Biotechnology*, Vols. 1–8, Verlag Chemie, 1981-1986.

7. Moody, G. W., and Baker, P. B., (Eds.), *Bioreactors and Biotransformations*, Elsevier, Amsterdam, 1987.

8. Wise, D. L., (Ed.), *Organic Chemicals from Biomass*, Benjamin Cummings, Menlo Park, CA, 1983.

5

The Chirality Pool

Das eine Molekül hat ein anderes geboren.

Emil Fischer

When planning the synthesis of an optically active product the easiest (albeit less challenging) approach is to use an optically active raw material available from the chirality pool. Since many of these inexpensive raw materials originate from large scale fermentation processes, it is appropriate to follow our discussion of fermentation with a discussion of the chirality pool. Examples of relatively inexpensive carbohydrates, amino acids, terpenes, alkaloids, etc., that are available from the chirality pool are collated in Table 5-1. Many of these materials have prices comparable with bulk petrochemicals and find a variety of industrial applications as chiral substrates, chiral auxiliaries, and as sources of chiral ligands for asymmetric catalysis. (See chapter 8.)

I. CARBOHYDRATES

A. The Monosaccharides

The most abundant source of optically pure compounds in nature comprises the carbohydrates, which generally have the D configuration. The monosaccharides are the simplest members and the most important from the viewpoint of industrial synthesis [1–3]. The D-series of C_3–C_6 aldoses are shown in Figure 5-1. Glucose

Table 5-1 The Chirality Pool

Product		World production (tons per annum)	Price (approx.) $/kg
Carbohydrates	D-Sucrose	100,000,000	0.6
	D-Glucose	5,000,000	0.9
	D-Sorbitol	650,000	1.0
	D-Lactose	180,000	1.0
	D-Fructose	50,000	2.0
	D-Mannitol	10,000	4.0
	D-Gluconic acid	25,000	4.0
	L-Ascorbic acid	35,000	10
	D-Isoascorbic acid	5,000	10
	Xylitol	3,000	10
Hydroxy acids	L-Lactic acid	25,000	3.0
	L-Tartaric acid	10,000	6.0
	L-Malic acid	10	40
Amino acids	L-Glutamic acid	400,000	2.0
	L-Lysine	75,000	5.0
	L-Phenylalanine	5,000	10
	D-Phenylglycine	6,000	15
	L-Aspartic acid	8,000	3.0
	L-Valine	300	30
	D-Valine	100	30
	L-Proline	200	40
	L-Threonine	200	40
	D-Alanine	100	35
Terpenes	(−)-α-Pinene	25,000	1.0
	(−)-β-Pinene	10,000	1.5
	(+)-3-Carene	500	2.0
	(+)-Limonene	10,000	2.0
	(−)-α-Phellandrene	100	5.0
	(−)-Menthol	6,000	15
	(−)-Carvone	500	20
	(+)-Camphor	1,000	10
Alkaloids	Ephedrine	500	60
	Pseudoephedrine	400	85
	Quinine	500	100
	Quinidine	250	150
	Cinchonidine	50	200

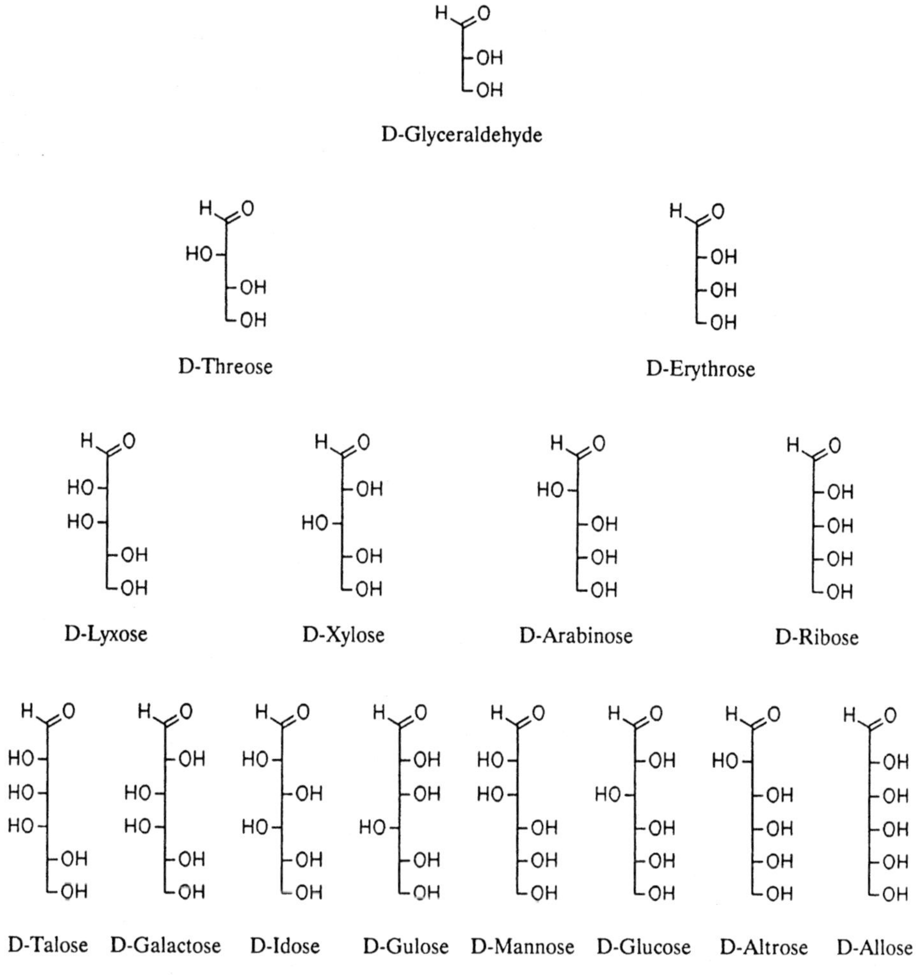

Figure 5-1. Fischer-projection formulas C$_3$–C$_6$ aldoses of the D-series.

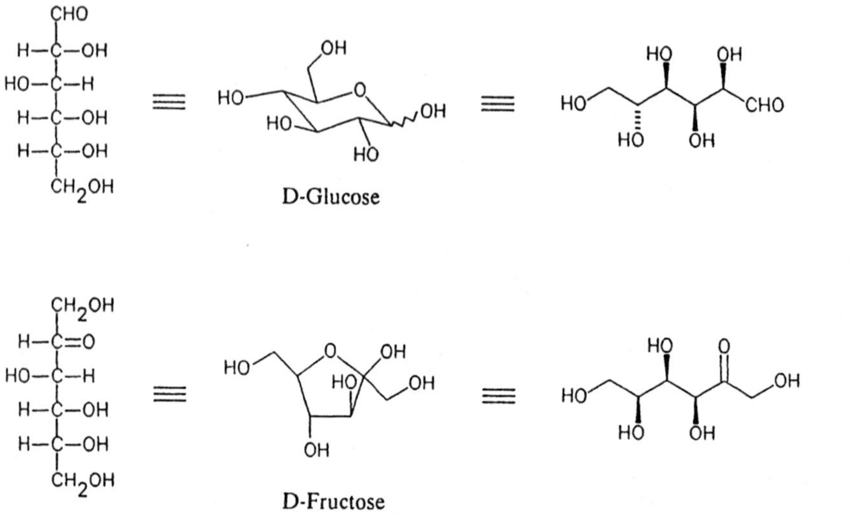

Figure 5-2. Structures of the two most important monosaccharides, glucose and fructose.

(Figure 5-2) is by far the most important and is available from the hydrolysis of starch or, together with fructose, from sucrose hydrolysis. D-Galactose is available, together with glucose, from the hydrolysis of lactose a byproduct of cheese manufacture. The second most important monosaccharide is D-fructose (Figure 5-2), a 2-ketose available from glucose isomerization or from the hydrolysis of inulin, a polyfructoside present in many plants and vegetables (e.g., the Jerusalem artichoke or topinamboer).

B. Hydrogenation Products

D-Sorbitol (Figure 5-3) is produced in 97–99% yield by catalytic hydrogenation of D-glucose over a Raney nickel catalyst, or as a 75:25 mixture with D-mannitol by hydrogenation of invert sugar (the 1:1 mixture of glucose and fructose obtained from sucrose hydrolysis). A method has also been described [4] for the production of D-mannitol in 62–66% yield from D-glucose by using glucose isomerase in combination with a copper hydrogenation catalyst, with both immobilized on silica.

D-Sorbitol finds wide application as a food additive and is the raw material in the Reichstein-Grussner process (Figure 5-4) for the manufacture of L-ascorbic acid. (See also chapter 4.) Indeed, the latter process [5] constitutes a prime example of an industrial synthesis of an optically active product by utilization of an inexpensive raw material available from the chirality pool. Since this process is operated on a large scale (> 25,000 tons) all of the intermediates are, in principle, readily available, inexpensive chiral synthons.

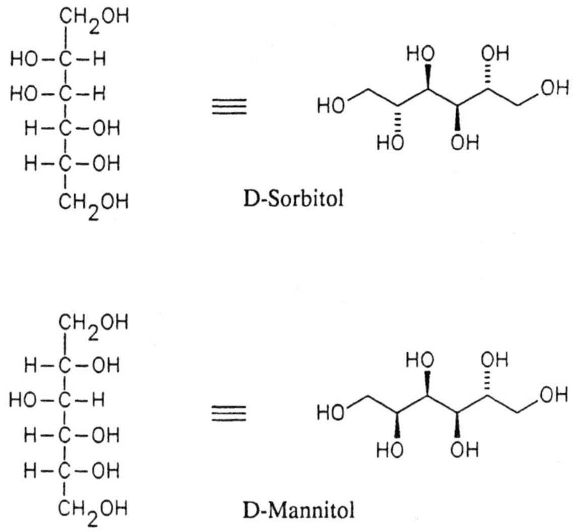

CH$_2$OH
HO−C−H
HO−C−H
H−C−OH
H−C−OH
CH$_2$OH

≡

D-Sorbitol

CH$_2$OH
H−C−OH
HO−C−H
H−C−OH
H−C−OH
CH$_2$OH

≡

D-Mannitol

Figure 5-3. Structures of D-sorbitol and D-mannitol.

In addition to their conversion to polyols by catalytic reduction, monosaccharides can also be converted to chiral aminopolyols by reductive amination with H$_2$ and NH$_3$ (or RNH$_2$) over a Raney nickel catalyst [6]; an example of this reaction is the reductive amination of glucose with ammonia or methylamine to yield glucamine or N-methylglucamine, respectively (Reaction 5-1).

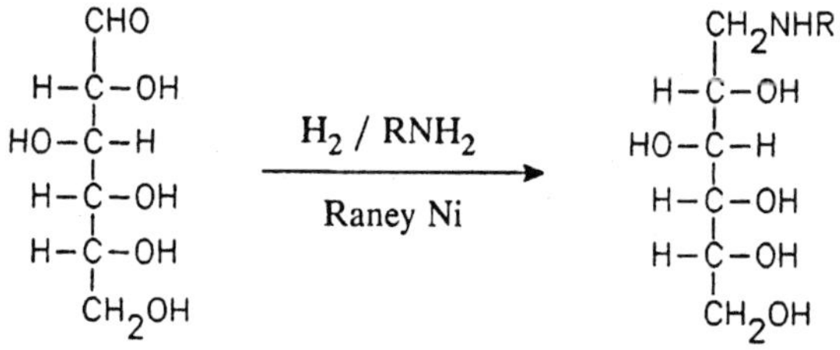

CHO
H−C−OH
HO−C−H
H−C−OH
H−C−OH
CH$_2$OH

$\xrightarrow{\text{H}_2 \text{ / RNH}_2}$ Raney Ni

CH$_2$NHR
H−C−OH
HO−C−H
H−C−OH
H−C−OH
CH$_2$OH

(5-1)

Glucamine (R = H)
N-Methylglucamine (R = Me)

C. Oxidation Products of D-Glucose

As noted in the previous chapter D-glucose can be converted to a variety of derivatives by microbial oxidation (Figure 5-5). Many of these transformations can also be achieved by oxidation with molecular oxygen in alkaline media over noble metal catalysts [2,3,7,8] or by enzymatic oxidation [2,3]. The various oxidation

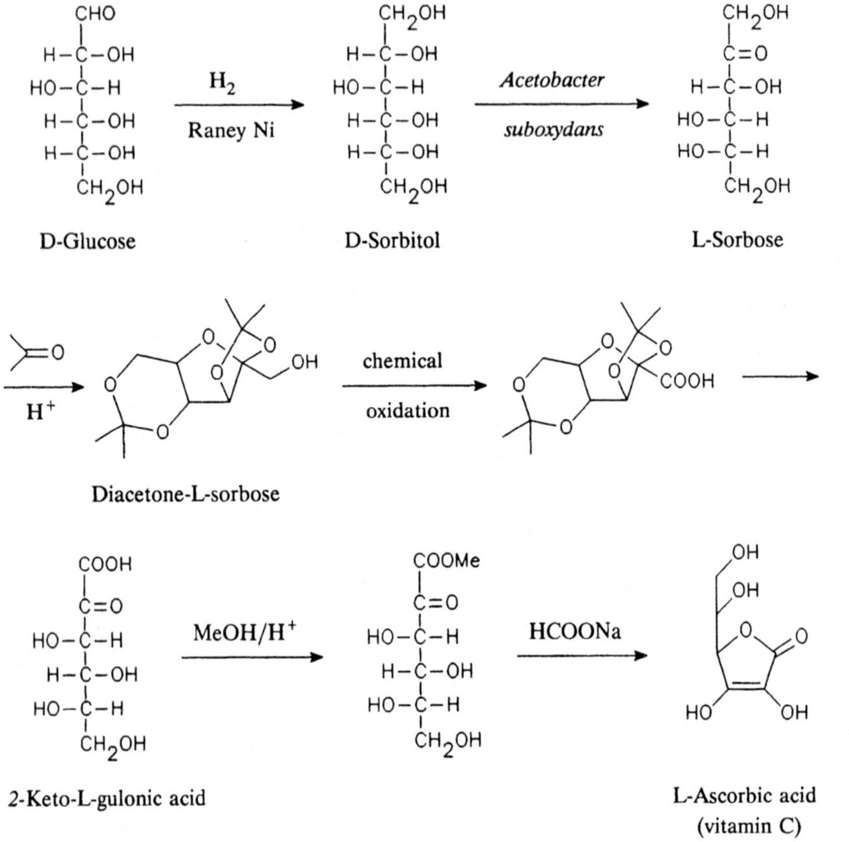

D-Glucose   D-Sorbitol   L-Sorbose

Diacetone-L-sorbose

2-Keto-L-gulonic acid   L-Ascorbic acid (vitamin C)

Figure 5-4. The Reichstein-Grussner process for the production of vitamin C [5].

products can be further converted to commercially interesting products; for example, 2-keto-D-gluconic acid is the precursor of D-isoascorbic acid (erythrobic acid, isovitamin C) (Reaction 5-2):

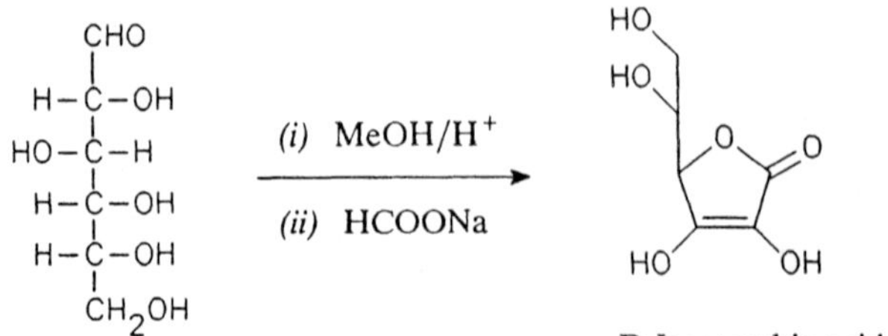

D-Isoascorbic acid

(5-2)

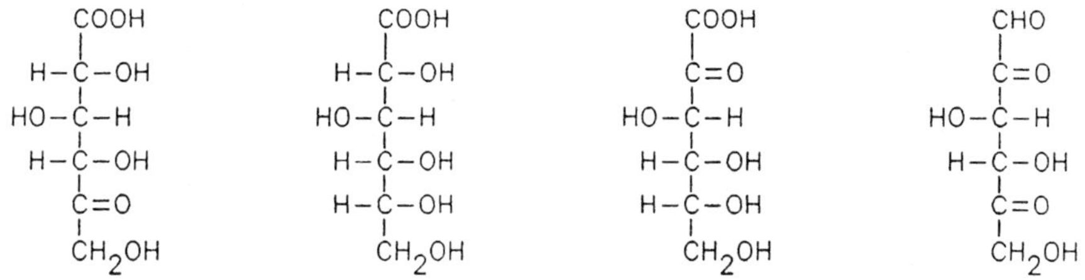

```
   COOH            COOH            COOH             CHO
 H−C−OH          H−C−OH           C=O              C=O
 HO−C−H          HO−C−H          HO−C−H           HO−C−H
 H−C−OH          H−C−OH          H−C−OH           H−C−OH
   C=O           H−C−OH          H−C−OH             C=O
  CH2OH           CH2OH           CH2OH            CH2OH
```

5-Keto-D-   D-gluconic   2-Keto-D-   2,5-Diketo-D-
gluconic acid (*f*)   acid (*e,f,c*)   gluconic acid (*f*)   gluconic acid (*f*)

```
    CHO            CHO             CHO             COOH
    C=O          H−C−OH          H−C−OH           H−C−OH
  HO−C−H         HO−C−H          HO−C−H           HO−C−H
  H−C−OH         H−C−OH          H−C−OH           H−C−OH
  H−C−OH         H−C−OH          H−C−OH           H−C−OH
   CH2OH          CH2OH           COOH             COOH
```

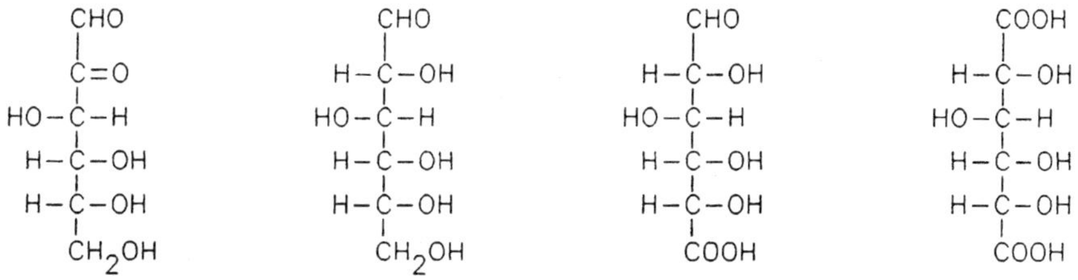

2-Keto-D-   6-Aldehyde-D-   D-Glucuronic   D-Glucaric
glucose (*e,f*)   glucose (*e*)   acid (*f*)   acid (*f,c*)

Figure 5-5. Oxidation products of D-glucose produced by fermentation (*f*), catalytic oxidation over noble metals (*c*) or enzymatic transformation (*e*).

D. C₃-Chirons from Protected Carbohydrate Precursors

Although carbohydrates have been widely used as chiral building blocks (chiral synthons or simply chirons) in organic synthesis [9–11], very few methods have found industrial application. This is largely due to the fact that carbohydrates are 'over-functionalized' with hydroxyl groups of similar or identical chemical reactivities. Moreover, they contain an abundance of asymmetric centers, more than are required for target chiral molecules (which are generally not carbohydrates). Consequently, carbohydrate-based syntheses tend to be circuitous and/or involve the use of expensive and/or environmentally unacceptable reagents. Hence, in order to exploit the vast synthetic potential of the carbohydrate chirality pool, there is a definite need to develop simple catalytic methods for their conversion [12].

There has been a growing interest [13–18] in the use of those C_3-chirons that have an asymmetric center at the central carbon atom for the synthesis of optically active compounds of biological importance. For example, the protected glycerol, glyceraldehyde, and glyceric acid derivatives (Figure 5-6) are versatile chirons amenable to a variety of synthetic transformations [13,16].

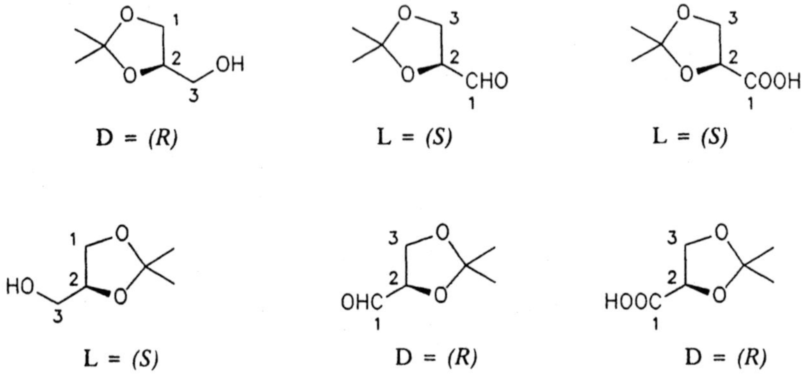

D = (R)   L = (S)   L = (S)

L = (S)   D = (R)   D = (R)

Figure 5-6. Examples of protected C₃-chirons.

Some potential applications of *(R)*-isopropylideneglycerol (*(R)*-IPG) are outlined in Figure 5-7 [17]; however, note that many of these applications are not yet economically viable due to the relatively high price of *(R)*-IPG. The corresponding protected glyceric acids have been less commonly used but are, in principle, useful C₃- chirons in, for example, the synthesis of phospholipids [19].

A suitable starting material for the synthesis of C₃-chirons is D-mannitol. Since the latter has a twofold axis of symmetry, cleavage at the C-3 to C-4 bond affords two identical chiral molecules without any wastage of carbon atoms. Indeed, one study [20] describes the high yield (97%) conversion of protected D-mannitol derivatives to the corresponding glyceric acid using sodium hypochlorite in the presence of a homogeneous or a heterogeneous ruthenium catalyst at room temperature (r.t.) as in Reaction 5-3.

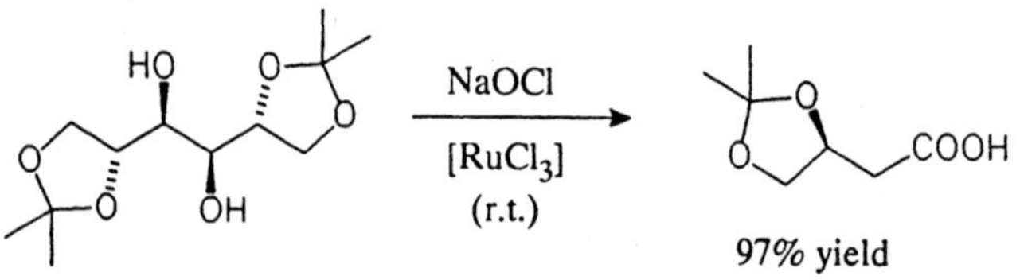

$$\text{(5-3)}$$

High yields were obtained only when the pH was carefully maintained at 8 in order to avoid deprotection. The intermediate formation of the corresponding protected glyceraldehyde was observed but the reaction could not be stopped at this stage. Hence, for the synthesis of the protected glyceraldehyde from D-mannitol, one still has to rely on oxidative cleavage with stoichiometric quantities of sodium periodate [21] or lead tetraacetate [22].

Alternatively, the L-*(S)* and D-*(R)* protected glyceraldehydes may be prepared from the protected L-ascorbic or D-isoascorbic acids, respectively, in a two-step

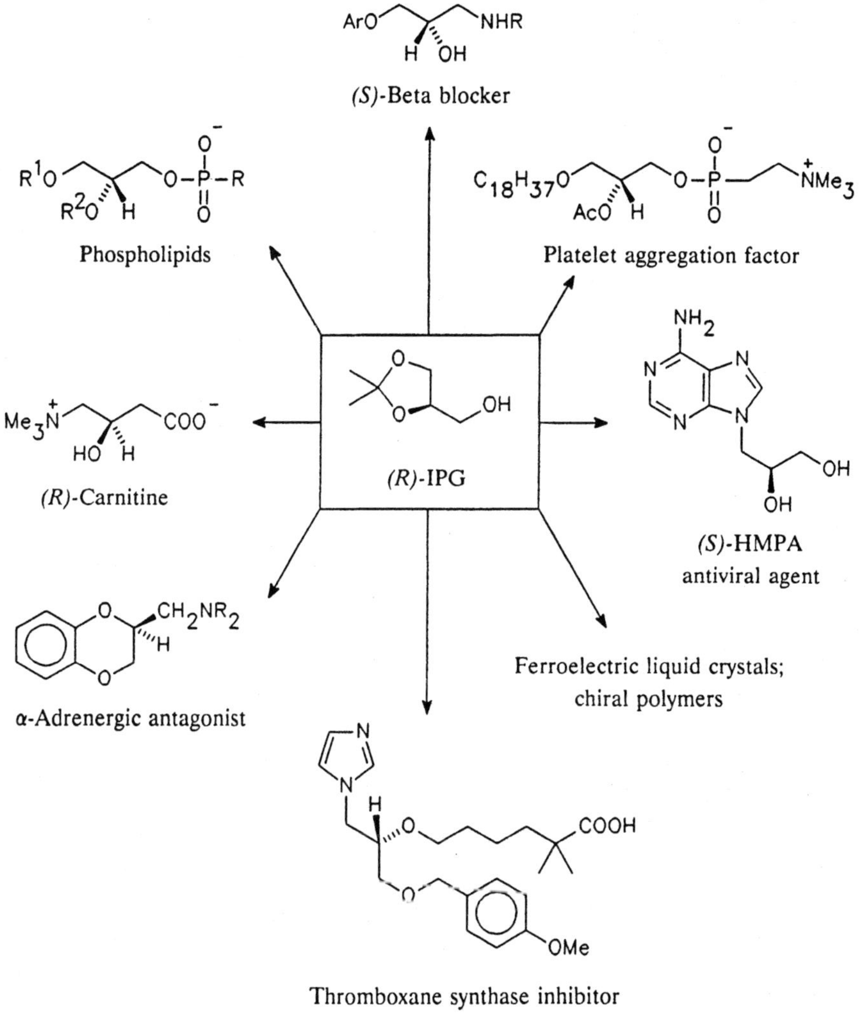

Figure 5-7. Potential applications of *(R)*-isopropylideneglycerol (*(R)*-IPG); see ref. [17].

procedure [19,23] involving oxidation with H_2O_2 followed by NaOCl (Figure 5-8). Overall yields were higher (about 60%) when the cyclohexylidene derivative was used (instead of isopropylidene) in a two-phase system, thus limiting over-oxidation of the aldehyde products. Addition of a ruthenium catalyst in the second step gives the corresponding glyceric acids in high overall yield (79–84%). The protected glycerols are obtained by catalytic reduction of the corresponding glyceraldehydes using Ru/C or Pd/C catalysts [24].

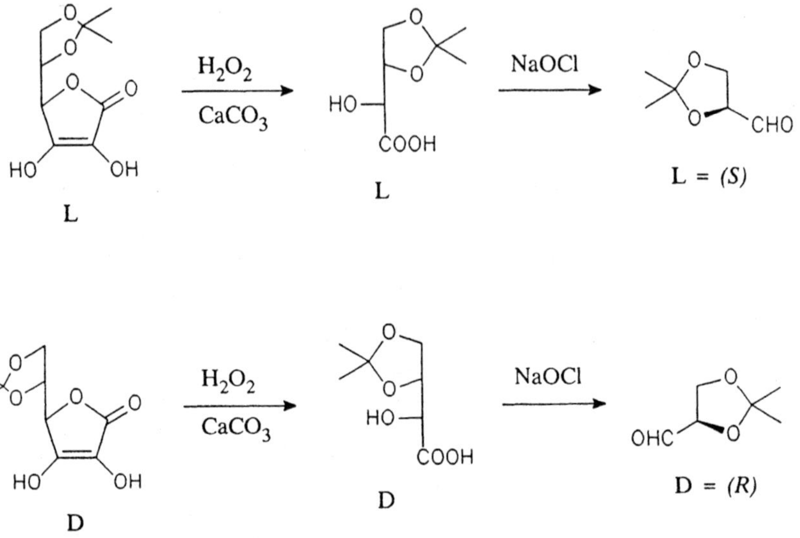

Figure 5-8. Synthesis of protected glyceraldehydes by oxidative degradation of L-ascorbic and D-ascorbic acid.

The above methods would appear to have synthetic potential as alternatives to classical methods that use stoichiometric amounts of lead tetraacetate or sodium periodate. However, they also have to be compared with other approaches such as the enantiospecific microbial oxidation of racemic 1,2-isopropylideneglycerol (Reaction 5-4) which affords *(R)*-IPG together with the *(R)*-isomer of the protected glyceric acid.

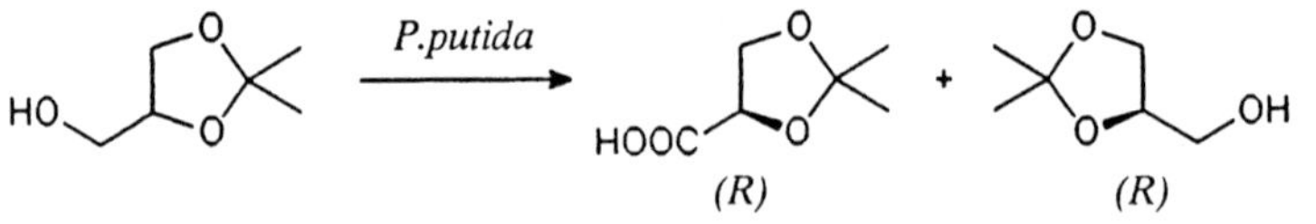

Moreover, these C_3 chirons have to be compared with other C_3 chirons such as chiral glycidyl derivatives that are prepared by enzymatic hydrolysis (see chapter 7) or assymetric epoxidation (see chapter 8.)

E.   C₃ Chirons from Unprotected Carbohydrates

An alternative approach is to degrade unprotected carbohydrates directly to unprotected C_3 chirons, which, if needed, can be subsequently protected. In this approach a combination of selective degradation methods is needed and only one C_3 chiron is produced per molecule of substrate. For example, in basic media,

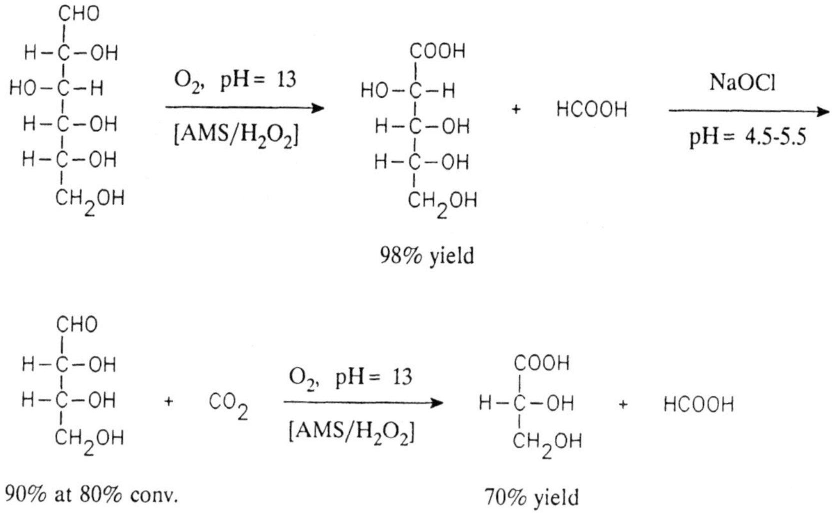

Figure 5-9. One-pot procedure for conversion of glucose to D-glyceric acid.

sodium anthraquinone-2-sulfonate (AMS) catalyzes the oxidative cleavage of aldoses to the next lower aldonic acids with molecular oxygen [26]. In one report [27] this has been combined with the known hypochlorite oxidation of aldonic acids to the next lower aldoses [28] to give an elegant three-step, one-pot procedure for the conversion of glucose to D-glyceric acid (Figure 5-9). This should be compared with the synthesis of the latter by enzymatic decarboxylation of L-tartaric acid with cell-extracts of a *Pseudomonas* species [29].

Similarly, by reversing the sequence of the oxidation procedure D-gluconic acid is converted to D-glyceraldehyde via D-erythronic acid (Figure 5-10) [27]. The latter is an interesting chiral building block (C_4 with two asymmetric centers) in its own right. It is also available from oxidative degradation of D-isoascorbic acid with H_2O_2 in basic media. Similarly, oxidative degradation of L-ascorbic acid with H_2O_2 under basic conditions yields L-threonic acid and the latter has been used as a chiral building block in a synthesis (Figure 5-11) of (R)-carnitine [30]; however, the overall yield was only 20%.

In principle, the above methodology of alternating oxidations with AMS/H_2O_2 and NaOCl is applicable to the selective degradation of other carbohydrates and provides access to a wide variety of C_3, C_4 and C_5 chirons using inexpensive reagents and simple procedures. Therefore, this methodology has enormous potential as an economical route to valuable chiral intermediates from readily available carbohydrates. Surprisingly, D-fructose is readily cleaved by hypochlorite at pH = 9 to give D-erythrose and glycolic acid, providing a very short route (Reaction 5-5) to this C_4 chiron [27].

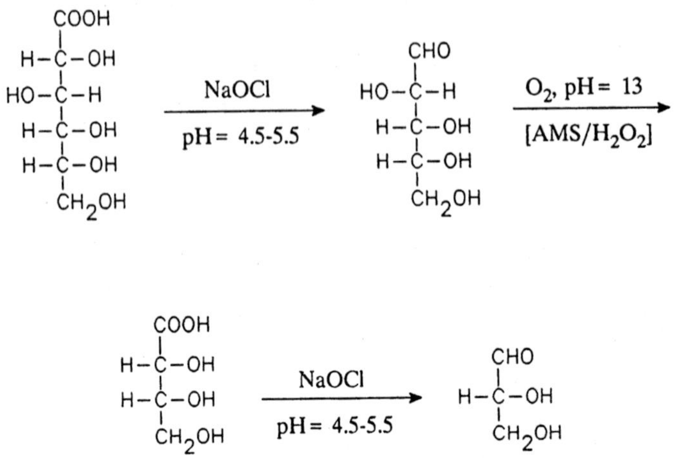

Figure 5-10. One-pot procedure for conversion of D-gluconic acid to D-glyceraldehyde.

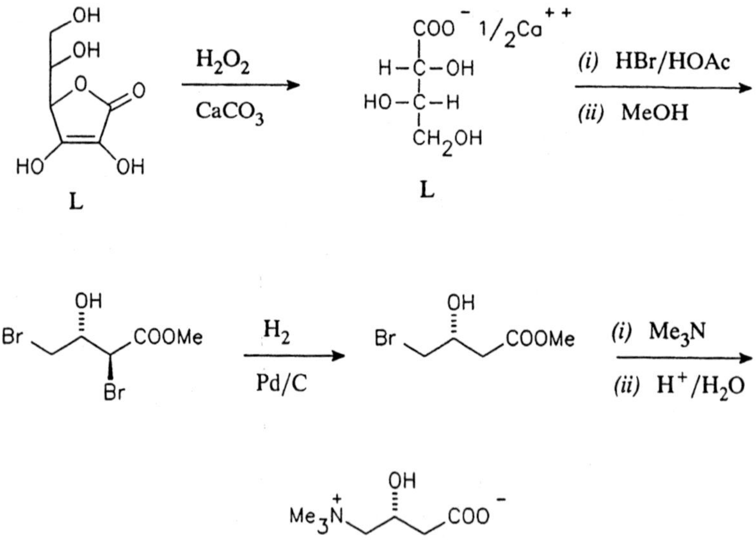

(R)-Carnitine

Figure 5-11. Synthesis of *(R)*-carnitine from L-ascorbic acid.

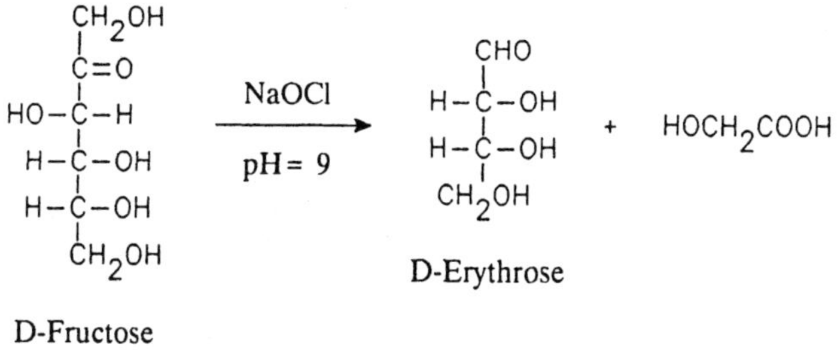

$$(5\text{-}5)$$

D-Fructose → D-Erythrose + HOCH$_2$COOH

II. CHIRAL HYDROXY ACIDS

As mentioned in the previous chapter, several chiral hydroxy acids are readily available from the fermentation of glucose or from selective microbial oxidations of aliphatic acids. The oldest and most well known are L-lactic acid and L-tartaric acid.

A. Lactic Acid

Both isomers of lactic acid are produced commercially by fermentation. The L = (S) isomer is the more important one and is widely applied as a food additive. The D = (R) isomer is gaining in importance as a chiral intermediate for the synthesis of optically active α-aryloxypropionic acid herbicides [31,32] by a double inversion sequence (Figure 5-12). The L isomer can also be used as the chiral building block by conversion to the mesylate followed by a single inversion to the final product, as illustrated in the synthesis of (R)-flamprop-isopropyl (Figure 5-13) [33].

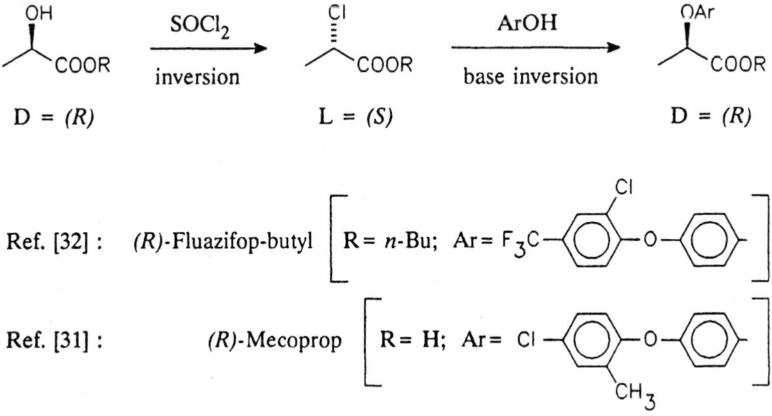

Ref. [32] : (R)-Fluazifop-butyl   R = n-Bu; Ar = F$_3$C—〈Cl〉—O—〈 〉—

Ref. [31] : (R)-Mecoprop   R = H; Ar = Cl—〈CH$_3$〉—O—〈 〉—

Figure 5-12. Optically active herbicides from D-lactic acid.

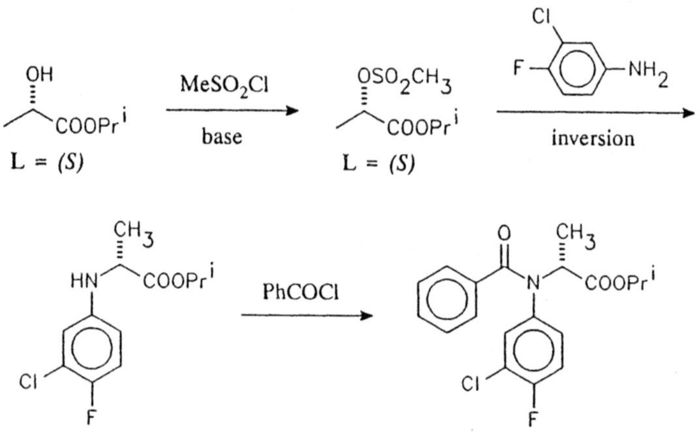

(R)-Flamprop-isopropyl

Figure 5-13. Optically active herbicide from L-lactic acid.

B. Tartaric Acid

(R,R)-L-Tartaric acid is produced in large quantities from cream of tartar, a waste product of wine manufacture. It is widely used as a resolving agent in industrial processes. (See chapter 6.) Its ready availability and low price also make it an interesting chiral synthon [34]. For example, it is used in the Zambon synthesis of (S)-naproxen [35] which has been described in chapter 3.

The enantiomeric (S,S)-D-tartaric acid is also a natural product, occurring in the African bush plant, *Bauhinia* [34]. It is currently produced industrially by resolution of racemic tartaric acid and is about 10–20 times more expensive than its enantiomer. It has been used in a synthesis of the antibiotic fosfomycin (Figure 5-14) [36]. However, considering the number of steps and the moderate stereoselectivity, it is doubtful whether this method can compete with the resolution of the racemic epoxide by crystallization of diastereomeric salts with optically active amines [37].

C. Other α-Hydroxy Acids

Other readily available chiral α-hydroxy acids are (S)-L-malic acid and both enantiomers of mandelic acid. The former is available from fermentation and more recently by catalytic asymmetric synthesis [38] (Figure 5-15). Both D- and L-mandelic acid are commercially available from classical resolution of the racemate.

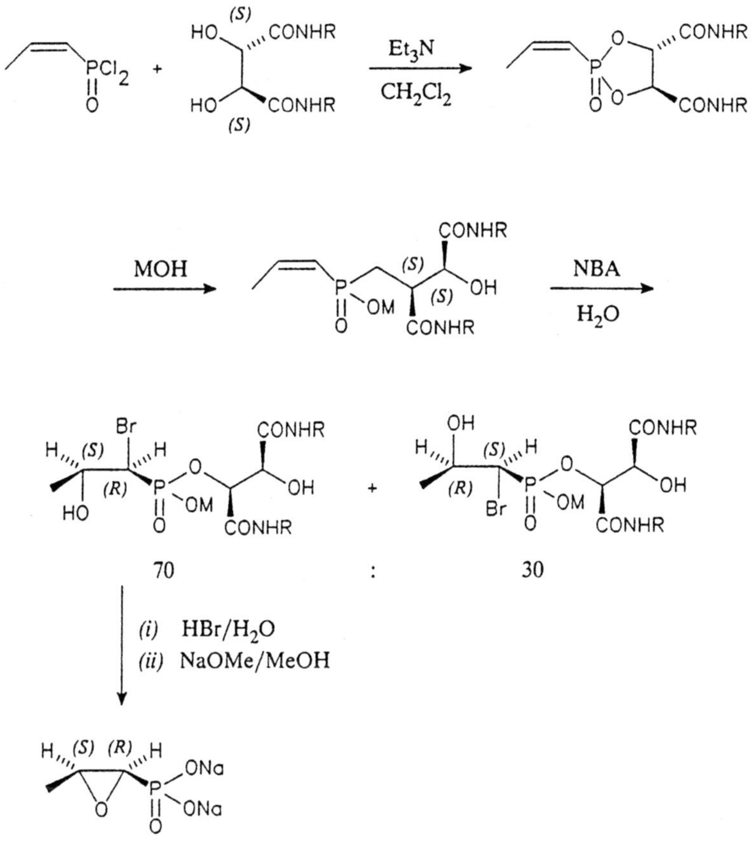

FOSFOMYCIN

Figure 5-14. Synthesis of fosfomycin using tartaric acid as a chiral auxiliary.

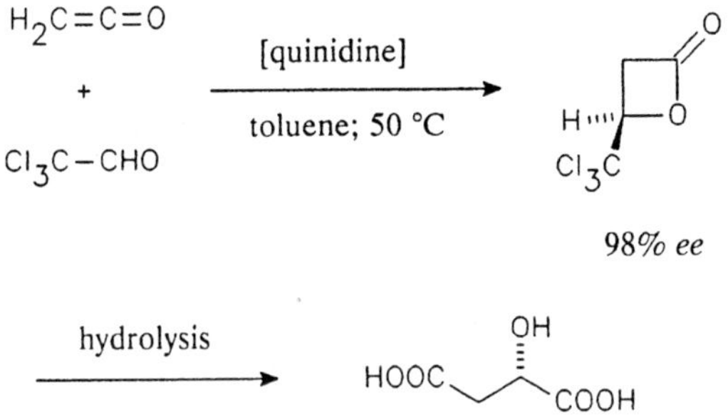

(S)-L-Malic acid

Figure 5-15. *(S)*-Malic acid via catalytic asymmetric synthesis.

Carbapenem intermediate

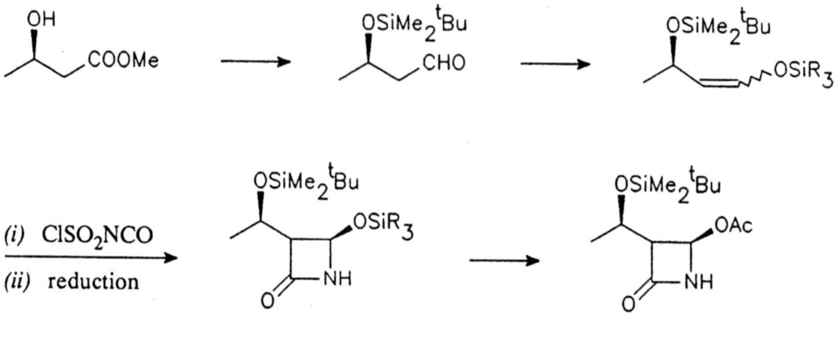

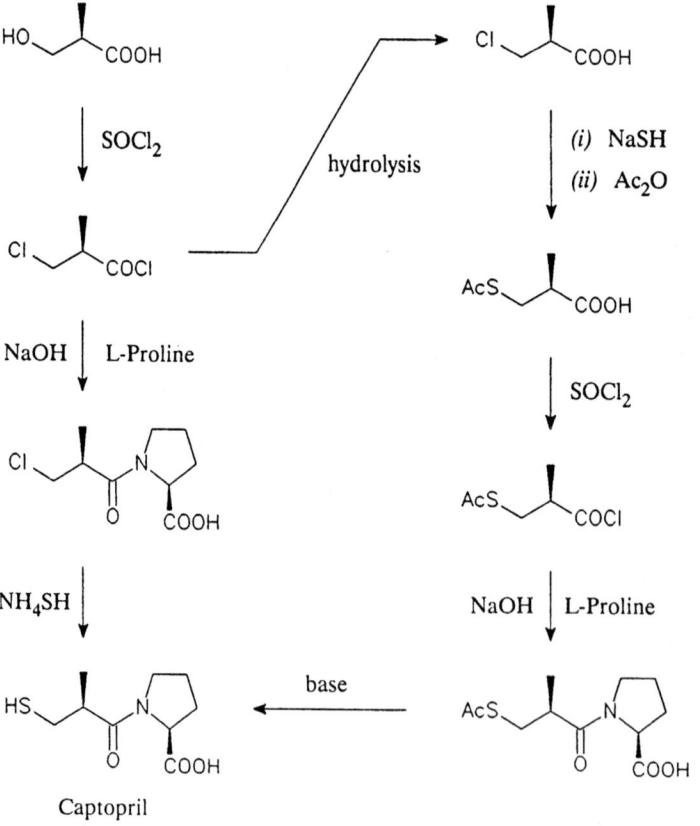

Figure 5-16. Synthesis of a carbapenem intermediate from *(R)*-β-hydroxybutyric acid.

Captopril

Figure 5-17. Routes to captopril from *(R)*-β-hydroxybutyric acid.

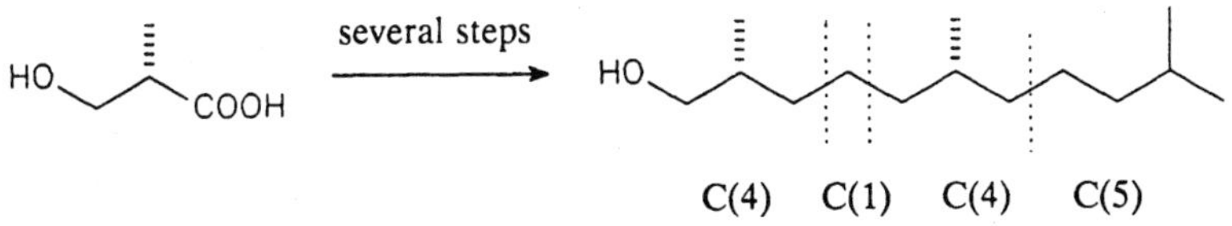

C(4)   C(1)   C(4)   C(5)

Figure 5-18. Construction of the side chain of vitamin E from *(S)*-β-hydroxybutyric acid.

D. Chiral β-Hydroxy Acids

(R)-β-Hydroxybutyric acid is produced by Kanegafuchi by microbial oxidation of *n*-butyric acid. (See chapter 4.) An alternative source is depolymerization of the commercially available poly-*(R)*-hydroxybutyrate, marketed by ICI as biopol[TM]. It is the key chiral synthon in a route developed by Kanegafuchi [39] to an intermediate for carbapenems (Figure 5-16).

Kanegafuchi has also commercialized the production of *(R)*-β-hydroxyiso-butyric acid by fermentation of isobutyric or methacrylic acid with mutant strains of *Candida rugosa* (see preceding chapter). It is the chiral synthon in routes to the ACE-inhibitor captopril (Figure 5-17) [40]. The key intermediate in the commercial production of captopril is the *(R)*-β-acetylmercaptoisobutyric acid, which is also produced by Andeno by classical resolution of the racemate [41]; the latter is produced by the addition of thioacetic acid to methacrylic acid [41]. The *(S)* enatiomer of β-hydroxyisobutyric acid is also, in principle, available from fermentation of isobutyric or methacrylic acid with various microorganisms, such as *Pseudomonas putida* [42]. It has been used to synthesize phytol, the side-chain of vitamin E [42,43] (Figure 5-18).

III. AMINO ACIDS

A. Natural Amino Acids

A variety of α-amino acids are readily available in bulk from fermentation and other processes (see chapter 4) and they probably constitute the most important class of compounds within the chirality pool. They have a relatively simple structure, with one or two asymmetric centers, and are amenable to a variety of chemical transformations [44]. Commercial availability has followed market needs, a classic example being the development of low cost production of L-aspartic acid and L-phenylalanine to meet the demand for the artificial sweetener aspartame. Similarly, L-proline, available from fermentation, is the key raw material for a variety of ACE-inhibitor drugs [45] including Captopril, Enalapril, and Lisinopril (Figure 5-19). For the latter two products, L-alanine and L-lysine are also key raw materials, respectively.

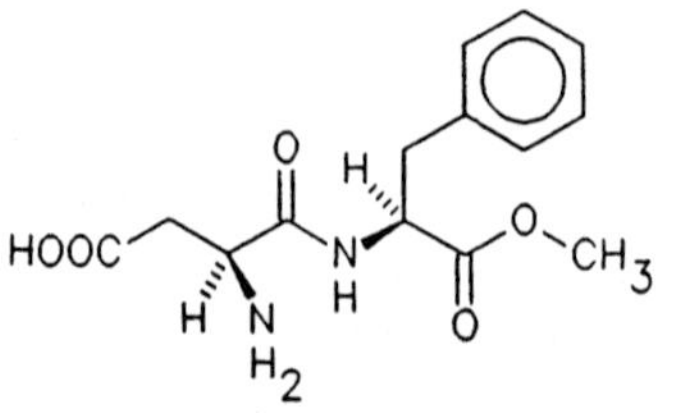

Aspartame
(L-phenylalanine/L-aspartic acid)

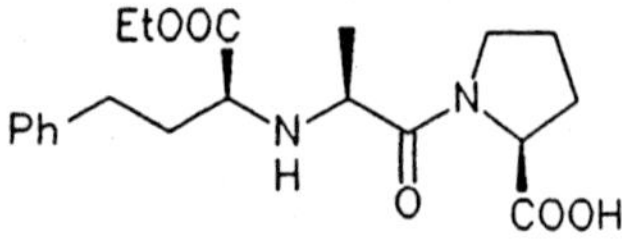

Enalapril
(L-alanine/L-proline)

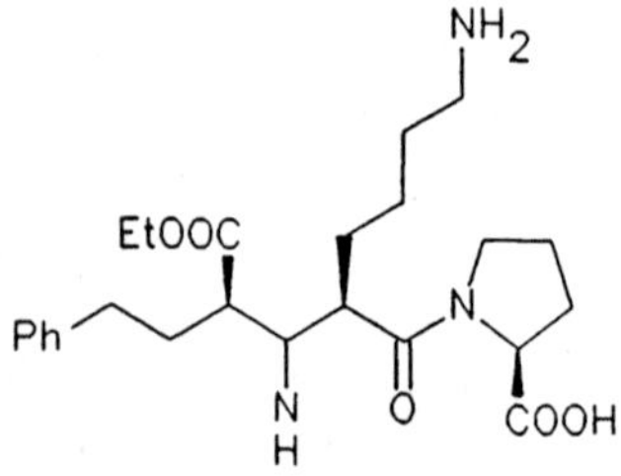

Lisinopril
(L-lysine/L-proline)

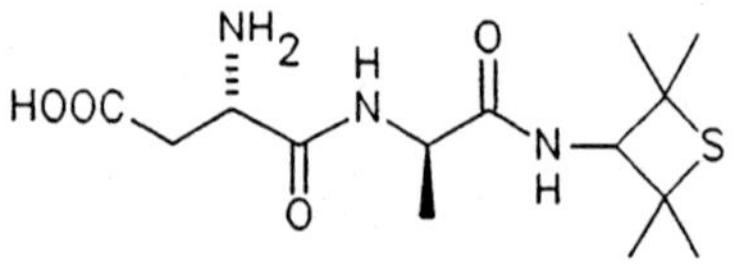

Alitame
(D-alanine)

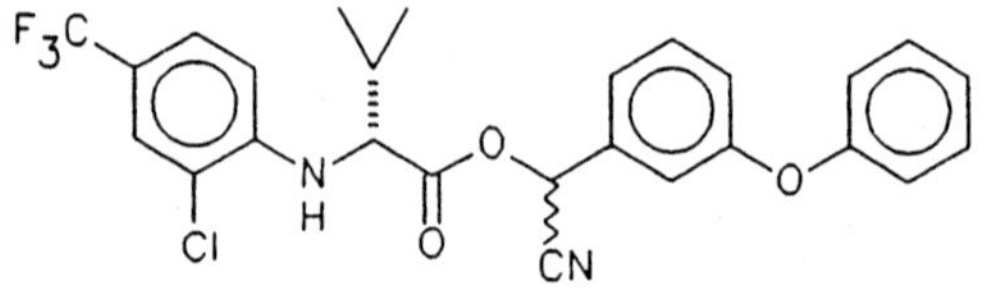

Fluvalinate
(D-valine)

Figure 5-19. Commercial products derived from amino acids as chiral synthons.

Another relevant commercial example is the Bristol-Myers Squibb synthesis [46] of the monobactam antibiotic, Azthreonam, using L-threonine as the chiral building block (Figure 5-20). This route to the monobactam nucleus is commercially more attractive than fermentation since the latter tends to give complex mixtures of monobactams that are difficult to separate. Similarly, the unsubstituted 3-aminomonobactamic acid nucleus (3-AMA) can be prepared analogously starting from L-serine [47]. Interestingly, 3-AMA can also be prepared [48] from 6-APA, another readily available chiral synthon (Figure 5-21).

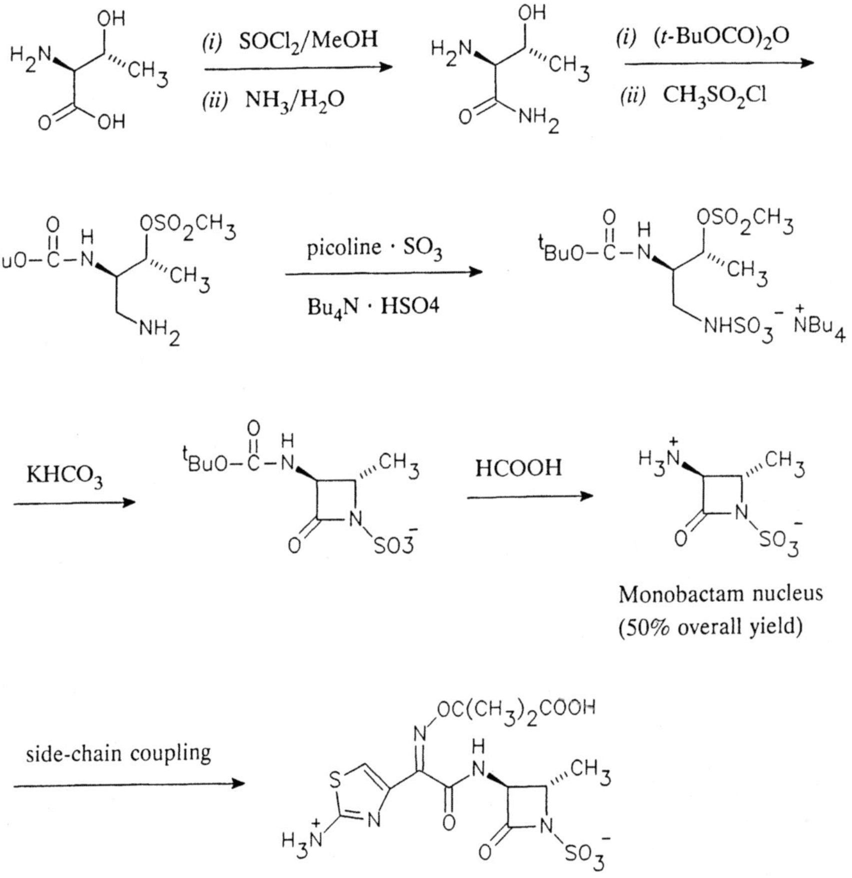

Azthreonam

Figure 5-20. Synthesis of Azthreonam nucleus from L-threonine.

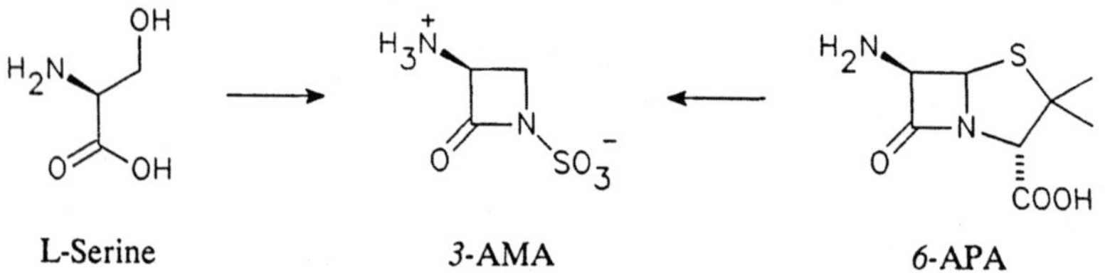

Figure 5-21. Synthesis of 3-aminomonobactamic acid (3-AMA) from L-serine or 6-APA

Various readily available amino acids, for example, L-glutamic acid [49], L-aspartic acid [50], and L-threonine [51] have been used as chiral synthons for the synthesis of carbapenem antibiotics, such as thienamycin.

B. Unnatural Amino Acids

In addition to the readily available natural L-amino acids, many D-amino acids are becoming increasingly available. For example, production of D-alanine has increased to meet the market demand for it as the raw material for Pfizer's new dipeptide sweetener, Alitame (Figure 5-19). Similarly, D-valine is a key raw material in the commercial synthesis of the optically active pyrethroid, Fluvalinate (Figure 5-19). L-Valine is a key raw material that is added to the fermentation broth in the commercial process for the synthesis of the cyclic peptide, cyclosporin. (See chapter 4 for the structure of cyclosporin.) Both D- and L-valine are manufactured [52] using the DSM aminopeptidase technology (see chapter 7) and L-valine is also available by fermentation.

Other D-amino acids that are readily available are D-phenylglycine and D-p-hydroxyphenylglycine, the side chains of ampicillin and amoxicillin, respectively. Interestingly, the L-isomer of phenylglycine, which is in principle available from the large scale manufacture of D-phenylglycine by diastereomer crystallization (see chapter 6), has been used as a chiral auxiliary in the synthesis of carbapenem antibiotics.

IV. TERPENES

A variety of optically active terpenes are readily available from natural sources [53]; the most important are the monoterpenes, several examples of which are illustrated in Figure 5-22. The most abundant source of monoterpenes is oil of turpentine, several hundred thousand tons of which are produced annually from conifers. The exact composition of oil of turpentine is very dependent on its geographical origin. Typically, turpentine from the southeastern United States

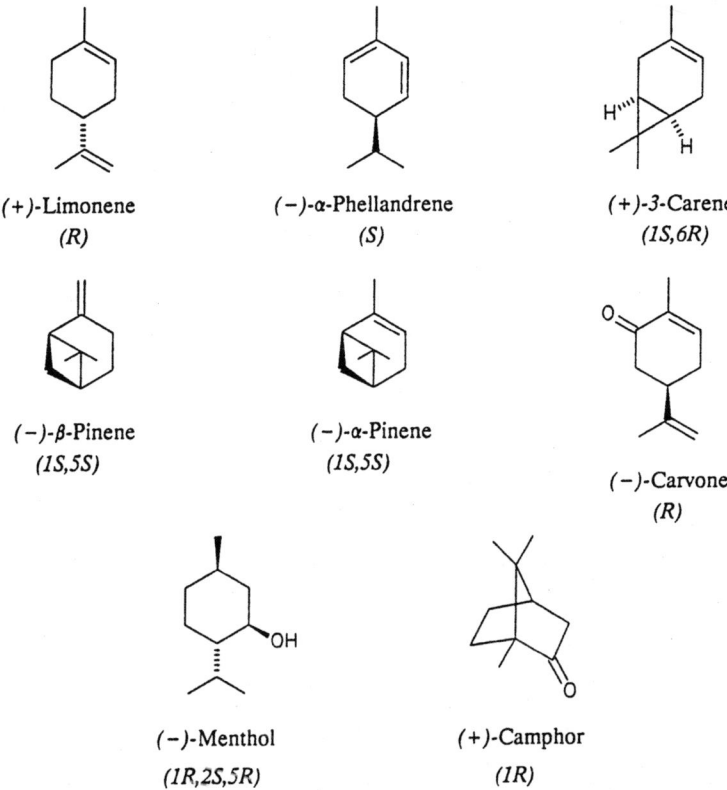

(+)-Limonene
(R)

(−)-α-Phellandrene
(S)

(+)-3-Carene
(1S,6R)

(−)-β-Pinene
(1S,5S)

(−)-α-Pinene
(1S,5S)

(−)-Carvone
(R)

(−)-Menthol
(1R,2S,5R)

(+)-Camphor
(1R)

Figure 5-22. Optically active terpene chirons.

contains 60–70% of α-pinene, 20–25% of β-pinene and 6–12% of others. Turpentine from the western United States, on the other hand, contains 2–25% of 3-carene in addition to α- and β-pinene. The optical purity of various components is also dependent on the origin of the turpentine. Turpentine from India and Pakistan contains (+)-3-carene of high optical purity. Natural β-pinene contains 92–95% of the (−)-isomer while the (+)-content of α-pinene is often not more than 35%.

Crude turpentine is widely used as a solvent (e.g., for paints and varnishes) and the purified monoterpenes are mainly used as raw materials for the synthesis of other terpenoid products. For example, β-pinene is converted to linalool, nerol, and geraniol (Figure 5-23) which in turn serve as the raw materials for other terpenoid products. (+)-Limonene is a byproduct of the citrus industry and is the starting material for the production of (−)-carvone (Figure 5-22). (+)-Camphor is obtained from the wood of the camphor tree. (−)-Menthol is obtained from natural

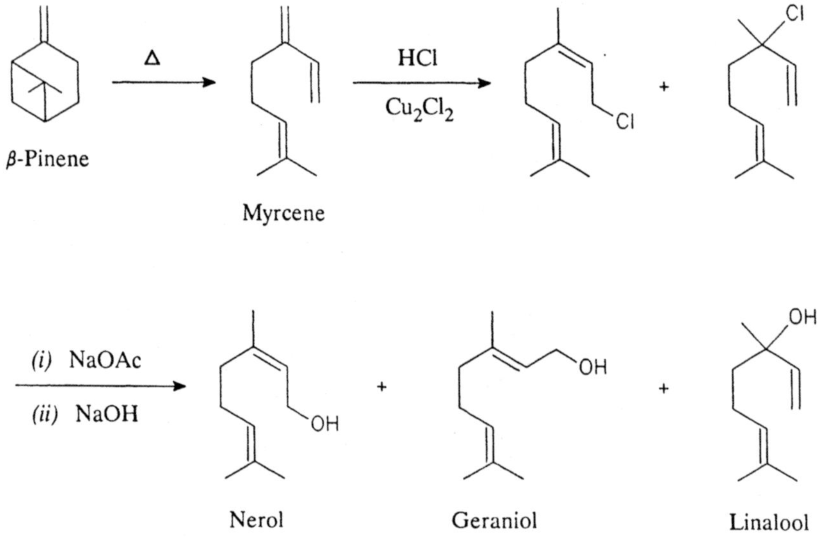

Figure 5-23. Commercial products derived from β-pinene.

sources but substantial amounts are currently being produced by catalytic asymmetric synthesis. (See chapter 8.)

In addition to being used as raw materials for other terpenoids, the monoterpenes also find employment as precursors of resolving agents (e.g., camphorsulphonic acid) and chiral ligands for asymmetric catalysis. In general, they have been only sporadically employed as chiral synthons. This can be attributed to the fact that, in common with carbohydrates, their structures are too elaborate for nonterpenoid target molecules. Moreover, they are not always readily available in high optical purity (see above) and, being liquids with few functional groups, are difficult to purify. Thus, if carbohydrates are over-functionalized we may view terpenes as being under-functionalized.

Consequently, terpenes have generally been used as chiral auxiliaries rather than chiral synthons (for an explanation of the difference see later). An illustrative example is the synthesis [54] of (R)-carnitine from (+)-β-pinene (Figure 5-24). The route is circuitous (11 steps) with an inherent low overall yield and employs stoichiometric amounts of expensive and/or environmentally unfriendly reagents. The degradation of the pinene structure involves, in successive steps, the use of m-chloroperbenzoic acid, dichromate, and periodate. Moreover, the pinene structure is 'sacrificed' in the synthesis. The commercial utility of such a route is, to say the least, questionable.

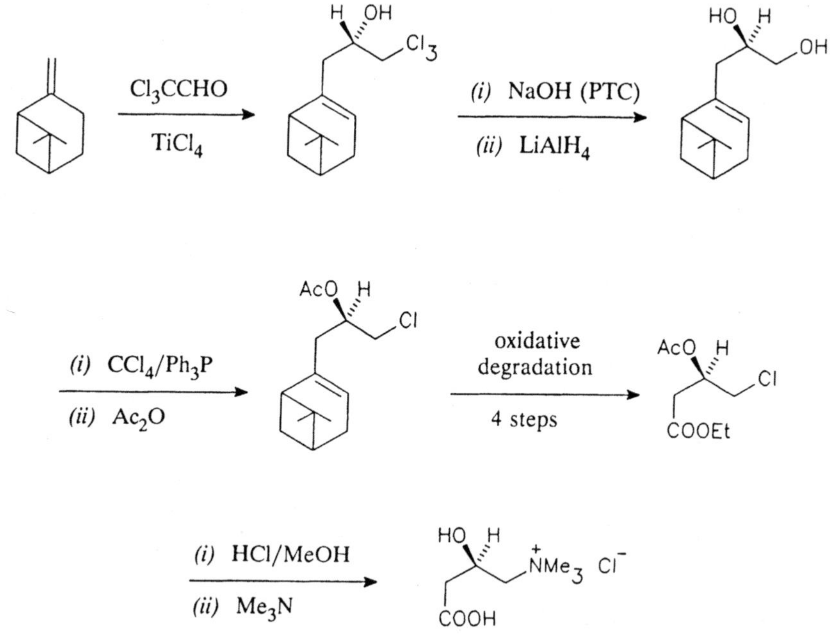

(R)-Carnitine · HCl

Figure 5-24. Synthesis of *(R)*-carnitine from *(+)*-β-pinene.

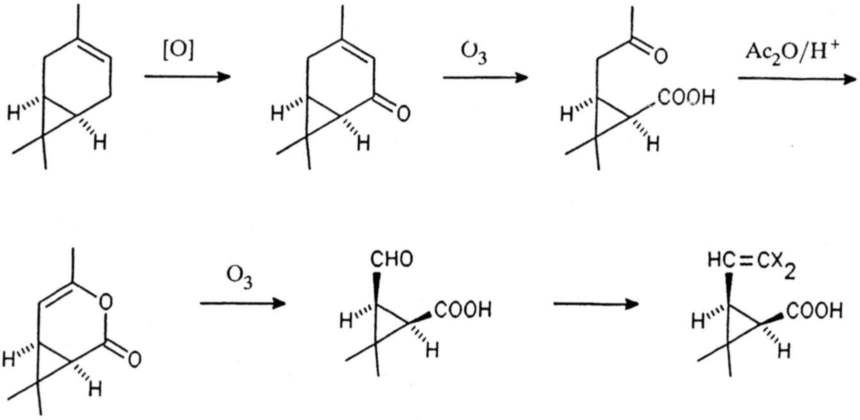

(1R,cis)-Caronaldehyde X= Cl, Br

Figure 5-25. Conversion of *(+)*-3-carene to intermediates for optically active pyrethroids.

A structural feature of monoterpenes that has stimulated interest is the presence of a geminal dimethyl group in conjunction with optical activity. This could make them, in principle, interesting precursors for optically active pyrethroids. (See chapter 2 for structures.) Examination of the configuration of *(+)-3-carene* reveals that it should be a suitable starting material for the optically active cyclopropane compounds that are key components of several optically active pyrethroids. For example, *(+)-3-carene* has been converted in 4 steps to *(1R,cis)*-caronaldehyde which then can be converted to the *(1R,cis)*-dihalovinylcyclopropene carboxylic acids (Figure 5-25) [55].

However, neither this nor the 3-carene-based routes appear to be competitive with alternative commercial routes to these products [55]. Indeed, despite its esthetically pleasing chiral structure, the major (and perhaps only) application of *(+)-3-carene* is as a solvent for paints and varnishes.

V. ALKALOIDS

If the monoterpenes have a problem with possessing an elaborate structure then the alkaloids are in serious trouble. Indeed, even the most inexpensive and readily available cinchona alkaloids—represented by quinine, quinidine, cinchonine, and cinchonidine (Figure 5-26)—are unlikely to find employment as chiral building blocks. They are widely used for small scale resolutions of racemic acids by diastereomeric salt crystallization. Because of their relatively high price and molecular weight they are only sporadically applied for large scale resolutions. They are, however, finding increasing use as chiral ligands for asymmetric catalysis or as asymmetric catalysts themselves in base-promoted reactions. (See chapter 8.)

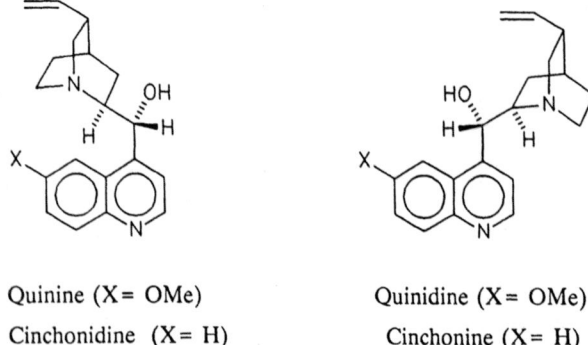

Quinine (X= OMe)   Quinidine (X= OMe)
Cinchonidine (X= H)   Cinchonine (X= H)

Figure 5-26. Structures of cinchona alkaloids.

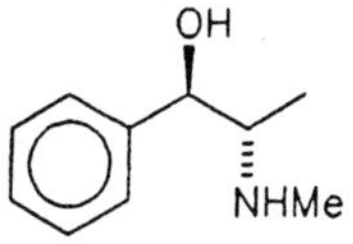

(1R,2S)-Ephedrine 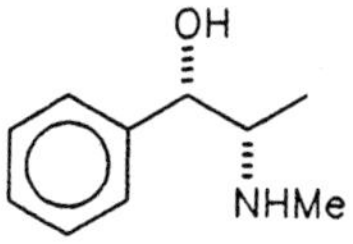   (1S,2S)-Pseudoephedrine

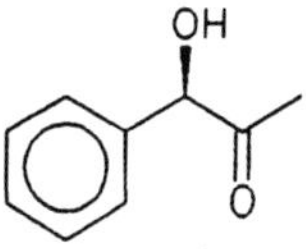

(R)-PAC 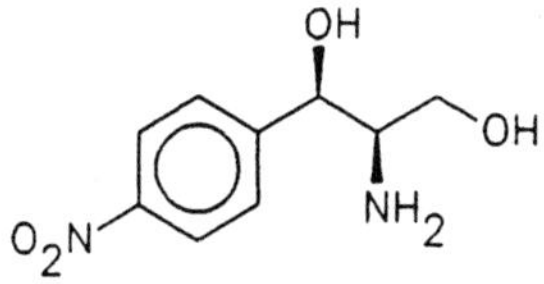   (1R,2R)-Chloramphenicol

Figure 5-27. Structures of miscellaneous optically active products.

VI. MISCELLANEOUS OPTICALLY ACTIVE PRODUCTS

In addition to the various classes of natural products discussed in this chapter, there is an extensive chirality pool comprising the miscellaneous optically active products and their precursors that are produced on a large scale. Examples include ephedrine, pseudoephedrine, and their precursors (R)-phenylacetylcarbinol ((R)-PAC), and chloramphenicol (Figure 5-27). For many products that are produced by racemate resolution (e.g., D-phenylglycine and chloramphenicol), the wrong isomer is also, in principle, available.

VII. CHIRAL SYNTHON VERSUS CHIRAL AUXILIARY

The distinction between a chiral synthon and a chiral auxiliary is simply that in the former the structure of the raw material is largely incorporated into the product while in the latter it is not; the chiral auxiliary orchestrates the asymmetric induction at newly formed stereogenic centers without being incorporated in the final product. The attractiveness of using a chiral auxiliary depends to a large extent on whether or not it is 'sacrificed' or is easily recovered and recycled. For example, in the Zambon process for (S)-naproxen (see chapter 3), L-tartaric acid is used as a chiral auxiliary but is recovered and recycled. In contrast, the synthesis of the carbacephem intermediate uses L-phenylglycine as a chiral auxiliary which is consumed (Figure 5-28) [56]; only the NH_2 group of L-phenylglycine remains in the product.

Recovery and recycling of a chiral auxiliary can be compared to that of a resolving agent in a classical resolution process. Hence, by analogy, not only the

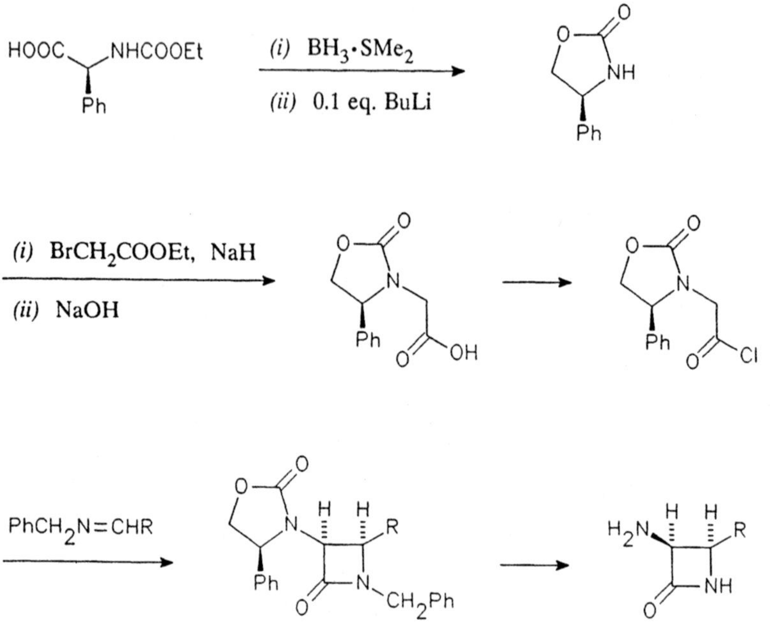

Figure 5-28. L-Phenylglycine as a chiral auxiliary.

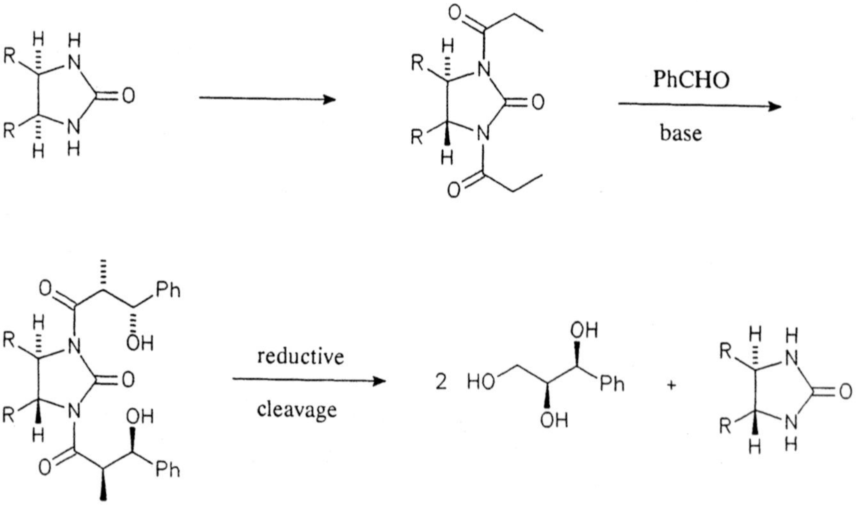

Figure 5-29. Asymmetric aldol-type condensation with "double" chiral auxiliary.

efficiency of this process is important for the economics but also the molecular weight of the chiral auxiliary (which should be as low as possible). The report [57] of a "double" chiral auxiliary for asymmetric aldol condensations provides an elegant means for minimizing the amount of the auxiliary required (Figure 5-29).

REFERENCES

1. Lichtenthaler, F. W. (Ed.), *Carbohydrates as Organic Raw Materials*, Verlag Chemie, Weinheim, 1991.
2. Röper, H., in *Carbohydrates as Organic Raw Materials*, Lichtenthaler, F. W. (Ed.), Verlag Chemie, Weinheim, pp. 267-288. 1991.
3. a) Röper, H., and Kod, H., *Starch/Stärke*, 40, 453-464 (1988). b) Röper, H., and Kod, H., *Starch/Stärke*, 42, 342-349 (1990).
4. Makkee, M., Kieboom, A. P. G., and Van Bekkum, H., *Carbohydrate Research*, 138, 237-245 (1985).
5. Crawford, T. C., and Crawford, S. A., *Advan. Carbohydrate Chem.*, 37, 79-155 (1980), and references cited therein.
6. Kelkenberg, H., *Tenside Surfactants Detergents*, 25, 8-13 (1988).
7. Van Bekkum, H., in *Carbohydrates as Organic Raw Materials*, Lichtenthaler, F. W. (Ed.), Verlag Chemie, Weinheim, pp. 289-310, 1991.
8. Sheldon, R. A., *Stud. Surf. Sci. Catal.*, 59, 33-54 (1991).
9. Lichtenthaler, F. W., in *New Aspects Org. Chem. I., Proc.4th Int. Kyoto Conf.*, Yoshida, Z., Shiba, T., and Ohshiro, Y., (Eds.), Kodansha, Tokyo, 1989, pp. 351-384.
10. Vasella, A., in *Modern Synthetic Methods, Vol. 2,*, Scheffold, R. (Ed.), Otto Salle Verlag, Frankfurt, 1980, pp. 173-265.
11. Hanessian, S., *Total Synthesis of Natural Products: The Chiron Approach*, Pergamon Press, Oxford, 1986.
12. Emons, C. H. H., *Ph.D. Thesis*, Eindhoven University of Technology, The Netherlands, 1992.
13. Jurczak, J., Pikul, S., and Bauer, T., *Tetrahedron*, 42, 447-488 (1986).
14. Takano, S., *Pure Appl. Chem.*, 59, 353 (1987).
15. Golding, B., *Chem. Ind.*, (London), 617-621 (1988).
16. Altenbach, H. J., *Nachr. Chem. Tech. Lab.*, 36, 33-37 (1988).
17. Kooreman, H. J., Van Nistelrooij, H., and Pryce, R. J., in *Proc. Chiral Synthesis Symp. and Workshop*, Spring Innovations, Stockport, UK, 1989, pp. 37-38.
18. Peters, U., Bankova, W., and Welzel, P., *Tetrahedron*, 43, 3803-3816 (1987).
19. a) Bhatia, S. K., and Hajdu, *Tetrahedron Lett.*, 28, 271-271 (1987). b) Bhatia, S. K., and Hajdu, *Tetrahedron Lett.*, 28, 3767-3770 1987). c) Bhatia, S. K., and Hajdu, *J. Org. Chem.*, 53, 5034-5039 (1988).
20. Emons, C. H. H., Kuster, B. F. M., Vekemans, J. A. J. M., and Sheldon, R. A., *Tetrahedron Asymmetry*, 2, 359-362 (1991).
21. Jackson, D. Y., *Synth. Commun.*, 18, 337 (1988).
22. a) Baer, E., and Fischer, H. O. L., *Biol. Chem.*, 128, 463 (1939). b) Baer, E., and Fischer, H. O. L., *Biol. Chem.*, 128, 475 (1939).
23. Mizuno, Y., and Sugimoto, K., European Patent, 143.973 (1984).

24. Emons, C. H. H., Kuster, B. F. M., Vekemans, J. A. J. M., and Sheldon, R. A., *Chimica Oggi (Chemistry Today)*, Nov./Dec. 59-65, (1992).

25. Bertola, M. A., Koger, H. S., Phillips, G. T., Marx, A. F., and Claassen, V. P., European Patent, 244.912 (1987).

26. Hendriks, H. E., Kuster, B. F. M., and Marin, G. B., *Carbohydrate Res.*, **214**, 71 (1991).

27. Emons, C. H. H., Kuster, B. F. M., Vekemans, J. A. J. M., and Sheldon, R. A., in *Proc. Chiral 92 Symp.*, Spring Innovations, Stockport, UK, 1992, pp. 39-46.

28. a) Whistler, R. L., and Schwegler, R., *J. Am. Chem. Soc.*, **81**, 5190 (1959). b) Whistler, R. L., and Yagi, K. J., *J. Org. Chem.*, **26**, 1050 (1961).

29. Furuyoshi, S., Kawabata, N., Tanaka, H., and Soda, K., *Agric. Biol. Chem.*, **53**, 2101-2105 (1989).

30. Bock, K., Lundt, I., and Pederson, C., *Acta Chem. Scand.*, **B37**, 341 (1983).

31. a) Eveleens, W., Spaans, J., and Wissink, H. G., European Patent Appl., 9.285 (1980) to AKZO; *CA:* **93**, 95006p (1980). b) Gras, G., German Patent 3.024.265 (1981) to Rhone-Poulenc; *CA:* **94**, 191956q (1981).

32. Crosby, J., *Tetrahedron*, **47**, 4789-4846 (1991).

33. Scott, R. M., and Armitage, G.D., German Patent 2.946.652 (1980) to Shell; *CA:* 185974g (1980).

34. Seebach, D., in *Modern Synthetic Methods*, Scheffold, R. (Ed.), Otto Salle Verlag, Frankfurt, 1980, pp. 91-171.

35. a) Giordano, C., Castaldi, G., Cavicchioli, S., and Villa, M., *Tetrahedron*, **45**, 4243-4252 (1989). b) Cavicchioli, S., Giordano, C., and Uggeri, F., *J. Org. Chem.*, **52**, 3018-3027 (1987).

36. Giordano, C., and Graziano, C., *J. Org. Chem.*, **54**, 1470-1473 (1989).

37. a) Glamkowski, E. J., Gal, G., Purick, R., Davidson, A. J., and Sletzinger, M., *J. Org. Chem.*, **35**, 3510 (1970). b) Girotra, N. N., and Wendler, N. L., *Tetrahedron Lett.*, 5911 (1984).

38. Wynberg, H., and Staring, E. G. J., *J. Am. Chem. Soc.*, **104**, 166 (1982).

39. a) Ohashi, T., and Hasegawa, J., *J. Synth. Org. Chem., Japan*, **45**, 331-345 (1987). b) Ohashi, T., *Proc. Chiral 90 Symp.*, Spring Innovations, Stockport, UK, 1990, pp. 65-71.

40. Shimazaki, M., Hasegawa, J., Kan, K., Nomura, K., Nose, Y., Kondo, H., Ohashi, T., and Watanabe, K., *Chem. Pharm. Bull.*, **30**, 3139-3146 (1982).

41. Sheldon, R. A., Porskamp, P. A., and Ten Hoeve, W., in *Biocatalysts in Organic Synthesis*, Tramper, J., Van der Plas, H. C., and Linko, P. (Eds.), Elsevier, Amsterdam, 1985, pp. 59-80.

42. Leuenberger, H. G. W., in *Biocatalysts in Organic Synthesis*, Tramper, J., Van der Plas, H. C., and Linko, P. (Eds.), Elsevier, Amsterdam, 1985, pp. 99-118, and references cited therein.

43. Cohen, N., Eichel, W. F., Lopresti, R. J., Neukom, C., and Sauci, G., *J. Org. Chem.*, **41**, 3505-3511 (1976).

44. Coppola, G. M., and Schuster, H. F., *Asymmetric Synthesis – Construction of Chiral Molecules using Amino Acids*, Wiley, New York, 1987.

45. Sheldon, R. A., Zeegers, H. J. M., Houbiers, J. P. M., and Hulshof, L. A., *Chimica Oggi (Chemistry Today)*, May, 1991, pp. 35-47.

46. a) Floyd, D. M., Fritz, A. W., and Cimarusti, C. M., *J. Org. Chem.*, **47**, 176-178 (1982). b) Cimarusti, C. M., and Sykes, R. B., *Chem. Brit.*, 302-303 (1982).

47. Slusarchyk, W. A., *Heterocycles*, **21**, 191-209 (1984).

48. Cimarusti, C. M., Applegate, H. E., Chang, H. W., Floyd, D. M., Koster, W. H., Slusarchyk, W. A., and Young, M. G., *J. Org. Chem.*, **47**, 180-182 (1982).

49. a) Ohta, T., Kimura, T., Sato, N., and Nozoe, S., *Tetrahedron Lett.*, **29**, 4303-4305 (1988). b) Ohta, T., Sato, N., Kimura, T., Nozoe, S., and Izawa, K., *Tetrahedron Lett.*, **29**, 4305-4308 (1988).

50. Salzmann, T. N., Ratcliffe, R. W., Christensen, B. G., and Bouffard, F. A., *J. Am. Chem. Soc.*, **102**, 6161-6163 (1980).

51. a) Shiozaki, M., Ishida, N., Hiraoka, T., and Yanagisawa, H., *Tetrahedron Lett.*, **22**, 5205-5208 (1981). b) Yanagisawa, H., Ando, A., Shiozaki, M., and Hiraoka, T., *Tetrahedron Lett.*, **24**, 1037-1040 (1983).

52. a) Sheldon, R. A., Schoemaker, H. E., Kamphuis, J., Boesten, W. H. J., and Meijer, E. M., in *Stereoselectivity of Pesticides*, Ariëns, E. J., Van Rensen, J. J. S., and Welling, W. (Eds.), Elsevier, Amsterdam, 1988, pp. 409-451. b) Meijer, E. M., Kamphuis, J., Van Balken, J. A. M., Hermes, H. F. M., Van den Tweel, W. J. J., Kloosterman, M., Boesten, W. H. J., and Schoemaker, H. E., in *Trends in Drug Research*, Claassen, V. (Ed.), Elsevier, Amsterdam, 1989, pp. 363-382.

53. Erman, W. F., *Chemistry of the Monoterpenes, Parts A and B*, Marcel Dekker, New York, 1985.

54. Pellegata, R., Dosi, I., Villa, M., Lesma, G., and Palinisano, G., *Tetrahedron*, **41**, 5607 (1985).

55. Crosby, J., *Tetrahedron*, **47**, 4789-4846 (1991), and references cited therein.

56. Evans, D. A., and Sjogren, E. B., *Tetrahedron Lett.*, **26**, 3783-3786 (1985).

57. Davies, S. G., *Proc. 2nd CPhI Conf.*, Manufacturing Chemist, Morgan Grampion, London, 1992, p. 134.

ADDITIONAL READING

1. Blaser, H. U., The Chiral Pool as a Source of Enantioselective Catalysts and Auxiliaries, *Chem. Rev.*, **92**, 935-952 (1992).

6

Racemate Resolution via Crystallization

> The seed, which had come from God-only-knows-where, taught the
> atoms the novel way in which to stack and lock, to crystallize, to freeze.
> Kurt Vonnegut, in *Cat's Cradle*

I. INTRODUCTION

Notwithstanding the revolutionary advances that have been achieved in enzymatic
kinetic resolutions (see chapter 7) and catalytic asymmetric synthesis (see chapter
8), racemate resolution by crystallization techniques is still probably the most
important method for the synthesis of pure enantiomers [1–3]. For example, a large
number of commercially important drugs or their precursors (Table 6-1) are
primarily manufactured by processes involving 'classical resolution'. Moreover,
in the foreseeable future, these methods will continue to play an important role in
both small and large scale syntheses. There are several reasons for this [2]. First,
the rationale and efficiency of crystallization have been greatly improved through
a better understanding of the underlying properties upon which they are based (e.g.,
phase diagrams of enantiomer and diastereomer systems). Second, they are not
restricted to the resolution of racemates per se but can also be used for further
purification of partially resolved substances prepared by other techniques (e.g.,
asymmetric synthesis or biotransformation). Third, they provide the basis for
developing crystallization-induced asymmetric transformations whereby a theoretical yield of 100% of a pure enantiomer is possible in a single stage.

Table 6-1 Optically Active Pharmaceuticals Produced (Wholly or Partially) Using Crystallization Techniques

Product	Therapeutic class	Worldwide sales ($ millions) 1990
Amoxycillin	Antibiotic	2000
Ampicillin	Antibiotic	1800
Captopril	Cardiovascular	1520
Diltiazem	Calcium antagonist	960
Naproxen	Antiinflammatory	950
Cefalexin	Antibiotic	900
Cefadroxil	Antibiotic	300
Timolol	Cardiovascular	325
α-Methyldopa	Cardiovascular	225
Chloramphenicol	Antiinfective	80
Dextromethorphan	Antitussive	50
Ethambutol	Tuberculostatic	50

II. RESOLUTION BY DIRECT CRYSTALLIZATION OF ENANTIOMERIC MIXTURES

Whether or not a racemate can be separated by direct, preferential crystallization of one of the enantiomers depends primarily on the type of racemate one is dealing with.

A. Types of Racemates

For the purpose of our discussion racemates can be conveniently divided into two types: **racemic compounds** and **conglomerates** (a third type, pseudoracemates, is rarely encountered). The two classes of racemate are readily distinguished by reference to their melting point diagrams (Figure 6-1). Alternatively, powder X-ray or solid state IR spectra can be used; the spectra of enantiomers are identical with that of the racemic conglomerate but different from that of a racemic compound.

The most common type, accounting for about 90% of all racemates, is referred to as a racemic compound. In this case, the racemic crystals consist of a perfectly ordered array of R and S molecules—a crystalline 1:1 addition complex—and individual crystals contain equal amounts of both enantiomers. The second type is referred to as a conglomerate and consists of a mechanical mixture of crystals of the two enantiomers in equal amounts; although in bulk a conglomerate is optically neutral, individual crystals contain only one enantiomer.

In a conglomerate the 1:1 racemic mixture is simply a eutectic, having by definition a lower melting point than that of the enantiomers because each enantiomer

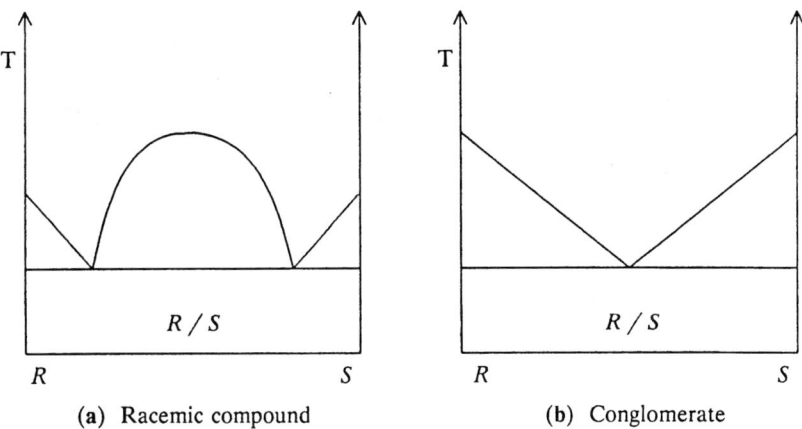

(a) Racemic compound   (b) Conglomerate

Figure 6-1. Melting point diagrams of racemates.

lowers the melting point of the other. From a practical viewpoint it is very important whether or not a racemate is a **conglomerate** since the latter is amenable to resolution by direct crystallization. Unfortunately, only about 5–10% of all racemates are conglomerates. As noted above they are readily recognized by their melting point diagrams, that are determined using a differential scanning calorimeter.

B. Resolution of Racemates by Direct Crystallization

Direct crystallization is widely used for industrial scale resolutions, for example in the manufacture of α-methyl-L-dopa [4] and chloramphenicol [5]. It is feasible only with conglomerates and depends on the correlation between the melting point diagram and the solubility diagram, that is, the mixture having the lowest melting point is the most soluble, and for a conglomerate this is the racemic mixture.

In the context of industrial scale resolutions, there are essentially two methods for the direct crystallization of conglomerates. The first involves circulating a supersaturated solution of the racemate simultaneously through two crystallization chambers or fluid beds that contain seed crystals of the respective enantiomers (Figure 6-2). Before recooling, the depleted solution is resaturated at a higher temperature in a make-up vessel in order to restore the original level of supersaturation required in the crystallization chambers. This is essentially the method used by Merck for the resolution of an intermediate in the industrial synthesis of the antihypertensive agent, α-methyl-L-dopa (Figure 6-3). The process is carried out on a scale of several hundred tons per annum.

The second method, first described by Gernez [6] in 1866, is referred to as **preferential crystallization** or **resolution by entrainment**. It consists of collect-

Figure 6-2. Separation of a conglomerate by simultaneous crystallization ($t_1 > t_2$).

ing alternate crops of each enantiomer in a single vessel. For example, seeds of the D-enantiomer are added to a supersaturated solution artificially enriched with the D-enantiomer. A crop of the D-enantiomer is then collected that is equal to approximately twice the amount of the original enrichment. An amount of fresh racemate equal to the weight of the collected crystals is then dissolved in the filtrate by warming. The resulting solution is cooled back to the operating temperature and crystallization of the L-enantiomer induced by seeding. In principle, this process can be repeated ad infinitum. In practice, however, the number of cycles is limited by the accumulation of impurities in the solution. It should be noted that such resolution processes are by definition metastable in that they involve crystallization from supersaturated solutions; therefore, they are very sensitive to random seeding by extraneous impurities and reproducibility can be a problem.

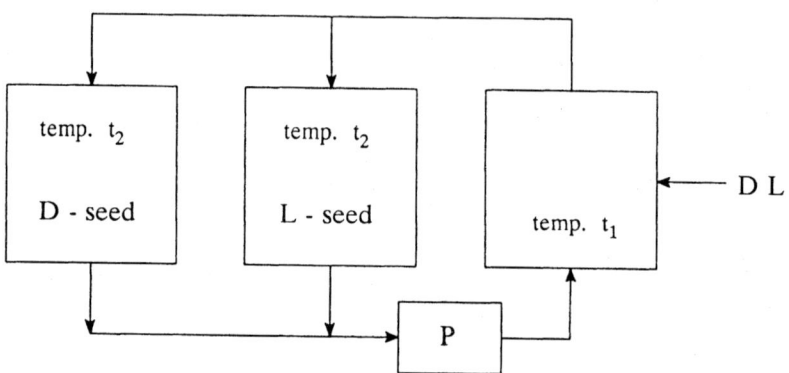

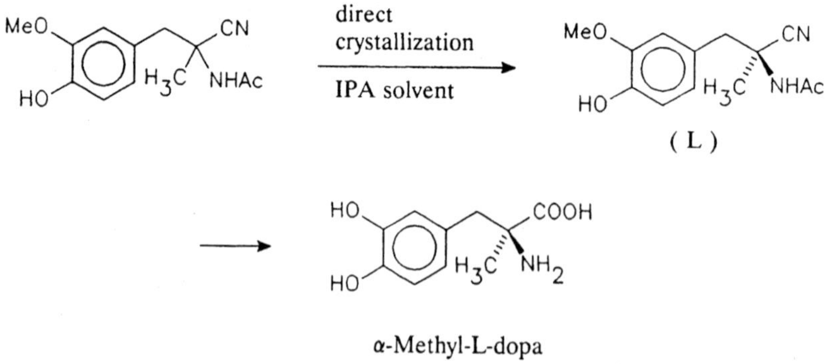

α-Methyl-L-dopa

Figure 6-3. α-Methyl-L-dopa synthesis.

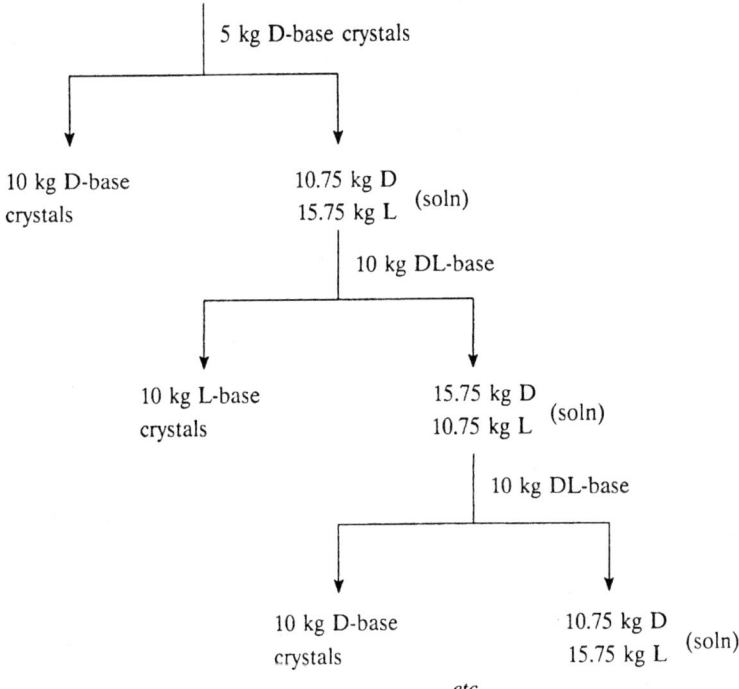

DL-threo chloramphenicol base

31.5 kg DL-base and 141.4 mol HCl in 300 ltr H_2O

Figure 6-4. Resolution by preferential crystallization.

An industrially relevant example of the separation of a racemate by preferential crystallization is the resolution of the chloramphenicol intermediate (Figure 6-4). This process has been used for more than 30 years by Roussel-Uclaf for the commercial production of large quantities of chloramphenicol [5]. Another relevant example is the resolution of L-glutamic acid by direct crystallization [7]. This process, introduced in the early 1960s by Ajinomoto and operated on a scale in excess of 10,000 tons per annum, competed successfully for many years with the fermentation process but has since been abandoned.

C.   Direct Crystallization of Salts with Achiral Acids or Bases

The chance that a particular racemate is a conglomerate, and hence amenable to resolution by direct crystallization, is less than one in ten. Fortunately, the chance can be increased by salt formation with achiral acids or bases. The occurrence of conglomerates among salts is estimated to be 2–3 times that of covalent compounds [1]. Virtually all of the naturally occurring α-amino acids, for example, are resolvable via preferential crystallization, either directly or as salts. Various strategies have been discussed [8,9] for achieving efficient separations of racemic amino acids. DL-Dopa, for example, can be resolved by preferential crystallization of its hydrochloride [8]. Similarly, a practical process has been described [9] for the resolution of DL-lysine by preferential crystallization of its p-aminobenzene-

D-form (g/100ml H$_2$O)

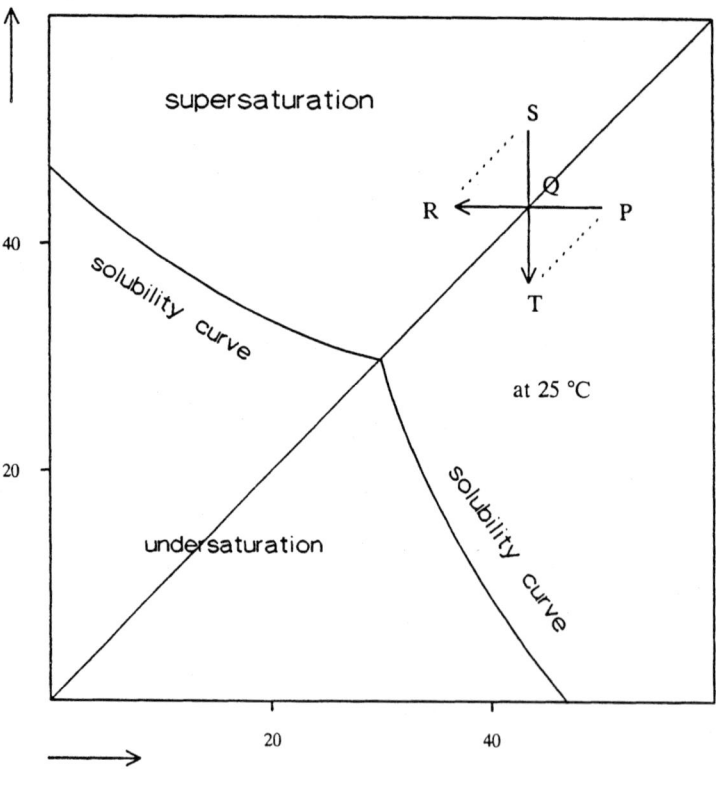

L-form (g/100ml H$_2$O)

Figure 6-5. Phase diagram of the reciprocal resolution of D- and L-lysine p-amino-benzenesulfonate (reproduced with permission from ref. [9]).

sulfonate salt. Figure 6-5 illustrates, by reference to the phase diagram, how this resolution is achieved.

A solution of composition P containing an excess of the L-lysine-aminobenzenesulfonate is prepared at an elevated temperature, cooled to 25°C and seeded with L-crystals. As the L-isomer crystallizes the composition of the solution moves through Q to point R. At this point crystals of the L-isomer are filtered off. Subsequently, an amount of racemate equal to that of the L-isomer separated is dissolved in the mother liquor by heating. The composition of the solution moves to the point S. Cooling the mixture to 25°C and seeding with the D-isomer results in crystallization of the latter whereby the composition of the solution moves to T. Filtration of the D crystals followed by dissolution of fresh racemate returns the composition to point P. The entire cycle can be then repeated indefinitely giving both D- and L-isomer reciprocally.

Simple amines can also often be resolved via direct crystallization of their salts with achiral acids. Thus, racemic α-methylbenzylamine (**I**) and 1-phenyl-2-(*p*-tolyl)-ethylamine (**II**) have been resolved by preferential crystallization of their salts with cinnamic acid (Figure 6-6) [10].

Many chiral carboxylic acids can similarly be resolved by preferential crystallization of their salts with achiral amines (Figure 6-6). For example, Nohira and coworkers [11] have reported the resolution of the synthetic pyrethroid intermediate (**III**) and the related compounds (**IV**) and (**V**) by preferential crystallization of their diethylamine salts. Similarly, naproxen (**VI**) and *cis*-benzoylaminocyclohexane-carboxylic acid (**VII**) have been resolved as their salts with ethylamine [12] and benzylamine [13], respectively.

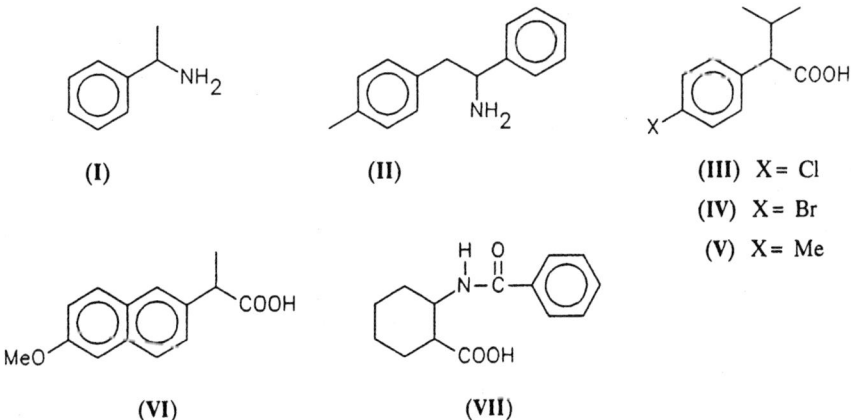

(I)	
(II)	
(III) X = Cl	
(IV) X = Br	
(V) X = Me	
(VI)	
(VII)	

Figure 6-6. Amines and carboxylic acids resolved by preferential crystallization of their salts.

D.   Crystallization-Induced Asymmetric Transformation of a Racemate

Preferential crystallization of a conglomerate is a particularly attractive method for industrial synthesis when it is accompanied by spontaneous in situ racemization of the enantiomer that remains in excess in solution; this allows for a theoretical yield of 100%. Such a process is called a **crystallization-induced asymmetric transformation** and may occur with racemates or diastereomeric mixtures. In the former case, the process is often referred to as a **deracemization**. An elegant example has been reported by Okada and coworkers [14] and is illustrated in Figure 6-7; such a process is a 'spontaneous resolution' and the enantiomer obtained is a matter of chance. Racemization occurs spontaneously in this example, while in other cases, in situ racemization can be induced by the addition of an external catalyst. For example, it is well known [15–17] that the racemization of α-amino acids can be induced by catalytic amounts of aldehydes via the reversible formation of Schiff bases (Figure 6-8).

Racemization via Schiff base formation is particularly facile with benzylic amino acids. Combining this knowledge with the fact that many amino acids can be resolved by preferential crystallization of their salts with aromatic sulfonic acids led to the development of an elegant method for the direct resolution of *p*-hydroxyphenylglycine [18]. Thus, seeding of supersaturated solutions of arenesulfonate salts of DL-*p*-hydroxyphenylglycine, containing a catalytic amount of an aldehyde, resulted in preferential crystallization of the D-isomer. The mother liquor was shown to be optically inactive, demonstrating that in situ racemization had occurred.

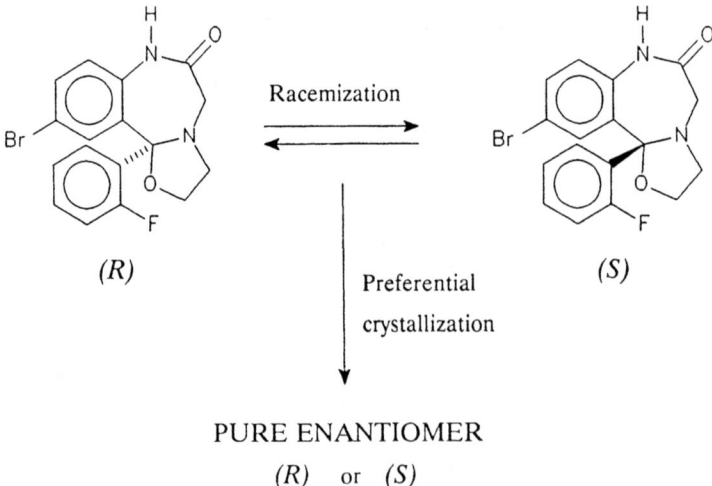

Figure 6-7. Asymmetric transformation of a racemate.

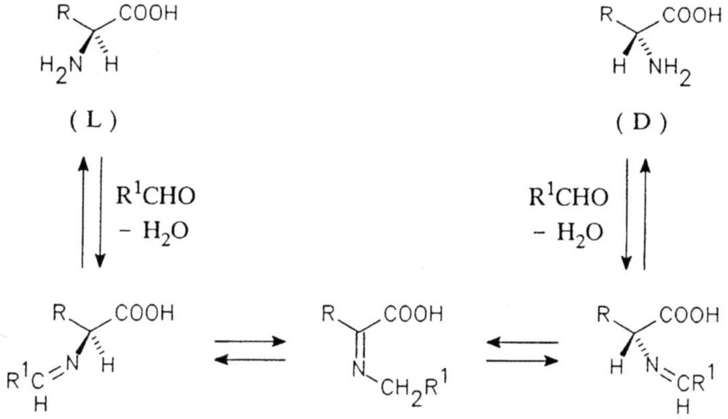

Figure 6-8. Aldehyde-catalyzed racemization of amino acids.

This method (Reaction 6-1) does not appear to be able to compete, however, with other commercial processes (diastereomeric salt crystallization or enzymatic resolution) for D-*p*-hydroxyphenylglycine, probably because of low productivities.

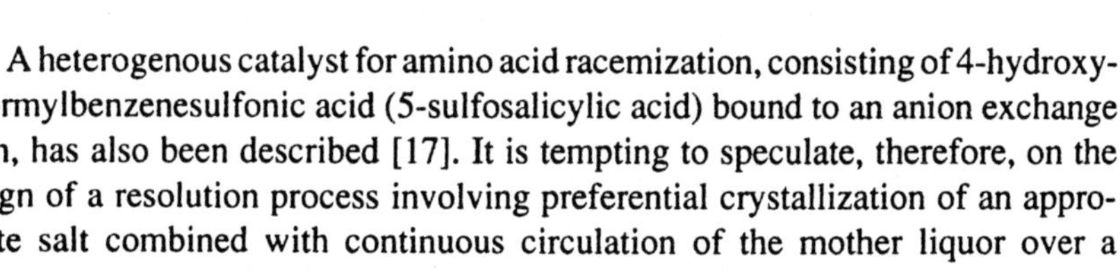

A heterogenous catalyst for amino acid racemization, consisting of 4-hydroxy-3-formylbenzenesulfonic acid (5-sulfosalicylic acid) bound to an anion exchange resin, has also been described [17]. It is tempting to speculate, therefore, on the design of a resolution process involving preferential crystallization of an appropriate salt combined with continuous circulation of the mother liquor over a heterogeneous racemization catalyst, but such a process has not yet been described in the literature.

Another industrially relevant example of crystallization-induced asymmetric transformation is shown in Figure 6-9. The synthetic pyrethroid (**VIII**) undergoes base-catalyzed epimerization at the CH group adjacent to the cyano group. When a solution of (**VIII**) is allowed to crystallize in the presence of a catalytic amount of freshly calcined KF, the *(α-S,2R,3R)*-isomer is obtained in 98.6% yield [19].

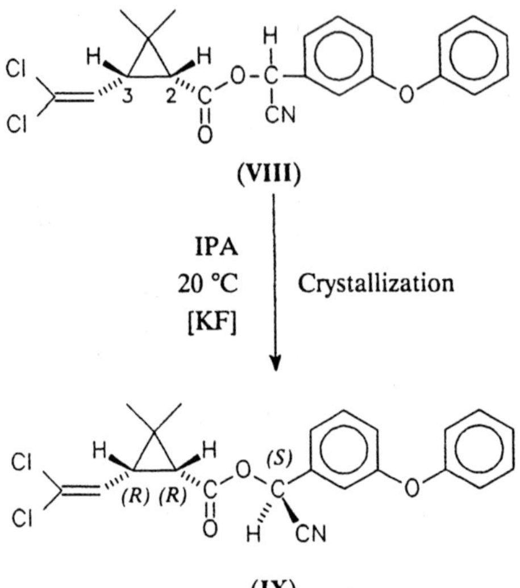

(VIII)

IPA
20 °C   Crystallization
[KF]

(IX)

Figure 6-9. Asymmetric transformation of a synthetic pyrethroid.

E.   Enrichment of Partially Resolved Enantiomers

Although, as noted above, racemic compounds cannot be resolved by direct crystallization it is possible to purify partially resolved samples by crystallization. The latter may be available from other methods (e.g., enzymatic kinetic resolution or catalytic asymmetric synthesis.) The feasibility and yield of such a process is dependent on the shape of the phase diagram as shown in Figure 6-10. The situation for a conglomerate is illustrated in Figure 6-10a where Q_A and Q_R are the quantities of solvent required to dissolve one mole (or 1 g) of the pure enantiomer or racemate, respectively, at a particular temperature.

A mixture of composition M, containing an excess of the D-enantiomer, will completely dissolve in a quantity of solvent Q_B, corresponding to the point K where the solubility curve is dissected by a vertical line originating from point M. At values of Q between Q_B and Q_R a two-phase system results that consists of crystals of the pure D-enantiomer and a solution containing both D- and L-forms. Accordingly, to obtain a pure enantiomer from a partially resolved conglomerate, in principle, one merely has to add a quantity of solvent necessary to dissolve the amount of racemate present in the sample. In practice, however, the whole sample is usually dissolved by heating the mixture, followed by cooling to yield crystals of the pure D-isomer.

The situation for a racemic compound is depicted in Figure 6-10b. In this case the most soluble mixture, having the eutectic composition E, is not the racemate.

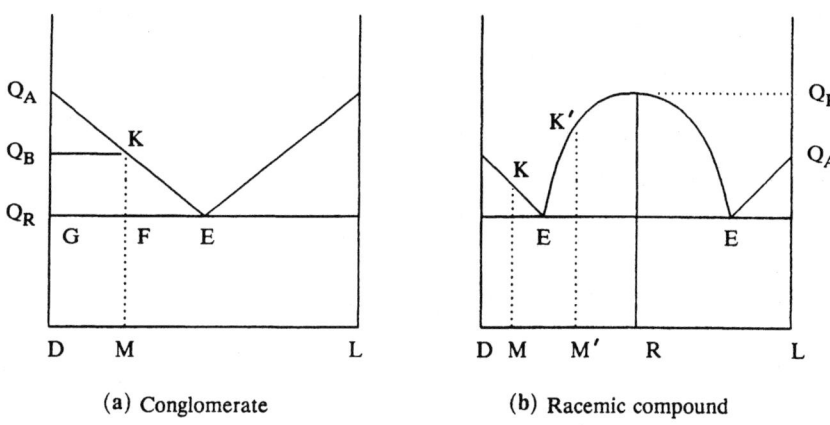

(a) Conglomerate   **(b)** Racemic compound

Figure 6-10. Phase diagrams for a racemic compound and a conglomerate.

Consequently, recrystallization of a partially resolved mixture, under equilibrium conditions, will yield either a pure enantiomer or the racemate depending on its composition. Thus, mixtures of composition M and M′ will afford the pure D-enantiomer and racemate, respectively. In order to know whether a partially resolved racemic compound can be further enriched, it is necessary to know the exact shape of the phase diagram.

Having shown that the feasibility of enrichment of a partially resolved mixture is dependent on the shape of the phase diagram, let us now consider the effect of the latter on the yield of this purification process. Three typical situations are illustrated in Figure 6-11 for a sample with an *ee* of 80%. For a conglomerate (Figure 6-11a) the theoretical yield is given by EF/EG = 80%, so in principle, one can collect the total amount of enantiomer present in excess. For a racemic compound, this figure drops rapidly as the composition of the eutectic moves towards the edges of the diagram. Thus, for a eutectic composition corresponding to 0.6 (Figure 6-11b) and 0.85 (Figure 6-11c) the maximum yields (given by EF/EG) are 75% and 30%, respectively.

A few examples of the rational application of the above principles to the preparation of optically active substances have been described in the literature [2]. Just as in the case of racemate resolutions, the formation of salts with achiral acids or bases can be have a beneficial effect, by altering the shape of the phase diagram. For example, 2-phenoxypropionic acid is a racemic compound with an unfavorable eutectic (Figure 6-12). Its *n*-propylamine salt, on the other hand, has a more favorable eutectic and mixtures with greater than 37% *ee* can be readily purified by crystallization [20]. Such partially resolved mixtures are obtainable from other techniques, for example, diastereomeric salt crystallization or enzymatic hydrolysis.

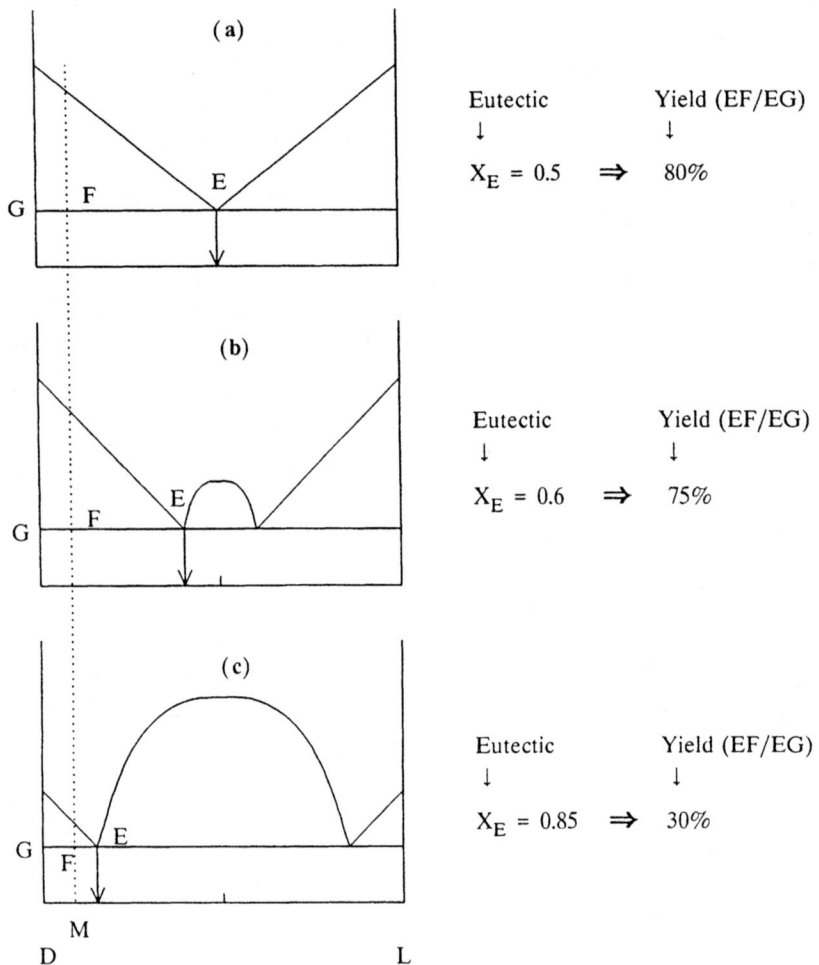

Figure 6-11. Influence of the shape of the phase diagram on the maximum theoretical yield of pure enantiomer obtainable from a partially resolved mixture.

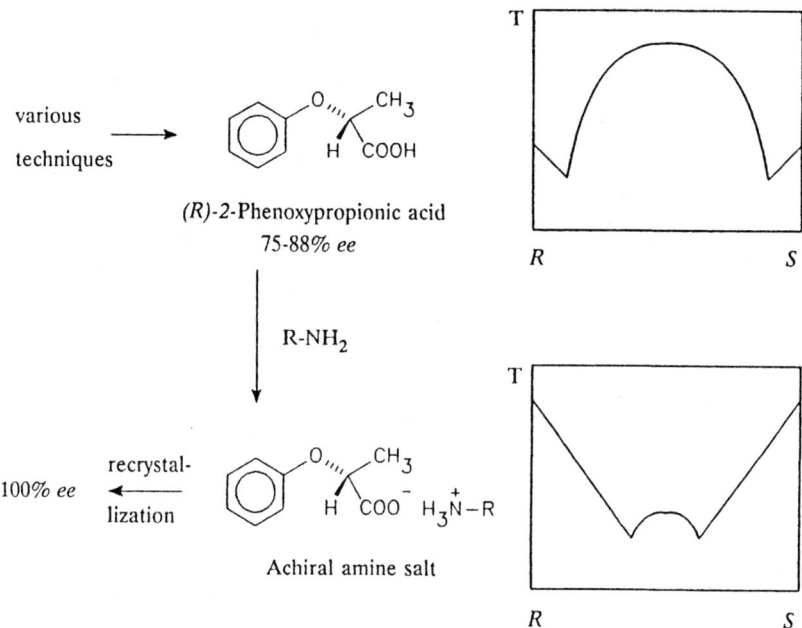

Figure 6-12. Strategy for the purification of partially resolved 2-propionic acid [20].

III. RESOLUTION BY DIASTEREOMER CRYSTALLIZATION

Derivatization of a racemate by reaction with an optically pure compound yields a mixture of diastereomers whose physical properties differ. Such mixtures are, therefore, amenable to separation by physical methods such as crystallization. The most attractive method involves the formation of diastereomeric salts between acids and bases and, in this case, the resolving agent can be readily recovered and recycled. This method is often referred to as **classical resolution**. Although it was first practiced by Pasteur more than a hundred years ago, and despite its low-tech image, it is still the most widely used resolution method. Through an improved understanding of the underlying principles that govern such processes and combined with an effective modus operandi [2], it is now possible to perform such resolutions rationally and with a high probability of success. We shall now discuss these underlying principles in more detail.

A.   Types of Diastereomeric Mixtures

When a racemic acid A is combined with an optically pure base B a mixture of two diastereomeric salts is formed (Reaction 6-2):

$$(dl) - A \quad + \quad (l) - B \quad \Rightarrow \quad (d)\text{-}A\ (l)\text{-}B \quad + \quad (l)\text{-}A\ (l)\text{-}B$$

$$\underset{\text{Racemate}}{} \qquad \underset{\text{Resolving agent}}{} \qquad \underset{\text{Diastereomers}}{\overset{\textbf{n salt} \qquad\qquad \textbf{p salt}}{}} \qquad (6\text{-}2)$$

The letter **p** (positive) is used to designate the diastereomer resulting from the two enantiomers with the same sign of rotation and **n** (negative) those of unlike signs [1,2]. In this convention, no account is taken of the absolute configuration of the asymmetric centers.

Analogous to the direct crystallization of enantiomeric mixtures, a knowledge of the shape of the phase diagrams of diastereomeric mixtures is an essential guide for their successful resolution. In contrast with racemates, diastereomers usually form eutectic mixtures or solid solutions rather than double salts (the equivalent of racemic compounds). Just as with racemates the maximum yield is determined by the position of the eutectic (Figure 6-13).

The maximum yield of pure diastereomer obtainable from crystallization is given by ME/PE (Figure 6-13) and approaches the theoretical maximum of 50% when the eutectic (E) is close to one of the pure components. The diastereomeric mixture (Figure 6-13a) in Figure 6-13 has an unfavorable eutectic, while Figure 6-13b has a favorable one. Furthermore, it should be noted that solid solutions exhibiting a large solubility difference between the **p** and **n** salts may also be amenable to resolution [3]. The feasibility of resolutions of diastereomeric mixtures can also be predicted by reference to their ternary phase diagrams with a particular solvent (Figure 6-14).

A concentrated solution (given by A) of an equimolar mixture of **p** and **n** salts (labeled M) deposits a mixture of composition A_S. Since the solid deposited is not yet pure, the remaining mother liquor must have a composition corresponding to the

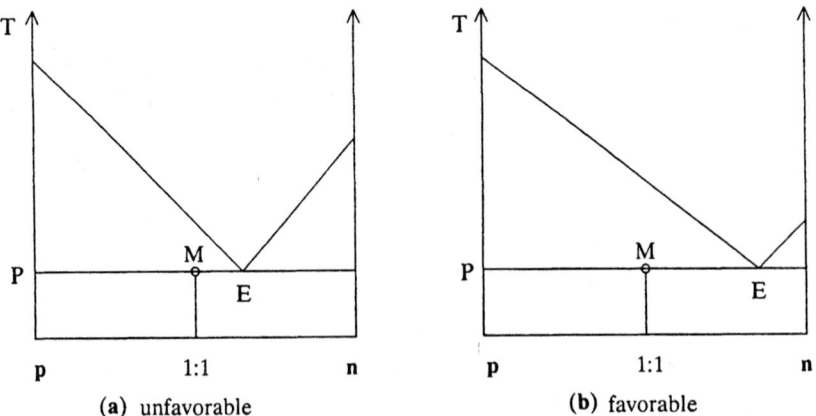

Figure 6-13. Phase diagrams for diastereomeric mixtures exhibiting a eutectic.

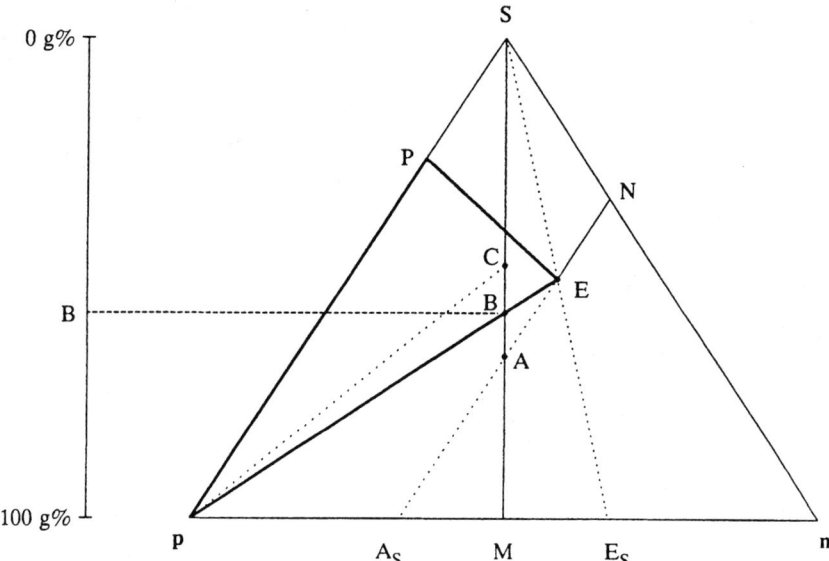

Figure 6-14. Ternary phase diagram of a diastereomeric mixture (N and P represent the solubilities of the pure **n** and **p** salts).

eutectic E. The enantiomeric composition is then given by E_S. A more dilute solution (C) deposits crystals of pure **p** salt. The maximum yield of P is obtained when the composition of the mixture is B, at the point at which the line connecting **p** with E dissects the line M-S. In all cases, the solution is unsaturated above the isotherm PEN.

Crystallization of a mixture having a composition within the triangle described by PEp always results in isolation of the pure **p** salt. The maximum theoretical yield of pure **p** salt obtainable from a 1:1 mixture is given by the formula:

$$R_{max.} = \frac{0.5 - ES}{1 - ES}$$

In other words, the closer E_S is to the vertex **n**, the higher the maximum yield.

Rational strategies for the systematic optimalization of classical resolutions have been described elsewhere [1–3] and the reader should refer to these texts for details. In industrial practice, a chemical yield of > 40% of material of high enantiomeric purity (> 95%) that is obtainable in a single crystallization is usually a requirement for an economically viable resolution. Unfortunately, no adequate theory exists that can predict the solubility of any organic compound, let alone differences between two diastereomeric salts (but see section III-C) and the selection of a resolving agent still belongs in the realm of trial and error, with only experience to act as a guide.

B.   Sources of Resolving Agents

When embarking on a search for an efficient classical resolution, one is initially faced with the question of where to look for a suitable and available resolving agent. One obvious source is the chirality pool, that is, the various acids and bases that occur naturally and/or are manufactured on a large scale by fermentation. In addition, several synthetic resolving agents are readily available (e.g., α-methylbenzylamine). Examples of several commonly used resolving agents, both natural and synthetic are shown in Figures 6-15 and 6-16. An advantage of synthetic resolving agents is the fact that both enantiomers are available.

Information regarding successful resolutions is generally fragmented in the literature but there are a few compilations of procedures [21,22]. The Japanese company Yamakawa has recently compiled a large number of resolutions employing mandelic acid [23] or α-methylbenzylamine [24].

NATURAL and SEMI-SYNTHETIC

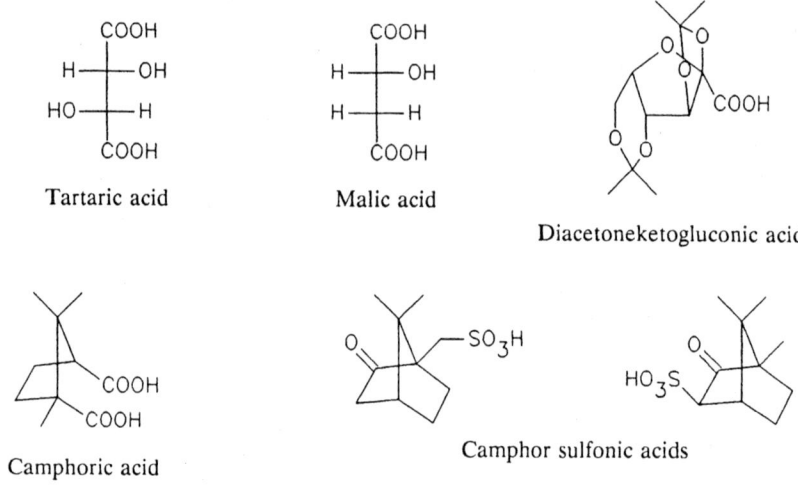

SYNTHETIC

Figure 6-15. Commonly used acidic resolving agents.

NATURAL

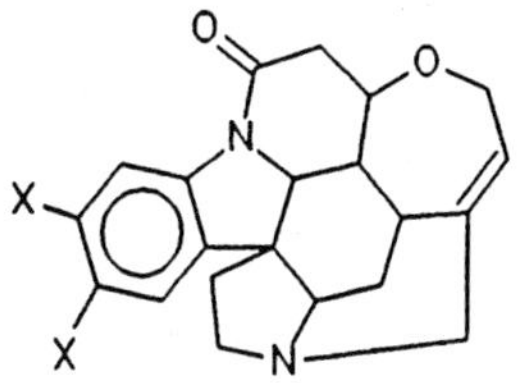

Brucine (X = OMe)
Strychnine (X = H)

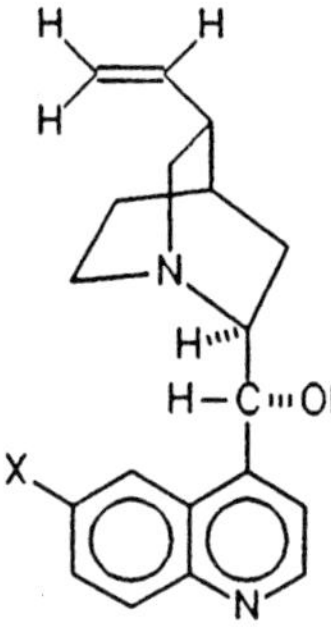

Quinine (X = OMe)
Cinchonidine (X = H)

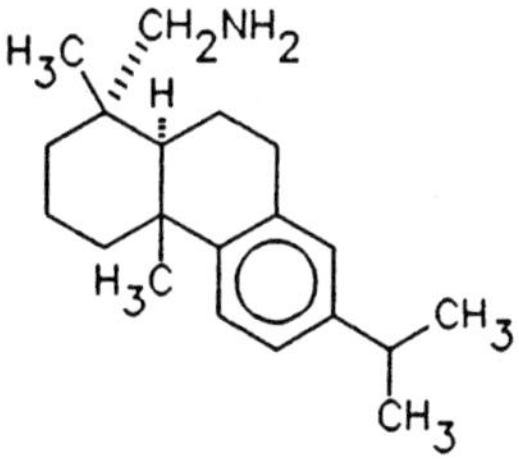

Dehydroabietylamine

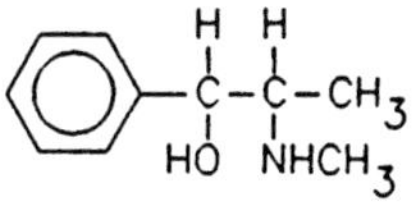

Ephedrine

SYNTHETIC

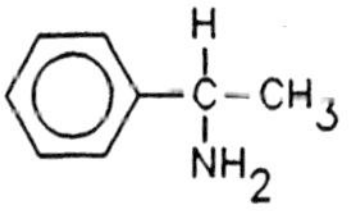

α-Methylbenzylamine

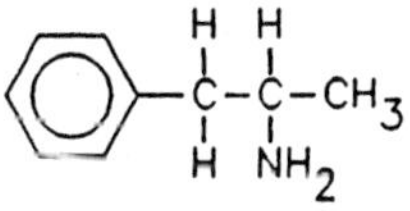

Amphetamine

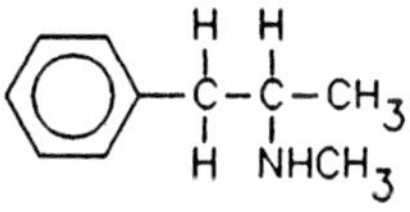

Deoxyephedrine

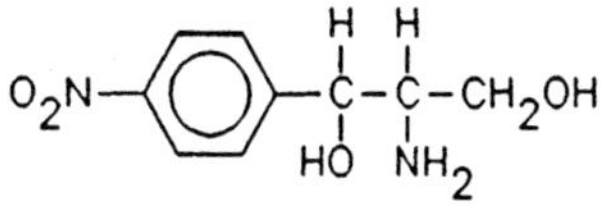

Chloramphenicol intermediate

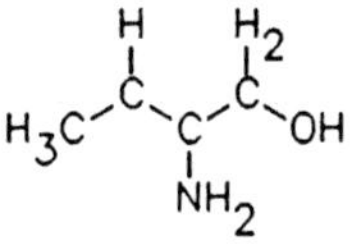

2-Amino-1-butanol

Figure 6-16. Commonly used basic resolving agents.

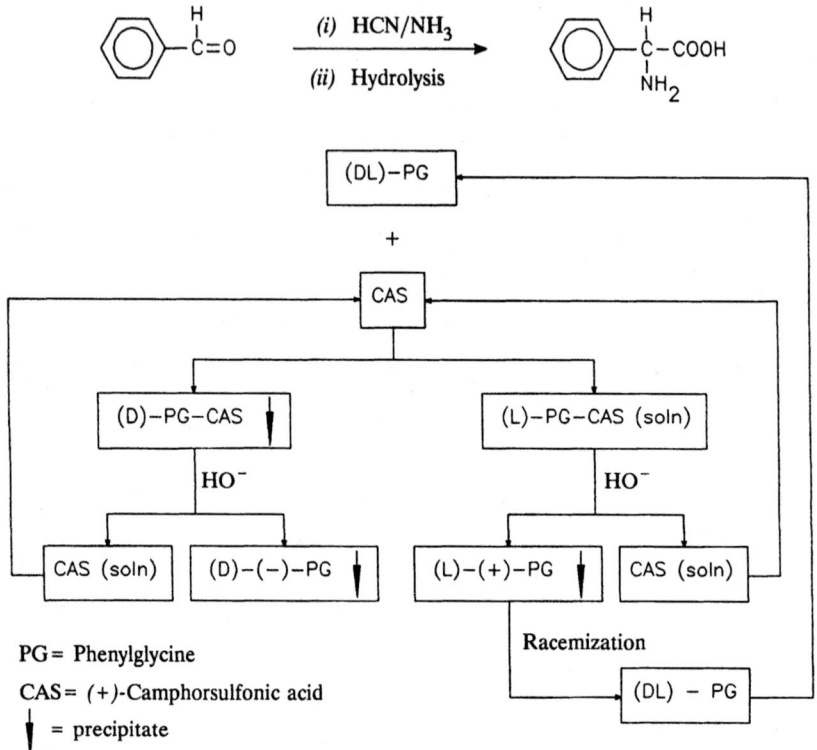

Figure 6-17. Andeno process for D-phenylglycine (D-PG).

Figure 6-18. Industrial route to *(S,S)*-ethambutol.

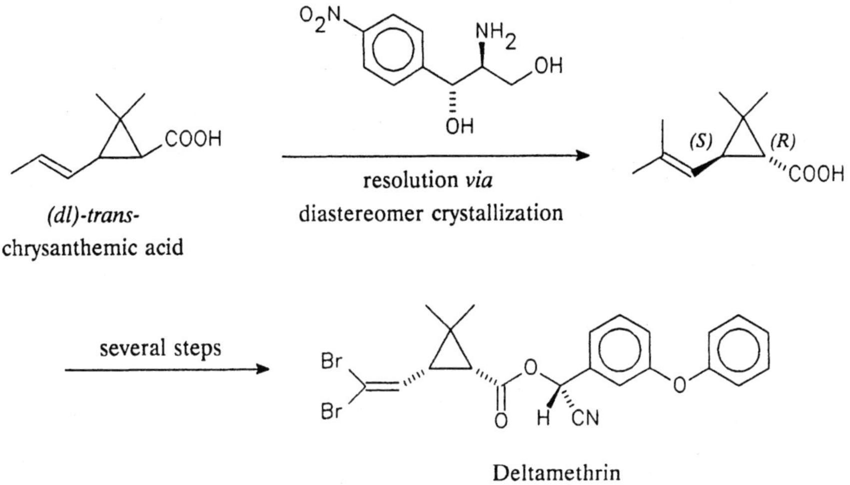

Figure 6-19. Roussel-Uclaf synthesis of deltamethrin.

A good example of an industrial process employing classical resolution is the manufacture of the important antibiotic intermediate D-phenylglycine. The latter is produced on more than a thousand tons per annum scale [25] by Andeno using optically pure camphor sulfonic acid in aqueous medium as the resolving agent (Figure 6-17); the L-isomer is racemized in a separate step.

In another example, (S)-2-aminobutanol, the key intermediate in the manufacture of the tuberculostatic drug ethambutol, is resolved with L-tartaric acid in methanol as solvent (Figure 6-18) [26]. Similarly, processes have been described for the resolution of naproxen using cinchonidine [27], dehydroabietylamine [28], and N-methyl-D-glucamine [29]. Which of these resolving agents is actually used by Syntex for the industrial production of naproxen is not clear.

Optically pure intermediates in existing large-scale industrial processes often constitute a good source of synthetic resolving agents. Sometimes a particular company may use an intermediate from one process as a resolving agent in another. A good example of this is the use of the chloramphenicol intermediate as the resolving agent in the Roussel-Uclaf synthesis [30] of the optically active pyrethroid insecticide deltamethrin (Figure 6-19).

C. Criteria for Good Resolving Agents

Although classical resolutions are not predictable there are certain empirical guidelines, based largely on experience, that can be used as criteria for selecting existing or developing new resolving agents [31]. These are:

1. The asymmetric center should be as close as possible to the functional group responsible for salt formation.
2. The diastereomeric salt should preferably have a tight, rigid structure; this usually occurs when several polar functional groups that readily form hydrogen bonds are present.
3. Strong acids and bases generally give better results than weak acids and bases.
4. The resolving agent must be chemically and optically stable under the conditions generally encountered in resolution processes (i.e., elevated temperatures at strongly acidic or basic pH).
5. Both enantiomers should be readily available; this is often an advantage of synthetic resolving agents.
6. The resolving agent should be easily recovered from the crystallization step.
7. For high productivities, the molecular weight of the resolving agent should be as low as possible; high molecular weight in combination with relatively high price is a serious economic disadvantage of the alkaloid bases, for example.

D. Designer Resolving Agents

Notwithstanding the broad range of commercially available resolving agents, there remains a definite need for better, synthetic resolving agents whose price and availability is not dependent on the precarious harvesting of natural products. Consequently, a steady stream of new, tailor-made resolving agents are being offered commercially, presumably developed by applying these criteria.

For example, a new group of 'designer resolving agents', the chiral phosphoric acids have been described [32]. They are easily prepared from readily available, inexpensive raw materials (Figure 6-20). Their pK_a's of 2–3 permit resolution of a broad range of amines and underivatized amino acids. They are highly crystalline

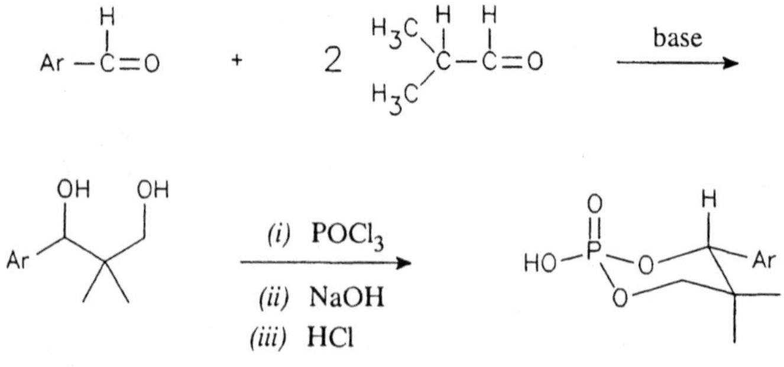

Ar = phenyl; o-chloorphenyl; $2,4$-dichloorphenyl; o-methoxyphenyl

Figure 6-20. Designer chiral phosphoric acid resolving agents.

and thermally stable under both acidic and basic conditions and have low water solubility, thus allowing for easy recovery and recycling. In short, they satisfy the optimum criteria very well. Other tailor-made resolving agents include amine (**X**) and acid (**XI**) available from Chiron and Kiralchem, respectively.

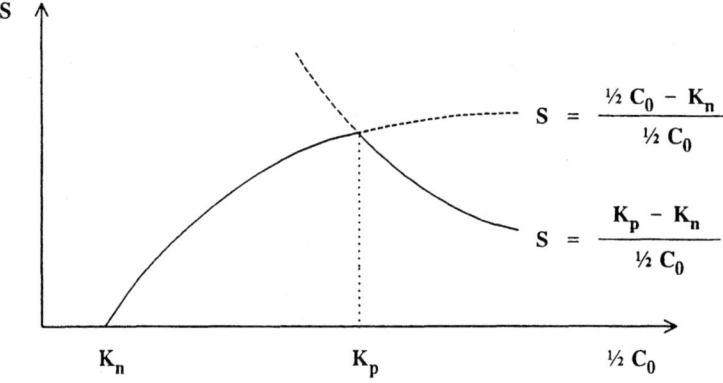

$$(X) \qquad\qquad (XI)$$

E. Resolving Capability; The Fogassy Parameter, S

The resolving capability of a resolving agent can be expressed by the parameter **S**, introduced by Fogassy [33,34]; **S** is the product of the chemical yield **K** (**K** = 1 for 50% yield) and the enantiomeric purity **t** (**t** = 1 for 100% purity) of the material. In principle, **S** cannot exceed 1 since no more than 50% yield of material of 100% *ee* can be obtained. **S** can also be related to the solubility difference of the **p** and **n** salt according to the following approximation [34]:

$$\mathbf{K} \times \mathbf{t} = \frac{\mathbf{K_p} - \mathbf{K_n}}{\frac{1}{2}\mathbf{C_0}}$$

where $\mathbf{K_p}$ and $\mathbf{K_n}$ are the solubilities of the **p** and **n** salt, respectively, and $\mathbf{C_0}$ is the initial concentration. Figure 6-21 illustrates the relationship between **S** and $\mathbf{C_0}$ for the case where the **p** salt is the more soluble diastereomer. The theoretical maximum value of **S** is attained when $\frac{1}{2}\mathbf{C_0} = \mathbf{K_p}$, that is, when saturation of the **p** salt is achieved.

S

$$S = \frac{\frac{1}{2} C_0 - K_n}{\frac{1}{2} C_0}$$

$$S = \frac{K_p - K_n}{\frac{1}{2} C_0}$$

K_n $\qquad\qquad$ K_p $\qquad\qquad$ $\frac{1}{2} C_0$

Figure 6-21. A plot of S versus $\frac{1}{2}C_0$.

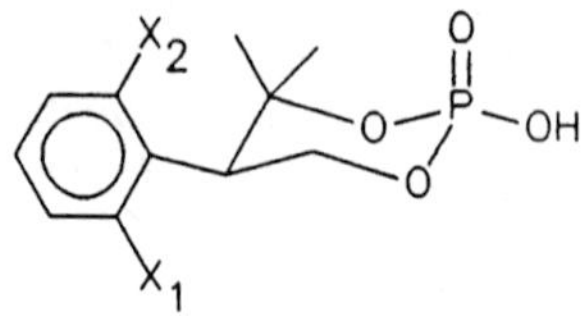

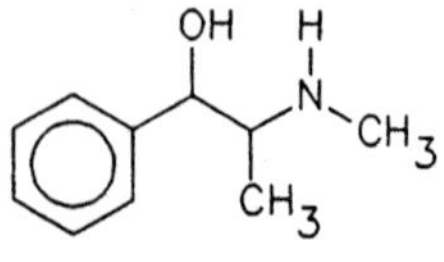

Phencyphos $(X_1 = H, X_2 = H)$
Fluocyphos $(X_1 = H, X_2 = F)$
Chlocyphos $(X_1 = H, X_2 = Cl)$
Brocyphos $(X_1 = H, X_2 = Br)$
Dichlocyphos $(X_1 = Cl, X_2 = Cl)$

Ephedrine

Figure 6-22. Resolution of ephedrine with chiral phosphoric acids.

As noted previously, the chiral phosphoric acid resolving agents were designed largely on the basis of experience and intuition. Subsequently, the same compounds have formed the basis for studies aimed at relating physical properties of diastereomeric salts to resolving capability, the ultimate goal being the rational design of resolving agents. To this end the physical properties were studied of the pairs of diastereomeric salts formed from the chiral phosphoric acids and the alkaloid ephedrine and compared with observed values of S in four different solvents (Figure 6-22) [35].

The results indicated a good correlation between the resolving capability, S and $\Delta\Delta H_f$, the difference in heats of fusion of the two diastereomers. In contrast, no correlation was observed between S and melting points or density differences. The apparent relationship between S and $\Delta\Delta H_f$ is perhaps not so surprising when one considers that the heat of fusion is the energy needed to break the network of non-bonding interactions existing in the solid state. Clearly, the extent of such interactions will play an important role in determining the solubility of a diastereomeric salt.

F. The Marckwald Principle and Reciprocal Resolutions

The idea that the two enantiomers of a resolving agent provide access to both enantiomers of a racemic substrate was first proposed by Marckwald [36]. The Marckwald principle may be applied if both enantiomers of a resolving agent are available. In this case the mother liquors of a first separation can be treated with the other enantiomer of the resolving agent (Figure 6-23).

When a racemic acid can be resolved by an optically active amine, it is frequently (but not always) the case that the racemic amine can itself be resolved by the optically active acid. This is known as the **principle of reciprocal resolu-**

(dl)- A   +   *(d)*-B   *(dl)*-A   +   *(l)*-B

      ⇓   ⇓

(d)-A *(d)*-B   +   *(l)*-A *(l)*-B   *(d)*-A *(l)*-B   +   *(l)*-A *(l)*-B

   p *(+)*   **n** *(−)*   **n** *(+)*   **p** *(-)*

    ⇓   ⇓

less soluble salt   less soluble salt

    ⇓   ⇓

   (d)-A   *(l)*-A

mirror image relationship

Figure 6-23. Resolution of an acid *(dl)*-A by *(d)*-B and *(l)*-B according to the Marckwald principle.

tions and is illustrated in Figure 6-24. This principle is often put to use in industrial resolutions since the required resolving agent *(d)*-B can be initially prepared by resolution of its racemate using optically pure product, *(d)*-A. It should be noted, however, that reciprocal systems are not mirror images of one another and that success is not guaranteed. In other words, if the diastereomer mixture p(+),n(−)

(dl)- A   +   *(d)*-B   *(dl)*-B   +   *(d)*-A

      ⇓   ⇓

(d)-A *(d)*-B   +   *(l)*-A *(d)*-B   *(d)*-B *(d)*-A   +   *(l)*-B *(d)*-A

   p *(+)*   **n** *(−)*   **p** *(+)*   **n** *(+)*

    ⇓   ⇓

less soluble salt   less soluble salt

    ⇓   ⇓

   (d)-A   *(l)*-B

no mirror image relationship

Figure 6-24. The principle of reciprocal resolutions.

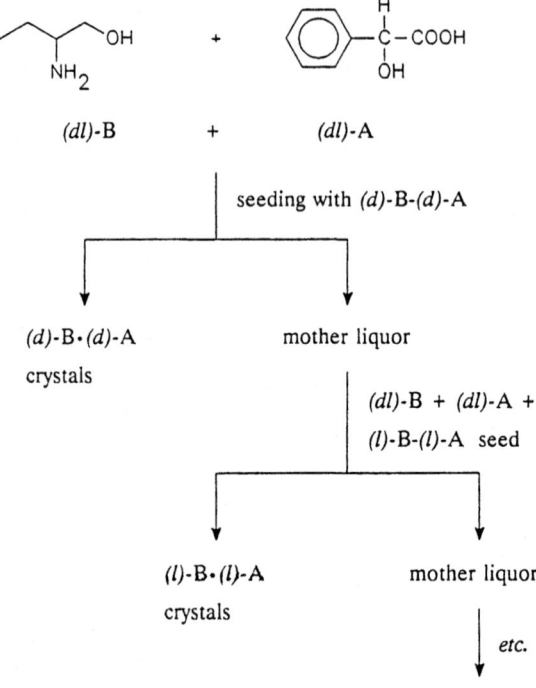

Figure 6-25. Mutual resolution of two racemates according to the Marckwald principle.

forms a eutectic, there is nothing to prevent the mixture **p(+),n(+)** from forming a double salt or solid solution. Fortunately, this does not happen very often in practice.

A further variant on the use of the Marckwald principle allows the mutual resolution of *(dl)*-A and *(dl)*-B from the same solution. It has been applied to the resolution of mixtures of 2-amino-1-butanol and mandelic acid, for example (Figure 6-25) [37].

G. Resolution with Less than One Equivalent of Resolving Agent

Ideally, in the resolution of a racemate via diastereomeric salt formation, only one of the diastereomers crystallizes from the solution. It should be possible, therefore, to accomplish the separation by using half an equivalent of resolving agent. This modification is known as 'the method of half quantities' and was first described by Marckwald [36] for the resolution of tartaric acid with cinchonidine. In practice, however, the economic optimum may be somewhere between a half and one equivalent of resolving agent.

The **method of Pope and Peachey** [38] is a variation on the above theme. It involves the use of a half-quantity of resolving agent in combination with a half-quantity of an achiral acid (or base). The method is illustrated in Reactions 6-3 and 6-4 for the resolution of a racemic amine and acid using HCl and KOH as the achiral acid and base, respectively.

$$(d)\text{-}B(l)\text{-}B + (d)\text{-}A + HCl \Rightarrow (l)\text{-}B(d)\text{-}A{\downarrow} + (d)\text{-}B{\cdot}HCl$$

(in solution)

(6-3)

$$(d)\text{-}A(l)\text{-}A + (d)\text{-}B + KOH \Rightarrow (l)\text{-}A(d)\text{-}B{\downarrow} + (d)\text{-}A^-K^+ + H_2O$$

(in solution)

(6-4)

Several examples have been described [1–3] of the successful application of this method to resolutions. Indeed, it may be widely applied in industrial practice since details of such industrial resolution processes are generally not described. A further variation on this theme consists of the direct combination of a salt, for example, the hydrochloride of the racemic base with half an equivalent of a salt of the optically active acid (Reaction 6-5).

$$(d)\text{-}B(l)\text{-}B{\cdot}2HCl + NH_4^+(d)\text{-}A^- \Rightarrow (l)\text{-}B(d)\text{-}A{\downarrow} + (d)\text{-}B{\cdot}HCl + NH_4Cl$$

(6-5)

An analogous resolution can easily be envisaged for a racemic acid.

IV. ASYMMETRIC TRANSFORMATIONS OF DIASTEREOMERIC SALTS

The theoretical once-through yield of a resolution via diastereomer crystallization is 50% and, in general, the unwanted enantiomer has to be racemized and recycled in order for the process to be economically viable. Once-through yields of > 50% are possible, however, when the crystallization is carried out under conditions where the diastereomer remaining in solution undergoes spontaneous epimerization, often referred to as **diastereomer interconversion**. The latter is the diastereomeric equivalent of racemization for a racemate. The overall process constitutes a **crystallization-induced asymmetric transformation of a diastereomeric mixture** and allows for a theoretical yield of 100%, analogous to crystallization-induced asymmetric transformation of a racemate. (See section II-D.) An industrial synthesis of dextropropoxyphene, for example, involves an asymmetric transformation of the diastereomeric salts formed from an aminoketone and dibenzoyl-L-tartaric acid (Figure 6-26) [39].

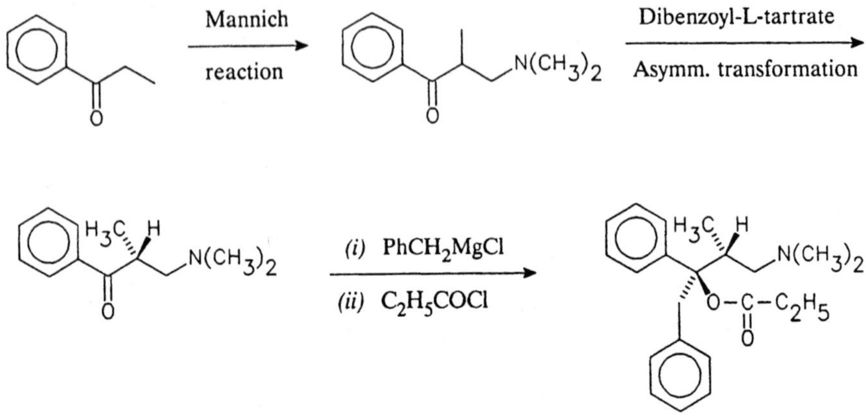

Figure 6-26. Industrial synthesis of dextropropoxyphene via an asymmetric transformation.

As mentioned previously, amines and amino acids are susceptible to race-mization via transient, reversible Schiff base formation in the presence of catalytic amounts of carbonyl compounds. Advantage can be taken of this phenomenon in designing asymmetric transformations of diastereomeric salts. Strictly speaking, they involve racemization of a small amount of free amine (or amino acid) present in the solution. An elegant example of such a process is illustrated in Figure 6-27.

In this process [40], the addition of a catalytic amount of an aldehyde promotes the in situ racemization of small amounts of the free amine via Schiff base formation. This is made possible by using less than one equivalent (92% m) of the (+)-camphorsulfonic acid (CAS) resolving agent. The desired (S)-amine crystal-lizes as its (+)-CAS salt in virtually quantitative yield (theoretical yield = 92%). This extremely efficient one-pot procedure has been used by Merck for the multikilo scale synthesis of an intermediate for a candidate cholecystokinin antagonist [40].

Similarly, catalytic amounts of aldehydes can be used to promote asymmetric transformations of diastereomeric salts of racemic amino acids. For example, the previously mentioned asymmetric transformation of DL-p-hydroxyphenylglycine (DL-HPG) salts with achiral arenesulfonic acids (see section II-D) has its direct equivalent in the asymmetric transformation of diastereomeric salts of DL-HPG with (+)-1-phenylethanesulfonic acid (PES) (Figure 6-28). The required D-HPG salt with (+)-PES was obtained in 85% yield, based on DL-HPG, when the crystallization was carried out in the presence of a catalytic amount of salicylalde-hyde and acetic acid [41].

For many proteinogenic amino acids that are produced on a large scale by fermentation, the L-isomer is often less expensive than the racemate. In such cases,

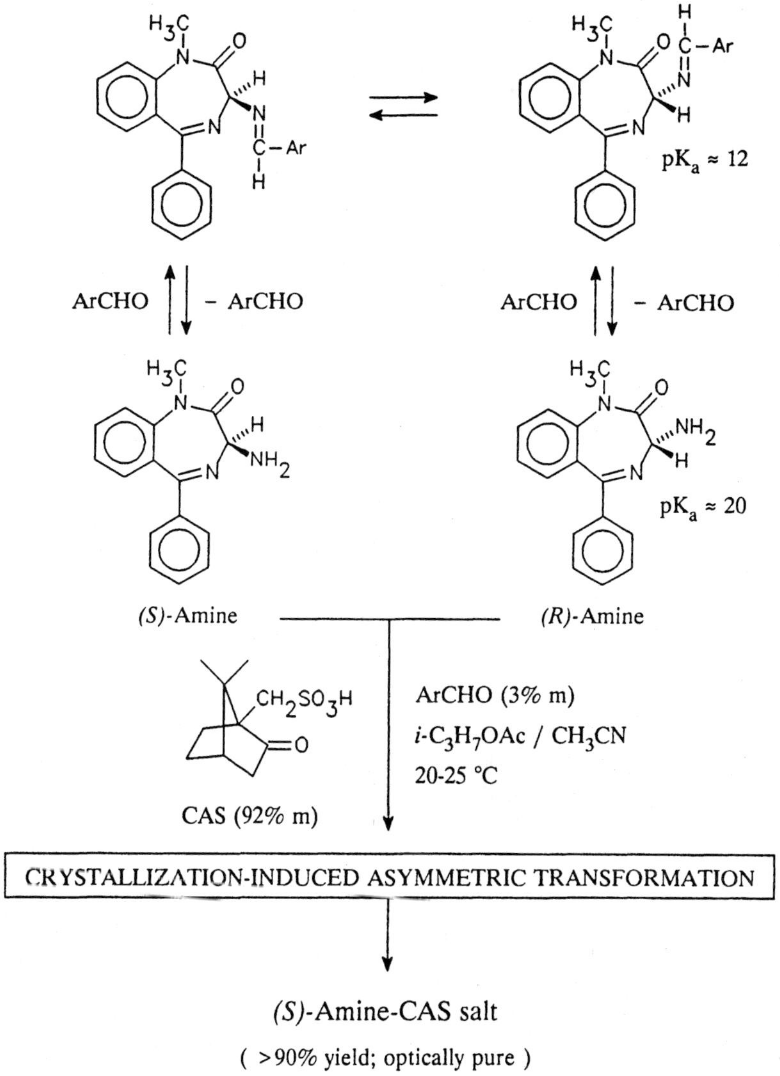

ArCHO = *3,5*-Dichlorobenzaldehyde

Figure 6-27. Crystallization-induced asymmetric transformation of diastereomeric salts via Schiff base formation.

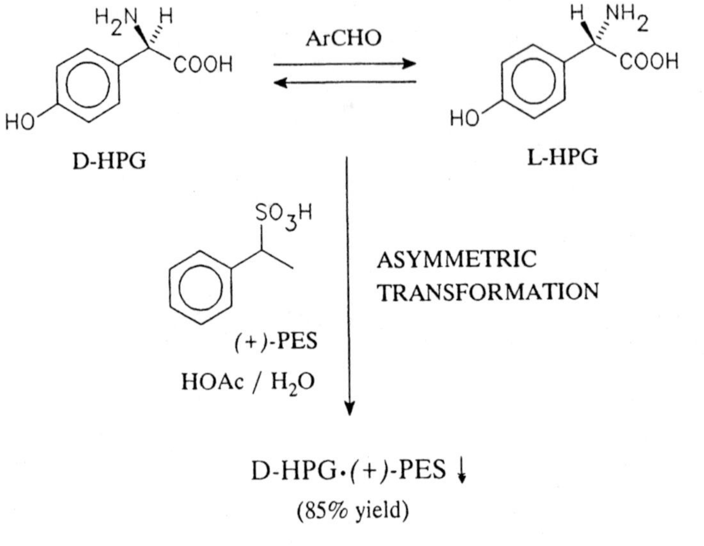

ArCHO = Salicylaldehyde

Figure 6-28. Crystallization-induced asymmetric transformation of p-hydroxyphenyl-glycine (HPG) salts with (+)-phenylethanesulfonic acid (PES).

diastereomer interconversion can be used to convert the readily available L-isomer to the D-isomer. A case in point is the synthesis of (R)= D-proline from the readily available (S)= L-proline [42]. Independent experiments showed that the L= (2S,3S)-tartaric acid salt of D-proline is less soluble than the corresponding salt of L-proline.

Consequently, heating equimolar amounts of L-tartaric acid and L-proline at 85°C in the presence of 10% m n-butyraldehyde in n-butyric acid, followed by cooling in an ice-bath, deposited crystals of the D-proline-L-tartaric acid salt in 93–95% yield and 93–95% optical purity (Reaction 6-6). Subsequent recrystallization from aqueous ethanol afforded optically pure D-proline in 85% overall yield.

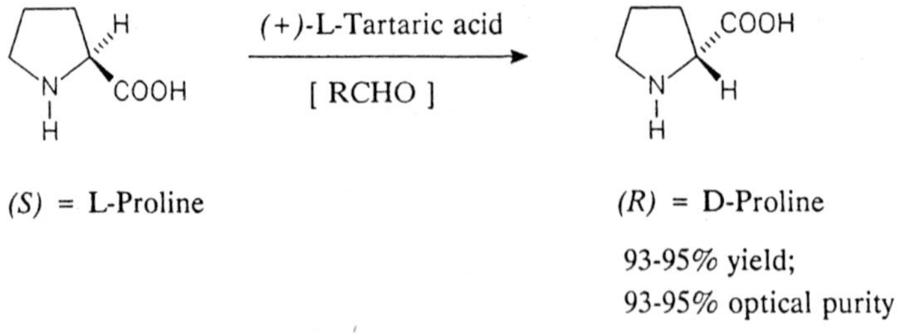

(S) = L-Proline

(R) = D-Proline

93-95% yield;
93-95% optical purity   (6-6)

Similarly, asymmetric transformations have been reported for diastereomeric salts of amino acid amides [43] and amino acid esters [44] in the presence of a catalytic amount of a carbonyl compound (Reactions 6-7 and 6-8, respectively). The efficient synthesis of the commercially important antibiotic intermediates, D-phenylglycine and D-p-hydroxyphenylglycine, by asymmetric transformation of the diastereomeric salts of the corresponding amino amides with, for example, L-mandelic acid appears to be economically attractive [43].

$$
\underset{\substack{\text{NH}_2}}{\overset{\substack{\text{H}}}{\text{Ar}-\text{C}-\text{COOEt}}} \quad \xrightarrow[\text{[PhCHO]}]{\text{L-Tartaric acid}} \quad \underset{\substack{\text{H} \quad \text{NH}_2}}{\text{Ar} \diagdown \text{COOEt}} \tag{6-7}
$$

$$
\underset{\substack{\text{NH}_2}}{\overset{\substack{\text{H}}}{\text{Ar}-\text{C}-\text{CONH}_2}} \quad \xrightarrow[\text{[PhCHO]}]{\text{L-Mandelic acid}} \quad \underset{\substack{\text{H} \quad \text{NH}_2}}{\text{Ar} \diagdown \text{CONH}_2} \tag{6-8}
$$

Ar = Phenyl, p-Hydroxyphenyl

V. THE FUTURE: RATIONAL DESIGN OF RESOLUTIONS

The chiral phosphoric acid resolving agents were designed on the basis of intuition and empirical guidelines derived from years of resolution experience. The ultimate goal is, of course, the rational design of resolution processes. In other words, being able to correlate resolving capability with molecular structure. Bearing in mind the considerable progress that has been made in the rational design of drugs and enzyme inhibitors with the aid of computer assisted modeling (CAMM) of drug-receptor interactions and enzyme inhibition, one might expect analogous studies of pairs of diastereomeric salts to provide the necessary clues. However, the design of resolving agents represents a higher order of complexity compared to drug design since the modelling of resolving agents must predict the crystal packing pattern, something that is far more complex than the modelling of drug-receptor interactions. Several studies have attempted to correlate the crystal structures of pairs of diastereomeric salts with the outcome of resolutions [35,45–49]. The overall conclusion appears to be that the insoluble diastereomer has a dramatically different crystal packing arrangement to the soluble one. For example, the crystal structures of the two pairs of diastereomeric salts formed from ephedrine and the chiral phosphoric acids (**XII**) and (**XIII**) were analyzed [48] in detail with the aim of providing an answer to the questions:

Can the observed resolution behavior be rationalized from the observed crystal structures ?

Can any relationship between resolving capability and crystal structure be expressed in a quantitative model?

Ephedrine

(XII); X = H

(XIII); X = Cl

Of the four salts, only the diastereomer formed from *(1R,2S)*-ephedrine and *(R)*-**(XIII)** was insoluble. Interestingly, the *(R)*-**(XIII)** salt did not exhibit the isostructural crystal packing pattern shown by the other three salts. The insoluble salt showed far more favorable van der Waals interactions as a result of more efficient packing. Apparently, the presence of the *ortho*-chloro substituent in **(XIII)** triggers, upon crystallization, a completely different hydrophobic packing pattern with the *(1R,2S)*-base than with *(1S,2R)*-base.

Furthermore, it was concluded that the observed [49] correlation between solubility and heat of fusion data, on the one hand, and lattice energy differences on the other, opens the way to predicting resolution efficiency (by simulation, modelling, and lattice energy calculations) in a series of related compounds. However, one cannot help feeling that these studies merely confirm, in hindsight, what was intuitively recognized, namely that the less soluble salt is the one with the higher heat of fusion and that the latter correlates with a larger number of nonbonded interactions in the crystalline state. This still does not provide the answer to the crucial question: how can the number of nonbonded interactions be predicted from a knowledge of molecular structures. Thus, an objective analysis of the current state of the art leads to the inevitable conclusion that although considerable progress has been made we are still a long way from the goal of rational design of resolutions. For the time being we shall still have to rely on our old friends, intuition and experience.

REFERENCES

1. Jacques, J., Collet, A., and Wilen, S., *Enantiomers, Racemates and Resolutions*, Wiley, New York, 1981.
2. Collet, A., in *Chiral Separations by HPLC*, Krstulovic, A.M., (Ed.), Ellis Horwood, Chichester, 1989, pp 81-104.
3. Wilen, S.H., Collet, A., and Jacques, J., *Tetrahedron*, **38**, 1-12 (1978)
4. Reinhold, D.F., Firestone, R.A., Gaines, W.A., Chemerda, J.M., and Sletzinger, M., *J. Org. Chem.*, **33**, 1209-1213 (1968).

5. Amiard, G., *Experientia*, **15**, 1-7 (1959)
6. Gernez, D., *Compt. Rend.*, **63**, 843 (1866).
7. Wakamatsu, H., *Food Eng.*, Nov 1968, p. 92.
8. a) Asai, S., *Ind. Eng. Chem. Process Des. Dev.*, **24**, 1105-1109 (1985). b) Asai, S., *Ind. Eng. Chem. Process Des. Dev.*, **22**, 429-432 (1983).
9. Yamada, S., Yamamoto, M., and Chibata, I., *J. Agr. Food Chem.*, **21**, 889-894 (1973).
10. Nohira, H., Kai, M., Nohira, M., Nishikawa, J., Hoshiko, T., and Saigo., K., *Chem. Lett.*, 951-952 (1981).
11. Nohira, H., Terunuma, D., Kobe, S., Asakura, I., Miyashita, A., and Ito, T., *Agric. Biol. Chem.*, **50**, 657-680 (1986).
12. European Patent, 298.395 to Industria Chimica Profarmaco.
13. Nohira, H., Watanabe, K., and Kurokawa, M., *Chem. Lett.*, 299-300 (1976).
14. a) Okada, Y., Takebayashi, T., Hashimoto, M., Kasuaga, S., Sato, S., and Tamura, C., *J. Chem. Soc., Chem. Commun.*, 784-785 (1983). b) Okada, Y., and Takebayashi, T., *Chem. Pharm. Bull.*, **36**, 3787-3792 (1988).
15. Pugniere, M., San Juan, C., and Previero, A., *Biotechnol. Lett.* **7**, 31-36 (1985).
16. Yamada, S., Hongo, C., Yoshioka, R., and Chibata, I., *J. Org. Chem.*, **48**, 843-846 (1983).
17. Belokon, Y.N., Tararov, V.I., Saveleva, T.F., and Belikov., V.M., *Makromol. Chem.*, **187**, 1077-1086 (1986).
18. Chibata, I., Yamada, S., Hongo, C., and Yoshioka, R., European Patent Appl., 0.070.114 (1982) to Tanabe.
19. Kral, V., Dvorak, D., Zavada, J., Stibor, I., Mostecky, J., and Votava, V., Czechoslovakian Patent, 257.094 (1989); *CA*: **112**, 35450 (1990).
20. Gabard, J., and Collet, A., *Nouv. J. Chim.*, **10**, 685 (1986).
21. Newman, P., *Optical Resolution Procedures for Chemical Compounds*, Optical Resolution Information Center, New York, 1978; covers the literature up to 1976.
22. Wilen, S.H., in *Tables of Resolving Agents and Optical Resolutions*, Eliel, E.L. (Ed.), University of Notre Dame Press, Notre Dame, Indiana, 1972.
23. *Optical Resolution of Amines with Mandelic Acid, Yamakawa Technical Bulletin No. 1*, Yamakawa Chemical Industry Co., Tokyo, Japan, 1991.
24. *Optical Resolution of Acids with α-Methylbenzylamine, Yamakawa Technical Bulletin No. 2*, Yamakawa Chemical Industry Co., Tokyo, Japan, 1991.
25. Sheldon, R. A., *Chem. Ind.*, 212-219 (1990).
26. Samant, R. S., and Chandalla, S. B., *Ind. Eng. Chem. Process Des. Dev.*, **24**, 426-429 (1985).
27. Fried, J. H., and Harrison, I. T., German Patent 2.007.177 (1970) to Syntex; *CA* 73: 120417v (1970).
28. British Patent 1.296.493 (1972) to Syntex.
29. British Patent 2.025.968 (1979) to Syntex.
30. Davies, S.G., Brown, J.M., and Fleet, G., *Chem. Britain*, **25**, 259 (1989).
31. Sheldon, R. A., Porskamp, P. A., and Ten Hoeve, W., in *Biocatalysts in Organic Synthesis*, Tramper, J., Van der Plas, H. C., and Linko, P., (Eds.), Elsevier, Amsterdam, 1985, pp. 59-80.
32. Ten Hoeve, W., and Wynberg, H., *J. Org. Chem.*, **50**, 4508-4514 (1985).
33. Fogassy, E., Lopata, A., Faigl, F., Darvas, F., Acs, M., and Toke, L., *Tetrahedron Lett.*, **21**, 647-650 (1980).

34. a) Fogassy, E., Faigl, F., Acs, M., and Grofsik, A., *J. Chem. Res. (S)*, 346-347 (1981);
    b) Fogassy, E., Faigl, F., Acs, M., and Grofsik, A., *J. Chem. Res. (M)*, 3981-3996 (1981).

35. Sheldon, R. A., Hulshof, L. A., Bruggink, A., Leusen, F. J. J., Van der Haest, A. D., and Wynberg, H., *Chemistry Today*, **9** (Jul/Aug), 23-29 (1991).

36. Marckwald, W., *Ber. Dtsch. Chem. Ges.*, **29**, 42-43 (1896).

37. Nohira, H., Fujii, H., Yajima, M., and Fujimura, R., European Patent, 0.036.265 (1986).

38. Pope, W. J., and Peachey, S. J., *J. Chem. Soc.*, **75**, 1066 (1899).

39. Pohland, A., Peters, L. R., and Sullivan, H. R., *J. Org. Chem.*, **28**, 2483 (1963).

40. Reider, P. J., Davis, P., Hughes, L. D., and Grabowski, E. J. S., *J. Org. Chem.*, 955-957 (1987).

41. Yoshioka, R., Tohyama, M., Ohtsuki, O., Yamada, S., and Chibata, I., *Bull. Chem. Soc., Japan*, **60**, 649 (1987).

42. Shiraiwa, T., Shinjo, K., and Kurokawa, H., *Chem. Letts.*, 1413-1414 (1989).

43. Boesten, W. H. J., Netherlands Patent, 90.00386 and 90.00387 (1990) to Stamicarbon.

44. Clark, J. C., Phillipps, G. H., and Steer, M. R., *J. Chem. Soc., Perkin Trans. I*, 475-481 (1976).

45. a) Gould, R. O., and Walkinshaw, M. D., *J. Am. Chem. Soc.*, **106**, 7840-7842 (1984). b) Gould, R. O., Kelly, R., and Walkinshaw, M. D., *J. Chem. Soc., Perkin Trans. II*, 847-852 (1985).

46. a) Fogassy, E., Acs, M., Faigl, F., Simon, K., Rohonczy, J., and Ecsery, Z., *J. Chem. Soc., Perkin Trans. II*, 1881-1886 (1986). b) Fogassy, E., Faigl, F., Acs, M., Simon, K., Koszda, E., Podanyi, B., Czugler, M., and Reck, G., *J. Chem. Soc., Perkin Trans. II*, 1385-1392 (1988).

47. a) Czugler, M., Csöregh, I., Kalman, A., Faigl, F., and Acs, M., *J. Mol. Struct.*, **196**, 157-170 (1989). b) Faigl, F., Simon, K., Lopata, A., Kozsda, E., Hargitai, R., Czugler, M., Acs, M., and Fogassy, E., *J. Chem. Soc., Perkin Trans. II*, 57-63 (1990).

48. Leusen, F. J. J., Bruins Slot, H. J., Noordik, J. H., Van der Haest, A. D., Wynberg, H., and Bruggink, A., *Recl. Trav. Chim., Pays-Bas*, **110**, 13-18 (1991).

49. a) Van der Haest, A. D., Wynberg, H., Leusen, F. J. J., and Bruggink, A., *Recl. Trav. Chim., Pays-Bas*, **109**, 523-528 (1990). b) See also: Leusen, F. J. J., Bruins Slot, H. J., Noordik, J. H., Van der Haest, A. D., Wynberg, H., and Bruggink, A., *Recl. Trav. Chim. Pays-Bas*, **111**, 111-118 (1992).

ADDITIONAL READING

1. Mason, S. F., *Molecular Optical Activity and Chiral Discrimination*, Cambridge University Press, Cambridge, 1982.

2. Van der Haest, A. D., *Classical Resolutions; Design of Resolving Agents and Studies of Diastereomeric Salts*, Ph.D. Thesis, University of Groningen, The Netherlands, 1992.

3. Kozma, D., Pokol, G., and Acs, M., Calculation of the Efficiency of Optical Resolutions on the Basis of the Binary Phase Diagram for the Diastereomeric Salts, *J. Chem. Soc., Perkin Trans. II*, 435-439 (1992).

7

Enzymatic Transformations

If we wish to catch up with Nature, we shall need to use the same methods as she does, and I can foresee a time in which physiological chemistry will not only make greater use of natural enzymes but will actually resort to creating synthetic ones.

Emil Fischer, 1902

Enzymes are the ubiquitous components of living cells, where they catalyze and regulate reactions of essential biochemical pathways. In common with all catalysts, they accelerate the attainment of chemical equilibria but cannot mediate a thermodynamically unfavorable reaction. For the purpose of the present discussion we restrict enzymatic transformations to processes involving either cell-free enzymes or whole cells under nonfermenting conditions, that is, where no growth of the organism occurs.

The majority of enzymatic transformations that have been used for the industrial synthesis of optically active compounds have involved hydrolytic processes which, in general, are kinetic resolutions. This contrasts with the chemocatalytic methods (see chapter 8) which are predominantly asymmetric syntheses. Consequently, we have elected to divide the material on the basis of methodology, i.e., bio- versus chemocatalysis rather than type of reaction—kinetic resolution versus asymmetric synthesis. The enzymatic processes dealt with in this chapter will be mainly (but not exclusively) kinetic resolutions.

I. ENZYME FUNDAMENTALS

A. Historical Development

The term enzyme (literally "in yeast") was coined by Kuhne [1] in 1876. A major controversy in the second half of the 19th century, associated with prominent scientists such as Pasteur and Liebig, was whether or not the process of fermentation was separable from the living cell. It was finally resolved in 1897 with the demonstration, by Buchner, that alcoholic fermentation could be sustained by a cell-free yeast extract. By the beginning of the 20th century the protein structure of enzymes was beginning to be recognized. Consensus on this point was brought closer by the first crystallization of an enzyme, urease, in 1926 by Sumner, who showed it to be a simple protein. Following the general acceptance of the protein structure of enzymes the way was clear for more precise analysis of their composition and structure. In 1965, this culminated in the first determination of the three dimensional structure of an enzyme, lysozyme, by X-ray crystallography [2]. Although the concept of enzymes as the catalysts of biological processes has been with us for well over a century their unique three-dimensional structures have been recognized for less than thirty years.

The use of the protease, chymotrypsin (CHT), to catalyze the enantioselective hydrolysis of an ester was reported in 1961 [3]. Interestingly, this involved a prochiral diester as the substrate (Reaction 7-1) thus allowing for the selective formation of one product, the *(R)*-monoester.

$$(7\text{-}1)$$

That this early example was not followed by the widespread application of enzymes in organic synthesis is largely attributable to the fact that conventional wisdom held that enzymes work only in aqueous media. Hence, it was widely assumed that enzymatic synthsis was incompatible with most of organic chemistry, which is carried out in nonaqueous media. This situation changed dramatically in the middle of the 1980s following the pioneering work of Klibanov [4] who recognized that most enzymes could function quite happily in organic solvents; this represented a quantum leap forward and it paved the way for explosive developments in the application of enzymes in enantioselective organic synthesis [6–32]. Add to this the revolutionary developments in both recombinant-DNA technology (microbial engineering) which allow for the production of any enzyme in industrial quantities, and in protein engineering to custom design enzymes, and one has the ideal ingredients for a commercially attractive technology.

B. Enzyme Nomenclature

Enzymes occur in myriad forms and catalyze a broad range of chemical reactions. They are classified according to the International Enzyme Commission into six main classes, according to the type of reaction catalyzed (Table 7-1). A given enzyme is assigned a code (the E.C. number), that consists of four numbers. The first indicates to which of the six classes the enzyme belongs. The second denotes the subclass, the third the sub-subclass and the fourth the serial number of the enzyme in its sub-subclass. For example, esterases belong to class 3 (hydrolases), subclass 1 (acting on ester bonds). They are further divided into sub-subclasses on the basis of the type of ester, for example, carboxylic esters (3.1.1) and thiolesters (3.1.2). The fourth number then denotes the specific type of carboxylic ester being hydrolysed. Lipases, for example, catalyze the hydrolysis of fatty acid triglycerides and are denoted by E.C. 3.1.1.3.

Large scale applications of enzymes in biotransformations have been largely confined to the relatively simple hydrolases and, to a lesser extent, lyases and isomerases, that do not require stoichiometric quantities of cofactors such as NAD(P)H or ATP. Because of their cofactor dependency oxidoreductases are generally used as whole cells rather than cell-free enzymes. Transferases and ligases generally require complex cofactors and have found little use in organic synthesis, although the latter are widely used in genetic engineering.

Table 7-1 The Six Major Classes of Enzymes

Enzyme class [*]	Reactions catalyzed
1. Oxidoreductases (650)	Redox reactions
2. Transferases (720)	Group transfer reactions (e.g., methyl, acyl, glycosyl, phosphate)
3. Hydrolases (636)	Hydrolysis of various functional groups
4. Lyases (255)	Additions to or formation of C=C, C=O, C=N bonds, etc., by elimination (e.g., decarboylase, dehydratase, aldolase)
5. Isomerase (120)	Structural and geometric rearrangements (e.g., racemase, epimerase)
6. Ligases (83)	Formation of C–O, C–S, C–N, and C–C bonds with consumption of ATP)

[*] Numbers in parentheses are known numbers according to the IUPAC compilation

C. Characteristic Features

A characteristic feature of an enzyme is its amino acid sequence, generally referred to as its **primary structure**. Methods of amino acid sequence analysis are now so well developed that determination of the primary structure of an enzyme is a relatively routine task. The **secondary structure** refers to the way in which the linear polypeptide chain coils into a more stable helical configuration, with the adjacent turns held in position by hydrogen bonds. The active site of an enzyme consists of a constellation of amino acid residues brought together spatially from different parts of the polypeptide chain. A given enzyme possesses a unique three dimensional structure that is produced by folding of the polypeptide chain. This so-called **tertiary structure** is essential for the catalytic activity of the enzyme. The primary structure is relatively stable but loss of the tertiary structure can readily occur by **denaturation** as a result of excessive heat, extreme pH, and the presence of organic solvents or detergents that disrupt hydrogen bonds. Hence, generally speaking, enzymes are more fragile than industrial (chemo)catalysts and require more care in handling.

Most, if not all, enzymes are globular proteins, that is, the molecule is highly compact. In many cases, several polypeptide chains bind together to form larger oligomers, referred to as the **quaternary structure**. Enzymes are further divided into intracellular and extracellular categories based on whether or not they are excreted through the cell wall. Extracellular enzymes are generally much more stable, thus ensuring a relatively long half-life in the environment (soil, water, etc.).

D. Popular Misconceptions

The widespread application of enzymes in organic synthesis has been hampered by a number of misconceptions regarding their properties as catalysts. Enzymes were considered to be highly labile catalysts, their use being restricted to a narrow pH range in aqueous solution at room temperature; these are conditions that are incompatible with most of organic chemistry. Moreover, a legacy of Fischer's 'lock-and-key' concept of enzyme catalysis (see chapter 2) was the perception that enzymes exhibit very narrow substrate specificity, that is, each substrate needs a different enzyme. Similarly, it was generally perceived that only one isomer (the 'natural' isomer) was amenable to production using enzymes and that the reaction was completely stereospecific. Other misconceptions prevalent among organic chemists are: enzymatic processes are carried out at high dilutions (no more so than with many chemocatalysts), work-up is difficult (not really a problem) and catalysts vary from batch to batch (so do chemocatalysts).

Fortunately, these misconceptions are now largely obsolete and it is widely accepted that most enzymes exhibit a broad substrate specificity and are more robust catalysts than was generally thought. Moreover, the identification of a wide range of 'extremophiles' (microbes capable of existing under relatively extreme conditions of temperature, pressure, pH, and salinity) and thermostable enzymes has significantly extended the scope of enzymes.

E. Advantages and Limitations

The main advantages and limitations of enzymes for use in industrial organic synthesis are summarized in Table 7-2. Although they are considerably more versatile and flexible than was previously imagined, their limited operational stability and flexibility with regard to reaction conditions (relative to most industrial catalysts) remain major drawbacks. Other major limitations are the relatively limited commercial availability and high price of most enzymes. Thus, of the more than 2500 known enzymes only about 250 have been commercially exploited and, of these, about 25 account for more than 80% of all applications. Indeed, the fact that so much has been achieved, when only such a small fraction of the known enzymes has been exploited, gives an idea of their tremendous potential.

F. Sources, Availability, and Prices

The market for enzymes is conveniently divided into three segments [33]: industrial enzymes, therapeutic applications, and in food and clinical diagnostics. The world market for industrial enzymes was about $500 million in 1990, of which 75% is accounted for by three major applications: detergents, dairy products (mainly cheesemaking), and starch processing (Table 7-3).

When one considers the relative contribution of the various applications to the total sales of enzymes, it is not surprising that enzyme manufacturers devote much more effort to developing better enzymes (e.g., by protein engineering) for detergents and cheesemaking than for organic synthesis. Nevertheless, applications in organic synthesis can benefit from the development of improved enzymes for other industries. Indeed, many of the enzymes currently used for enantioselective organic

Table 7-2 Advantages and Limitations of Enzymes

Advantages	Limitations
Very efficient catalysis	Availability and price
Mild conditions	Operational stability and lack of flexibility
No organic solvent needed	Substrate and/or product inhibition
Wide range of activity	Not all types of reaction accessible
High substrate selectivity, chemoselectivity, regioselectivity, and stereoselectivity	Cofactor regeneration sometimes needed
Often fewer steps	
Inexpensive source of complex chiral ligands	

Table 7-3 Industrial Enzymes

Enzyme type/ Conversion	Sales (1990) ($ million)	Main applications
1. Proteases (Protein hydrolysis)		
Alkaline bacterial proteases (e.g., subtilisin)	180	Detergents
Calf and microbial rennet (chymosin)	70	Cheesemaking, dairy
Plant proteases (papain, bromelain)	25	Food processing, beverages, brewing
Pancreatic proteases (trypsin, chymotrypsin)	10	Leather processing
2. Carbohydrate conversions		
Amylases (glucoamylases, α-amylases, etc.)	100	Starch processing, brewing, baking detergents
Pectinases	25	Food processing, winemaking
Glucose isomerase	50	Food and beverages
3. Lipases (fat hydrolysis)	10	Detergents, fat processing, cheesemaking, organic synthesis
4. Others	30	Various
TOTAL	500	

syntheses were developed for other applications (e.g., subtilisin for detergents and *Mucor miehei* lipase for cheesemaking). The above reasoning applies, of course, to the sales of enzymes. On the other hand, the sales of optically active pharmaceutical intermediates produced using enzymes amount to much more, but the enzyme manufacturer is primarily interested in the sales of enzymes. In this respect, the position of the enzyme manufacturer is no different from that of the conventional catalyst manufacturer. Thus, the major outlets for chemical catalysts are in emission control, oil refining, and polymer synthesis and not in the synthesis of fine organic chemicals.

The prices of major industrial enzymes (as 100% pure protein) are compared with precious metal catalysts and chiral ligands in Table 7-4. The important number is, of course, not the price of the enzyme per kilo but its price as a percentage of the product value. Fortunately, since enzymes generally exhibit a high activity, only a very small amount of enzyme is needed. For example, although chymosin (calf rennet) costs $5000/kg, the enzyme costs represent only 0.1–0.3% of the value of the cheese. Similarly, glucose isomerase costs $250/kg but one kilo of enzyme produces 50–100 tons of high fructose corn syrup and the enzyme costs represent 2–3% of product value. Thus, an important parameter is the productivity number (PN) of an enzyme, given by:

$$\text{Productivity number (PN)} = \frac{\text{amount of product (kg)}}{\text{dry wt of catalysts (kg)} \times \text{time (h)}}$$

Typical sources of enzymes are the liver (e.g., pig liver esterase and horse liver alcohol dehydrogenase) and pancreas (e.g., porcine pancreas lipase) of animals, plant extracts, and a wide variety of microorganisms. Although a few important enzymes are still obtained from animal (e.g., calf rennet) or plant (e.g., papain) sources, nowadays most industrial enzymes are of microbial origin (bacteria, yeasts, and molds).

As noted in chapter 4, microorganisms are orders of magnitude more efficient production units than plants or animals. Consequently, there is a marked trend towards the use of microorganisms for the production of all enzymes, including those originally derived from plants and animals. Using r-DNA technology, a

Table 7-4 Prices of Enzymes (1991)

Enzyme	Price ($/kg) of pure enzyme	Use	Enzyme cost % product value
Industrial enzymes			
Glucoamylase	30	Starch	1–2
α-Amylase	250	Processing	
Glucose isomerase	250	High fructose syrup	2–3
Alkaline protease	100	Detergents	1–4
Calf rennet (Chymosin)	5,000	Cheese	0.1–0.3
Microbial rennet	500		
Lipase-P	1,500	Fat hydrolysis	3–5
Penicillin acylase	2,000	6-APA manufacture	5–10
Diagnostic enzymes			
Horseradish peroxidase	300,000	Clinical analysis	?
Chloroperoxidase	> 1,000,000	Organic synthesis	?
Others			
Rabbit muscle Aldolase (RAMA)	640,000	Organic synthesis	?
Precious metal catalysts			
Platinum	16,000	Oil refining	< 5
Rhodium	160,000	Acetic acid manufacture	5–10
Chiral ligands			
BINAP	100,000	Catalytic asymmetric synthesis	5–10

Table 7-5 Whole Cells Versus Isolated Enzymes in Organic Synthesis

	Whole cell	Isolated enzyme
Advantages	Cheap	Simpler equipment
	Cofactors present	Simpler work-up
		Less contamination from other enzymes
		Higher tolerance to organic solvents
Disadvantages	High dilution	Expensive
	Complex (expensive) work-up	Addition of cofactors necessary where
	Side-reactions caused by other	required
	enzymes present	

cloned gene can now be overexpressed in a suitable host organism to generate large quantities of any enzyme by fermentation. Moreover, they can be selectively mutated using protein engineering (site-directed mutagenesis) to modify (redesign) the amino acid sequence and hence the catalytic properties of the enzyme. The only limiting factor is whether or not the potential revenues warrant the substantial development costs involved.

G. Whole Cells or Isolated Enzymes?

Any organic chemist embarking on the application of enzymes in (enantioselective) synthesis is initially faced with the choice of whether to use whole cells, that is, the microorganism themselves, or isolated cell-free enzymes. The pros and cons of using whole cells as opposed to isolated enzymes are listed in Table 7-5. In practice, very few whole cell systems (e.g., baker's yeast *Saccharomyces cerevisiae*) are commercially available and their preparation depends on access to fermentation facilities and expertise. Beginners are advised, therefore, to use commercially available, cell-free enzymes. Whole microbial cells are often used in the case of labile, intracellular enzymes and/or cofactor-dependent enzymes.

H. Where Are Biocatalysts Useful?

The use of biocatalysts is particularly advantageous in cases where it leads to a significantly shorter route (process simplification) or high regio- and/or stereo-selectivity that is unattainable by other means. A pertinent example is the one-step tyrosinase-catalyzed synthesis of L-tyrosine from phenol, pyruvic acid, and ammonia (Figure 7-1). The reaction is completely regio- and enantioselective, and it is difficult to envisage a chemocatalytic method that could compete with this elegant synthesis. Interestingly, this synthesis is overshadowed commercially by an even

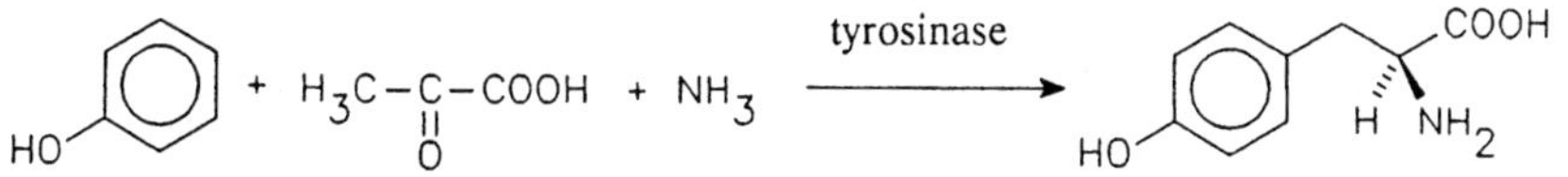

Figure 7-1. Enzymatic synthesis of L-tyrosine.

simpler route, namely the one-step synthesis of L-tyrosine by de novo fermentation of glucose. (See chapter 4.) Another relevant example is the porcine pancreas lipase (PPL)-catalyzed, regio- and enantioselective acylation of butane-1,2-diol (Reaction 7-2) reported by Klibanov and coworkers [35]. In short, a useful enzyme is one that combines high (stereo)selectivity with broad substrate specificity.

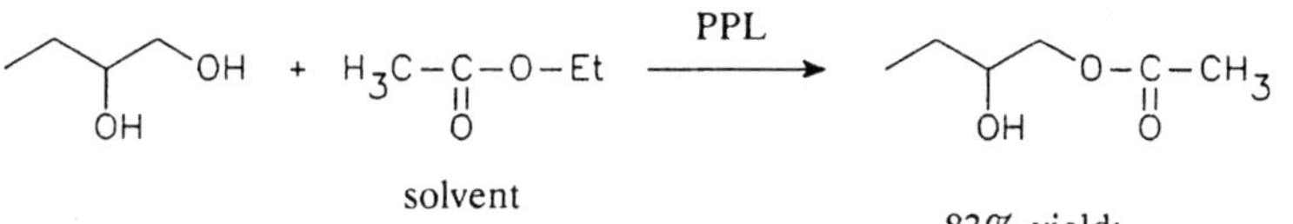

solvent

83% yield;
99% mono-acetate   (7-2)

I. Factors Governing Activity, Selectivity, and Stability

As with all catalytic processes, it is important to know which factors influence the activity. The efficiency of an enzymatic transformation is indicated by its molar activity (formerly turnover number) which is defined as the number of substrate molecules converted in one minute by one molecule of enzyme. It can be calculated from the specific activity of an enzyme if the molecular weight of the latter is known.

All enzymatic processes involve interaction between the enzyme, its substrate and its immediate environment (solvent, pH, temperature, pressure, and the presence of cofactors). All enzymes have an optimum temperature and pH range for activity, the two parameters often being interdependent. Temperature dependence exhibits an optimum at the point where thermal denaturation becomes important. Generally it is between 40–60°C, although some thermostable enzymes have optima up to 100°C.

For most enzymes the pH optimum lies in the range 5–7 but extreme values from 1.5–10.5 are known, for pepsin and alkaline phosphatase, respectively. Figure 7-2 shows some examples of activity as a function of pH.

Yields of enzymatic processes (e.g., peptide synthesis [36]) can also be influenced by changing the pressure. Advantage can be taken of the stability of many enzymes towards high pressure in order to carry out reactions in supercritical fluids. For example, cholesterol oxidase catalyzes the oxidation of cholesterol to 4-cholesten-3-one (Figure 7-3) at 123 bar and 31°C in supercritical CO_2, giving a

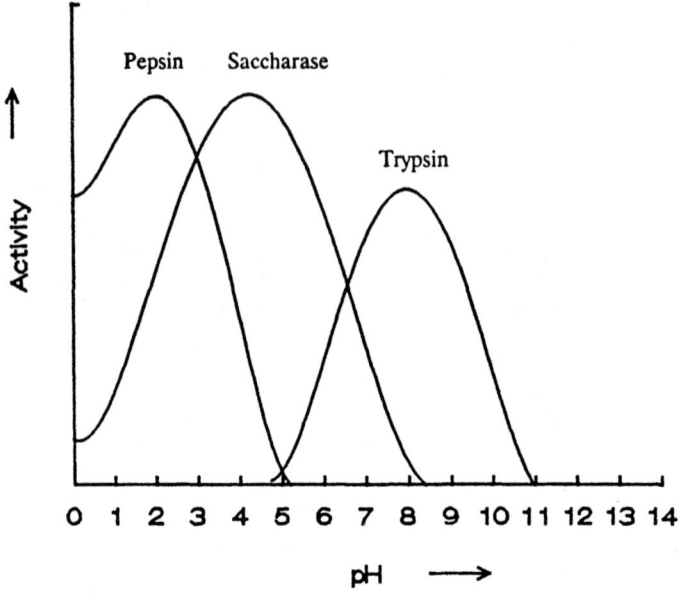

Figure 7-2. Activity of various enzymes as a function of pH.

100% conversion in one hour [37]. This procedure has several advantages: cholesterol is 50 times as soluble in supercritical CO_2 as in water; the solvent is inert, nontoxic, and nonflammable, and the enzyme, being insoluble, is easily recovered.

J. Aqueous Versus Organic Media

In common with other reactions, enzymatic processes are also influenced by the nature of solvent. Following the pioneering work of Klibanov [4,5] it is now widely accepted that many, if not all, enzymes can function in organic solvents containing

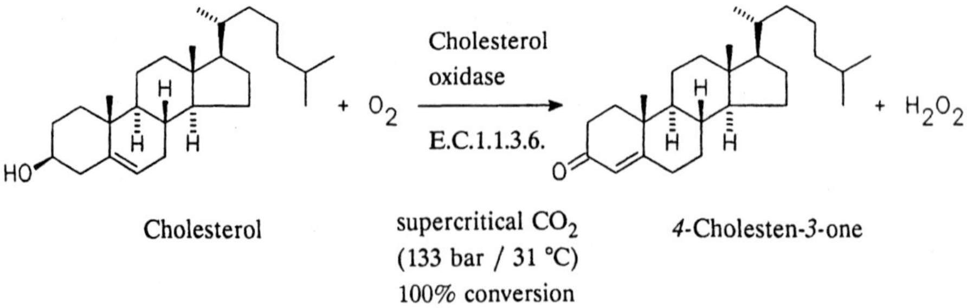

Cholesterol   Cholesterol oxidase   4-Cholesten-3-one
                    E.C.1.1.3.6.

$+ O_2$ $+ H_2O_2$

                    supercritical CO_2
                    (133 bar / 31 °C)
                    100% conversion

Figure 7-3. Enzymatic oxidation in supercritical CO_2.

very little water (< 0.1%; i.e., in nearly anhydrous media). Indeed, the nature of the solvent can dramatically effect the substrate specificity, activity, stability, and regio- and stereoselectivity of an enzymatic transformation [38–41]. Note, however, that most enzymes do not function in totally anydrous media.

Specifically, when a 'dry' enzyme powder is suspended in an organic solvent it displays acceptable activity if certain ground rules are observed [38]. Generally, hydrophobic solvents, such as hydrocarbons, should be used since they do not 'strip' the last traces of water which are essential for the functioning of the enzyme. In the case of more hydrophilic, water immiscible solvents (e.g., ethyl acetate), it is recommended to saturate them with water. Stripping of the last traces of water generally leads to the denaturation of enzymes in hydrophilic solvents. However, certain enzymes (e.g., many lipases) appear to bind the necessary water so tightly that they function in both hydrophobic and hydrophilic solvents.

For optimum activity in nonaqueous media, the enzyme powder should be freeze-dried from aqueous solution at the pH optimum for the particular enzyme. Apparently, the enzyme 'remembers' the pH of the aqueous solution it was last exposed to. In other words, the ionization state and the corresponding activity of the enzyme is retained in the solid state and in organic solvents. Similarly, a water-immiscible organic solvent should be saturated with an aqueous buffer of the optimum pH rather than in plain water, prior to adding the enzyme powder.

The use of enzymes in organic media offers several advantages (these will be discussed in more detail later):

1. Enzymes are generally more thermally stable in nonaqueous media. For example, in water PPL is inactivated virtually instantaneously at 100°C while in hydrocarbons it has a half-life of tens of hours at this temperature [5]. The increased stability in organic media can be readily understood when one considers that many thermal deactivation mechanisms require water (e.g., hydrolysis of peptide bonds and deamidation reactions) [4,5].

2. Water 'lubricates' enzyme molecules, rendering them conformationally flexible. Consequently, dehydration causes enzymes to adopt a more rigid structure. Hence, they remain properly folded in organic media even at high temperatures and are resistant to denaturation. An interesting consequence of the enhanced rigidity is a change in substrate specificity. Hydrophobic interactions often provide a driving force for enzyme-substrate binding in aqueous media, for instance, lipophilic substrates are 'expelled' from water into the active center of the enzyme. However, in nonaqueous media this driving force no longer exists which leads to completely different substrate-binding effects. An illustrative example of such a dramatic change in substrate specificity is shown in Figure 7-4. Similarly, Klibanov [39–41] found that the nature of the solvent had a marked influence on the regio- and enantioselectivities of enzymatic processes. For example, the enantiomeric ratio of the subtilisin-

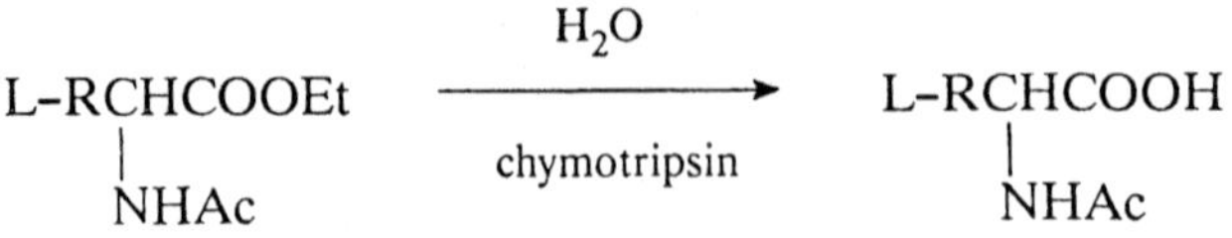

$$L-RCHCOOEt \quad \xrightarrow[\text{chymotripsin}]{H_2O} \quad L-RCHCOOH$$
$$\underset{\text{NHAc}}{|} \qquad\qquad\qquad\qquad\qquad \underset{\text{NHAc}}{|}$$

Solvent	Rel. rate phenylalanine/histidine
H_2O	> 100
n-octane	< 0.1

Figure 7-4. Dependence of substrate specificity on solvent.

catalyzed transesterification reaction (Reaction 7-3) varied from 3 in aceto-
nitrile to 61 in dioxane [39].

$$PhCH(CH_3)OH \ + \ CH_3CH_2CH_2COOCH=CH_2 \quad \longrightarrow$$

$$PhCH(CH_3)OOCCH_2CH_2CH_3 \ + \ CH_3CHO \qquad\qquad (7\text{-}3)$$

3. Another advantage of using enzymes in nonaqueous media is that it allows
    reactions to be carried out which are not feasible in water due to thermo-
    dynamically unfavorable equilibria (e.g., esterifications or peptide formation).
    Indeed, the transesterification shown in Reaction (7-3) is a perfect example of
    a transformation that is not feasible in water.
4. Last but not least, performing enzymatic transformations in organic solvents
    renders them more compatible with organic synthesis; this is important be-
    cause the majority of compounds of interest to (industrial) organic chemists
    are insoluble in water. The realization that enzymes can function perfectly well
    in organic solvents has been the single most important factor in eliminating
    the prejudice that most organic chemists had with respect to using enzymes.

K. Catalyst Immobilization

The immobilization of isolated enzymes or whole cells provides for facile recovery
and recycling of these often expensive catalysts. Moreover, immobilized biocatalysts are generally more stable and easier to handle than their 'free' counterparts
and more amenable to continuous processing [42–46]. The use of immobilized
microbial cells [44,46] circumvents the laborious and expensive isolation and
purification of intracellular enzymes. Moreover, the stability is improved by

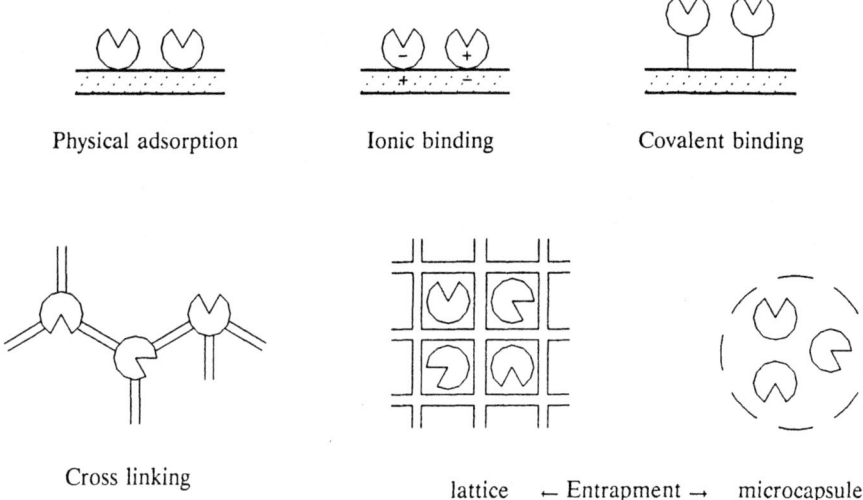

Figure 7-5. Techniques for immobilizing biocatalysts.

maintaining the enzyme in its natural environment. These advantages are particularly valid for multi-enzyme and/or cofactor dependent systems. Immobilization methods are divided into three categories: carrier binding, crosslinking, and entrapment (Figure 7-5).

Carrier binding is commonly used with free enzymes and entrapment with whole cells. It consists of attaching the biocatalyst to a water-insoluble support by physical adsorption, or by ionic or covalent binding. The most commonly used physical adsorbents are silica, alumina, activated charcoal, and various polysaccharides. Chibata and coworkers [42] were the first to apply immobilized enzymes to industrial organic synthesis. Fungal amino acylase (E.C. 3.5.1.14) was immobilized on DEAE-Sephadex (diethylamino ethylcellulose) by ionic binding. This catalyst forms the basis of the Tanabe process for the enantioselective hydrolysis of N-acylamino acids [42].

Enzymes immobilized by covalent binding are more stable towards leaching than those involving physical adsorption or ionic binding. It is achieved by reaction with amino, hydroxyl or carboxyl groups in the polypeptide chain. For example, oxirane activated acrylic beads (Eupergit C) react with free amino groups (Reaction 7-4).

$$\text{polymer}-O\diagdown\!\!\diagup^{O} + H_2N-\boxed{E} \longrightarrow \text{polymer}-O\diagdown\overset{HO}{\diagup}\diagdown\overset{H}{N}-\boxed{E}$$

$-\boxed{E}$ = Immobilized enzyme by covalent binding

$$(7\text{-}4)$$

Crosslinked enzymes are insoluble macromolecules produced by reaction of enzymes with bifunctional reagents, the most popular of which is glutaraldehyde (Reaction 7-5).

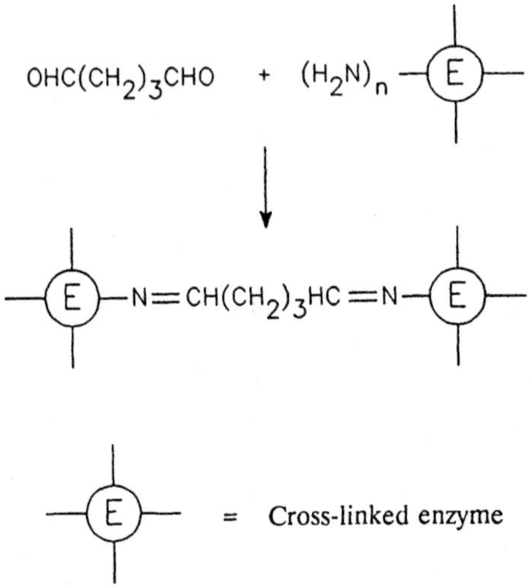

$$OHC(CH_2)_3CHO \quad + \quad (H_2N)_n - \boxed{E} -$$

$$- \boxed{E} - N = CH(CH_2)_3HC = N - \boxed{E} -$$

$$- \boxed{E} - \quad = \quad \text{Cross-linked enzyme} \qquad (7\text{-}5)$$

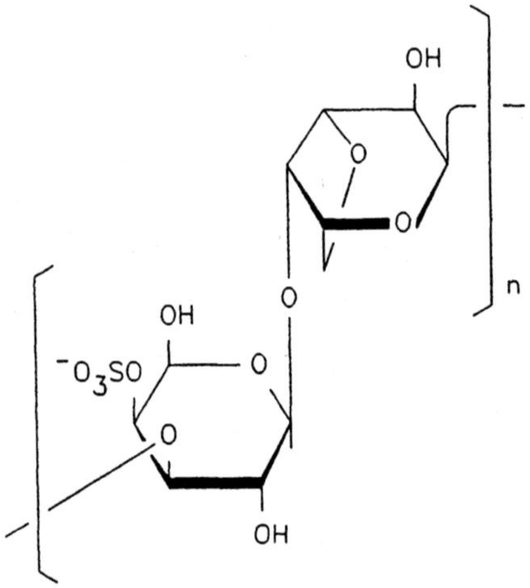

κ-Carrageenan (n = 250 - 2000)

Figure 7-6. Structure of κ-carrageenan.

Entrapment has the advantage that the biocatalyst is not subjected to structural modification and the catalyst is protected from proteases of high molecular weight. Several natural polysaccharides, such as alginate, agar, and κ-carrageenan (Figure 7-6) are excellent gel materials and are widely used for the entrapment of biocatalysts. Chibata and coworkers [44,47] screened a broad range of gel materials and found κ-carrageenan to be the best. Aqueous κ-carrageenan is mixed with the biocatalyst and dropped into a solution of the gelling agent (e.g., calcium chloride). This method has replaced entrapment in synthetic polyacrylamide gels (lattice type) in the industrial production [44] of L-aspartate and L-malate using immobilized microbial cells.

L. Membrane Bioreactors

The use of membrane bioreactors [48–50], whereby the biocatalyst is entrapped within a membrane or separated from the environment by it, is a special example of the entrapment technique.

The two different types of membrane bioreactors are illustrated in Figure 7-7. In the first type (Figure 7-7a), the reactor is divided into two compartments by a permselective membrane, possibly in the form of hollow fibers. The enzymatic hydrolysis of a racemic ester is confined to one compartment and the permselective membrane allows passage of the *(S)*-carboxylic acid and the alcohol products but rejects the *(R)*-ester. This type is employed commercially by Degussa, Akzo and Bend Research [49]. In the second type (Figure 7-7b), developed by Sepracor [50], the enzyme is physically entrapped within a microporous membrane (usually an asymmetric hollow fiber membrane). A slight pressure gradient on the outside of the membrane maintains the organic/aqueous phase boundary at the outside of the surface of the hydrophilic membrane. The enzyme is thus effectively entrapped within the microporous membrane by taking advantage of two physical properties, its large size and insolubility in organic solvents. Physical entrapment has the added advantage that once the enzyme is deactivated it can be flushed from the system, by reversing the pressure differential and back flushing, and replaced with fresh enzyme.

II. ENZYMATIC SYNTHESIS OF AMINO ACIDS

Since L-amino acids are the basic building blocks of proteins, it is perfectly appropriate that biocatalytic methods play a major role in the industrial synthesis of optically pure amino acids. Thus, both fermentation (see chapter 4) and enzymatic transformations are widely used. Since the latter generally involve hydrolytic processes we shall begin with a discussion of the various types of hydrolytic enzymes (hydrolases).

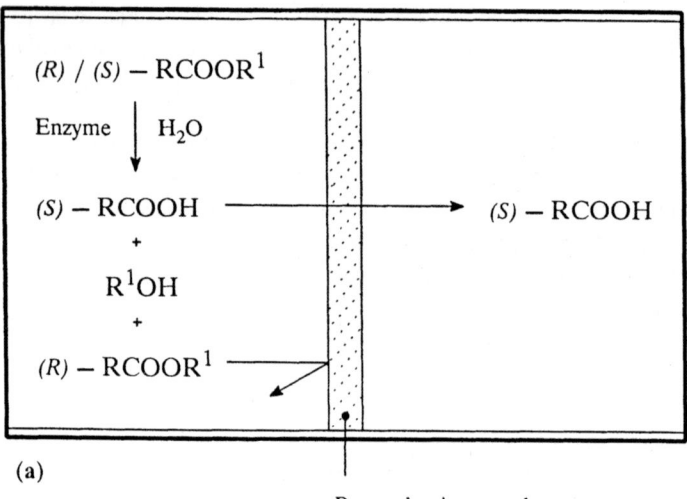

(a)

Permselective membrane

(b)

Enzyme in microporous membrane

Figure 7-7. Membrane bioreactors.

Table 7-6 Commercially Available Proteases

Enzyme class	E.C. No.	Source
1. **Serine proteases** (alkaline)	3.4.21	
Chymotripsin, trypsin		Animal pancreas
Subtilisins		*Bacillus subtilis*
		Bacillus lichenifornis
		Bacillus amyloliquefaciens
2. **Thiol proteases**	3.4.22	
Papain		Papaya
Bromelain		Pineappple
3. **Carboxyl proteases** (acid)	3.4.23	
Rennin (chymosin)		Calf stomach
Pepsin		Pig stomach
Microbial rennin		*Mucor* sp.
		Aspergillus sp.
4. **Metalloproteases** (neutral)	3.4.24	
Thermolysin		*Bacillus thermoproteolyticus*

A. Types of Hydrolytic Enzymes

Hydrolases are by far the most important class of enzymes from the viewpoint of industrial applications. They comprise a variety of subclasses, the most important of which are the proteases (polypeptide hydrolysis), amidases (amides in general), glycosidases (starch, cellulose, etc.), and lipases and esterases (carboxylic ester hydrolysis). Proteases constitute the commercially most important family of enzymes and the four major types of proteases are listed in Table 7-6. Classification of the different types is on the basis of the mechanism of action involved.

A stimulating factor in the explosive development of enzymatic methods was the realization that enzymes are able to mediate transformations quite different from those that they catalyze in nature. This is perfectly illustrated by the hydrolases. It has long been known [3] that proteases such as chymotrypsin are able to catalyze ester hydrolysis. In a complementary manner lipases have been shown to catalyze peptide bond formation.

B. Acylases

The oldest of the enzymatic kinetic resolution methods for the industrial production of amino acids is the Tanabe acylase process [42,44]. The L-specific aminoacylase (E.C. 3.5.1.4) from *Aspergillus oryzae* is used for the manufacture of several L-amino acids such as L-methionine and L-valine (Figure 7-8). The Tanabe process

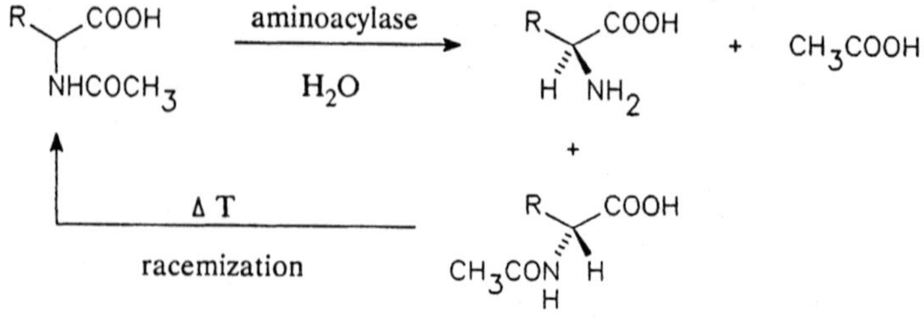

$R = (CH_3)_2CH; \ CH_3SCH_2CH_2; \ etc.$

Figure 7-8. Tanabe acylase process for amino acid resolution.

employs a packed column of the aminoacylase immobilized on DEAE-Sephadex in continuous operation. Degussa, on the other hand, employs the free acylase in a membrane bioreactor. The acylase method is attractive for L- and subtractive for D-amino acids and is used commercially only for the production of L-amino acids. The process is efficient in enzyme use and racemization of the unwanted isomer is straightforward.

A closely related enzyme is penicillin acylase (E.C. 3.5.1.11) which catalyzes the selective deacylation of Pen-G to 6-APA [51]. The enzymatic route to 6-APA (Figure 7-9) has been commercialized as an environmentally friendly alternative

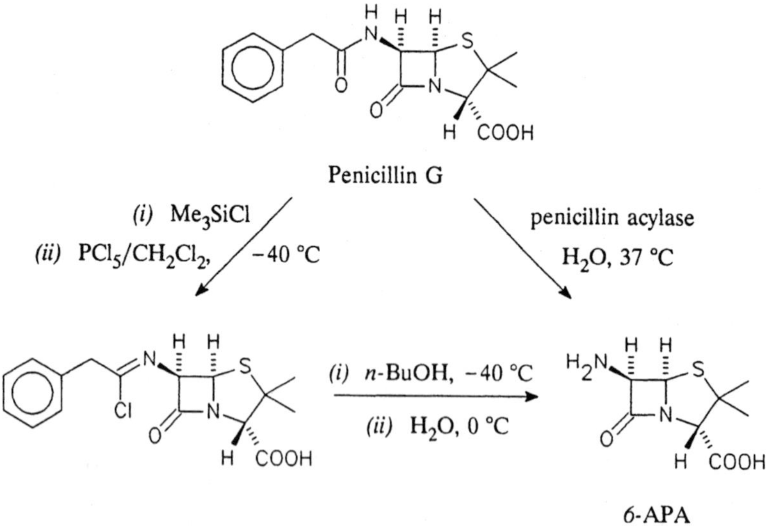

Figure 7-9. Enzymatic versus chemical deacylation of penicillin G.

to the chemical method for deacylation; it obviates the need for chlorinated hydrocarbon solvent and produces less inorganic salts. Interestingly, penicillin acylase also catalyzes the hydrolysis of phenylacetyl derivatives of simple amines, albeit in modest enantioselectivities [52].

C. Ester Hydrolysis

Enantiospecific hydrolysis of amino acid esters is catalyzed by amino acid esterase (E.C. 3.1.1.43) and a variety of proteases having esterase activity (e.g., chymotripsin, papain, and different subtilisins) [28]. A major disadvantage of the esterase method is the competing nonenzymatic hydrolysis, which results in low optical purities. Enantiospecific hydrolysis of N-acetyl amino acids [53] is catalyzed by subtilisin (E.C. 3.4.21.14) (Reaction 7-6). The method is, however, circuitous as it involves a double derivatization and is not commercially applied.

$$\text{(7-6)}$$

D.   Amino Amide Hydrolysis

The DSM process [28] for amino acid resolution involves the enantiospecific hydrolysis of racemic amino acid amides mediated by an L-specific aminopeptidase (amidase). This process was discussed in chapter 3 (Figure 3-23) and it is used for the manufacture of several D- and L-amino acids, such as D- and L-valine. The substrates are readily prepared from simple raw materials. Strecker reaction of the appropriate aldehyde with hydrogen cyanide and ammonia gives the corresponding amino nitrile. The latter is converted in high yield to the racemic amino amide by alkaline hydrolysis in the presence of a catalytic amount of acetone.

The resolution step is performed with soluble, permeabilized whole cells of *Pseudomonas putida* ATCC 12633. The method combines virtually complete enantiospecificity with a broad substrate specificity, allowing for the resolution of a wide range of amino acids (Table 7-7). A further advantage accrues from the fact that the substrate is a precursor of the racemic amino acid; in other words, it is a precursor process as opposed to a derivative process. (See chapter 3.)

An elegant, simple method is used for the separation of the L-amino acid and D-amino acid amide products. The addition of one equivalent of benzaldehyde gives a water-insoluble Schiff base of the D-amino amide which is then readily separated; subsequent acid hydrolysis yields the optically pure D-amino acid. If the L-amino acid is the desired product, the D-N-benzylidene amino amide is racemized under mild conditions; after hydrolysis, the racemic amino amide and benzaldehyde are recycled.

Table 7-7 Substrate specificity of the Aminopeptidase from *Pseudomonas putida* ATCC 12633 (Selected Examples)

$$R-\underset{\underset{NH_2}{|}}{\overset{\overset{H}{|}}{C}}-CONH_2 \xrightarrow[\text{aminopeptidase}]{\text{L-specific}} \underset{H}{\overset{R}{\underset{NH_2}{\diagdown}}}\overset{COOH}{} \quad + \quad \underset{H_2N}{\overset{R}{\diagdown}}\overset{CONH_2}{\underset{H}{}}$$

R	R
CH_3	$PhCH_2$
CH_3CH_2	$PhCH_2CH_2$
$CH_3CH_2CH_2$	Ph
$(CH_3)_2CH$	$o\text{-}ClC_6H_4$
$(CH_3)_2CHCH_2$	$p\text{-}HOC_6H_4$
$CH_2=CH_2CH_2$	$p\text{-}HOC_6H_5CH_2$
$CH_3SCH_2CH_2$	$2\text{-}C_{10}H_7$ (naphthyl)
$CH_3SCH_2CH_2CH_2$	$2\text{-}C_4H_3S$ (thienyl)

Both D- and L-amino acids are widely used as pharmaceutical and agro-chemical intermediates. L-Valine, for example, is used in the manufacture of cyclosporin (see chapter 4), while D-valine is a key raw material for the broad spectrum pyrethroid insecticide, fluvalinate, developed by Zoecon (see chapter 2). The DSM method has also been used to produce L-homophenylalanine [54], a possible precursor to a variety of ACE-inhibitor drugs such as enalapril (see chapter 2). The racemic amino amide is readily prepared from cinnamaldehyde (Figure 7-10).

Related amidases (e.g., lactamases) can be used for the enantiospecific hydrolysis of cyclic amides (i.e., lactams). For example, Enzymatix scientists have developed whole cell microbial biocatalysts that mediate the enantiospecific hydrolysis of the bicyclic γ-lactam shown in Figure 7-11 [54]. Enantiocomplementary microorganisms were discovered that produce either enantiomer in > 98% *ee*. The (−)-lactam is used as a chiral synthon for the synthesis of carbocyclic nucleosides (e.g., carbovir) that exhibit potent antiviral activity. Unfortunately, the amino acid coproduct cannot be converted to the racemic lactam which is a serious limitation to the method.

E. The Hydantoin Method

All the methods discussed above for the synthesis of α-amino acids employ L-specific enzymes and therefore involve two extra steps (i.e., they are subtractive) for D-amino acid production. However, several commercially important amino acids

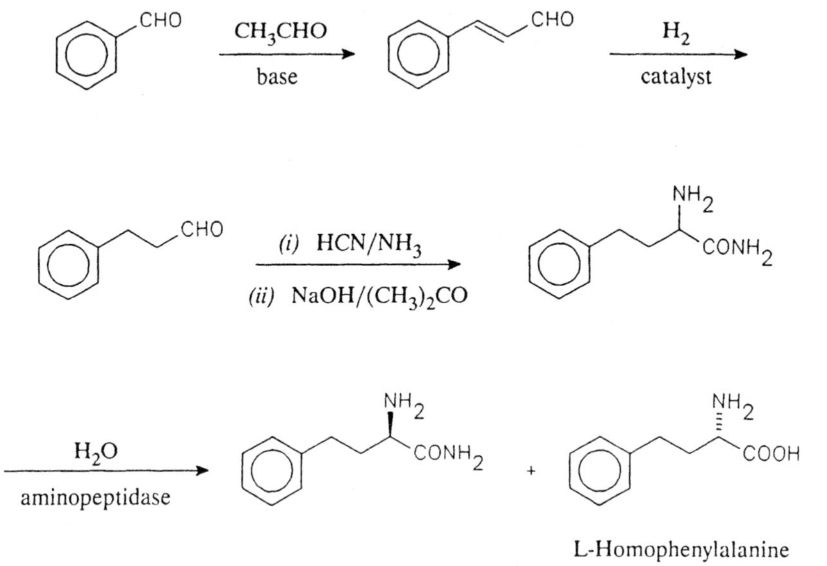

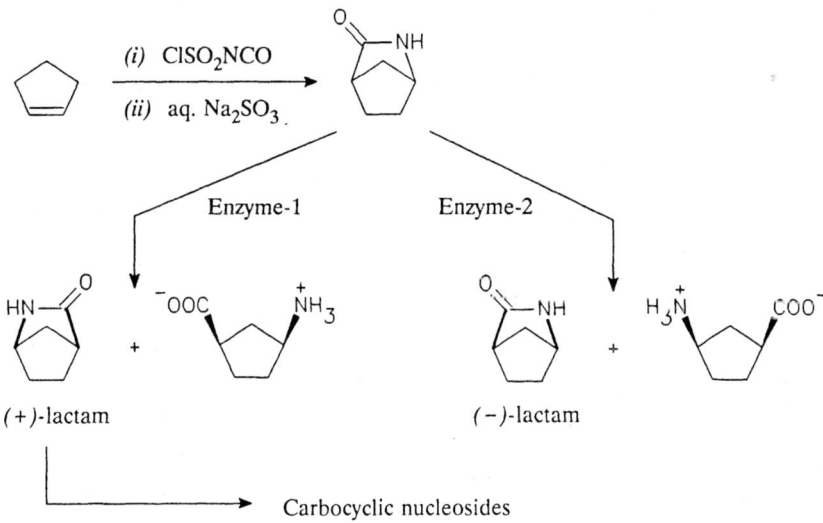

Figure 7-10. Synthesis of L-homophenylalanine.

Figure 7-11. Enantiospecific hydrolysis of a bicyclic γ-lactam.

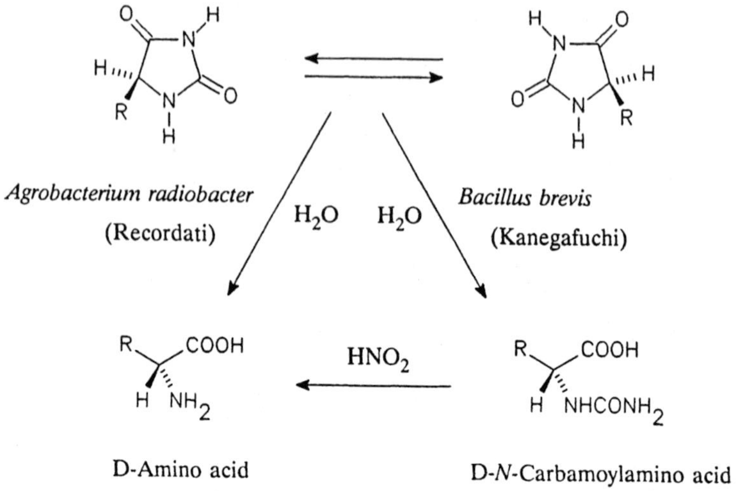

Figure 7-12. Enantiospecific hydrolysis of hydantoins.

are D-isomers, the most prominent of which are D-phenylglycine and D-p-hydroxy-phenylglycine, key raw materials for the semisynthetic penicillins, ampicillin and amoxycillin, respectively. An elegant enzymatic method that is attractive for D-amino acids, and utilizes a 'precursor' substrate is the hydantoinase (dihydro-pyrimidinase; E.C. 3.5.2.2)-mediated, D-specific hydrolysis of 5-substituted hydantoins (Figure 7-12).

The racemic hydantoins, easily prepared from the corresponding aldehydes via the Bucherer-Berg reaction, are converted by whole cells of *Bacillus brevis*, containing a D-specific hydantoinase, to a mixture of D-N-carbamoyl amino acid and L-hydantoin [55]. An additional advantage derives from the fact that the latter undergoes spontaneous racemization under the alkaline conditions, which provides for a maximum theoretical yield of 100%. The D-amino acid is obtained by subsequent treatment of the D-N-carbamoyl compound with nitrous acid. This process is operated commercially by Kanegafuchi for the production of D-p-hydroxyphenylglycine.

In an even more elegant approach, Recordati uses whole cells of *Agrobacterium radiobacter* that not only contain D-specific hydantoinase but also a second enzyme able to catalyze the hydrolysis of the D-N-carbamoyl amino acid, thus allowing for a one step conversion in a theoretical yield of 100% [56].

F. Enzymatic Racemization

In the above example, spontaneous racemization occurs under the alkaline conditions due to the acidic character of the C-H bond in the hydantoin ring. If such a

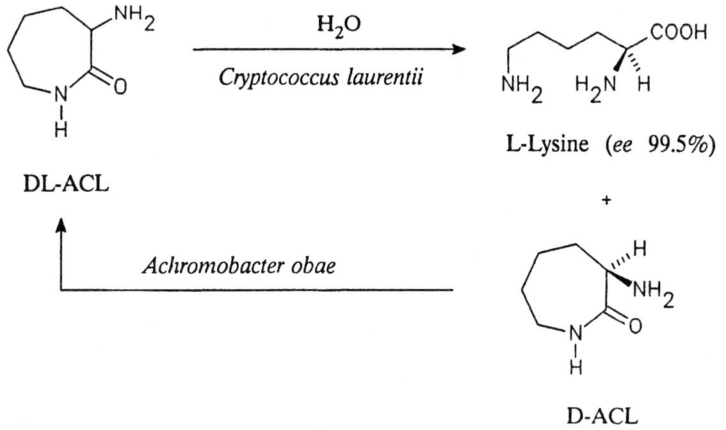

Figure 7-13. L-lysine production by enzymatic dynamic kinetic resolution.

spontaneous racemization is not feasible, one may be able to effect in situ racemization by employing a racemase. For example, the bacterium *Cryptococcus laurentii* contains an enzyme which catalyzes the enantiospecific hydrolysis of α-amino-ε-caprolactam (ACL) to L-lysine (Figure 7-13). In the presence of a second bacterium, *Achromobacter obae*, the remaining D-α-amino-ε-caprolactam undergoes an ACL racemase-catalyzed racemization [57]. A 10% solution of racemic ACL is converted to L-lysine in almost quantitative yield in 25 hours using whole cells of *C. laurentii* and *A. obae* in batch operation. The method was used by Toray in Japan to produce more than 4000 tons per annum of L-lysine but has been discontinued in favor of fermentation.

A similar process has been described for the production of L-cysteine by enantiospecific hydrolysis of DL-2-amino-2-thiazoline-4-carboxylic acid (ATC) [58]. Three enzymes are involved, all of which are present in *Pseudomonas thiazolinophilium*. Ring opening is catalyzed by an L-specific ATC hydrolase; a second enzyme converts L-S-carbamoylcysteine intermediate into L-cysteine and a third enzyme racemizes the D-ATC (Figure 7-14). A molar yield of L-cysteine of 95% is obtained at a product concentration of 31 g l⁻¹.

G.   Asymmetric Synthesis from Prochiral Precursors

Ammonia lyases catalyze the enantioselective addition of ammonia to α,β-unsaturated carboxylic acids. For example aspartic acid, a key raw material for the artificial sweetener, aspartame, is produced commercially by Tanabe from fumaric acid and ammonia (Figure 7-15). The process employs aspartase-containing whole cells of *E. coli*, immobilized by entrapment in a polyacrylamide gel, in a continu-

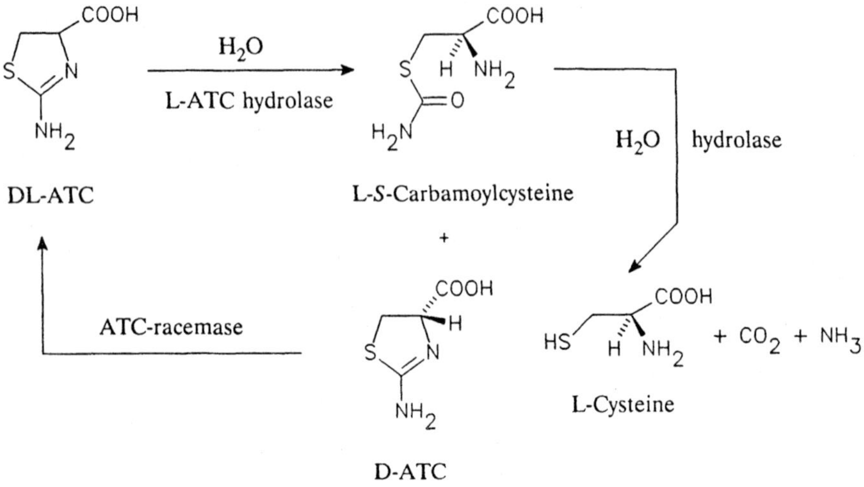

Figure 7-14. L-Cysteine production by enzymatic dynamic kinetic resolution.

ous, fixed-bed operation [59,60]. The second key raw material for aspartame, L-phenylalanine, can be produced by the analogous phenylalanine ammonia lyase-catalyzed addition of ammonia to *trans*-cinnamic acid [61]. The process was commercialized by Genex but production has been discontinued in favor of L-phenylalanine production by de novo fermentation.

Amino acid dehydrogenases catalyze the enantioselective reductive amination of α-keto acids to L-amino acids, with consumption of an equivalent of the cofactor

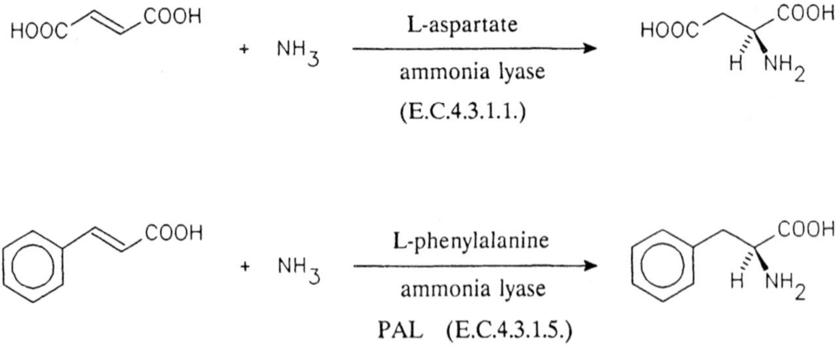

Figure 7-15. L-Amino acids by lyase-catalyzed addition of ammonia to unsaturated acids.

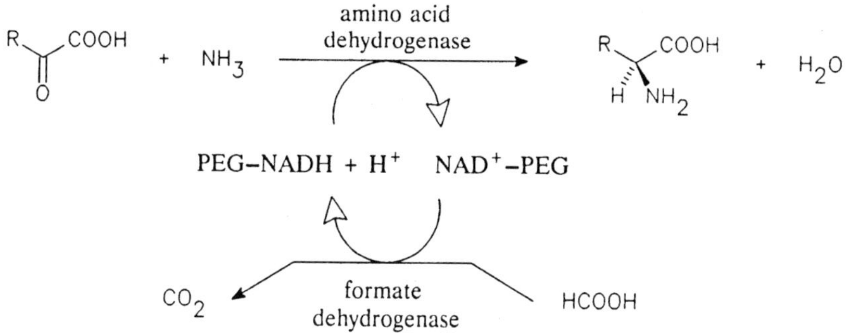

Figure 7-16. Enantioselective reductive amination of α-keto acids with in situ cofactor regeneration in a membrane bioreactor.

R	Enzyme	Product
CH_3	L-alanine dehydrogenase	L-Alanine
$(CH_3)_2CHCH_2$	L-leucine dehydrogenase	L-Leucine
$(CH_3)_2CH$	L-leucine dehydrogenase	L-Valine
$(CH_3)_3C$	L-leucine dehydrogenase	L-*tert*-Leucine
$C_6H_5CH_2$	L-phenylalanine dehydrogenase	L-Phenylalanine

NADH (Figure 7-16). Commercial exploitation of these enzymes depends on finding an economical method for regenerating the NADH and this has been achieved by employing formate dehydrogenase (E.C. 1.2.1.43) from *Candida boidinii* [62]. A second problem is separation of substrate and products from the NADH. This has been elegantly solved by chemically binding the latter to polyethylene glycol (Mw. 20,000) and immobilizing the PEG-NAD(H) conjugate on one side of an ultrafiltration membrane. Such a continuously operated membrane bioreactor has been commercialized by Degussa for the production of L-alanine from pyruvic acid.

A further variation on this theme is the coupling of an α-hydroxy acid dehydrogenase with an amino acid dehydrogenase in a membrane bioreactor, to effect conversion of an α-hydroxy acid to an L-amino acid by in situ formation of the α-keto acid [63]. Since reduction equivalents are provided by the α-hydroxy acid there is no net consumption of NADH. Thus, L-lactic acid can be converted to L-alanine by employing L-lactic acid dehydrogenase in conjunction with L-ala-

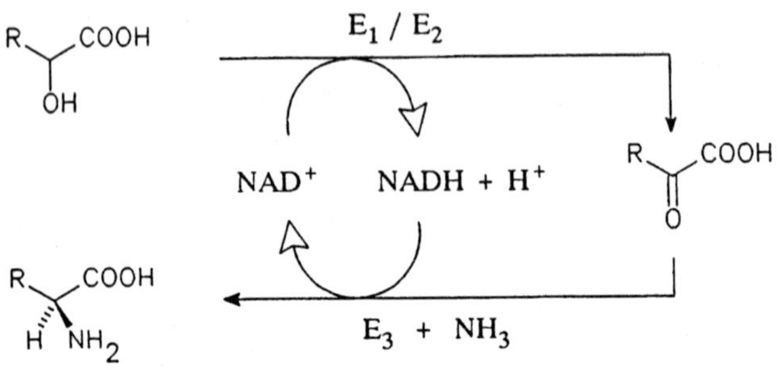

E_1 / E_2 = D- and L-hydroxy acid dehydrogenase

E_3 = L-amino acid dehydrogenase

Figure 7-17. L-Amino acids from α-hydroxy acids.

nine dehydrogenase. Similarly, racemic lactic acid can be used as the substrate by using both D- and L-lactate dehydrogenase in the same system (Figure 7-17).

H. Enzymatic Deracemization

The system illustrated in Figure 7-17 reflects what appears to be a general trend in enzymatic transformations, namely the use of coupled multienzyme systems that circumvent cofactor consumption. Another elegant example is the enzymatic deracemization of a racemic amino acid using a combination of four enzymes (Figure 7-18) [64]. Although such multienzyme systems have not yet achieved commercial status, their use in combination with membrane bioreactors would appear to have industrial potential.

I. Enzymatic C-C Coupling

Synthesis of L-amino acids via carbon-carbon bond formation is catalyzed by a variety of pyridoxal-phosphate-dependent lyases and transferases. In contrast to redox cofactors such as NAD(P)H, pyridoxal phosphate is a true cocatalyst, that is, it is not consumed in the process. One example of C-C coupling is the tyrosinase-catalyzed synthesis of L-tyrosine mentioned earlier (Figure 7-1); other commercially relevant examples are illustrated in Figure 7-19.

Using a crude extract of SHMT from *Klebsiella aerogenes*, for example, reaction of glycine with formaldehyde yields L-serine at a concentration of 450 g l^{-1} and a productivity of 9 g l^{-1} h^{-1} [65]. On the other hand, it appears to be more economical to make L-tryptophan and L-tyrosine via de novo fermentation.

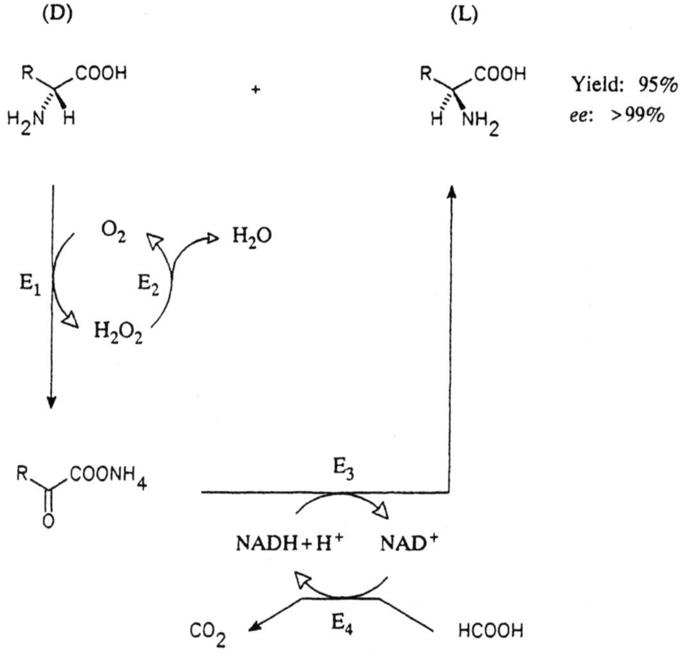

(D)   (L)

R⎯COOH   +   R⎯COOH   Yield: 95%
H₂N⎯H   H⎯NH₂   ee: >99%

E_1 O_2 $\rightarrow$ H_2O
                    E_2

                 H_2O_2

R⎯COONH₄ E_3

        NADH+H⁺   NAD⁺

        CO_2 E_4 HCOOH

Alanine   R = CH_3
Leucine   R = $(CH_3)_2CHCH_2$
Methionine R = $CH_3SCH_2CH_2$

E_1 = D-amino acid oxidase   (E.C. 1.4.3.3)
E_2 = catalase   (E.C. 1.11.1.6)
E_3 = leucine dehydrogenase   (E.C. 1.4.1.9)
E_4 = formate dehydrogenase   (E.C. 1.2.1.2)

Figure 7-18. Enzymatic deracemization of an amino acid.

J.   Optically Active Precursors

Optically active amino acids can also be made by enzymatic transformation of other optically active amino acids. For example, L-alanine is produced by decarboxylation of L-aspartic acid catalyzed by aspartate β-dacarboxylase (E.C. 4.1.1.12), a pyridoxal-phosphate-dependent enzyme. In the industrial process, operated by Tanabe [66], whole cells of *Pseudomonas dacunhae* are used to produce L-alanine

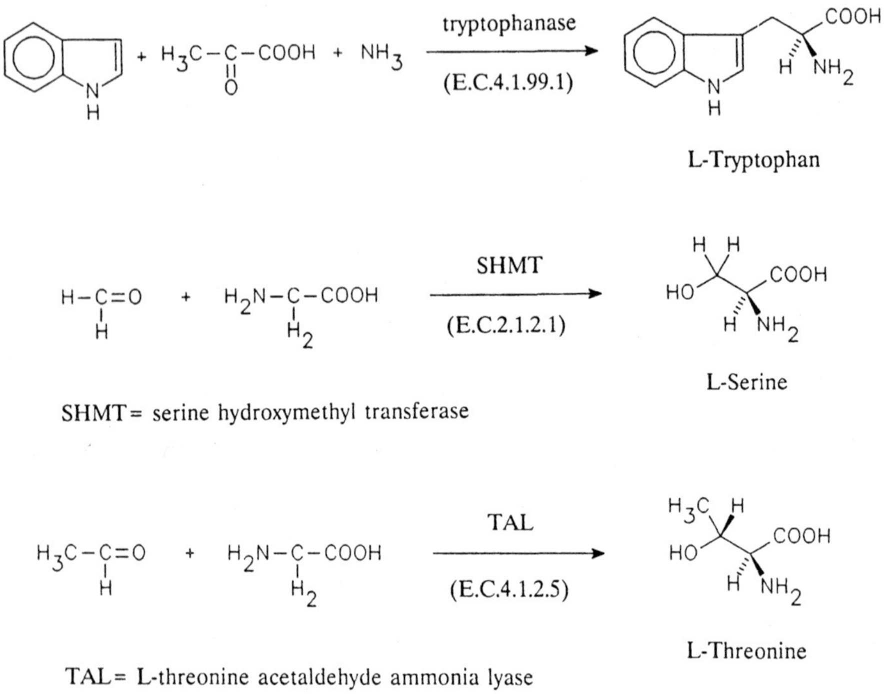
tryptophanase
(E.C.4.1.99.1)
L-Tryptophan
SHMT
(E.C.2.1.2.1)
L-Serine
SHMT= serine hydroxymethyl transferase
TAL
(E.C.4.1.2.5)
L-Threonine
TAL= L-threonine acetaldehyde ammonia lyase

Figure 7-19. Amino acid synthesis via enzymatic C-C bond formation.

at concentrations of 400 g l^{-1} (Reaction 7-7). Reaction (7-7) is actually an example of the use of an inexpensive raw material from the chirality pool.

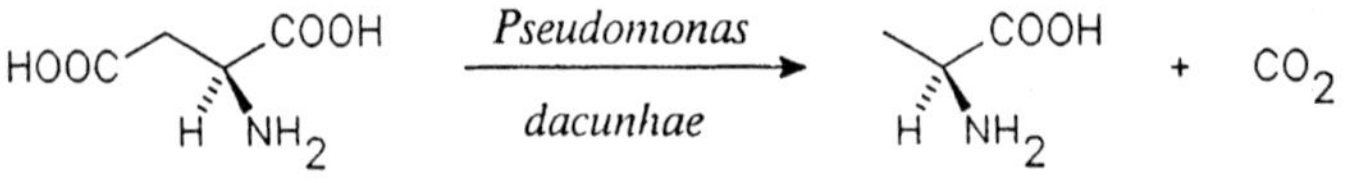
Pseudomonas
dacunhae

(7-7)

K. Optically Active α-Hydroxy Acids

Analogous to the formation of L-aspartic acid from fumaric acid and ammonia, enzymatic addition of water to fumaric acid affords L-malic acid (Reaction 7-9). The reaction is catalyzed by the lyase, fumarase (E.C. 4.2.1.2) and is performed industrially by Tanabe [42-44] using immobilized cells of *Brevibacterium ammoniagenes*.

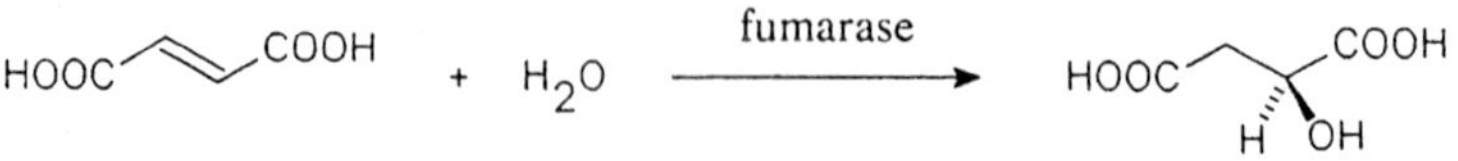
fumarase

(7-8)

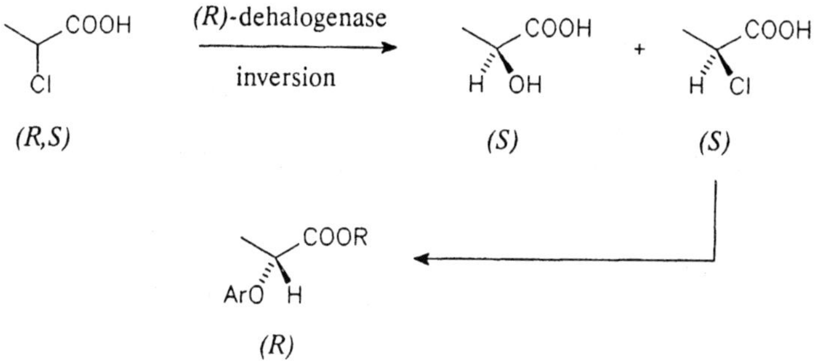

Figure 7-20. Preparation of *(S)*-2-chloropropionic acid using an *(R)*-specific dehalogenase.

As mentioned previously, optically active α-hydroxy carboxylic acids can also be prepared by enantioselective reduction of α-keto acids with continuous regeneration of the NADH cofactor (as a PEG conjugate) in a membrane bioreactor [67]. Both L- [68] and D-hydroxy acids [69] can be made in this way.

L.   Dehalogenases

Yet another enzymatic transformation that affords optically active α-hydroxy acids is the enantioselective hydrolysis of α-chloro acids mediated by dehalogenases [70]. Both $R(=D)$ and $S(=L)$ specific dehalogenases have been reported for α-chloropropionic acid as substrate, by ICI [70] and Unitka [71], respectively (Figure 7-20). In both cases the enzyme-mediated hydrolysis proceeds with inversion of configuration, suggesting an S_N2 mechanism. The *(R)*-specific hydrolysis has been developed by ICI into a commercial process for the production of *(S)*-2-chloropropionic acid, a key intermediate for the synthesis of optically active α-phenoxypropionate herbicides such as ICI's fusilade. Obviously, the economics of this process are very dependent on the value of the *(S)*-lactic acid coproduct.

It is worth noting, that dehalogenases have, up till now, received only cursory attention. If these reactions could be carried out in anhydrous organic media this would allow for enantiospecific substitution of chlorine by other nucleophiles (e.g., RNH_2, RSH) which could provide new methods for the synthesis of amino acids and peptides, for example.

III.   PEPTIDE SYNTHESIS

The commercial importance of peptides has been increasing [72,73]. Notwithstanding the significant advances in automated chemical synthesis, enzymatic methods for their synthesis remain attractive for two reasons: minimal protection is required

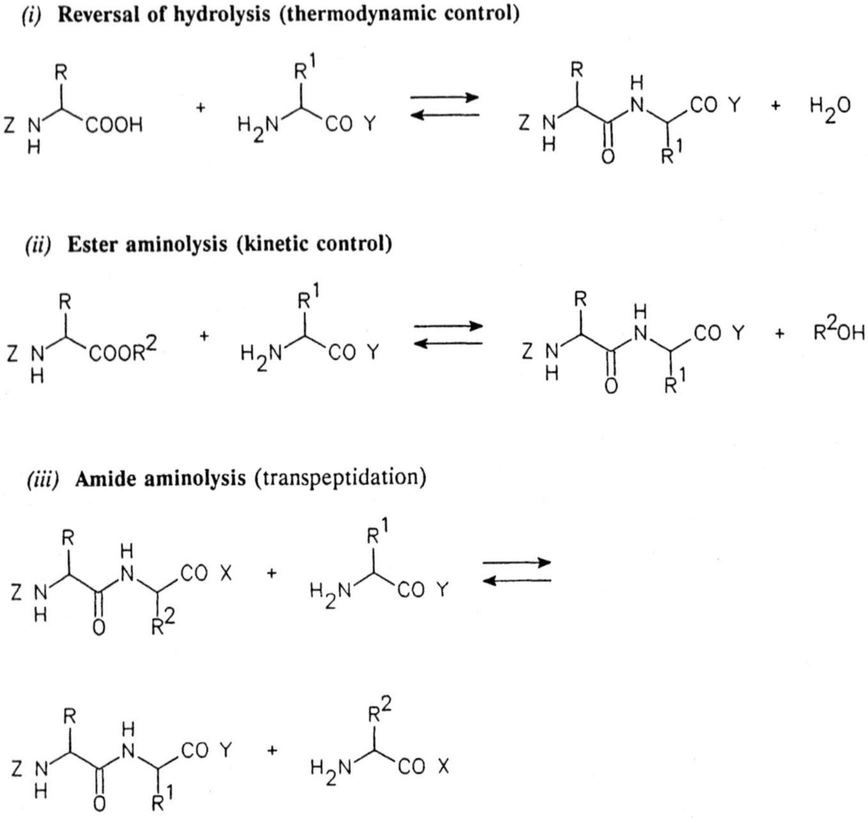

(i) Reversal of hydrolysis (thermodynamic control)
Z N H
COOH
H2N
CO Y
H2O
(ii) Ester aminolysis (kinetic control)
COOR2
R2OH
(iii) Amide aminolysis (transpeptidation)
CO X

Figure 7-21. Strategies for enzymatic peptide synthesis.

and no racemization is observed. There are three strategies for the enzymatic synthesis of peptides [73], as outlined in Figure 7-21.

A. Reversal of Hydrolysis

In order to implement the 'reversal of hydrolysis' strategy in an aqueous medium, the system must be designed such that the product is continuously removed, for example, by precipitation or extraction into a water immiscible organic solvent [73]. Obviously, the use of an immobilized enzyme is impractical when the product precipitates. Moreover, the use of a biphasic (aqueous-organic) system leads to channeling of the two phases in a fixed bed of immobilized

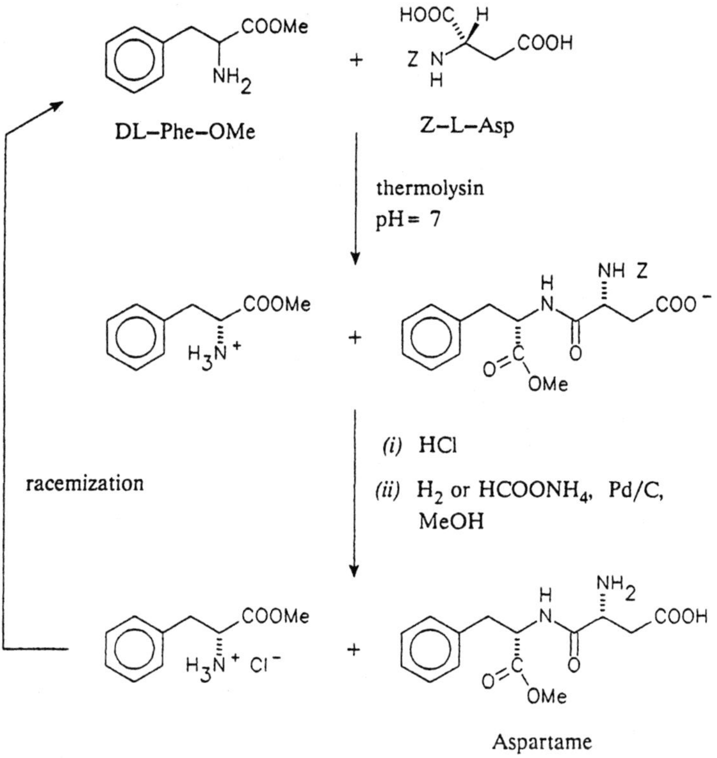

Figure 7-22. DSM-Toyo Soda process for aspartame.

enzyme. Hence, a 'water-free' organic solvent is preferred with the use of an immobilized enzyme [74].

In the DSM-Toyo Soda process for the synthesis of aspartame [74–76], for example, the protease, thermolysin (E.C. 3.4.24.4) catalyzes the regio- and enantiospecific coupling of protected L-aspartic acid with L-phenylalanine methyl ester. In the commercial process (Figure 7-22), racemic phenylalanine methyl ester is used as the substrate to give a 1:1 salt between protected aspartame and the D-phenylalanine methyl ester. This salt is sparingly soluble in water and precipitates from the reaction mixture. Subsequent removal of the amine component by treatment with acid is followed by catalytic hydrogenation to remove the benzyloxycarbonyl protecting group. Although an alternative coupling procedure using immobilized thermolysin in ethyl acetate has been described [74,75], it is generally believed that the commercial process employs the soluble enzyme in aqueous solution, as described above.

Covalent attachment of polyethyleneglycol (PEG) to enzymes renders them soluble in nonpolar organic solvents [77,78]. The technique is relatively simple and has been used to solubilize thermolysin [79], chymotripsin [80], and papain [81]

100% yield   (7-9)

in nonpolar solvents, thus facilitating their use in peptide formation. PEG-thermo-lysin, for example, gives a quantitative yield of protected dipeptide (Reaction 7-9) in benzene containing 2% diisopropylamine and 2% methanol [79].

B. Aminolysis of Esters

The aminolysis (kinetic control approach) requires the use of an ester as an acyl donor and is limited to those enzymes (e.g., serine and cysteine proteases) that form an acylenzyme intermediate. When proteases are used, product formation (esterase activity) competes with its consecutive hydrolysis (amidase activity). Hence, to maximize yields the amidase activity has to be eliminated or substantially reduced with respect to the esterase activity. This is feasible with proteases since the rate determining step is different. Thus, formation of the acylenzyme inter-mediate is rate determining in amide hydrolysis and deacylation in ester hydrolysis. Various approaches have been used to control the esterase versus amidase activity, for example, the use of water miscible organic solvents [82–84], changing the pH [82], or covalent modification of the active site to afford so-called 'damaged proteases', (e.g., thiolsubtilisin) that show esterase without amidase activity [85]. Other studies have shown that dramatic increases in yield can be obtained in an aqueous medium by decreasing the temperature from ambient to −25°C [86]. This so-called 'enzymes-on-the-rocks' technique appears to have tremendous synthetic potential (Figure 7-23).

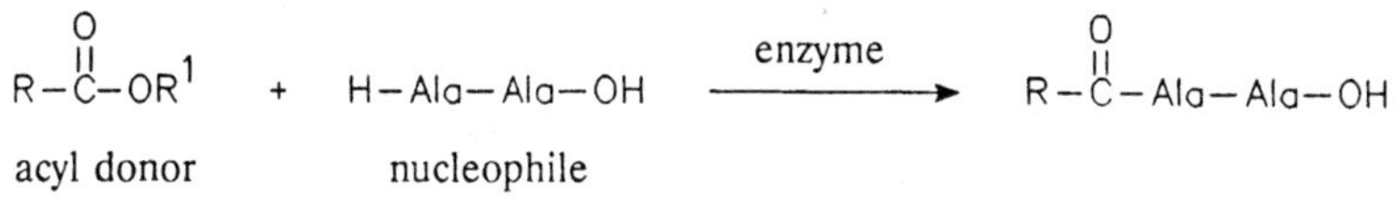

$$\underset{\text{acyl donor}}{R - \overset{\overset{O}{\|}}{C} - OR^1} \; + \; \underset{\text{nucleophile}}{H - Ala - Ala - OH} \; \xrightarrow{\text{enzyme}} \; R - \overset{\overset{O}{\|}}{C} - Ala - Ala - OH$$

Acyl donor	Enzyme	Peptide yield (%)	
		25 °C	−25 °C
Mal–Tyr–OMe	α-chymotrypsin	10	94
Mal–Phe–Ala–OEt Cl	papain	42	79
Z–Glu–OMe	V8-protease	5	76

Mal = maleyl = HOOC⎓CO-

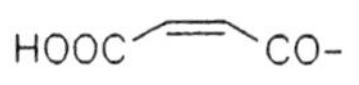

Z = benzyloxycarbonyl = ⟨O⟩—CH$_2$-O-CO-

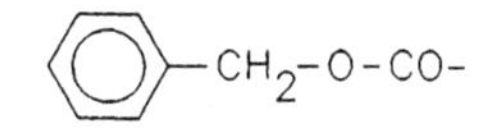

Figure 7-23. 'Enzymes on the rocks.'

Yet another approach to minimizing hydrolysis with respect to aminolysis is to operate in low-water systems. For example, chymotrypsin is very active as a suspension in dichloromethane in the presence of a small amount of water (0.2%) [6]. Alternatively, lipases and other esterases may be used for peptide formation via ester aminolysis in essentially anhydrous organic solvents [87,88].

C. Transpeptidation

Transpeptidation is used industrially in the production of human insulin from porcine insulin by trypsin-catalyzed replacement of a terminal alanine residue with threonine [73] (Figure 7-24).

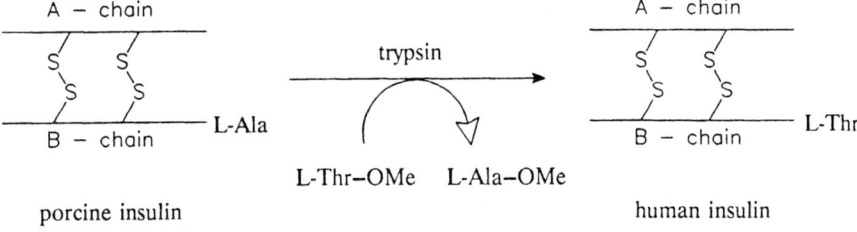

porcine insulin   human insulin

Figure 7-24. Human insulin from porcine insulin by transpeptidation.

IV.   ENZYMES IN ORGANIC MEDIA

The transformations discussed in the preceding sections were predominantly operations in aqueous media, many involving water as a reagent in a hydrolytic process. As mentioned earlier, the use of enzymes in nonpolar solvents offers several advantages. For example, hydrolytic processes such as ester or peptide hydrolysis can be carried out in a synthetic (condensation) mode, something that is obviously not feasible in aqueous solution. Various approaches have been used for carrying out enzymatic transformations in nonpolar, organic media. Covalent modification of enzymes with polyethylene glycol (PEG), for example, renders them soluble in organic media [77,78]; alternatively, enzymes can be solubilized by reverse micelle formation [89]. The simplest method, however, is to suspend the enzyme in the organic solvent [4].

A.   Lipases and Esterases

Two types of enzymes that have received increasing attention in the context of organic synthesis are esterases and lipases. Lipases (E.C. 3.1.1.3) are a special example of the carboxyl esterases (E.C. subgroup 3.1.1). They are ubiquitous in animals, plants, and microorganisms, where their function is to catalyze the hydrolysis of triglycerides to glycerol and fatty acids. A distinguishing feature of lipases is that, in contrast to other esterases, they act very efficiently on emulsions of water insoluble substrates (i.e., at lipid-water interfaces) [90]. Industrial applications include fat conversion, accelerated cheese ripening and detergents. Moreover, the enantiospecificity and commercial availability of many lipases offers interesting possibilities for the synthesis of optically active compounds [26,27,30,31]. A further advantage is that many of these transformations can be carried out in concentrated solution, sometimes with undiluted substrate. The main sources of commercial lipases are porcine pancreas and a variety of bacteria and yeast (Table 7-8).

Table 7-8   Sources of Lipases

Source (species)	Suppliers
Porcine pancreas (PPL)	Sigma, Amano, Boehringer Mannheim
Candida cylindracea (CCL) (= *Candida rugosa*)	Sigma, Amano, Boehringer Mannheim
Mucor miehei (Lipozyme)	Novo
Aspergillus oryzae (Lipolase)	Novo
Pseudomonas fluorescens (PFL)	Amano
Rhizopus arrhizus	Sigma, Boehringer Mannheim

Another esterase that is widely used in organic synthesis is pig liver esterase (PLE) [25]. The natural function of PLE is to mediate the hydrolysis of the various esters in the porcine diet and so it is not surprising that it exhibits wide substrate specificity.

In low-water media, lipases and esterases catalyze esterification, transesterification, and interesterification (Figure 7-25). These transformations can be employed for the kinetic resolution of chiral alcohols or carboxylic acids (esters). For example, a transesterification can be carried out using a racemic alcohol together with an achiral ester, such as ethyl acetate, as both acylating agent and solvent [99].

The interesterification procedures can, in some cases, lead to higher enantioselectivities than the corresponding hydrolysis or (trans)esterification [92]. This derives from the fact that the substrate must visit the active site of the enzyme twice during the process. This leads to "double-sieving" of the substrate through two

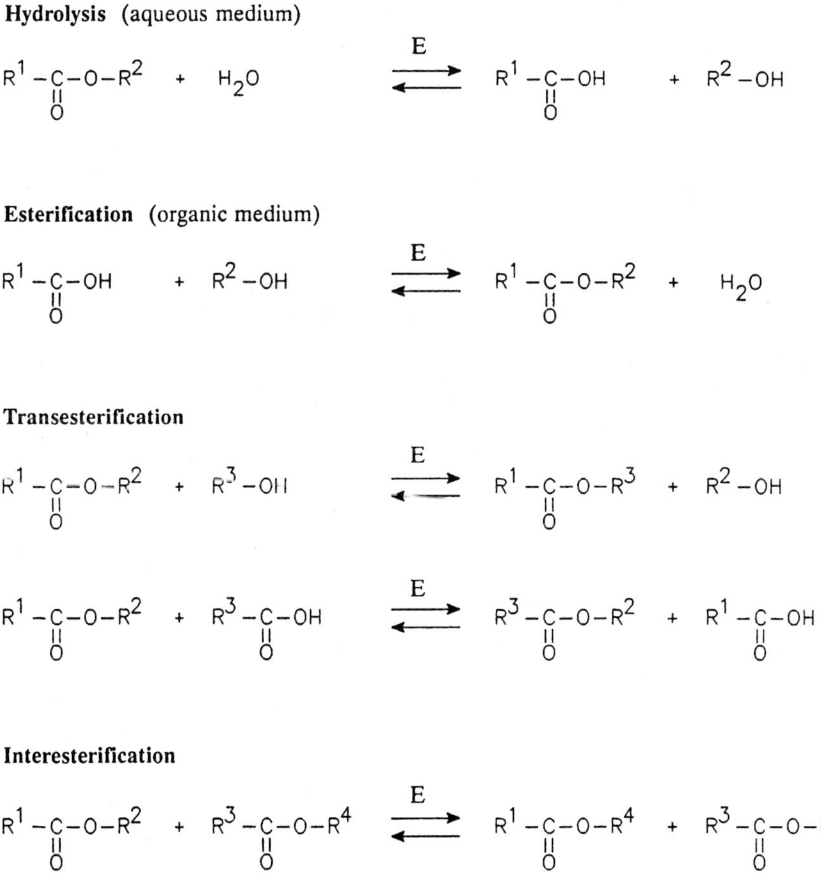

Hydrolysis (aqueous medium)

$$R^1-\underset{\underset{O}{\|}}{C}-O-R^2 \quad + \quad H_2O \quad \underset{\longleftarrow}{\overset{E}{\longrightarrow}} \quad R^1-\underset{\underset{O}{\|}}{C}-OH \quad + \quad R^2-OH$$

Esterification (organic medium)

$$R^1-\underset{\underset{O}{\|}}{C}-OH \quad + \quad R^2-OH \quad \underset{\longleftarrow}{\overset{E}{\longrightarrow}} \quad R^1-\underset{\underset{O}{\|}}{C}-O-R^2 \quad + \quad H_2O$$

Transesterification

$$R^1-\underset{\underset{O}{\|}}{C}-O-R^2 \quad + \quad R^3-OH \quad \underset{\longleftarrow}{\overset{E}{\longrightarrow}} \quad R^1-\underset{\underset{O}{\|}}{C}-O-R^3 \quad + \quad R^2-OH$$

$$R^1-\underset{\underset{O}{\|}}{C}-O-R^2 \quad + \quad R^3-\underset{\underset{O}{\|}}{C}-OH \quad \underset{\longleftarrow}{\overset{E}{\longrightarrow}} \quad R^3-\underset{\underset{O}{\|}}{C}-O-R^2 \quad + \quad R^1-\underset{\underset{O}{\|}}{C}-OH$$

Interesterification

$$R^1-\underset{\underset{O}{\|}}{C}-O-R^2 \quad + \quad R^3-\underset{\underset{O}{\|}}{C}-O-R^4 \quad \underset{\longleftarrow}{\overset{E}{\longrightarrow}} \quad R^1-\underset{\underset{O}{\|}}{C}-O-R^4 \quad + \quad R^3-\underset{\underset{O}{\|}}{C}-O-R^2$$

Figure 7-25. Reactions catalyzed by lipases and esterases.

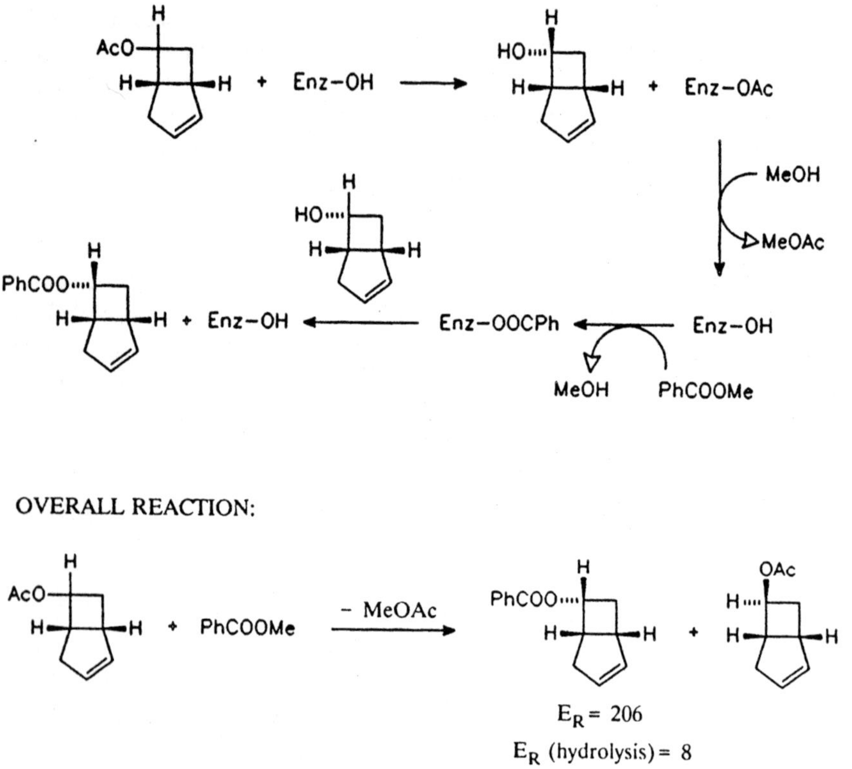

OVERALL REACTION:

$E_R = 206$

E_R (hydrolysis) = 8

Figure 7-26. Double enantioselection in an enzymatic interesterification.

enantioselective steps, as shown for the resolution of the racemic bicyclic acetate in Figure 7-26 with *Mucor miehei* lipase [92].

B. Kinetic Resolution of Carboxylic Acids (Esters)

A great many carboxylic acids can be resolved using lipases or esterases [25–31] in either a hydrolytic or synthetic (esterolytic) mode. Rates are often lower in the latter case and, hence, more enzyme is required. Klibanov and coworkers have described the lipase-mediated enantiospecific hydrolysis of α-halopropionic acids in hexane as solvent (Figure 7-27) [93]. The transformation is of commercial interest in connection with the manufacture of the biologically active *(R)*-isomers of α-phenoxypropionic acid herbicides. (See chapter 2.) Chemie Linz (Austria) has reportedly scaled this process up to pilot plant scale. The required product is the *(S)*-α-halopropionic acid that is converted via an inversion into the *(R)*-α-phenoxy propionic acid. Alternatively, the latter can be prepared [94] by lipase-catalyzed

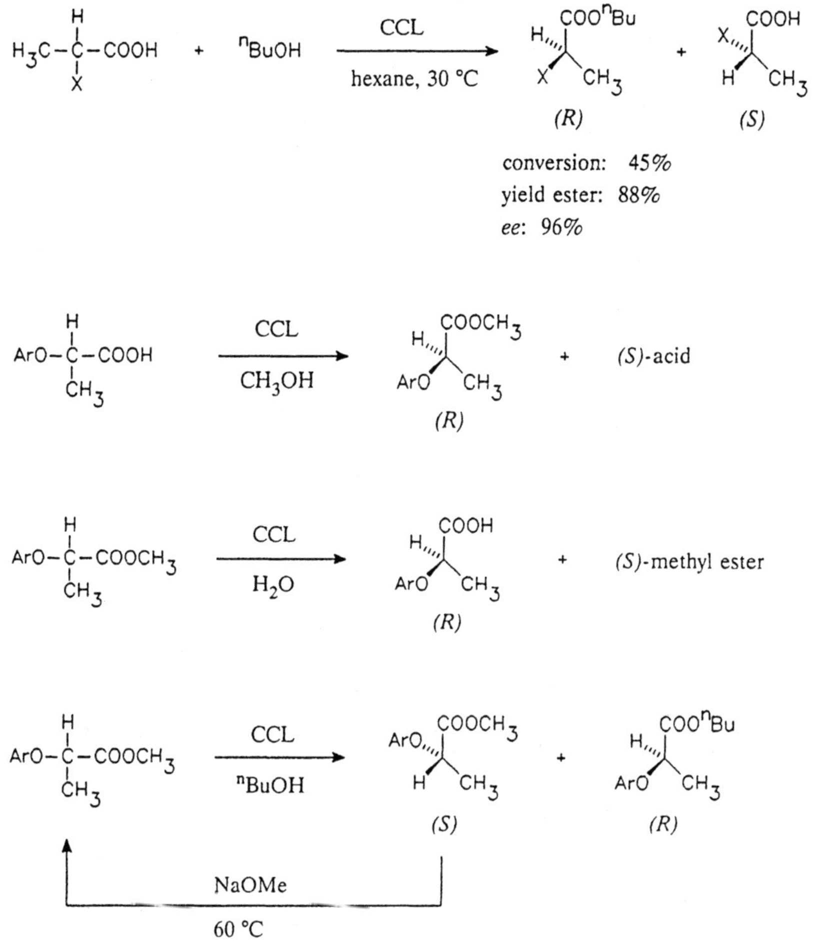

Figure 7-27. Lipase-catalyzed kinetic resolutions of α-halo and α-aryloxypropionates.

hydrolysis, esterification, or transesterification (Figure 7-27). In the transesterification, the unwanted (S)-enantiomer is readily racemized by heating with a catalytic amount of sodium methoxide [94].

The therapeutically active (S)-enantiomers of the structurally related α-arylpropionic acid class of antiinflammatory drugs have been similarly prepared using lipases [30]. The preparation of (S)-naproxen by CCL-catalyzed hydrolysis of the methyl ester (Figure 7-28), for example, was reported by Sih and coworkers [30,95].

A continuous process has been described for the enantiospecific hydrolysis of the ethoxyethyl ester of naproxen using a column packed with CCL immobilized

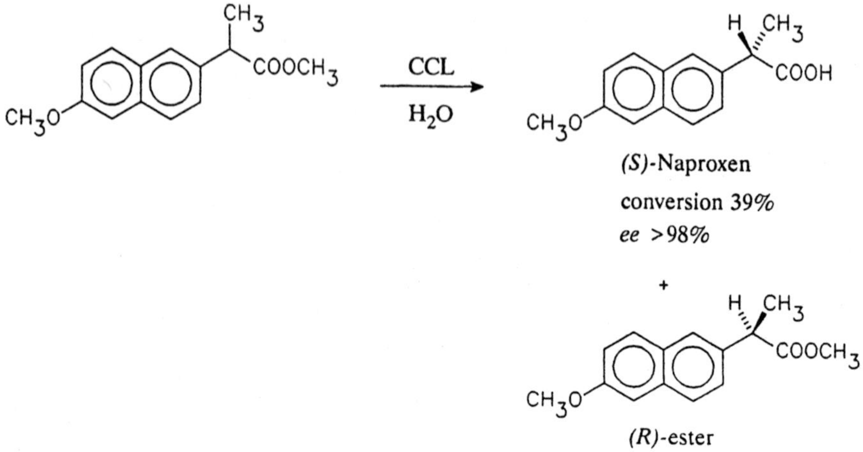

(S)-Naproxen
conversion 39%
ee >98%

+

(R)-ester

Figure 7-28. Synthesis of (S)-naproxen via enzymatic hydrolysis.

on an ion exchange resin [96]. Naproxen and ibuprofen methyl esters were also shown [97] to undergo (S)-specific hydrolysis in the presence of an esterase isolated from a *Bacillus subtilis* and cloned into *E. coli*. The same esterase also catalyzed the (S)-specific hydrolysis of some α-aryloxypropionic acid esters, which is quite surprising since they have the opposite configuration to the α-arylpropionic acids [97] as shown below:

(S)

(S)

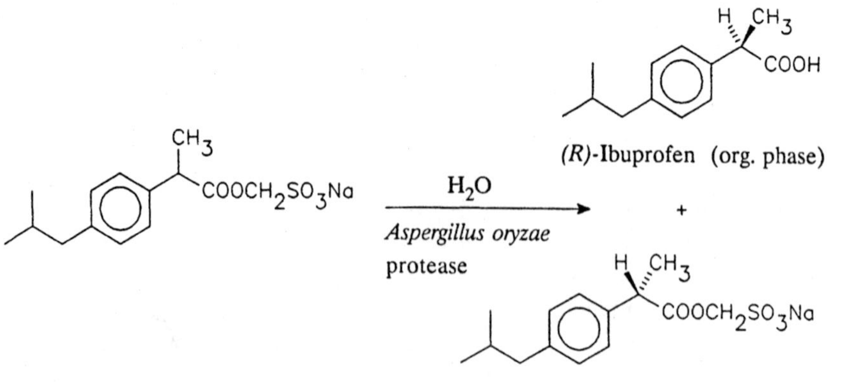

(R)-Ibuprofen (org. phase)

+

H_2O

Aspergillus oryzae protease

(S)-ester (aqueous phase)　　**(7-10)**

Candida cylindraceae lipase-catalyzed *(S)*-specific hydrolysis of ibuprofen esters in a multiphase membrane reactor has been described by Sepracor workers [50,98]. The same authors reported [50] an interesting variation on this theme, namely the protease-catalyzed, *(R)*-specific hydrolysis of the water soluble ester of ibuprofen containing the bisulfite adduct of formaldehyde as the alcohol moiety. Reaction in a two-phase, aqueous-organic medium affords the *(R)*-acid dissolved in the organic phase, together with unreacted ester and product alcohol in the aqueous phase; this is the reverse of the normal situation (Reaction 7-10).

Much effort has been devoted [30] to finding an enzyme for the enantiospecific hydrolysis of racemic β-acetylmercaptoisobutyric acid esters to the *(S)*-acid, a key intermediate in the synthesis of the ACE-inhibitor, captopril. Here the problem is not only enantiospecificity but also chemoselectivity (i.e., ester versus thioester hydrolysis). A variety of microbial lipases give only statistical mixtures of hydrolytic products and high chemo- and enantioselectivities are obtained only with sterically hindered acylthio groups [30]. Interestingly, transesterification with *n*-propanol in the presence of PPL or lipase P [99] gives the *(S)*-ester via selective thiotransesterification (Figure 7-29).

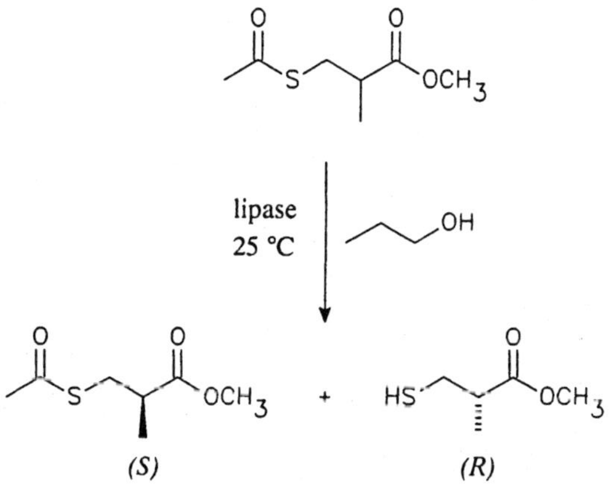

Enzyme	Time (h)	Conv. (%)	(S) - ester Yield (%)	ee (%)
Lipase P	24	60	31	93
PPL	72	51	42	95

Figure 7-29. Enzymatic thiotransesterification.

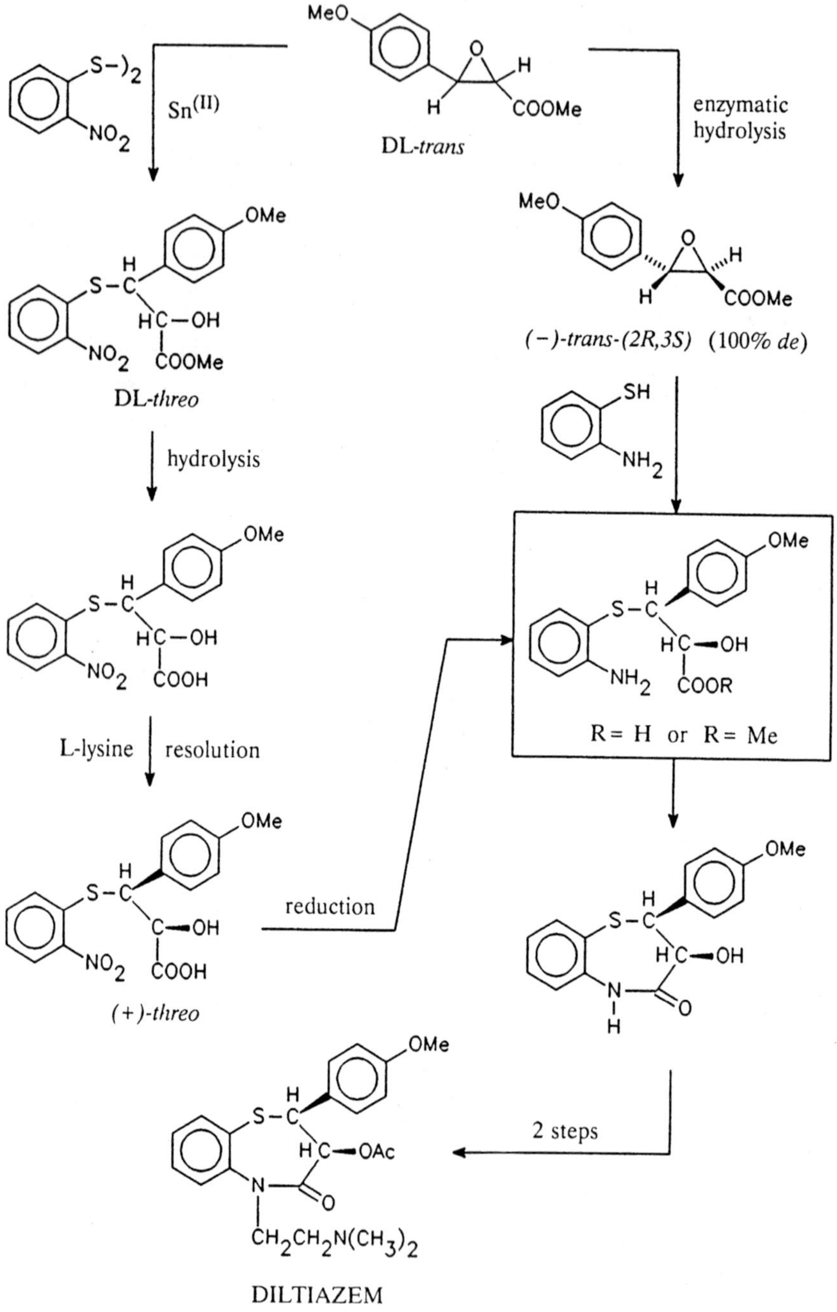

Figure 7-30. Two routes to diltiazem.

The synthesis of a key intermediate for diltiazem by enantiospecific hydrolysis of a glycidate ester has been developed by DSM-Andeno [100] and by Sepracor [50]. This has led to the development of a shorter, more economical route to diltiazem than the existing route via a late classical resolution (Figure 7-30).

C. Kinetic Resolution of Alcohols

A variety of chiral alcohols can be resolved via lipase-catalyzed hydrolysis of the corresponding acetates [101,102] or transesterification with ethyl acetate as acylating agent and solvent [91]. Optically active glycidyl derivatives, which are of interest as intermediates to optically active beta blockers, have been synthesized [103,104] by PPL-catalyzed hydrolysis of racemic glycidyl butyrate (Reaction 7-11).

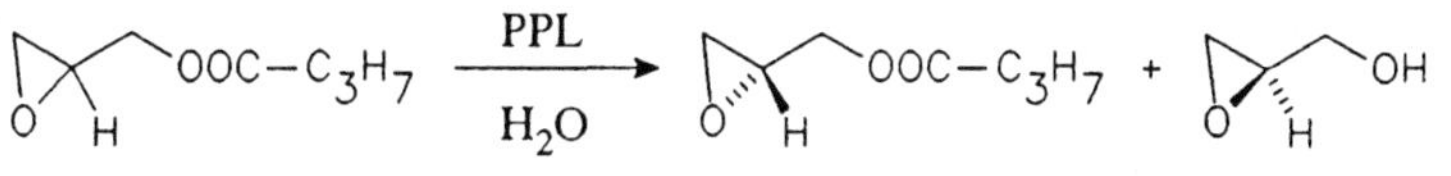

$$\text{(7-11)}$$

The products are readily converted to *(R)-* and *(S)-*glycidyl tosylates of high enantiomeric purity that are attractive intermediates for a variety of commercially interesting products [104].

Another reaction of considerable interest is the enantiospecific hydrolysis of the acetate of racemic *m*-phenoxybenzaldehyde cyanohydrin to yield a key intermediate for the optically active pyrethroid insecticide *(S,S)-*fenvalerate (Figure 7-31).

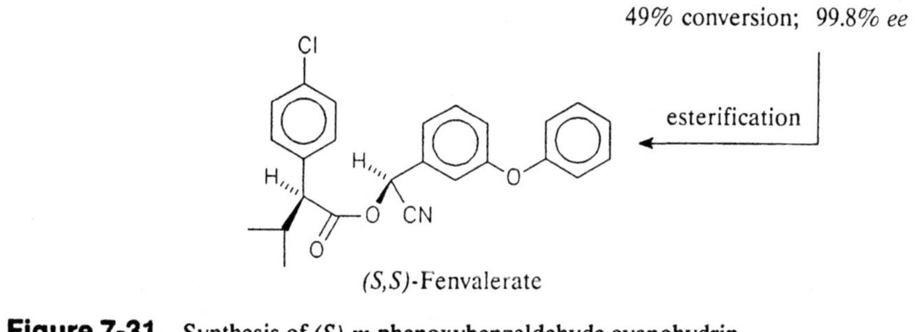

Figure 7-31. Synthesis of *(S)-m*-phenoxybenzaldehyde cyanohydrin.

A variety of commercial microbial lipases were investigated [105] for this transformation and the best results were obtained with a lipase from *Arthrobacter sp.*

D. Intramolecular Transesterification

Methyl esters of hydroxy carboxylic acids undergo enantiospecific intramolecular transesterification in the presence of lipases in organic solvents [106–108], to give optically active lactones (e.g., Reaction 7-12).

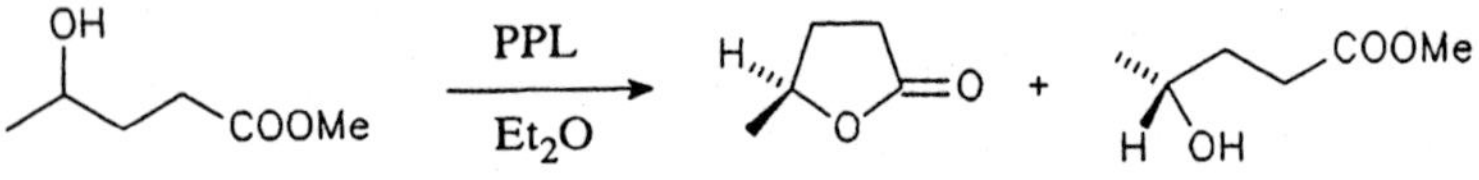

$$\begin{array}{cc} \text{(S)-lactone} & \text{(R)-hydroxy ester} \\ 36\% \text{ conversion} & 60\% \text{ conversion} \\ >94\% \; ee & >94\% \; ee \end{array} \qquad (7\text{-}12)$$

By carrying out the reaction to low (36%) or high (60%) conversion, either the *(S)*-lactone or the *(R)*-hydroxy ester is obtained in high optical purity [108].

E. Irreversible Transesterification

An inherent problem in the kinetic resolution of alcohols by enzymatic transesterification is the reverse reaction of the achiral alcohol, derived from the ester component, with the optically active ester product. Thus, the forward reaction in Figure 7-32 yields the acetate of the *(R)*-alcohol, together with the *(S)*-alcohol as the remaining isomer of the substrate. The reverse reaction of the achiral alcohol with the *(R)*-acetate, on the other hand, affords the *(R)*-alcohol, leading to a decrease

$$\underset{(R,S)}{ROH} + EtOAc \underset{\longleftarrow}{\overset{E}{\rightleftharpoons}} \underset{(R)}{ROAc} + EtOH + \underset{(S)}{ROH}$$

$$ROH + R^1COOC(R^2){=}CH_2 \xrightarrow{E} \underset{(R)}{R^1COOR} + R^1R^2C{=}O + \underset{(S)}{ROH}$$

E = Enzyme: lipase from *Pseudomonas sp.*

Ester	EtOAc	:	$CH_2{=}C(CH_3)OAc$	:	CH_2CHOAc	:	$CH_2{=}CHOOCBu^n$
Rel. rate	1	:	100	:	400	:	2000

Figure 7-32. Enzyme-catalyzed irreversible transesterification.

in optical purity. An elegant means to circumvent this problem is to employ an enol ester whereby the product alcohol tautomerizes to a carbonyl compound, thus rendering the transformation 'irreversible' [109,110]. Indeed, not only does this lead to improved enantiospecificities but also to dramatic increases in rates of reaction [110].

The same goal has also been achieved by using oxime esters [111] or anhydrides [112] as irreversible acyl transfer agents. In the latter case, enantiospecificities are improved considerably by removing the acid that is formed [113]. The anhydride method was used, for example, for the preparation of the ACE-inhibitor intermediate, (R)-2-hydroxy-4-phenylbutanoic acid (Reaction 7-13) [112].

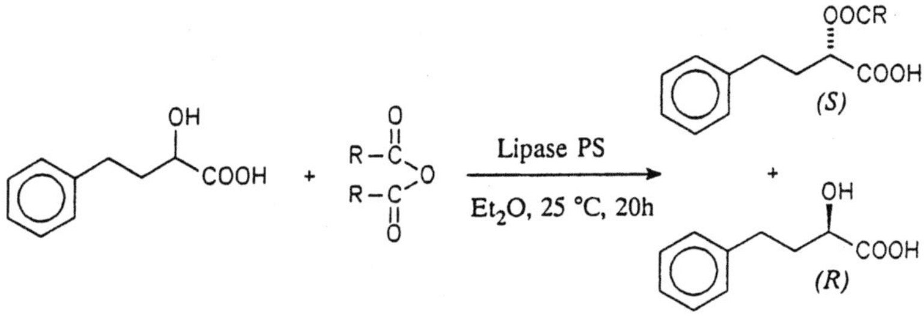

39% yield
60% conversion
>99% ee   (7-13)

A further refinement on the theme of irreversible transesterification is the use of an acyl donor containing an ionic group [114]. This renders the product ester soluble in water, thus facilitating its separation from the optically active alcohol (Reaction 7-14).

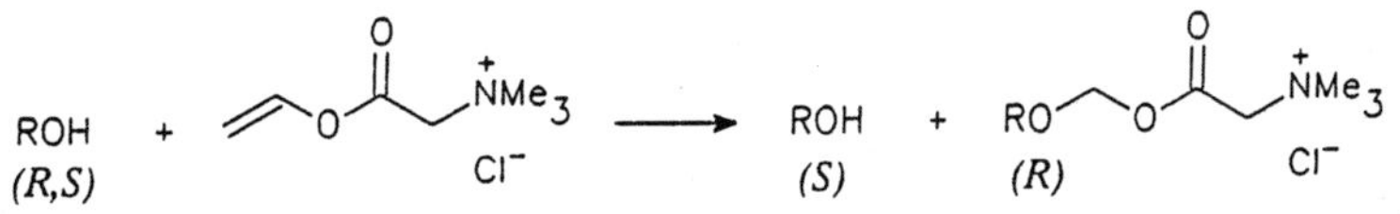

separable by solvent extraction   (7-14)

F. Amine Resolution

All of the lipase and esterase catalyzed reactions discussed above involve rate-limiting formation of an acylenzyme intermediate via the so-called serine protease mechanism (Figure 7-33). The active site of the enzyme is an aspartic acid-histidine-serine triad. Acylation of the hydroxyl group of the serine residue by the

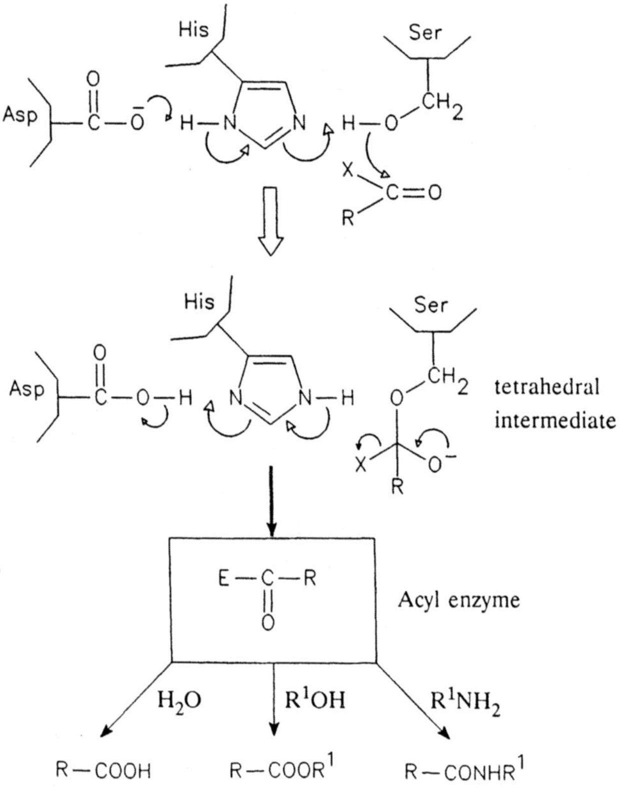

Figure 7-33. Serine protease mechanism.

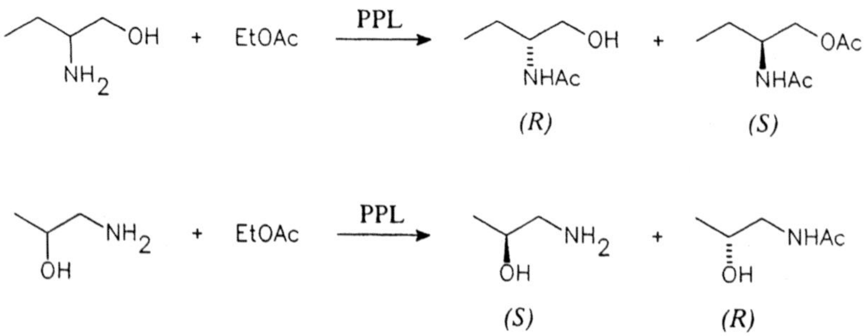

Figure 7-34. PPL-mediated enantioselective acylation of amino alcohols.

substrate is facilitated by a 'relay mechanism' involving the aspartate and histidine residues. The acylenzyme intermediate subsequently reacts with nucleophiles present in the reaction medium to yield the products, such as carboxylic acids or esters.

Similarly, an amine is acylated to give the corresponding amide and this can be utilized to resolve amines [115–118]. For example, chiral amino alcohols are resolved [115] by PPL-catalyzed acetylation using ethyl acetate both as the acyl donor and the solvent (Figure 7-34). The transformation is of commercial interest in connection with the synthesis of *(S)*-2-amino-1-butanol, a key intermediate for the antitubercular drug ethambutol. (See chapter 2.) Other authors [119,120] have reported that lipase-catalyzed hydrolysis or transesterification of the *N*-acylated derivative are superior methods, ones that require much lower amounts of enzyme (Reaction 7-15) [120]; however, optical purities are modest (66% *ee* at 60% conversion).

$$(7\text{-}15)$$

G. *Meso-* and Prochiral Diesters

Pig liver esterase [25,121,122] and, to a lesser extent, lipases [31,123] have been widely used in the application of the enzymatic *meso* trick whereby a *meso*-diester undergoes enantioselective hydrolysis or transesterification via enantiotopic group differentiation. This approach can be used to prepare optically active half esters of *meso*-dicarboxylic acids or *meso*-diols in theoretical yields of 100%. An example of the latter is the synthesis of the prostaglandin intermediates (Figure 7-35) by enantioselective hydrolysis of the *meso*-diacetate in the presence of PLE [124], PPL [125], or electric eel acetylcholine esterase, EECE [126].

Similarly prochiral diesters undergo enantioselective lipase-mediated hydrolysis to give optically active dicarboxylic acids or diols in theoretical yields of 100%. For example, lipase-catalyzed hydrolysis of the prochiral diacetate of 2-*O*-benzylglycerol gives the *(S)*-monoacetate (Figure 7-36) [13,109]. Alternatively, the *(R)*-enantiomer was obtained by transesterification of the corresponding *meso*-diol using the same enzyme. Thus, both enantiomers can be prepared from the same raw material and the same enzyme by varying the mode of operation, with the same enantiotopic group differentiation being observed in both cases.

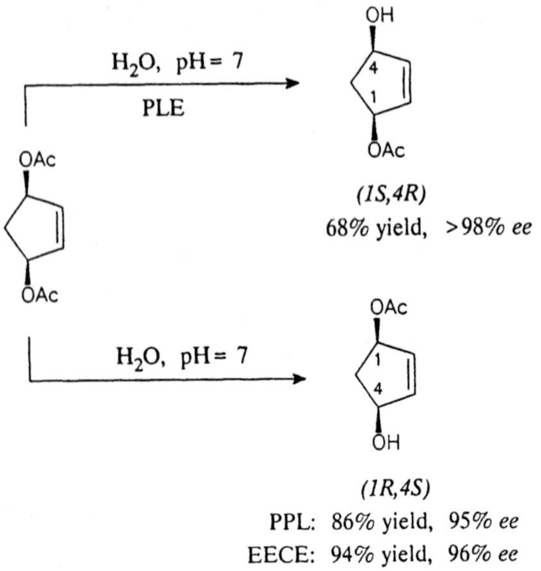

Figure 7-35. Enantiotopic group differentiation in a *meso*-diacetate.

Ohno and coworkers [121,122] have developed a so-called **symmetrization-asymmetrization concept** for the synthesis of chiral synthons using PLE. The target molecule is first subjected to retrosynthetic analysis in order to generate a prochiral or *meso*-diester. The symmetrical diester is subjected to PLE-mediated asymmetric hydrolysis to generate an optically active half-ester and the latter is subsequently converted by conventional organic synthesis to the target molecule.

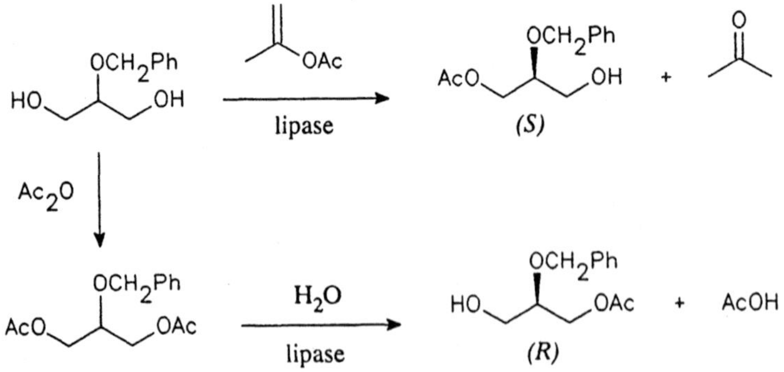

Figure 7-36. Enantiotopic group differentiation in a prochiral diol and diacetate.

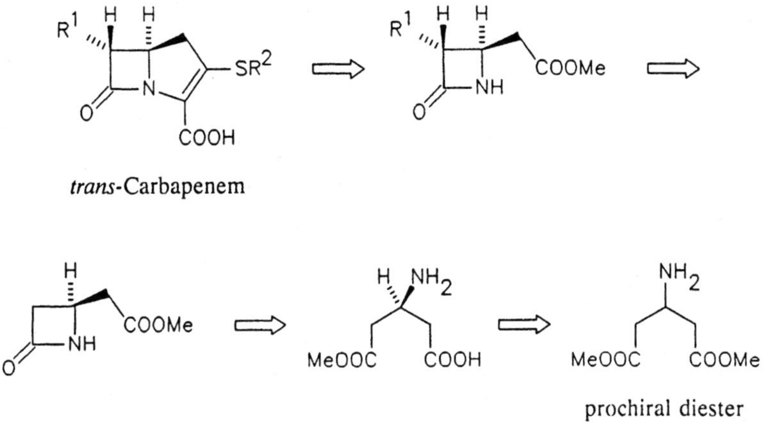

trans-Carbapenem

prochiral diester

Figure 7-37. The symmetrization-asymmetrization concept.

The concept is illustrated for the synthesis of carbapenem antibiotics (e.g., thienamycin) in Figure 7-37.

As noted earlier, a derivative of the target prochiral β-amino glutarate had been shown to undergo enantioselective chymotrypsin-mediated hydrolysis in 1965 (Reaction 7-1) [3]. Ohno and coworkers [121,122,127], however, showed that PLE-mediated hydrolysis of *N*-acetylated β-amino glutarates was much more efficient. Interestingly, the stereochemistry of the product could be controlled by varying the nature of the acyl group (Reaction 7-16).

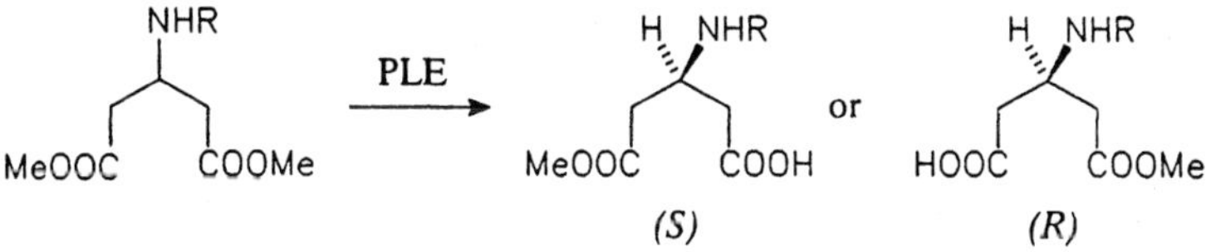

R	Yield (%)	Configuration	ee
H	94	*(R)*	41
CH₃CO	81	*(R)*	93
BOC = *t*-BuOCO	93	*(S)*	90
Z = PhCH₂OCO	93	*(S)*	93

(7-16)

These results provide interesting insights into the topography of the active site of PLE [121,122].

V. OXIDOREDUCTASES

A major distinction between the hydrolases (discussed in preceding sections) and the oxidoreductases is that the latter generally require the consumption of stoichiometric amounts of expensive cofactors [e.g., NAD(P) and FAD(P)]. Consequently, enzymatic redox processes are generally carried out as fermentations, that is, with growing whole cell systems in order to circumvent the cofactor regeneration problem. Baker's yeast-mediated reductions and microbial hydroxylations (discussed in chapter 4) are typical examples of such technologies. In this section we shall be concerned with the use of isolated enzymes in redox transformations where, if necessary, alternative methods for cofactor regeneration have to be employed.

Oxidoreductases are classified according to the Enzyme Commission on the basis of the organic transformation that is brought about. However, for the purpose of the present discussion, it is convenient to classify these enzymes according to the oxidant and stoichiometry involved:

Dehydrogenases

$$RH + D \rightleftharpoons R(-H) + DH_2$$

Monooxygenases

$$RH + O_2 + DH_2 \longrightarrow ROH + D + H_2O$$

Dioxygenases

$$RH + O_2 \longrightarrow RO_2H$$

Oxidases

$$RH + O_2 \longrightarrow R(-H) + H_2O_2$$

Peroxidases

$$RH + H_2O_2 \longrightarrow ROH + H_2O$$

D, DH_2 = cofactor

A. Dehydrogenases

Dehydrogenases are the only class of oxidoreductase enzymes that has received much attention from the viewpoint of synthetic applications of isolated enzymes. Dehydrogenases are, of course, the actual catalysts in many oxidations and reductions that involve whole cell systems (e.g., yeast reductions; see chapter 4). The

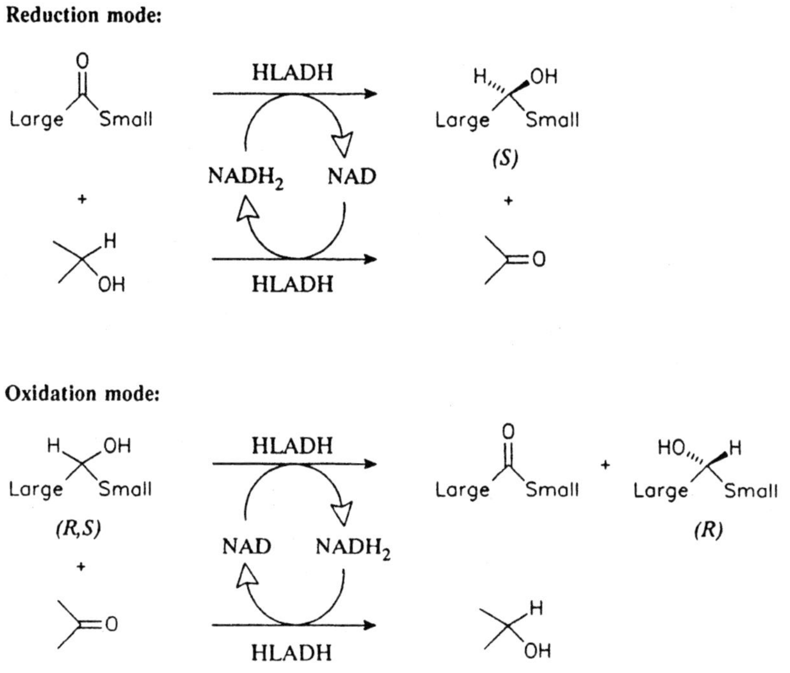

Reduction mode:

Oxidation mode:

Figure 7-38. Enantioselective reduction of prochiral ketones and enantiospecific oxidation of chiral alcohols with HLADH.

E.C. compilation describes 176 different dehydrogenases that can be classified on the basis of the cofactor required: nicotinamide base (NAD), flavin adenine dinucleotide (FAD) and pyrroloquinoline quinone (PQQ).

The most studied and well-known dehydrogenase is undoubtedly horse liver alcohol dehydrogenase (HLADH) [128,129] and it has, appropriately, the number E.C. 1.1.1.1. HLADH accommodates a broad range of substrates from acetaldehyde to bicyclic ketones, and is NADH-dependent. Reactions with HLADH (and other dehydrogenases) in a reduction mode can be used for the enantioselective reduction of a prochiral ketone (theoretical yield 100%) to give the *(S)*-alcohol via attack at the *Re*-face of the carbonyl group (Figure 7-38). In an oxidation mode, the reaction can be used for the kinetic resolution of a chiral alcohol. The *(S)*-enantiomer is selectively oxidized to leave the *(R)*-enantiomer as remaining substrate.

The simplest solution to the cofactor recycling problem is to carry out the reaction as a catalytic hydrogen transfer. An appropriate alcohol cosubstrate, for example, ethanol or isopropanol, is added in excess, often as the solvent, and its oxidation provides for regeneration of the cofactor (Figure 7-38).

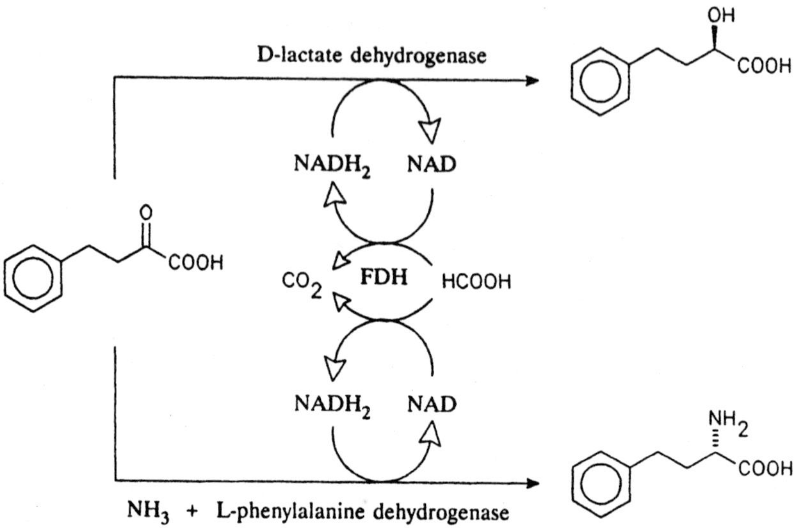

Figure 7-39. Enzymatic reduction and reductive amination of 2-oxo-4-phenylbutanoic acid.

HLADH and NAD have been co-immobilized on an ion exchange resin and used as a catalyst for the reduction of aldehydes to primary alcohols using an alcohol cosubstrate in organic solvents [130]. High activities were observed in n-alkanes, diisopropyl ether, and isopropyl acetate. Turnover numbers of 10^5–10^6 were observed based on the NAD charged, which translate to cofactor costs of 0.10–0.01 US$ per kg product, based on a price of 1000 US$ per kg for NAD. However, HLADH is a very expensive, relatively unstable enzyme and the enzyme costs were calculated to be about 1000 US$ per kg product. Hence, it is the cost of the HLADH and not the cofactor that is the limiting factor. Yeast alcohol dehydrogenase (YAD) is 400 times less expensive than HLADH but it loses more than 99% of its activity during immobilization and subsequent application in organic solvents [130]. The availability of cheaper, more stable alcohol dehydrogenases is obviously a prerequisite for commercial viability. In this context the isolation of thermostable alcohol dehydrogenases from thermophilic microorganisms, such as *Thermoanaerobium brockii* [131] and *Sulfolobus solfataricus* [132] would appear to be a step in the right direction [133].

Another solution to the cofactor regeneration problem is the use of the formate/formate dehydrogenase (FDH) system. In this approach the FDH (E.C. 1.2.1.43) catalyzes the oxidation of formate to CO_2 with simultaneous regeneration of the NADH cofactor [62,134]. The enzyme is readily available from *Candida boidinii* cultivated on methanol. The CO_2 coproduct is easily removed and does not interfere with the reaction. Wandrey and coworkers [62,63,134,135] pioneered the use of the formate/FDH system coupled with dehydrogenases in ultrafiltration membrane bioreactors. In order to prevent loss of NADH by leaking out of the system, the cofactor is modified by covalent attachment to polyethyleneglycol (PEG) of molecular weight 20,000. Thus, enzymes and the enlarged cofactor are trapped on one side of the membrane while the product passes through. The technique has been successfully applied [134–136] to the enantioselective reduction and reductive amination of a variety of α-keto acids, using α-hydroxy acid dehydrogenases and amino acid dehydrogenases, respectively. The latter were mentioned earlier in connection with amino acid synthesis.

The concept has also been applied to the synthesis of the ACE-inhibitor intermediates, (R)-2-hydroxy-4-phenylbutanoic acid and (S)=L-homophenyl-alanine (Figure 7-39) [136].

B. Oxidases and Oxygenases

Oxidases and mono- and dioxygenases utilize molecular oxygen as the oxidant. Reactions catalyzed by oxidases are actually oxidative dehydrogenations that result in the coproduction of H_2O_2 (or sometimes H_2O) and dehydrogenated substrate. Oxygenase-catalyzed reactions, on the other hand, result in incorporation of either one (monooxygenases) or both (dioxygenases) of the oxygens into the substrate.

Although oxidases mediate the same oxidative transformations as dehydrogenases but without the requirement of cofactor regeneration, they have found almost no synthetic applications. The most well-known examples, glucose, alcohol, and cholesterol oxidases, are used commercially in clinical analysis as biosensors.

Monooxygenases, such as the cytochrome P450 dependent variety, are ubiquitous in animals, plants, and microorganisms, where they play an important role in cellular metabolism and the detoxification of xenobiotics. Although they catalyze a variety of synthetically interesting transformations (e.g, olefin epoxidation, hydroxylation of aromatics and alkanes, and Baeyer-Villiger oxidation of ketones to lactones) they have found almost no applications. Some of the problems associated with these enzymes are: the requirement for cofactor regeneration, lack of stability outside the cell, and less than optimum (enantio)selectivities. Transformations involving monooxygenases are, therefore, generally carried out as precursor fermentations in, for example, steroid hydroxylation and olefin epoxidation. (See chapter 4.) Similarly, dioxygenases have not yet been applied to commercially relevant syntheses and transformations involving these enzymes are carried out as precursor fermentations.

C. Peroxidases

Peroxidases have synthetic potential since they can, in principle, mediate the same types of transformations as monooxygenases, that is, the incorporation of one

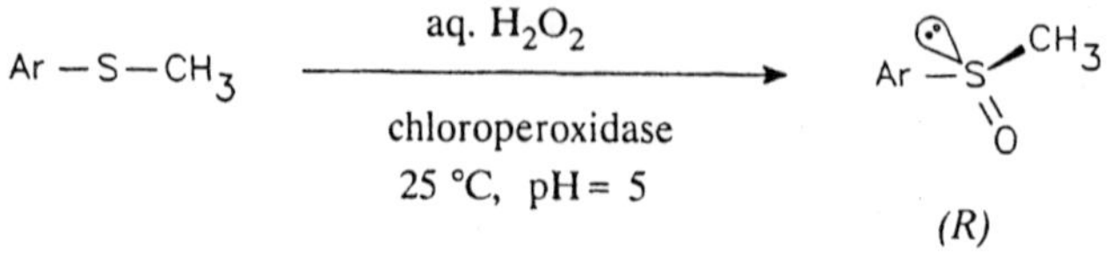

$$Ar-S-CH_3 \quad \xrightarrow[\substack{\text{chloroperoxidase} \\ 25\ ^\circ C,\ pH = 5}]{\text{aq. } H_2O_2} \quad Ar-\overset{\oplus}{\underset{O}{\overset{\ominus}{S}}}-CH_3$$

(R)

Ar	Yield (%)	ee (%)
p-CH$_3$C$_6$H$_4$	98	91
o-CH$_3$C$_6$H$_4$	27	33
p-CH$_3$OC$_6$H$_4$	72	90
o-CH$_3$OC$_6$H$_4$	24	27
C$_6$H$_5$	100	98
2-pyridyl	100	99

Figure 7-40. Chloroperoxidase-catalyzed enantioselective sulfoxidation.

oxygen atom into the substrate but without the consumption of a cofactor. The necessary reducing equivalents are derived from the hydrogens of H_2O_2. Moreover, peroxidases are generally extracellular enzymes and are considerably more robust than the intracellular monooxygenases.

Chloroperoxidase (CPO; E.C. 1.11.1.10), for example, is particularly interesting because of its close structural similarity to cytochrome P450. It is an extracellular enzyme produced by the marine fungus *Caldariomyces fumago*. In nature, CPO catalyzes the oxidative chlorination of organic compounds via the intermediacy of hypochlorite [138]. Based on the structural similarity to cytochrome P450 dependent monooxygenases CPO might be expected to catalyze synthetically interesting, (enantio)selective oxidations with H_2O_2 or ROOH. Indeed, one report [139] describes CPO-catalyzed enantioselective oxidations of prochiral sulfides with H_2O_2 (Figure 7-40).

VI. C-C BOND FORMING

Catalytic methods for the (enantio)selective formation of C-C bonds constitute a primary goal in modern organic synthesis. Although attention has been mainly focussed on chemocatalytic methods (see chapter 8), there are a few classes of synthetically useful enzymes capable of mediating C-C bond formation. One of these, the pyruvate decarboxylase-mediated acyloin condensation, has already been discussed in chapter 4 as these transformations are carried out as precursor fermentations with baker's yeast. Two other examples, cyanohydrin formation and aldol condensations, are discussed in this section.

A. Asymmetric Cyanohydrin Formation

Oxynitrilases occur widely in nature where they catalyze the last step in the conversion of cyanogenic glycosides to aldehydes and HCN [140]. Pioneering work of Becker and Pfeil [141] in the 1960s demonstrated that the *(R)*-oxynitrilase (E.C. 4.1.2.10), present in bitter almonds, catalyzes the enantioselective formation of *(R)*-cyanohydrins by reaction of HCN with aldehydes (Figure 7-41). Subsequently, an *(S)*-specific oxynitrilase (E.C. 4.1.2.11) was isolated from sorghum extract [142,143] and the synthetic potential of these enzymes has been explored by several groups [142–146]. The two enzymes differ in structure and substrate specificity; *(S)*-oxynitrilase is effective only with aromatic aldehydes while *(R)*-oxynitrilase accommodates aromatic, heterocyclic, and aliphatic aldehydes.

Achieving high enantioselectivities in aqueous media is often difficult due to competing nonenzymatic addition of HCN to the substrate and racemization of the product in aqueous buffer. Dramatic improvements in enantioselectivity are seen when the reactions are carried out in a water-immiscible solvent such as ethyl acetate or diisopropyl ether [143,144]. In these solvents the enzymatic reaction is hardly affected whereas the chemical reaction was much slower. For example, on changing from aqueous ethanol to ethyl acetate as solvent, the *ee* of the *(R)*-cyanohydrin obtained from *m*-phenoxybenzaldehyde increased from 10.5 to 98% [144].

(R)-Oxynitrilase is commercially available, highly efficient, and milligram quantities are enough to catalyze the formation of kilograms of cyanohydrin. It has been immobilized on cellulose-based supports and used in a fixed bed for continuous production in organic solvents [144].

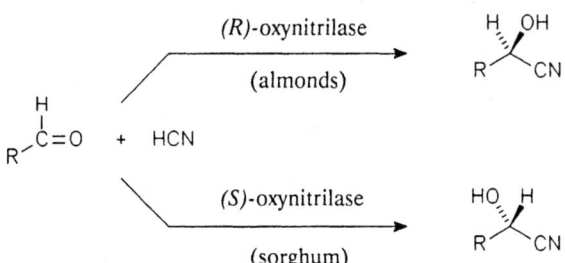

Figure 7-41. Enantioselective enzymatic cyanohydrin formation.

Table 7-9 *(R)*-Oxynitrilase-Catalyzed Cyanation [144] and Transhydrocyanation [147] of Aldehydes

| | Method | | | |
| | Cyanation | | Transhydrocyanation | |
Aldehyde	Yield (%)	ee (%)	Yield (%)	ee (%)
PhCHO	95	99	72	92
3-PhOC$_6$H$_4$CHO	99	98		
PhCH$_2$CHO	95	40	83	88
(CH$_3$)$_3$CCHO	78	73	58	92
CH$_3$(CH$_2$)$_2$CHO	75	96		
CH$_3$(CH$_2$)$_5$CHO			65	92
CH$_3$S(CH$_2$)$_2$CHO	97	80	60	92
H$_3$C⌒⌒CHO	68	97		
⌒⌒⌒⌒CHO			46	99

An alternative means of circumventing the nonenzymatic process has been described: enzymatic transhydrocyanation [147]. For instance, *(R)*-oxynitrilase-catalyzed reaction of acetone cyanohydrin with a wide variety of aldehydes yields *(R)*-cyanohydrins of consistently high enantiomeric purity. Results with typical aldehydes using the standard method in ethyl acetate and the transhydrocyanation method are compared in Table 7-9.

Optically pure cyanohydrins are potentially key chiral intermediates that can be readily converted to a variety of interesting products. For example, acid-catalyzed hydrolysis of *(S)*-cyanohydrins yields the corresponding *(S)*-α-hydroxy-carboxylic acids without racemization [143]. Alternatively, they can be reduced to optically active hydroxy amines [145]. Finally, the scope of the reaction has been further extended by the discovery that *(R)*-oxynitrilase catalyzes the enantio-selective addition of HCN to aliphatic ketones in diisopropyl ether as solvent. Acid-catalyzed hydrolysis afforded the corresponding α-hydroxy acids without racemization (Reaction 7-17) [148].

$$R \overset{O}{\underset{CH_3}{\bigwedge}} \xrightarrow[\text{(R)-oxynitrilase}]{\text{HCN}} \underset{R}{\overset{H_3C_{,,} OH}{\bigwedge}} CN \xrightarrow{H_2O/H^+} \underset{R}{\overset{H_3C_{,,} OH}{\bigwedge}} COOH \quad (7\text{-}17)$$

Yet another approach to the synthesis of optically active cyanohydrins is the lipase-mediated hydrolysis of cyanohydrin acetates. Several groups [149–151] have described the synthesis of acyl derivatives of *(R)* or *(S)* cyanohydrins by lipase-mediated hydrolysis, esterification, and transesterification reactions. For example, lipase-mediated hydrolysis of a racemic cyanohydrin acetate typically givess the *(R)*-cyanohydrin in > 98% *ee* (Reaction 7-18) [150].

$$(7\text{-}18)$$

Although the theoretical yield is 50%, the cyanohydrin coproduct is easily reconverted to the aldehyde starting material. In a further elaboration of this approach, advantage was taken of the reversibility of cyanohydrin formation to devise an elegant method for obtaining a theoretical yield of 100% in one step (Figure 7-42) [152]. Aromatic aldehydes are reversibly converted into the corresponding cyanohydrins by transhydrocyanation with acetone cyanohydrin, cata-

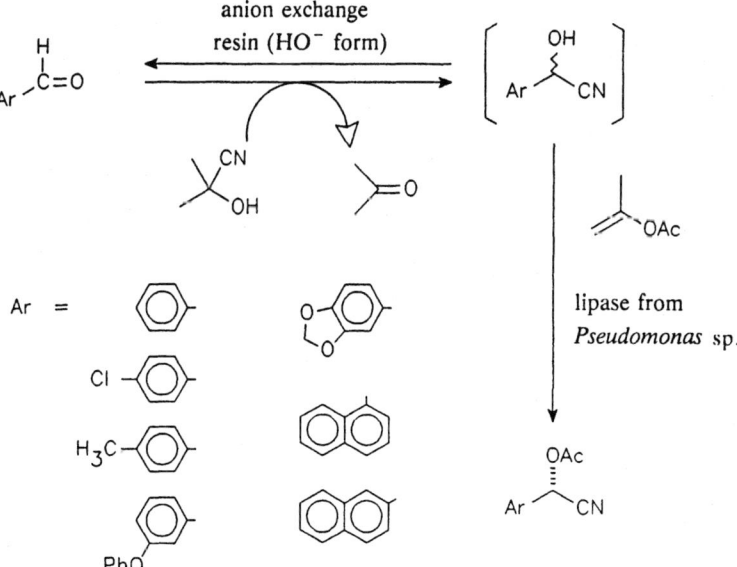

Figure 7-42. Lipase-catalyzed dynamic kinetic resolution of cyanohydrins.

lyzed by an anion exchange resin (HO⁻ form). The resulting racemic cyanohydrin is subjected to lipase-mediated, in situ, irreversible transesterification with isopropenyl acetate, yielding *(S)*-cyanohydrin acetate with high *ee*. Because of the reversible nature of the base-catalyzed transhydrocyanation, the remaining *(R)*-cyanohydrin undergoes in situ racemization. The overall result is an enzymatic second-order asymmetric transformation. *m*-Phenoxybenzaldehyde, for example, gives the *(S)*-cyanohydrin acetate in 80% yield and 89% *ee*. The success of the method also depends on the fact that acetone cyanohydrin, due to its steric bulk, cannot be accommodated by the lipase and is not acylated.

B. Asymmetric Aldol Condensation

The aldol condensation occupies a central position in organic chemistry and catalytic asymmetric aldol condensations remain an important synthetic challenge. There has been increased attention on a group of enzymes, **the aldolases**, that catalyze such reactions in vivo [153–158]. Aldolases (E.C. 4.1.2) are present in all organisms and more than 20 have been isolated. They fall into two distinct groups. Type I aldolases, found predominantly in higher plants and animals, require no metal cofactor and catalyze the aldol condensation through Schiff base formation between aldehyde substrate and a lysine residue (Figure 7-43). These enzymes are deactivated by NaBH₄ but are unaffected by enthylenediaminetetraacetic acid (EDTA). Type II aldolases are of microbial origin and are Zn²⁺-dependent enzymes. They are deactivated by EDTA but are unaffected by NaBH₄.

The best-known aldolase is the fructose 1,6-diphosphate aldolase from rabbit muscle (FDP aldolase, E.C. 4.1.2.13), also commonly known as rabbit muscle aldolase (RAMA). This enzyme is commercially available and catalyzes the condensation of dihydroxyacetone phosphate (DHAP) with a wide range (more than 75)

TYPE I   TYPE II

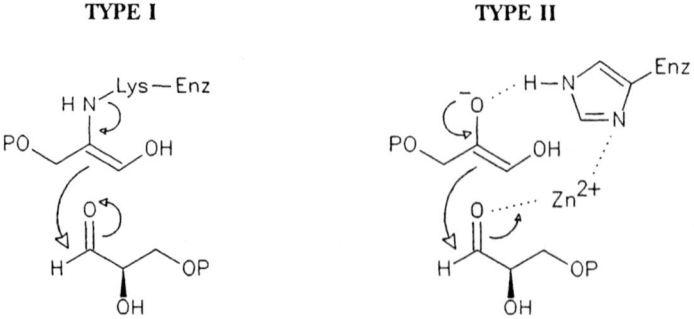

Figure 7-43. Mechanisms of Type I and Type II aldolases.

of aldehydes substrate (Reaction 7-19) [155]. The stereochemistry about the newly-formed C-C bond is strictly defined (D-*threo*).

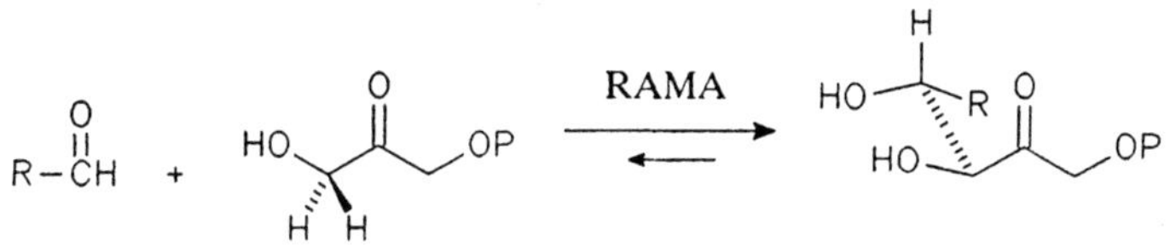

$$R-\overset{O}{\overset{\|}{C}}H \quad + \quad \text{(structure)} \quad \xrightarrow[\longleftarrow]{\text{RAMA}} \quad \text{(structure)} \tag{7-19}$$

Unfortunately, RAMA degrades rapidly, with a half-life of two days at ambient temperature. Wong and coworkers [153] have developed a 300-fold overproducing strain of *E. coli* that produces the Zn^{2+}-dependent *E. coli* FDP aldolase. This enzyme has a half-life of 60 days at ambient temperature.

A problem associated with FDP-aldolase-catalyzed condensations is the need for preparation of DHAP. Its synthesis is by no means trivial and it has limited stability. One solution to this problem is to use a mixture of dihydroxyacetone and a catalytic amount of inorganic arsenate instead of DHAP [160]. Dihydroxy-acetone reacts spontaneously with arsenate in aqueous solution to give the mono-arsenate. The latter is a mimic of DHAP and is accepted by the FDP aldolase. After the reaction the arsenate moiety is regenerated, thus creating a catalytic cycle (Figure 7-44).

Aldolases and the related transketolase (E.C. 2.2.1.1) which catalyzes the condensation of a wide range of aldehyde substrates with hydroxypyruvic acid (Reaction 7-20) are being exploited for the synthesis of unusual monosaccharides and azasugars [153–158]. The latter are the focus of increasing attention because of their antiviral activity.

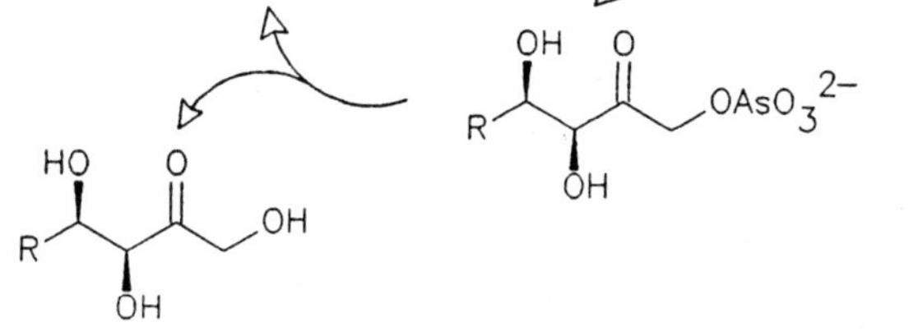

Figure 7-44. Replacement of DHAP with dihydroxyacetone and arsenate catalyst in FDP-aldolase-catalyzed reactions.

$$R-CHO + HO\overset{}{\underset{O}{\diagup}}COOH \xrightarrow{\text{transketolase}} R\overset{\overset{OH}{|}}{\underset{O}{\diagup}}OH + CO_2 \qquad (7\text{-}20)$$

In short, enzyme-mediated asymmetric aldol condensations appear to have considerable synthetic potential [161].

VII. FUTURE DEVELOPMENT

The application of isolated enzymes to the industrial synthesis of optically active compounds has expanded dramatically and will continue to do so in the foreseeable future. In the area of amino acid synthesis further improvements are still being made. For example, racemases have been discovered for N-acylamino acids [162] and amino acid amides [163], which in principle makes dynamic kinetic resolution of these substrates feasible, using acylases or amidases, respectively. The application of lipases and esterases will undoubtedly continue unabated. Moreover, we expect to see more emphasis on the application of oxidoreductases to asymmetric redox processes, such as oxidations with H_2O_2 or ROOH and reductions with H_2 or CO/H_2O. Thus, the observation that certain microorganisms [164,165], particularly thermophilic bacteria [166], are able to catalyze reductions with H_2 or CO/H_2O may have synthetic potential.

The scope of enzymatic transformations in general will increase as a result of:

1. The development of more stable enzymes derived from thermophilic microorganisms [167–169].
2. The use of r-DNA techniques to improve the production of enzymes (microbial engineering).
3. The use of protein engineering via site-directed mutagenesis or posttranslational chemical modification to generate tailor-made enzymes with novel properties and higher stabilities [170–172].
4. Increasing use of enzymes as catalysts for reactions they were never meant to catalyze in vivo.
5. Further development of the science of enzyme reactor engineering, for example, increased usage of membrane bioreactors and membrane enclosed enzymatic catalysis (MEEC) [173], possibly in combination with coupled, multienzyme systems.

In short, we have hardly scratched the surface with regard to the application of enzymatic methods to the synthesis of optically active compounds.

REFERENCES

1. Gutfreund, H., (Ed.), *Enzymes: 100 Years*, North Holland, Amsterdam, FEBS Letters, 62(Suppl.), 1976, and references cited therein.
2. Phillips, D. C., *Proc. Natl. Acad. Sci., USA*, **57**, 484-495 (1967).
3. Cohen, S. G., and Khedouri, E., *J. Am. Chem. Soc.*, **83**, 1093 (1961).
4. Klibanov, A. M., *Chemtech*, 354-359 (1986).
5. a) Zaks, A., and Klibanov, A. M., *Science*, **224**, 1249-1251 (1984). b) Ahern, J. J., and Klibanov, A. M., *Science*, **228**, 1280-1284 (1985).
6. Crout, D. H. G., and Cristen, M., in *Modern Synthetic Methods, Vol. 5*, Scheffold, R. (Ed.), Springer Verlag, Heidelberg, 1989, pp. 1-114.
7. Porter, R., and Clark, S., (Eds.), *Enzymes in Organic Synthesis*, Pitman, London, CIBA Foundation Symposium, No. 111, 1985.
8. Schneider, M. P. (Ed.), *Enzymes as Catalysts in Organic Synthesis*, Reidel, Dordrecht, 1986.
9. Tramper, J., Van der Plas, H. C., and Linko, P. (Eds.), *Biocatalysts in Organic Synthesis*, Elsevier, Amsterdam, 1985.
10. Davies, H. G., Green, R. H., Kelly, D. R., and Roberts, S. M., *Biotransformations in Preperative Organic Chemistry*, Academic Press, New York, 1989.
11. Abramowicz, D. A. (Ed.), *Biocatalysis*, Van Nostrand Reinhold, New York, 1990.
12. Whitaker, J. R., and Sonnet, P. E., *Biocatalysis in Agricultural Biotechnology*, American Chemical Society, Washington, 1989. ACS Symp. Ser., 389.
13. Wong, C. H., *Science*, **244**, 1145-1152 (1989).
14. Sonnet, P. E., *Chemtech*, 94-98 (1988).
15. a) Akiyama, A., Bednarski, M., Kim, M. J., Simon, E. S., Waldmann, H., and Whitesides, G. M., *Chem. Brit.*, 645-654 (1987). b) Akiyama, A., Bednarski, M., Kim, M. J., Simon, E. S., Waldmann, H., and Whitesides, G. M., *Chemtech*, 627-634 (1988).
16. Pratt, A. J., *Chem. Brit.*, 282-286 (1989).
17. Butt, S., and Roberts, S. M., *Chem. Brit.*, 127-134 (1987).
18. Soda, K., and Yonaha, K., in *Biotechnology, Vol. 7A*, Rehm, H. J. and Reed, G. (Eds.), VCH, Weinheim, 1987, pp. 605-602.
19. Chibata, I., Tosa, T., and Sato, T., in *Biotechnology, Vol. 7A*, Rehm, H. J. and Reed, G. (Eds.), VCH, Weinheim, 1987, pp. 653-684.
20. Jones, J. B., Perlman, D., and Sih, C. J., *Application of Biochemical Systems in Organic Synthesis*, Wiley, New York, 1976.
21. Sih, C. J., and Chen, C. S., *Angew. Chem., Int. Ed. Engl.*, **23**, 570 (1984).
22. Whitesides, G. M., and Wong, C. M., *Angew. Chem., Int. Ed. Engl.*, **24**, 617 (1985).
23. Jones, J. B., *Tetrahedron*, **42**, 3351 (1986).
24. Jones, J. B., in *Asymmetric Synthesis*, Morison, E. D. (Ed.), Academic Press, New York, 1986, pp. 309-344.
25. Ohno, M., and Otsuka, M., *Org. React.*, **37**, 1-55 (1989).
26. Klibanov, A. M., *Acc. Chem. Res.*, **23**, 114-120 (1989).
27. Chen, C. S., and Sih, C. J., *Angew. Chem., Int. Ed. Engl.*, **28**, 695-707 (1989).
28. a) Sheldon, R. A., Schoemaker, H. E., Kamphuis, J., Boesten, W. H. J., and Meijer, E. M., in *Stereoselectivity of Pesticides*, Ariëns, E. J., Van Rensen, J. J., and Welling, W (Eds.), Elsevier, Amsterdam, 1988, pp. 409-451. b) Meijer, E. M., Boesten, W. H. J., Schoemaker,

H. E., and Van Balken, J. A. M., in *Biocatalysts in Organic Synthesis*, Tramper, E. J., Van der Plas, H. C., and Linko, P. (Eds.), Elsevier, Amsterdam, 1985, pp. 135-156.

29. Dordick, J. S., *Enzym. Microb. Technol.*, **11**, 194-211 (1989).

30. Sih, C. J., Gu, Q. M., Fülling, G., Wu, S. H., and Reddy, D. R., *Dev. Ind. Microbiol.*, **29**, 221-229 (1988).

31. Xil, Z. F., *Tetrahedron Asymm.*, **2**, 733-750 (1991).

32. Elferink, V. H. M., Breitgoff, D., Kloosterman, M., Kamphuis, J., Van den Tweel, W. J. J., and Meijer, E. M., *Recl. Trav. Chim. Pays-Bas*, **110**, 63-74 (1991).

33. Gerhartz, W. (Ed.), *Enzymes in Industry. Production and Application*, VCH, Weinheim, 1990.

34. Frost, G. M., and Moss, D. A., in *Biotechnology, Vol. 7A*, Rehm, H. J. and Reed, G. (Eds.), VCH, Weinheim, 1987, pp. 65-211.

35. Klibanov, A. M., *Biochem. Biotechnol.*, **11**, 401 (1985).

36. Kunugi, S., Tanabe, K., Yamashita, K., Morikawa, Y., Ito, T., Kondo, T., Hirata, K., and Nomura, A., *Bull. Chem. Soc. Japan*, **62**, 514-518 (1989).

37. Randolph T. W., Clark, D. S., Blanch, H. W., and Prausnitz, J. M., *Science*, **238**, 387-390 (1990).

38. Klibanov A. M., *Trends Biochem. Soc.*, **14**, 141-144 (1989).

39. Fitzpatrick, P. A., and Klibanov, A. M., *J. Am. Chem. Soc.*, **113**, 3166-3171 (1991).

40. Rubio, E., Fernandez-Mayorales, A., and Klibanov, A. M., *J. Am. Chem. Soc.*, **113**, 695-696 (1991).

41. Sakurai, T., Margolin, A. L., Russell, A. J., and Klibanov, A. M., *J. Am. Chem. Soc.*, **110**, 7236-7237 (1988).

42. Chibata, I. (Ed.), *Immobilized Enzymes, Research and Development*, Kodansha, Tokyo, 1978.

43. Chibata, I., *Pure Appl. Chem.*, **50**, 667-675 (1978).

44. Chibata, I., Tosa, T. and Sato, T., in *Biotechnology, Vol. 7*, Rehm, H. J. and Reed, G. (Eds.), VCH, Weinheim, 1987, pp. 653-684.

45. Kennedy, J. F., and Cabral, J. M. S., in *Biotechnology, Vol. 7*, Rehm, H. J. and Reed, G. (Eds.), VCH, Weinheim, 1987, pp. 347-404.

46. Brodelius, P., and Vandamme, E. J., in *Biotechnology, Vol. 7*, Rehm, H. J. and Reed, G. (Eds.), VCH, Weinheim, 1987, pp. 405-464.

47. Takata, I., Tosa, T., and Chibata, I., *J. Solid-Phase Biochem.*, **2**, 225 (1977).

48. Matson, S. L., and Quinn, J. A., *Ann. N.Y. Acad. Sci.*, **469**, 152-165 (1986).

49. Van Eikeren, P., Brose, D. J., Muchmore, D. C., West, J. B., and Colton, R. H., *Proc. Chiral 92 Symp.*, Spring Innovations, Inc, Stockport, UK, 1992, pp. 63-69.

50. Young, J. W., and Bratzler, R. L., *Proc. Chiral 90 Symp.*, Spring Innovations, Inc, Stockport, UK, 1990, pp. 23-28.

51. a) Pessina, A., Luethi, P., Luisi, P. L., Prenosil, J., and Zhang, Y., *Helv. Chim. Acta*, **71**, 631-641 (1988); b) Verweij, J., and DeVroom, E., *Recl. Trav. Chim., Pays-Bas*, **112**, 1993, in press.

52. Fuganti, C., Grasselli, P., Seneci, P.F., and Servi, S., *Tetrahedron Lett.*, **27**, 2061-2062 (1986).

53. Schutt, H. Schmidt-Kastner, G., Arens, A., and Preiss, M., *Biotechnol. Bioeng.*, **27**, 420 (1985).

54. Evans, C. T., Roberts, S. M., Shoberu, K. A., and Sutherland, A. G., *J. Chem. Soc., Perkin Trans. I*, 589-592 (1992) and references cited therein.
55. Takahashi, J., Ohashi, T., Kii, Y., Kumagi, H., and Yamada, H., *J. Ferment. Technol.*, **57**, 328-332 (1979).
56. Olivieri, R. Fascetti, E., Angelini, L. and Degen, L., *Biotechnol. Bioeng.*, **23**, 2173 (1981).
57. a) Fukumura, T., *Agr. Biol. Chem.*, **40**, 1687 (1976). b) Fukumura, T., *Agr. Biol. Chem.*, **40**, 1695 (1976). c) Fukumura, T., *Agr. Biol. Chem.*, **41**, 1321 (1977). d) Fukumura, T., *Agr. Biol. Chem.*, **41**, 1327 (1977).
58. Sano, K., Yokozeki, K., Tamura, F., Yasuda, N., Noda, I., and Mitsugi, K., *Appl. Environ. Microbiol.*, **34**, 806-810.
59. Umemura, J., Takamatsu, S., Sato, T., Tosa, T., and Chibata, J., *Eur. J. Appl. Microbiol. Biotechnol.*, **20**, 291 (1984).
60. Chibata, I., Tosa, T., and Sato, T., in *Biotechnology, Vol. 7a*, Rehm, H.J., and Reed, G. (Eds.), VCH, Weinheim, 1987, pp. 653-684.
61. a) McGuire, J. C., US Patent 4598047 (1986) to Genex Corporation. b) Finkelmann, M. A. J., and Yang, H. H., US Patent 4584273 (1986) to Genex Corporation. c) Vollmer, P. J., US Patent 4584269 (1986) to Genex Corporation. d) See also: Yamada, S., Nabe, K., Izuo, N., Nakamichi, K., and Chibata, J. *Appl. Environ. Microbiol.*, **42**, 773 (1981).
62. a) Bückmann, A. F., Kula, M. R., Wichmann, R., and Wandrey, C., *J. Appl. Biochem.*, **3**, 301-315 (1981). b) Wichmann, R., Wandrey, C., Bückmann, A. F., and Kula, M. R., *Biotechnol. Bioeng.* **23**, 2789 (1981).
63. Wandrey, C., in *Enzymes as Catalyst in Organic Synthesis*, Schneider, M. P. (Ed.) Reidel, Dordrecht, 1986, pp. 263-284.
64. Nakajima, N., Esaki, N., and Soda, K., *J. Chem. Soc., Chem. Commun.*, 947-948 (1990).
65. a) Hsiao, H. Y., Wei, T., and Campbell, K., *Biotechnol. Bioeng.*, **28**, 857 (1986). b) Hsiao, H.Y., and Wei, T., *Biotechnol. Bioeng.*, **28**, 1510 (1986).
66. a) Chibata, I., Kakimoto, T., and Kato, J., *Appl. Microbiol.*, **13**, 638 (1965). b) See also: Jandel, A. S., Hustedt, H., and Wandrey, C., *Eur. J. Appl. Microbiol. Biotechnol.*, **15**, 59 (1982). c) See also: Fusee, M. C., and Weber, J. E., *Appl. Environ. Microbiol.*, **48**, 694 (1984).
67. Wichmann, R., Wandrey, C., Leuchtenberger, W., Kula, M. R., and Bückmann, A. F., United States Patent 4,326,031 (1982) to Degussa.
68. Schütte, H., Hummel, W., and Kula, M. R., *Eur. J. Appl. Microbiol. Biotechnol.*, **19**, 167 (1984).
69. a) Hummel, W., Schütte, H., and Kula, M. R., *Eur. J. Appl. Microbiol. Biotechnol.*, **18**, 75 (1983). b) Hummel, W., Schütte, H., and Kula, M. R., *Eur. J. Appl. Microbiol. Biotechnol.*, **21**, 7 (1985).
70. Taylor, S. C., in *Biocatalysis*, Abramowicz, D. A. (Ed.), Van Nostrand Reinhold, New York, 1990, pp. 157-165.
71. Motosugi, K., Esaki, N., and Soda, K., *Biotechnol. Bioeng.*, **26**, 805-807 (1984).
72. Jakubke, H.D., Kuhl, P., and Konnecke, A., *Angew. Chem., Int. Ed. Engl.*, 85-93 (1985).
73. a) Andersen, A. J., Fomsgaard, J., Thorbek, P., and Aasmul-Olsen, S., *Chimica Oggi (Chemistry Today)*, **9**(3), 17-23 (1991). a) Andersen, A. J., Fomsgaard, J., Thorbek, P., and Aasmul-Olsen, S., *Chimica Oggi, (Chemistry Today)*, **9**(4), 17-23 (1991).
74. Oyama, K., and Kihara, K., *Chemtech*, 100-104 (1984).

75. Oyama, K., Nishimura, S., Nonaka, Y., Kihara, K., and Hashimoto, T., *J. Org. Chem.*, **46**, 5241-5242 (1981).
76. Lindeberg, G., *J. Chem. Ed.*, **64**, 1062-1064 (1987).
77. Takahashi, K., Ajima, A., Yashimoto, T., Okada, M., Matshushima, A., Tamaura, Y., and Inada, Y., *J. Org. Chem.*, **50**, 3414-3415 (1985).
78. a) Nishio, T., Takahashi, K., Yoshimoto, T., Kodera, Y., Saito, Y., and Inada, Y., *Biotechnol. Lett.*, **9**, 771 (1986). b) Takahashi, K., Ajima, A., Yoshimoto, T., and Inada, Y., *Biochem. Biophys. Res. Commun.*, **125**, 761-766 (1984).
79. Ferjanic, A., Puigserver, A., and Gaertner, H., *Biotechnol. Lett.*, **10**, 101 (1988).
80. Matsushima, A., Okada, M., and Inada, Y., *FEBS Lett.*, **178**, 275 (1984).
81. Lee, H., Takahashi, K., Kodera, Y., Ohwada, K., Tsuzuki, T., Matsushima, A., and Inada, Y., *Biotechnol. Lett.*, **10**, 403 (1988).
82. Barbas, C. F., Matos, J. R., West, J. B., and Wong, C. H., *J. Am. Chem. Soc.*, **110**, 5162-5166 (1988).
83. Barbas, C. F., and Wong, C. H., *J. Chem. Soc., Chem. Commun.*, 533-534 (1987).
84. Chen, S. T., Hsiao, S. C., and Wang, K. T., *Biorg. Med. Chem. Lett.*, **1**, 445-450 (1991).
85. Wu, Z. P., and Hilvert, D., *J. Am. Chem. Soc.*, **111**, 4513-4514 (1989).
86. Schuster, M., Aaviksaar, A., and Jakubke, H. D., *Tetrahedron*, **46**, 8093-8102 (1990).
87. a) West, J. B., and Wong, C. H., *Tetrahedron Lett.*, **28**, 1629-1632 (1987). b) Matos, J. R., West, J. B., and Wong, C. H., *Biotechnol. Lett.*, **9**, 233 (1987).
88. a) Margolin, A. L., and Klibanov, A. M., *J. Am. Chem. Soc.*, **109**, 3802-3804 (1987). b) Kitaguchi, H., Tai, D. F., and Klibanov, A. M., *Tetrahedron Lett.*, **29**, 5487-5488 (1988).
89. Luisi, P. L., and Laane, C., *Trends Biotechnol.*, **4**, 153-161 (1986).
90. Alberghina, L., Schmid, R. D., and Verger, R. (Eds.), *Lipases: Structure, Mechanism, and Genetic Engineering*, GBF Monographs, Vol. 16, VCH, Weinham, 1991.
91. Janssen, A. J. M., Klunder, A. J. H., and Zwanenburg, B., *Tetrahedron*, **47**, 7645-7662 (1991) and references cited therein.
92. Macfarlane, E. L. A., Roberts, S. M., and Turner, N. J., *J. Chem. Soc., Chem. Commun.*, 569-571 (1990).
93. Kirchner, G., Scollar, M. P., and Klibanov, A. M., *J. Am. Chem. Soc.*, **107**, 7072-7076 (1985).
94. Cambou, B., and Klibanov, A. M., *Biotechnol. Bioeng.*, **26**, 1449-1454 (1984).
95. Gu, Q. M., Chen, C. S., and Sih, C. J., *Tetrahedron Lett.*, **27**, 1763 (1986).
96. Battistel, E., Bianchi, D., Cesti, P., and Pina, C., *Biotechnol. Bioeng.*, **38**, 659-664 (1991).
97. Mutsaers, J. H. G. M., and Kooreman, H. J., *Recl. Trav. Chim. Pays-Bas*, **110**, 185-188 (1991).
98. McConville, F. X., Lopez, J. L., and Wald, S. A., in *Biocatalysis*, Abramowicz, D. A. (Ed.), Van Nostrand Reinhold, New York, 1990, pp. 167-177.
99. Bianchi, D., and Cesti, P., *J. Org. Chem.*, **55**, 5657-5659 (1990).
100. Hulshof, L. A., and Roskam, J. H., European Patent Appl. 034714 (1989) to Stamicarbon.
101. Laumen, K., and Schneider, M. P., *J. Chem. Soc., Chem. Commun.*, 598-600 (1988).
102. Mori, K., and Bernotas, R., *Tetrahedron Asymm.*, **1**, 87-96 (1990).
103. Ladner, W. E., and Whitesides, G. M., *J. Am. Chem. Soc.*, **106**, 7250-7251 (1984).
104. Kloosterman, M., Elferink, V. H. M., Van Iersel, J., Roskam, J. H., Meijer, E. M., Hulshof, L. A., and Sheldon, R. A., *Trends Biotechnol.*, **6**, 251-256 (1988).

105. a) Hirohara, H., Mitsuda, S., Ando, E., and Komaki, R., in *Biocatalysts in Organic Synthesis*, Tramper, J., Van der Plas, H. C., and Linko, P. (Eds.), Elsevier, Amsterdam, 1985, pp. 119-134. b) Mitsuda, S., Yamamoto, H., Hirohara, H., and Nabeshima, S., *Agric. Biol. Chem.*, **54**, 2907-2912 (1990).

106. Yamada, H., Ohsawa, S., Sugai, T., Ohta, H., and Yoshikawa, S., *Chem. Lett.*, 1775-1776 (1989).

107. Makita, A., Nihira, t., and Yamada, Y., *Tetrahedron Lett.*, **28**, 805-808 (1987).

108. Gutman, A. L., Zuobi, K., and Boltansky, A., *Tetrahedron Lett.*, **28**, 3861-3864 (1987).

109. a) Wang, Y. F., and Wong, C. H., *J. Org. Chem.*, **53**, 3127 (1988). b) Wang, Y. F., Lalonde, J. J., Momongan, M., Bergbreiter, D. E., and Wong, C. H., *J. Am. Chem Soc.*, **110**, 7200-7205 (1988).

110. Degueil-Castaing, M., De Jeso, B., Drouillard, S., and Maillard, B., *Tetrahedron Lett.*, **28**, 953-954 (1987).

111. a) Chogare, A., and Sudesh-Kumar, J., *J. Chem. Soc., Chem. Commun.*, 1533-1535 (1989). b) Chogare, A., and Sudesh-Kumar, J., *J. Chem. Soc., Chem. Commun.*, 134-135 (1990).

112. a) Bianchi, D., Sugai, T., and Ohta, H., *Agric. Biol. Chem.*, **55**, 293-294 (1991). b) Bianchi, D. Cesti, P., and Battistel, E., *J. Org. Chem.*, **53**, 5531 (1988).

113. Berger, B., Rabiller, C. G., Königsberger, K., Faber, K., and Griengel, H., *Tetrahedron Asymm.*, **1**, 541-546 (1990).

114. McCague, R., and Evans, C. T., *Proc. Chiral 92 Symp.*, Spring Innovations, Inc, Stockport, UK, 1992, pp. 27-31.

115. Gotor, V., Brieva, R., and Rebolledo, F., *J. Chem. Soc., Chem. Commun.*, 957-958 (1988).

116. Kitaguchi, H. Fitzpatrick, P. A., Huber, J. E., and Klibanov, A. M., *J. Am. Chem. Soc.*, **111**, 3094-3095 (1989).

117. Brieva, R., Rebolledo, F., and Gotor, V., *J. Chem. Soc., Chem. Commun.*, 1386-1387 (1990).

118. a) Asensio, G., Andreu, C., and Marco, J. A., *Tetrahedron Lett.*, **32**, 4197-4198 (1991). b) Tuccio, B., Ferré, E., and Comeau, L., *Tetrahedron Lett.*, **32**, 2763-2764 (1991).

119. a) Francalanci, F., Cesti, P., Cabri, W., Bianchi, D., Martinengo, T., and Foa, M., *J. Org. Chem.*, **52**, 5079-5082 (1987). b) Francalanci, F., Cesti, P., Foa, M., and Martinengo, T., *Eur. Pat. Appl.*, 222561 (1987) to Montedison.

120. Bevinakalti, H. S., and Newadkar, R. V., *Tetrahedron Asymm.*, **1**, 583-586 (1990).

121. Ohno, M., in *Enzymes in Organic Synthesis*, Pitman, London, 1985, pp. 171-187. CIBA Foundation Symposium No. 111.

122. Ohno, M., Kobayashi, S., and Adachi, K., in *Enzymes as Catalysts in Organic Synthesis*, Schneider, M. P. (Ed.), Reidel, Dordrecht, 1986, pp. 123-142.

123. Schneider, M. P., *Performance Chemicals*, April, 1989, pp. 28-31.

124. Laumen, K., Reimerdes, E. H., and Schneider, M. P., *Tetrahedron Lett.*, **27**, 1255-1256 (1986).

125. Laumen, K., and Schneider, M. P., *J. Chem. Soc., Chem. Commun.*, 298-299 (1986).

126. Deardof, D. R., Mathews, A. J., McMeekin, D. S., and Craney, C. L., *Tetrahedron Lett.*, **27**, 1255-1256 (1986).

127. a) Ohno, M., Kobayashi, S., Limori, T., Wang, Y. F., and Izawa, T., *J. Am. Chem. Soc.*, **103**, 2405 (1981). b) Adachi, K., Kobayashi, S., and Ohno, M., *Chimia*, **40**, 311 (1986).

128. Lemiere, G. L., in *Enzymes as Catalysts in Organic Synthesis*, Schneider, M. P. (Ed.), Reidel, Dordrecht, 1986, pp. 19-34.

129. Jones, J. B., in *Enzymes in Organic Synthesis*, Pitman, London, 1985, pp. 3-21. CIBA Foundation Symposium No. 111.

130. a) Snijder-Lanbers, A. M., Vulfson, E. N., and Doddema, H. J., *Recl. Trav. Chim. Pays-Bas*, **110**, 226-230 (1991). b) See also: Grunwald, J., Wirz, B., Scollar, M. P., and Klibanov, A. M., *J. Am. Chem. Soc.*, **108**, 6732-6734 (1986).

131. a) Keinan, E., Hafeli, E. K., Seth, K. K., and Lamed, R., *J. Am. Chem. Soc.*, **108**, 162 (1986). b) Keinan, E., Seth, K. K., and Lamed, R., *J. Am. Chem. Soc.*, **108**, 3474 (1986).

132. a) Parvaresh, F., Vic, G., Thomas, D., and Legoy, M. D., *Ann. N.Y. Acad. Sci.*, **613**, 302-313 (1990); b) Parvaresh, F., Robert, H., Thomas, D., and Leroy, M. D., *Biotechnol. Bioeng.*, **39**, 467-473 (1992).

133. West, S., *Chimica Oggi (Chemistry Today)*, Jan/Feb, 1991, pp. 43-47.

134. Kula, M. R., and Wandrey, C., *Method Enzymol.*, **136**, 9-21 (1987).

135. Pabsch, K., Petersen, M., Rao, N. N., Alfermann, A. W., and Wandrey, C., *Recl. Trav. Chim. Pays-Bas*, **110**, 199-205 (1991).

136. Bradshaw, C. W., Wong, C. H., Hummel, W., and Kula, M. R., *Bioorg. Chem.*, **19**, 29-39 (1991).

137. Ortiz de Montellano, P. R. (Ed.), *Cytochrome P450; Structure, Mechanism and Biochemistry*, Plenum Press, New York, 1986.

138. Griffin, B.W., in *Peroxidases in Chemistry and Biology, Vol. II*, Everse, J., Everse, K. E., and Grisham, M. B., (Eds.), CRC Press, Ann Arbor, 1991, pp. 85-137.

139. a) Colonna, S., Gaggero, N., Manfredi, A., Casella, L., Gullottti, M., Carrera, G., and Pasta, P., *Biochemistry*, **29**, 10465-10468 (1990). b) Colonna, S., Gaggero, N., Casella, L., Carrera, G., and Pasta, P., *Tetrahedron Asymm.*, **3**, 95-106 (1992).

140. Xu, L. L., Singh, B. K., and Conn, E. E., *Arch. Biochem. Biophys.*, **250**, 322 (1986).

141. a) Becker, W., Freund, H., and Pfeil, E., *Angew. Chem., Int. Ed. Engl.*, **4**, 1079 (1965). b) Becker, W., and Pfeil, E., *J. Am. Chem. Soc.*, **88**, 4299 (1966).

142. a) Niedermeyer, U., and Kula, M. R., *Angew. Chem., Int. Ed. Engl.*, **102**, 423-424 (1990). b) Niedermeyer, U., Kula, M. R., Wandrey, C., Makryaleus, K., and Dranz, S., Eur. Patent 326.063 (1989) to Degussa. c) Kula, M. R., Niedermeyer, U., and Stürtz, I. M., Eur. Patent 350.908 (1989) to Degussa.

143. Effenberger, F., Hörsch, B., Förster, S., and Ziegler, T., *Tetrahedron Lett.*, **31**, 1249-1252 (1990).

144. Effenberger, F., Ziegler, T., and Förster, S., *Angew. Chem., Int. Ed. Engl.*, **26**, 458-459 (1987).

145. a) Kruse, C. G., Brussee, J., and Van der Gen, A., *Proc. Chiral 90 Symp.*, Spring Innovations, Stockport, UK, 1990, pp. 55-59. b) Brussee, J., Roos, E.C., and Van der Gen, A., Tetrahedron Lett., **29**, 4485-4488 (1988). c) Brussee, J., and Van der Gen, A., *Recl. Trav. Chim. Pays-Bas*, **110**, 25 (1991).

146. a) Brussee, J., Loos, W. T., Kruse, C. G., and Van der Gen, A., *Tetrahedron*, **46**, 979-986 (1990). b) Kruse, C. G., Geluk, H. W., and Van Scharrenburg, G. J. M., *Chimica Oggi (Chemistry Today)*, Jan/Feb, 1992, pp. 59-63.

147. Ognyanov, V. I., Datcheva, V. K., and Kyler, K. S., *J. Am. Chem. Soc.*, **113**, 6992-6996 (1991).

148. Effenberger, F., Hörsch, B., Weingart, F., Ziegler, T., and Kühner, S., *Tetrahedron Lett.*, **32**, 2605- 2608 (1991).

149. Effenberger, F., Gutterer, B., Ziegler, T., Eckhardt, E., and Aichholz, R., *Liebigs Ann. Chem.*, 47-54 (1991).

150. Van Almsick, A., Buddrus, J., Hönicke-Schmidt, P., Lamen, P., and Schneider, M. P., *J. Chem. Soc., Chem. Commun.*, 1391-1393 (1989).

151. a) Ohta, H., Miyamae, Y., and Tsuchihashi, G., *Agric. Biol. Chem.*, **53**, 215-222 (1989). b) Ohta, H., Hiraga, S., Miyamoto, K., and Tsuchihashi, G., *Agric. Biol. Chem.*, **52**, 3023-3027 (1988). c) Ohta, H., Kimura, Y., Sugaro, Y., and Sugai, T., *Tetrahedron*, **45**, 5469-5476 (1989).

152. Inagaki, M., Hiratake, J., Nishioka, T., and Oda, J., *J. Am. Chem. Soc.*, **113**, 9360-9361 (1991).

153. Wong, C. H., in *Biocatalysis*, Abramowicz, D. A. (Ed.), Van Nostrand Reinhold, New York, 1990, pp. 319-335.

154. Wong, C. H., in *Enzymes as Catalysts in Organic Synthesis*, Schneider, M. P. (Ed.), Reidel, Dordrecht, 1986, pp. 199-216.

155. Toone, E. J., Simon, E. S., Bednarski, M. D., and Whitesides, G. M., *Tetrahedron*, **45**, 5365-5422 (1989).

156. Drueckhammer, D. G., Hennen, W. J., Pederson, R. L., Barbas, C. F., Gautheron, T. K., and Wong, C. H., *Synthesis*, 499-525 (1991).

157. Toone, E. J., and Whitesides, G. M., in *Enzymes in Carbohydrate Synthesis*, ACS Symp. Ser., **466**, 1-22 (1991).

158. Whitesides, G. M., and Wong, C. H., *Angew. Chem., Int. Ed. Engl.*, **24**, 617 (1985).

159. Von der Osten, C. H., Sinskey, A. J., Barbas, C. F., Pedersen, R. L., Wang, Y. F., and Wong, C. H., *J. Am. Chem. Soc.*, **111**, 3924 (1989).

160. Drueckhammer, D. G., Durrwachter, J. R., Pederson, R. L., Crans, D. C., Daniels, L., and Wong, C. H., *J. Org. Chem.*, **54**, 70 (1989).

161. See for example: a) Kragl, U., Gygax, D., Ghisalbra, O., and Wandrey, C., *Angew. Chem.*, Int. Ed. Engl., **30**, 827-828 (1991). b) Nagy, J. O., and Bednarski, M. D., *Tetrahedron Lett.*, **32**, 3953-3956 (1991). c) Kajlmoto, T., Liu, K. C., Pederson, R. L., Zhong, Z., Ichikawa, Y., Porco, J. A., and Wong, C. H., *J. Am. Chem. Soc.*, **113**, 6187-6196 (1991).

162. European Patent Appl. 304.021 (1989) to Takeda.

163. European Patent Appl. 307.023 (1989) and 383.403 (1990) to DSM/Novo-Nordisk.

164. Simon, H., in *Biocatalysis*, Abramowicz, D. A. (ED.), Van Nostrand Reinhold, New York, 1990, pp. 217-242.

165. Fraisse, L., and Simon, H., *Arch. Microbiol.*, **150**, 381-386 (1988).

166. Adams, M.W.W., *CHEMTECH*, 692-699 (91991).

167. Sonnleitner, B., and Fiechter, A., *Trends Biotechnol.*, **1**(3), 74-80 (1983).

168. Zentgraf, B., and Ringpfeil, M., *Pure Appl. Chem.*, 1528-1533 (1991).

169. a) Daniel, R. M., Bragger, J., and Morgan, H. W., in *Biocatalysis*, Abramowicz, D. A. (Ed.), Van Nostrand Reinhold, New York, 1990, pp. 243-254. b) Zeikus, J. G., Lowe, S. E., and Saha, B. C., in *Biocatalysis*, Abramowicz, D. A. (Ed.), Van Nostrand Reinhold, New York, 1990, pp. 255-276.

170. Gray, C. J., *Genet. Eng. Biotechnol.*, **10**(5), 15-18 (1990).

171. Hilvert, D., *Trends Biotechnol.*, **9**, 11-17 (1991).

172. Ahern, T. J., *Pure Appl. Chem.*, 1538-1540 (1991).
173. a) Bednarski, M. D., Chenault, H. K., Simon, E. S., and Whitesides, G. M., *J. Am. Chem. Soc.*, **109**, 1283-1285 (1987). b) Bednarski, M. D., in *Catalysis of Organic Reactions*, Blackburn, D. W. (Ed.), Marcel Dekker, New York, 1990, pp. 61-69.

ADDITIONAL READING

1. Dordick, J. S., Principles and Applications of Nonaqueous Enzymology, in *Applied Biocatalysis*, Blanch, H. W., and Clark, D. S. (Eds.), Marcel Dekker, New York, 1991, pp. 1-51.
2. Roberts, S. M., and Turner, N. J., Some Recent Developments in the Use of Enzyme Catalysed Reactions in Organic Synthesis, *J. Biotechnol.*, **22**, 227-244 (1992).
3. Gupta, M. N., Enzyme Function in Organic Solvents, *Eur. J. Biochem.*, **203**, 25-32 (1992).
4. Jongejan, J. A., Van Tol, J. B. A., Geerlof, A., and Duine, J. A., Enantioselective Enzymatic Catalysis. 1. A Novel Method to Determine the Enantiomeric Ratio, *Recl. Trav. Chim Pays-Bas*, **110**, 247-254 (1991).
5. Van Tol, J. B. A., Jongejan, J. A., Geerlof, A., and Duine, J. A., Enantioselective Enzymatic Catalysis. 2. Applicability of Methods for Enantiomeric Ratio Determinations, *Recl. Trav. Chim Pays-Bas*, **110**, 255-262 (1991).
6. Holland, H. L., *Organic Synthesis with Oxidative Enzymes*, VCH, Weinheim, 1992.

8

Catalytic Asymmetric Synthesis

Abiological catalysts can be highly selective yet avoid the substrate restrictions usually imposed by the so-called lock-and-key-based selectivity of enzymatic catalysts.

Barry Sharpless, 1985

The enzymatic processes dealt with in chapter 7 were predominantly kinetic resolutions of racemates, usually involving an enantiospecific solvolysis. The subject matter of this chapter, on the other hand, has generally become synonymous with asymmetric syntheses catalyzed by chiral (transition) metal complexes. Note, however, that the subject also comprises chiral (Lewis) acids and bases as catalysts. Biocatalytic asymmetric syntheses are discussed in this chapter solely for the purpose of comparison with their chemocatalytic counterparts.

The reactions involved are generally redox transformations or C-C bond forming processes and they tend to complement the enzymatic hydrolytic processes of the preceding chapter. There are three different types of chemocatalysts: heterogeneous metal catalysts, homogeneous complexes and soluble chiral acids and bases. It is convenient to organize this chapter on the basis of reaction type rather than catalyst type. Most of the reactions discussed can also be carried out as kinetic resolutions [1] and, where relevant, examples are presented.

I.   HISTORICAL DEVELOPMENT

A.   Early Work

Cyanohydrin synthesis, the reaction that was the focus of Emil Fisher's seminal work on asymmetric induction, appears to be the first reaction to have been subjected to asymmetric catalysis. Bredig [2] showed that quinine catalyzes the enantioselective hydrocyanation of aldehydes. Subsequently, cinchona alkaloid bases have been shown to catalyze a wide variety of enantioselective processes [3].

As has been pointed out by Blaser [4], early studies of **chiral heterogeneous catalysts** were strongly influenced by research on the origins of chirality. Two different approaches were followed. In the first approach chiral solids, such as quartz [5] or biopolymers (e.g., silk fibroin [6], cellulose [7], and other polysaccharides [8]) were used as supports for metallic catalysts in asymmetric (de)hydrogenations. Enantioselectivities were poor to moderate, but the principle had been demonstrated.

The second approach involved the modification of heterogeneous metal catalysts by doping with naturally occurring chiral substances. The modification of a Pt hydrogenation catalyst by a cinchona alkaloid, for example, was first reported in 1939 [9]. Later, cinchona alkaloid-modified Pt catalysts were shown to mediate the asymmetric hydrogenation of α-keto esters with enantioselectivities in excess of 90% [4,10]. The most studied heterogeneous asymmetric hydrogenation catalyst is undoubtedly the Ni/tartrate/NaBr system developed by Izumi and coworkers [11], which catalyzes the enantioselective hydrogenation of β-diketones and β-keto esters.

The first reported example of **homogeneous asymmetric catalysis** by a soluble chiral metal complex dates from 1966 [12]; it involved an asymmetric cyclopropanation catalyzed by a chiral Schiff base complex of $Cu^{(II)}$ (Figure 8-1).

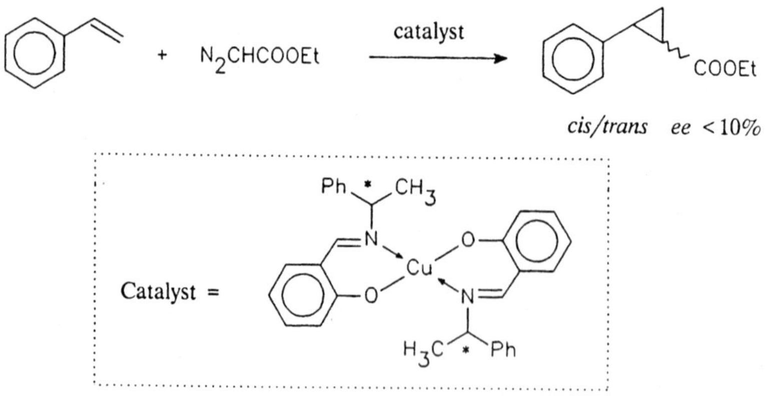

cis/trans   ee <10%

Figure 8-1. The first homogeneous asymmetric catalysis by a chiral metal complex.

The product was a mixture of *cis* and *trans* isomers and had an optical purity of < 10%. Subsequent systematic screening of chiral Schiff bases led to the development of cyclopropanation catalysts affording enantioselectivities up to 94%.

B.   The Monsanto L-Dopa Process

An important landmark in the history of asymmetric homogeneous catalysis was the development in the late 1960s of asymmetric hydrogenation catalysts based on rhodium complexes with chiral phosphine ligands. This resulted from the convergence of two prior developments. The first was the discovery by Wilkinson and coworkers of the homogeneous hydrogenation catalyst, RhCl(Ph$_3$P)$_3$ [13]; this was the first soluble catalyst that exhibited comparable activities to the well-known heterogeneous hydrogenations catalysts. The second was the development by the group of Horner [14] and Mislow [15] of methods for the synthesis of optically active phosphines.

In 1968, Monsanto scientists [16] described the catalytic asymmetric hydrogenation of α-phenylacrylic acid to α-phenylpropionic acid, with a 15% *ee*, using a rhodium-chiral phosphine catalyst. In the same year Horner and coworkers [14] reported the asymmetric hydrogenation of α-methoxystyrene in 6% *ee*. Both groups used the same chiral monodentate phosphine, methylpropylphenylphosphine (PPMP), with the chirality derived from the stereogenic phosphorus atom.

Further development of this discovery by the Monsanto group culminated in its application to the manufacture of L-Dopa [17,18]. The key step in this process is the asymmetric hydrogenation of an enamide (Figure 8-2). Interestingly, initial attempts with PPMP gave a product with only 28% *ee*. Replacement of the *n*-propyl group by *o*-anisyl (PAMP ligand) increased the *ee* to 60%. Subsequent replacement of the phenyl group by the bulkier cyclohexyl (CAMP ligand) resulted in an *ee* of 85%. Finally, application of a chelating bidentate analog of PAMP, so-called DIPAMP, gave a catalyst that was both more selective (95% *ee*) and more stable.

C.   Chiral Diphosphine Ligands

In 1971, Kagan and Dang [19] described an asymmetric hydrogenation catalyst containing a chiral diphosphine, so-called DIOP (Figure 8-3), derived from L-tartaric acid. The chirality of DIOP, in contrast to the above-mentioned ligands, is not due to a stereogenic phosphorus atom but to stereogenic centers in the carbon skeleton of the phosphine. Following these seminal discoveries a plethora of chiral bidentate phosphines was developed for use as chiral ligands in asymmetric catalysis by metal complexes [20–29]. Some of the more well known chiral phosphine ligands are illustrated in Figure 8-3.

From the viewpoint of scope of applications and industrial potential, BINAP, which appeared on the scene in the early 1980s [30], is the most important. At about the same time, Selke and coworkers [31] introduced the chiral diphosphinite ligand,

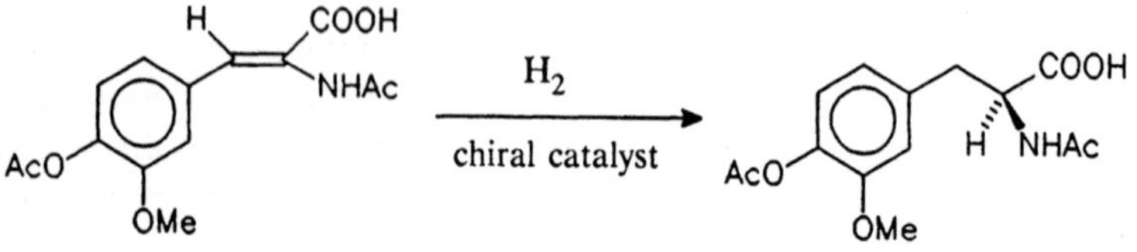

Catalyst = $[\,Rh(COD)L^*_2\,]^+\ BF_4^-$
COD = *1,5*-cyclooctadiene
L^* = chiral phosphine

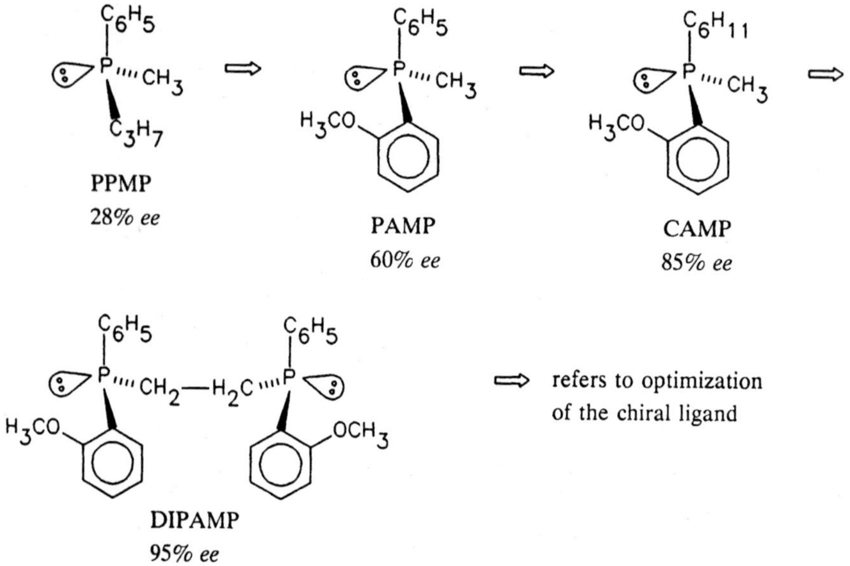

Figure 8-2. Optimization of the chiral ligand in the Monsanto L-dopa process.

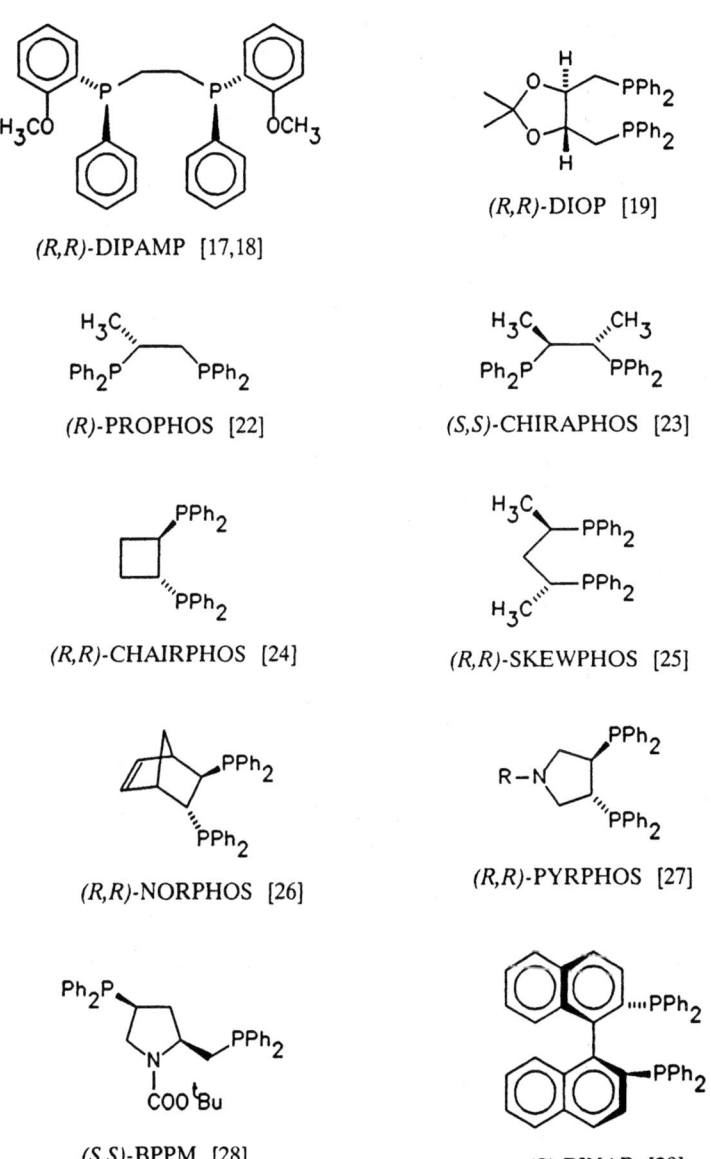

(R,R)-DIPAMP [17,18]

(R,R)-DIOP [19]

(R)-PROPHOS [22]

(S,S)-CHIRAPHOS [23]

(R,R)-CHAIRPHOS [24]

(R,R)-SKEWPHOS [25]

(R,R)-NORPHOS [26]

(R,R)-PYRPHOS [27]

(S,S)-BPPM [28]

(S)-BINAP [29]

Figure 8-3. Chiral bidentate phosphine and references (in brackets) to their synthesis.

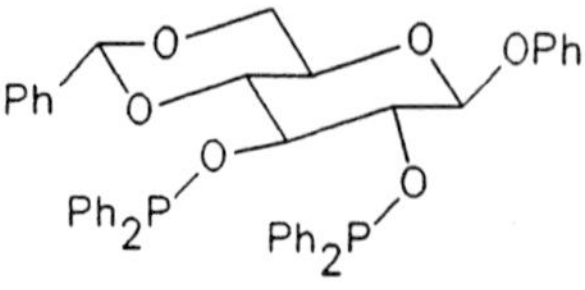

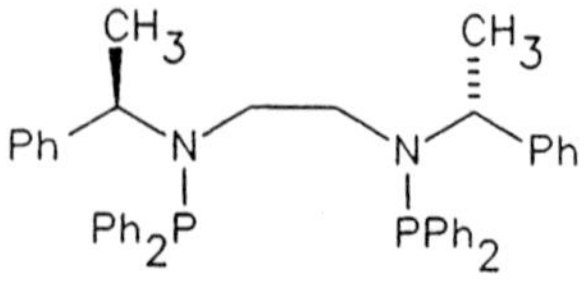

Ph–β–Glup   PNNP

Figure 8-4. Structures of Ph–β–glup and PNNP ligands.

Ph–β–glup, which was derived from glucose (Figure 8-4). This ligand formed the basis of an alternative L-dopa process, developed by VEB Isis Chemie [32], involving the same asymmetric hydrogenation as the Monsanto process.

Similarly, Enichem [33] developed a process for the manufacture of L-phenylalanine by asymmetric hydrogenation of Z-acetamido cinnamic acid using a rhodium complex of PNNP, a nitrogen analog of a diphosphinite, as the catalyst. Although the process is reportedly used in Italy it is doubtful whether it can compete with the synthesis of L-phenylalanine by de novo fermentation.

D.   Asymmetric Epoxidation

Another landmark in catalytic asymmetric synthesis was the discovery in 1980 of the catalytic asymmetric epoxidation of allylic alcohols by Katsuki and Sharpless [34]. The 'Sharpless epoxidation' [35] utilizes a soluble $Ti^{(IV)}$-tartrate catalyst and *tert*-butyl hydroperoxide (TBHP) as the oxidant and gives high enantioselectivities, often > 95%, for the epoxidation of a wide variety of allylic alcohols (see later). A modification of this system was subsequently applied by Kagan and coworkers [36] to the asymmetric oxidation of prochiral sulfides. However, although the Sharpless epoxidation has been widely applied in bench-scale organic synthesis, commercial applications are, as yet, limited to a few relatively small scale productions.

E.   The Takasago *l*-Menthol Process

From the viewpoint of industrial utilization, the real breakthrough in catalytic asymmetric synthesis was heralded by the discovery [30,37] of the Rh–BINAP-catalyzed enantioselective isomerization of prochiral allylic amines. This discovery led to the development, by Takasago [30] of a commercial process for the production of *l*-menthol based on the highly efficient asymmetric isomerization of diethyl-geranylamine to citronellaldiethylamine (Reaction 8-1). This process is currently used for the production of about 2000 tons per annum of *l*-menthol, making it by far the most important example of industrial catalytic asymmetric synthesis.

It was subsequently demonstrated by Noyori and coworkers [38–40] that ruthenium-Binap complexes have, in contrast to other systems, broad utility as

$$96\text{-}99\% \ EE \qquad (8\text{-}1)$$

highly efficient catalysts for the enantioselective reduction of a wide variety of substrates; several of these processes are currently in various stages of commercialization at Takasago [41]. These important discoveries made in the mid-1980s paved the way for extensive developments in the area of catalytic asymmetric synthesis [42–53]. The situation is reminiscent of the intensification of interest in enzymes that followed the demonstration that they can function in organic solvents.

Now that the problem of enantioselectivity has essentially been solved in many asymmetric hydrogenations, other problems are being addressed such as ease of recovery of the (expensive) catalyst. The development of water-soluble chiral ligands and the immobilization of expensive metal-chiral ligand catalysts pertain to this point.

II. BASICS OF TRANSITION METAL CATALYSIS

An in-depth treatment of transition metal catalysis is beyond the scope of this book. However, a rudimentary knowledge of the basic steps involved is useful for understanding asymmetric catalysis by transition metals.

A. The Catalytic Cycle

For every catalytic process we can draw a catalytic cycle. In many cases the substance added to the reaction is a catalyst precursor that is converted in situ to the active catalyst. The latter then enters the catalytic cycle (Figure 8-5). The activation step may involve ligand dissociation and/or a change in oxidation state of the metal. The active catalyst reacts with the substrate(s) to form a catalytic intermediate which subsequently decomposes into product(s) with simultaneous regeneration of the catalyst. A prerequisite for efficient catalysis is that both the rate of formation (k_1) and rate of decomposition (k_2) of the catalytic intermediate(s) are relatively fast; note that Figure 8-5 applies equally well to enzymatic catalysis.

This explains why in many transition metal-catalyzed processes (e.g., hydrogenation, carbonylation, hydroformylation), second-row elements (e.g., Pd, Rh, Ru) are often the best catalysts. First-row elements tend to be unreactive towards the substrate but form labile intermediates (i.e., $k_2 \gg k_1$). Third-row elements, on the other hand, tend to react readily with substrates but form intermediates that are

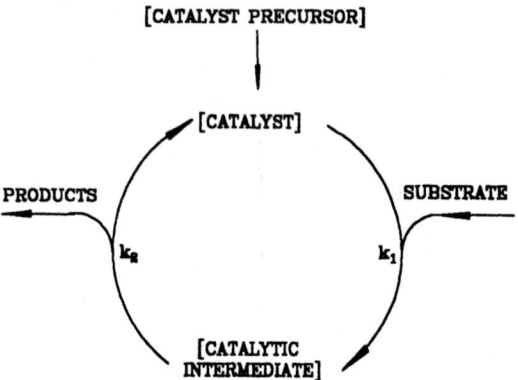

Figure 8-5. The catalytic cycle.

too stable (i.e., $k_1 \gg k_2$). Second-row elements tend to exhibit an optimum balance between the rate of intermediate formation and decomposition (i.e., $k_1 \approx k_2$ and both are reasonably high.)

For asymmetric catalysis it is also important to identify the step in which chiral discrimination occurs and it cannot be overemphasized that catalysis is purely a kinetic phenomenon. The isolation and characterization of intermediate complexes are only meaningful, therefore, when they can be related to kinetic measurements. This point is well illustrated by reference to the mechanism of asymmetric hydrogenation (see later).

B.   The Basic Steps

Catalytic reactions involving organometallic complexes of transition metals generally proceed through the following basic steps: ligand dissociation/coordination, oxidative addition, migratory insertion, and reductive elimination (Table 8-1). Ligand dissociation is followed by substrate coordination and activation. The latter may involve simple electron redistribution or transfer or cleavage via oxidative addition. In organometallic catalysis, substrate activation is generally followed by migratory insertion in a metal-hydride or metal-alkyl bond.

A convenient tool for understanding catalysis via organometallic intermediates is the 16/18 electron rule. An 18 electron complex possesses an inert gas configuration and must first undergo dissociation to create a vacant coordination site (coordinative unsaturation) necessary for substrate activation. The number of valence electrons counted for a particular metal is readily ascertained from its position in the periodic table (e.g., Mn=7, Fe=8, Co=9, etc.) and is independent of its oxidation state. Covalent (e.g., hydride, alkyl) and ionic (e.g., chloride) ligands

Table 8-1 Basic Steps in Transition Metal Catalysis

1. Ligand dissociation/coordination

   (a) Heterolytic: $M-L \quad \rightleftharpoons \quad M \; + \; L:$

               $L = CO, R_3P, etc.$

   (b) Homolytic: $M^{n+} - R \quad \rightleftharpoons \quad M^{(n-1)+} \; + \; R\cdot$

2. Oxidative addition and reductive elimination

   $$M^{n+} \; + \; A-B \quad \rightleftharpoons \quad \begin{matrix} A \\ \diagdown \\ B \diagup \end{matrix} M^{(n+2)+}$$

   $A-B = H_2, HCN, R_3Si-H, RX, RCH_2-H$

3. Insertion and extrusion

   (a) $M-H \; + \; H_2C=CH_2 \quad \underset{\text{dehydrogenation}}{\overset{\text{hydrogenation}}{\rightleftharpoons}} \quad \underset{\underset{H_2}{|}}{M-C-CH_3}$

   (b) $M-R \; + \; CO \quad \underset{\text{decarbonylation}}{\overset{\text{carbonylation}}{\rightleftharpoons}} \quad \underset{\overset{||}{O}}{M-C-R}$

count for one electron and coordinative ligands (e.g., CO, R_3P, R_3N) for two electrons. A mechanistic pathway in organometallic catalysis, generally speaking, proceeds via the alternating formation of 16 and 18 electron complexes. It should be pointed out, however, that although it is a useful tool there are many exceptions to the 16/18 electron rule. Moreover, it completely disregards homolytic pathways. A specific example that illustrates the above principle is the hydroformylation of olefins catalyzed by $HCo(CO)_4$ (Figure 8-6).

In the hydrogenation of olefins catalyzed by Wilkinson's complex, $Rh(Ph_3P)_3Cl$, the catalyst is a 16 electron complex. In contrast to what one would expect from the 16/18 electron rule, the complex undergoes initial dissociation of a Ph_3P ligand to give a reactive 14 electron species (Figure 8-7). This unexpected behavior is due to the steric bulk of the Ph_3P ligands that hinders the coordination of an extra ligand.

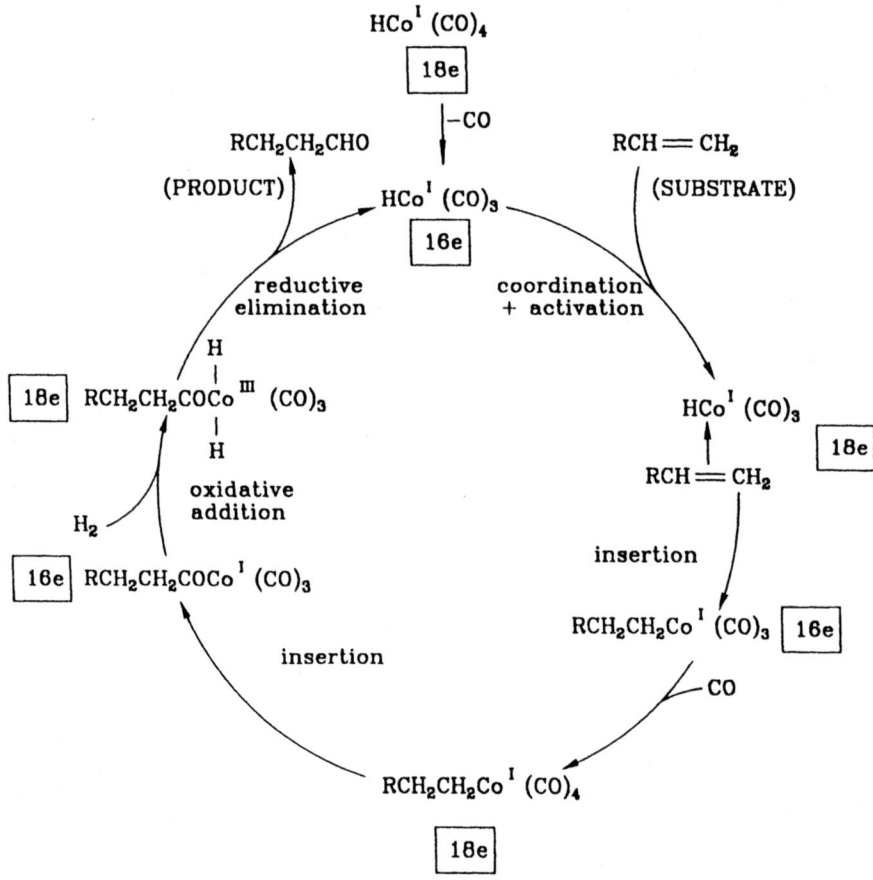

Figure 8-6. Catalytic cycle of cobalt-catalyzed hydroformylation of olefins.

The manifest ability of transition metals to catalyze myriad transformations of organic molecules stems from a combination of various characteristic features such as:

1. The accessibility of several oxidation states and coordination numbers for a particular metal.
2. The ability to coordinate groups in a specific array (template effect) that is conducive to high stereo- and/or regioselectivity.
3. The propensity for activation of small molecules (e.g., CO, H_2, HCN, etc.)
4. The ability to stabilize a wide variety of reactive intermediates (e.g., hydride, alkyl, aryl, and/or alkylidene) via coordination through σ- and/or π-bonding.

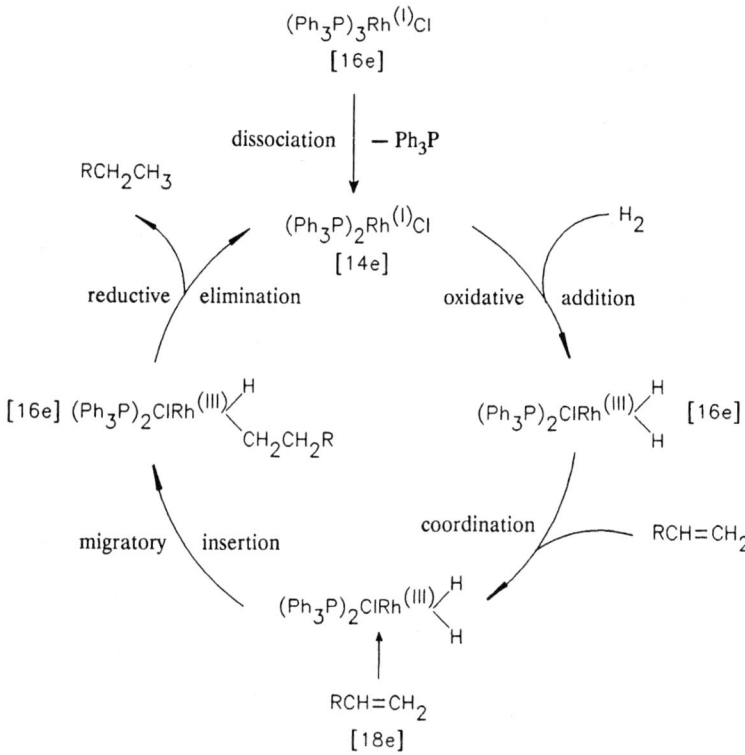

Figure 8-7. Catalytic cycle of RhCl(Ph3P)3-catalyzed hydrogenation.

C. Ligand-Accelerated Asymmetric Catalysis

A sine qua non for the observation of high enantioselectivity is that coordination of the chiral ligand results in a substantial rate acceleration. Sharpless [54] coined the term **ligand-accelerated catalysis** to describe this phenomenon. Thus, if the metal-chiral ligand complex (M-L*) readily dissociates in solution then high enantioselectivities will be observed only when M-L is a much more active catalyst than M (Reaction 8-2). Reactions catalyzed by metalloenzymes, for example, all involve substantial ligand accelerated catalysis derived from coordination of the protein chiral ligand to the metal.

$$M \ + \ L^* \ \rightleftharpoons \ M—L^*$$

| achiral | chiral |
| catalyst | catalyst |

(8-2)

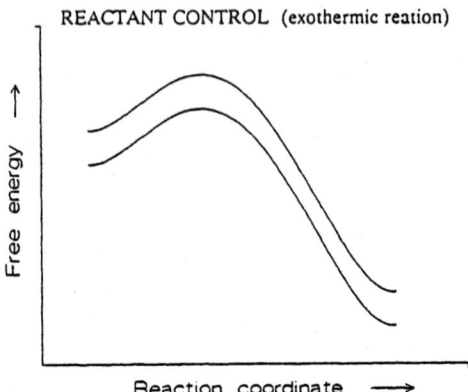

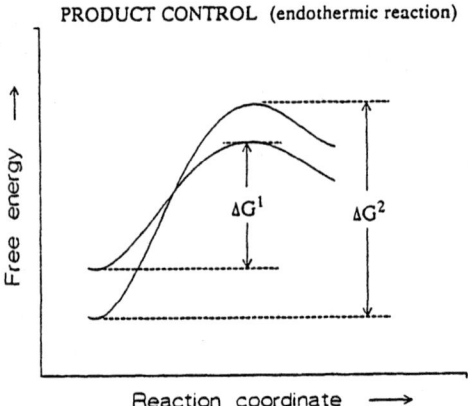

Figure 8-8. Reactant and product control in asymmetric catalysis.

D. Chiral Recognition and Enantiocontrol

In asymmetric catalysis, a chiral modifier (ligand) that is attached to the (metal) catalyst discriminates the enantiotopic features of a prochiral substrate. (See chapter 1.) In other words, the chiral catalyst reacts at different rates with, for example, the two enantiotopic faces of a prochiral olefin to form energetically

$$M \quad + \quad L^* \quad \rightleftharpoons \quad M—L^*$$

achiral   chiral
catalyst   catalyst

different diastereomeric transition states. In this case, the enantioselectivity is determined by the difference in free energy of these two diastereomeric transition states, a difference of 3 kcal/mole being large enough to give > 99% *ee*.

The enantioselection is determined in the first irreversible step involving diastereomeric transition states. Identification of this step provides the key to designing efficient asymmetric catalysts. What happens subsequent to the enantioselective step may have a bearing on other features of the reaction but not on the enantioselectivity.

A prerequisite for good enantiocontrol is generally assumed to be that the catalyst and substrate should be rigidly held. This is facilitated by the presence of secondary binding sites in the substrate (three-point contact model; see chapter 1), which serve to orient the latter in a unique disposition with regard to the chirality of the catalyst.

Since it is not possible to observe transition states directly, their structures and stabilities have to be inferred from those of the reactants, intermediates and products. A useful tool in this context is Hammond's postulate which states that the structure of the transition state resembles most closely the species closest to it in energy in the reaction profile. Accordingly, in an exothermic reaction the more stable transition state resembles more closely the diastereomers of the catalyst-substrate complex and the reaction is said to be under **reactant control**. For an endothermic reaction, on the other hand, it resembles more closely the catalyst-product diastereomers and the reaction is said to be under **product control**. In some cases, for instance in asymmetric hydrogenation catalyzed by chiral Rh complexes, the reaction profiles cross (Figure 8-8), which means that the enantiomer that is selectively formed is derived from the minor (i.e., less stable) diastereomer of the catalyst-substrate complex.

E. Chiral Ligands and Their Sources

The vast majority of chiral ligands used in catalytic asymmetric synthesis are based on raw materials available from the chirality pool; typical examples are tartaric acid (esters) and cinchona alkaloids. The former are used as chiral ligands in heterogeneous nickel-catalyzed asymmetric hydrogenations and homogeneous titanium-catalyzed asymmetric epoxidation. The latter are chiral ligands in heterogeneous palladium-catalyzed asymmetric hydrogenations and homogeneous osmium-catalyzed asymmetric dihydroxylation of olefins. Cinchona alkaloids are also asymmetric catalysts, in their own right, in a variety of base-catalyzed reactions [3]. It is no mere coincidence that chiral ligands for asymmetric catalysis are often based on the same substances that are used as resolving agents in classical resolutions; apparently the three dimensional structures of these molecules form an ideal basis for chiral recognition.

The majority of catalytic asymmetric syntheses to date involve the use of chiral bidentate phosphine ligands (Figure 8-3). They are generally synthesized from relatively inexpensive raw materials (e.g., tartaric acid and L-proline) that are available from the chirality pool. Most of them are commercially available, at least in small quantities. They are, however, expensive and often cost as much per kilo as the noble metals that they coordinate to. Therefore, a prerequisite for commercial viability is high turnover numbers (preferably > 10,000).

The chiral bidentate phosphines coordinate to transition metal ions (e.g., Ru, Rh, Pd) to form chelate rings varying in size from five for 1,2-diphosphines (e.g., DIPAMP, CHIRAPHOS) to seven for 1,4-diphosphines (e.g., DIOP, BINAP). A common feature of these chelate complexes is the arrangement of the four benzene rings in an alternating 'edge-face' array as illustrated for Rh–CHIRAPHOS below.

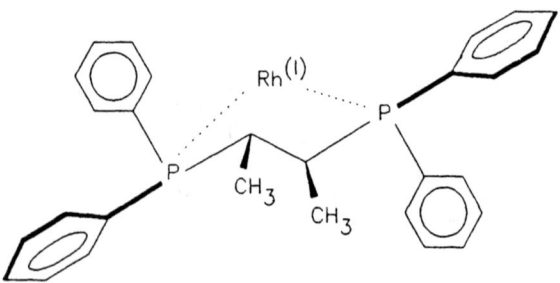

Structure of Rh$^{(I)}$CHIRAPHOS

Enantioselection results from the binding of a prochiral substrate (e.g., an olefin) to this chiral template. The preferred enantiomer is derived from the catalyst-substrate complex in which the bulky substituents of the substrate lie in front of the face-exposed benzene rings of the catalyst. It is worth noting that almost all of the chiral diphosphines in Figure 8-3 exhibit C_2 symmetry. Although this is not a prerequisite for efficient enantioselection, an attractive feature for C_2 symmetric ligands is that when bound to a metal the substrate experiences the same chiral environment regardless of the side from which it approaches.

Otsuka and Tani [30] and Noyori [38–40] have commented at length on the unique features of the axially dissymmetric diphosphine, BINAP, which make it such a highly effective ligand. An important feature is that conformational changes in the chelate ring cannot lead to achiral disposition of the four groups attached to the phosphorus atoms. This contrasts with the other 1,4-diphosphine, DIOP, in which conformational flexibility in the seven-membered chelate ring can lead to a loss of chirality (Figure 8-9). Thus, BINAP accommodates a wide variety of transition metals to form seven-membered chelate rings that are both pliable and conformationally unambiguous. Such a metal complex containing a large, pliable ligand has been compared to an enzyme [30].

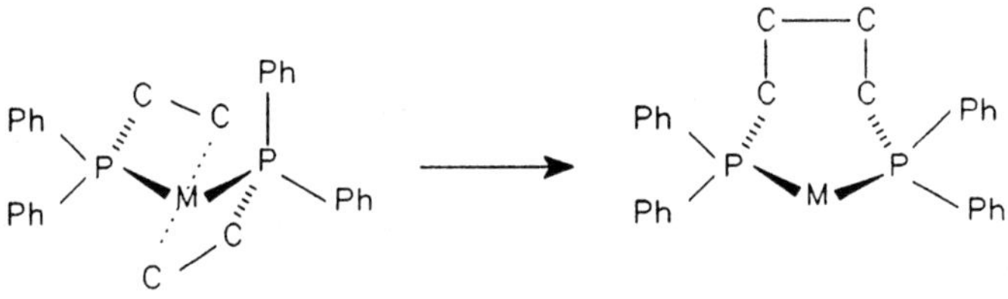

Figure 8-9. Loss of chirality via conformational change in DIOP-metal complex.

F.   Water-Soluble Chiral Ligands

An inherent disadvantage of homogeneous catalysts compared to their hetero-geneous counterparts is the problem of recovery and recycling of the catalysts. One approach to circumventing this problem [55] is to attach the catalysts to an insoluble support, similar to the immobilization of enzymes. Another approach which has received much attention is the use of metal complexes of water-soluble ligands [56]. For example, water-soluble phosphines are produced by introducing a highly polar functional group such as a sulfonate, carboxylate or tetraalkyl-ammonium salt. Rhodium complexes of tri(3-sulfonatophenyl)phosphine are highly soluble in water and can catalyze hydrogenations, carbonylations, hydro-formylations, etc., in two-phase, aqueous-organic systems [56]. After completion of the reaction, the catalyst in the aqueous phase is readily separated from the product(s) in the organic phase and recycled to the next batch.

This approach has also been successfully applied to asymmetric catalysis [57–61]. Sulfonated chiral phosphines, for example, form rhodium complexes that exhibit high enantioselectivities in catalytic hydrogenations of dehydroamino acids and which can be reused without loss of enantioselectivity. Some examples of water-soluble chiral ligands are shown in Figure 8-10.

III.   ASYMMETRIC C-H BOND FORMATION

In the majority of chiral compounds of industrial relevance one of the substitu-ents attached to a stereogenic carbon is a hydrogen atom. This hydrogen atom can be delivered to a prochiral carbon atom by hydrogenation of a C=X double bond (Figure 8-11). By the same token, a variety of double bond migration reactions result in the creation of stereogenic carbon centers containing a C-H bond (Figure 8-11). This probably explains why much of the effort in the area of catalytic asymmetric synthesis has been focussed on hydrogenations and allylic isomerizations.

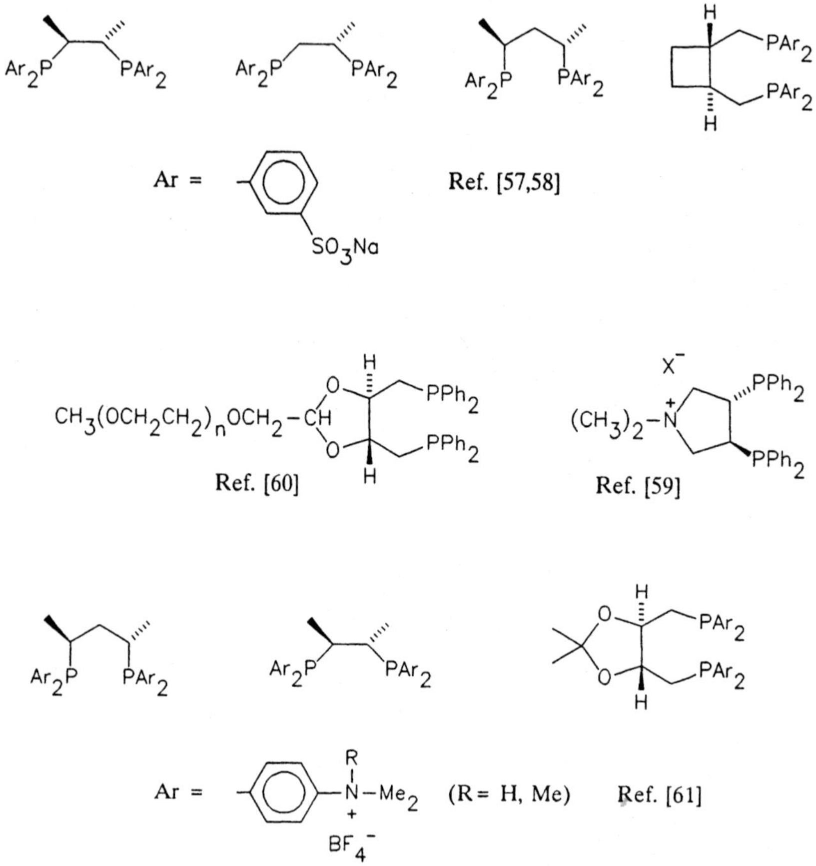

Figure 8-10. Structures of water-soluble chiral phosphines.

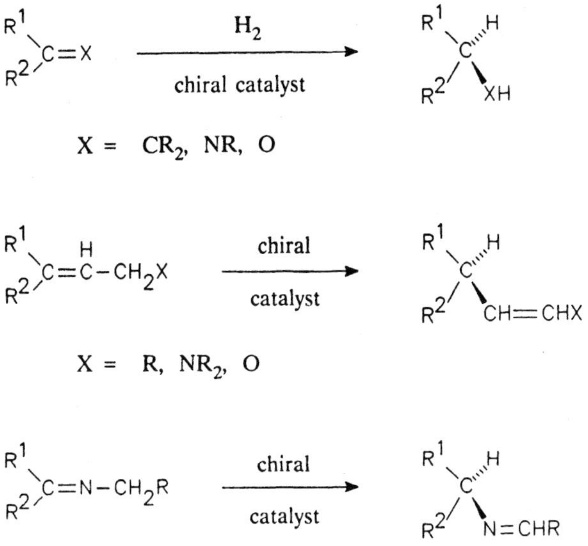

$$R^1_{R^2}C=X \xrightarrow[\text{chiral catalyst}]{H_2} R^1_{R^2}\overset{H}{\underset{XH}{C}}$$

$$X = CR_2, NR, O$$

$$R^1_{R^2}C=C\overset{H}{\underset{}{-}}CH_2X \xrightarrow[\text{catalyst}]{\text{chiral}} R^1_{R^2}\overset{H}{\underset{CH=CHX}{C}}$$

$$X = R, NR_2, O$$

$$R^1_{R^2}C=N-CH_2R \xrightarrow[\text{catalyst}]{\text{chiral}} R^1_{R^2}\overset{H}{\underset{N=CHR}{C}}$$

Figure 8-11. Asymmetric generation of C-H bonds by hydrogenation or isomerization.

A. Asymmetric Hydrogenation of Olefins

Asymmetric hydrogenation catalyzed by soluble transition metal complexes is undoubtedly the most studied technology within the area of catalytic asymmetric synthesis [62–67]. Most of the early work was concerned with the rhodium-catalyzed asymmetric hydrogenation of dehydroamino acid derivatives. Following the development of the DIPAMP ligand by Monsanto scientists several chiral bidentate diphosphines were applied with success to these reactions. (See Table 8-2 for a comparison of various systems.) The catalyst is a cationic Rh^I complex containing one bidentate chiral phosphine and a 1,5-cyclooctadiene (COD) ligand. Enantioselectivities for acetamido acrylic acid type substrates are very high and the hydrogenation conditions are moderate.

High enantioselectivities are obtained only with the Z-isomers of the dehydroamino acid which is attributed to the necessity for secondary binding of the oxygen atom of the acetamido group to the rhodium atom as shown in Figure 8-12. This secondary binding creates a rigid catalyst-substrate complex with a unique chiral ligand-olefin arrangement conducive to high enantioselection.

The mechanism of the rhodium-catalyzed reaction has been elucidated by Halpern and coworkers [68–70]. It is an interesting example of product control of enantioselection involving a crossover in the reaction profile. In other words, the stereochemistry of the product is controlled by the minor diastereomer of the

Table 8-2 Asymmetric Hydrogenations of Dehydroamino Acid Derivatives [31,42,62]

$$X = BF_4^-, \ ClO_4^-, \ etc. \ ; \qquad P{-}P = \text{chiral diphosphine}$$

Substrate	Chiral diphosphines *ee* (%)					
	DIPAMP	DIOP	CHIRAPHOS	BPPM	PYRPHOS	Ph−β−Glup
	94	73	91	98	-	98
	96	64	99	-	-	-
	95	81	89	91	99	97
	94	84	83	86	97	94

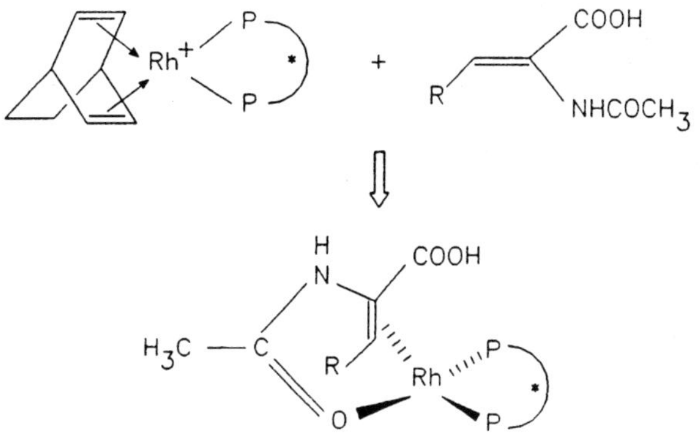

Figure 8-12. Secondary binding of the acetamido group to Rh in the catalyst-substrate complex.

catalyst-substrate complex. Subsequent oxidative addition of hydrogen to the latter is the first irreversible step and hence is the enantiocontrolling step in the process (Figure 8-13). In the case of dehydroamino acids the minor (i.e., less stable) diastereomer of the catalyst-substrate complex reacts much faster with hydrogen to afford the *(S)*-product.

Both the rates and enantioselectivities of these reactions are determined by the rate of oxidative addition of hydrogen and exhibit a complex dependence on hydrogen pressure and temperature. At low hydrogen pressures and/or high temperatures, the two substrate-catalyst diastereomers interconvert rapidly and yield the *(S)*-enantiomer. At high hydrogen pressures and/or low temperatures, interconversion between diastereomers is slow, which results in the formation of more of the *(R)*-enantiomer and a decrease in enantioselectivity. However, although the rhodium-based catalysts were very successful in the asymmetric hydrogenation of unsaturated acylamino acids and the related dehydro dipeptides [71–73], they proved to be much less effective with other prochiral olefins.

The $Ru(O_2CR)_2BINAP$ complexes, developed by Noyori and coworkers [38–40,74] and Takasago scientists [41], appear to have unprecedented broad utility as asymmetric hydrogenation catalysts. They are, for example, highly efficient catalysts for the asymmetric hydrogenation of a variety of α,β-unsaturated carboxylic acids that contain no additional functionality. *(S)*-Naproxen, for instance, was obtained [75] in 97% *ee* from the unsaturated precursor (Figure 8-14).

Similarly, allylic alcohols are another class of prochiral olefins that are hydrogenated with a high degree of selectivity using the Ru–BINAP catalysts [76]. For example, geraniol and nerol are hydrogenated at room temperature and 30–100 bar H_2 in methanol to give *(R)* or *(S)*-citronellol in virtually quantitative yield and

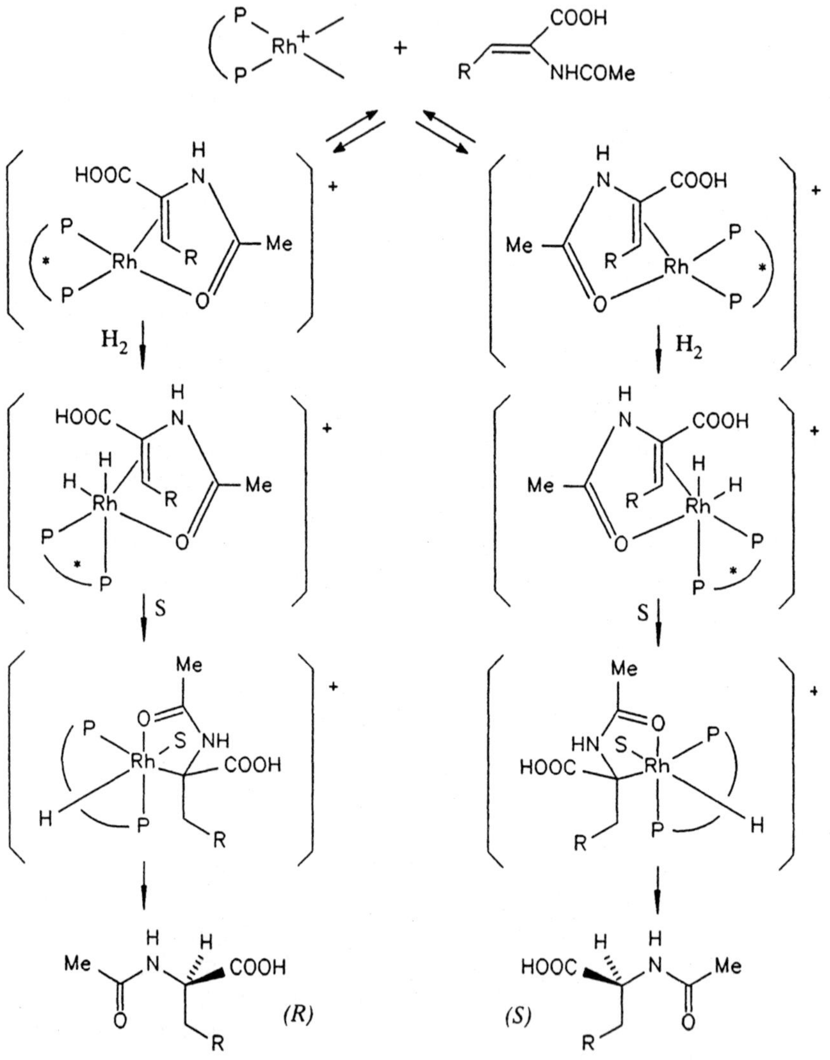

Figure 8-13. Mechanism of Rh-catalyzed asymmetric hydrogenation.

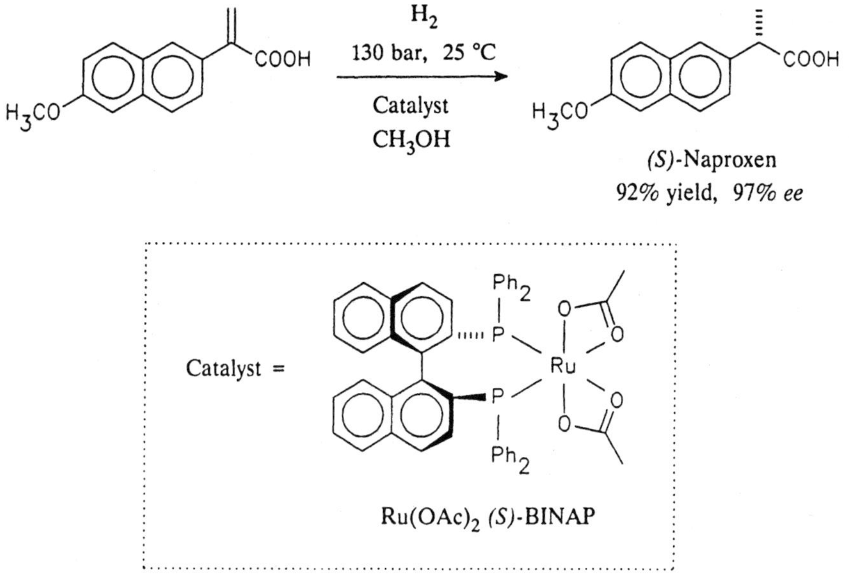

Figure 8-14. Synthesis of *(S)*-naproxen by catalytic asymmetric hydrogenation.

96–99% *ee* at substrate-catalyst ratios up to 50,000. These transformations (Figure 8-15) constitute elegant examples of the high degree of regio- and enantiocontrol, and are comparable to that observed in many enzymatic processes. This technique has also been successfully applied to the synthesis of a vitamin E intermediate and to the kinetic resolution of chiral allylic alcohols [77].

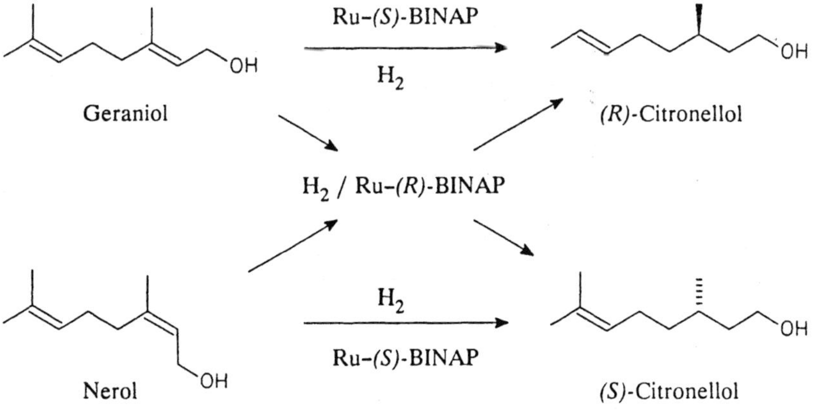

Figure 8-15. Synthesis of *(R)* and *(S)*-citronellol via Ru–BINAP-catalyzed hydrogenation.

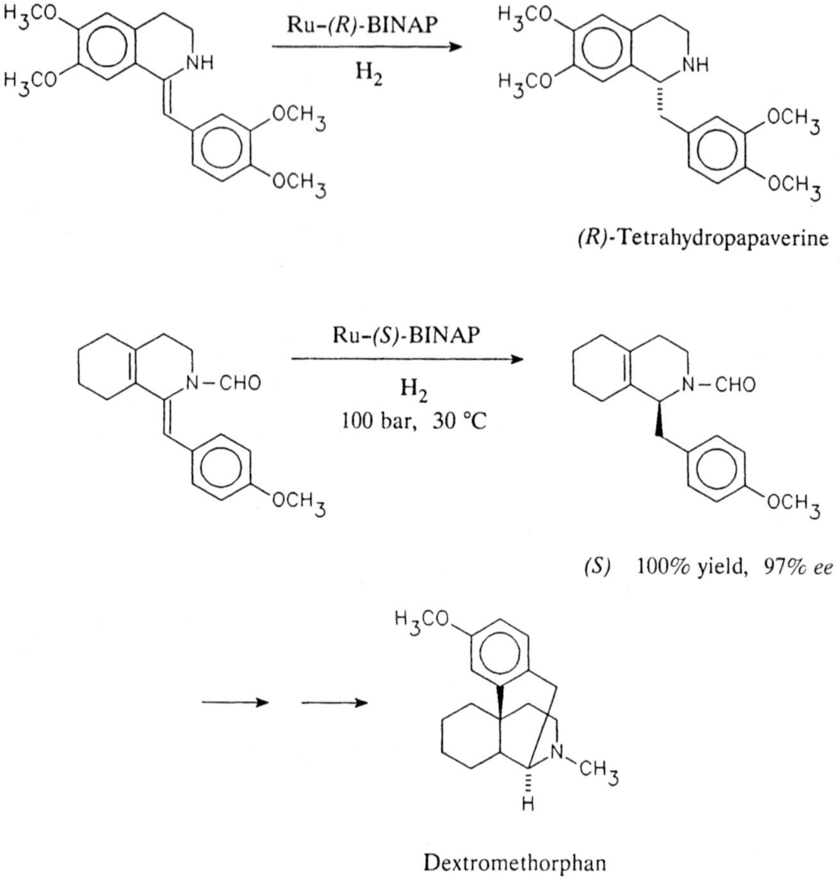

(R)-Tetrahydropapaverine

(S) 100% yield, 97% ee

Dextromethorphan

Figure 8-16. Asymmetric hydrogenation of Z-enamine and of Z-enamide.

A further application of the Ru–BINAP system is in the enantioselective hydrogenation of the Z-enamine and the Z-enamide [78] in 95–100% ee (Figure 8-16). These reactions are of commercial interest as key steps in the synthesis of a variety of alkaloids, such as tetrahydropapaverine and dextromethorphan. In contrast to the rhodium-catalyzed reactions discussed earlier, the mechanistic details of these Ru–BINAP-catalyzed asymmetric hydrogenations have yet to be elucidated.

B. Asymmetric Hydrogenation of C=O Bonds

The Ru–BINAP complexes have also been found to be excellent catalysts for the enantioselective reduction of a variety of functionalized ketones [41,79,80]. Inter-

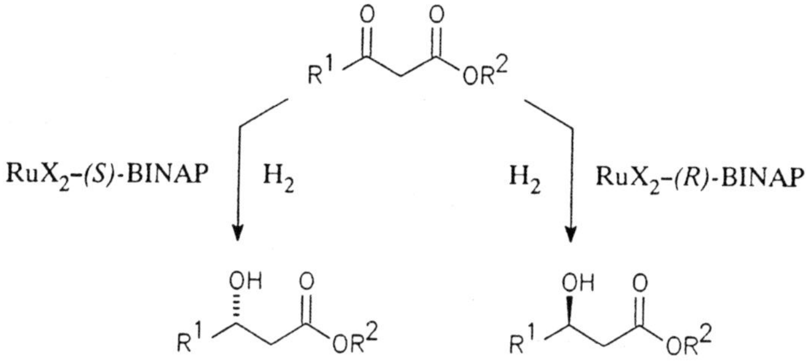

R^1	R^2	% ee
CH_3	CH_3	99
C_3H_7	CH_3	98
C_6H_5	CH_3	85

Figure 8-17. Asymmetric hydrogenation of β-keto esters.

estingly, the Ru–BINAP carboxylates that were singularly effective in olefin hydrogenation were ineffective in the hydrogenation of β-keto esters; on the other hand, the analogous Ru-halide complexes, RuX_2–BINAP, were excellent catalysts [79]. A variety of β-keto esters were hydrogenated in quantitative yield and about 98% ee using < 0.1% mol of catalyst in methanol at 25°C and 100 bar (Figure 8-17). Takasago [41] uses this technology for the manufacture of the chiral synthons (R)-and (S)-methyl-3-hydroxybutanoate in 98% yield and 99% ee.

When the β-keto ester contains a heteroatom (O, N, or Cl) at the γ or δ position the enantioselectivity is substantially decreased, which is probably due to the heteroatom competing with the ester group for coordination to Ru. Surprisingly, dramatic improvements were observed at higher reaction temperatures. Thus, the hydrogenation of ethyl-4-chloro-3-oxobutanoate gave the corresponding alcohol in 56% ee at room temperature but at 100°C the reaction was complete within 5 minutes and gave a product with 97% ee [71]. The (R)-enantiomer, obtained using (S)-BINAP as the chiral ligand, is of interest as a precursor of (R)-carnitine (Figure 8-18). The increase in enantioselectivity with increasing temperature is presumably due to a change in the enantioselection-determining step as was observed in Rh-catalyzed asymmetric hydrogenations. Similarly, ketones functionalized with dialkylamino, hydroxyl, alkylthio and even halogen were converted [79,80] to the corresponding alcohols in high enantioselectivity (Figure 8-19).

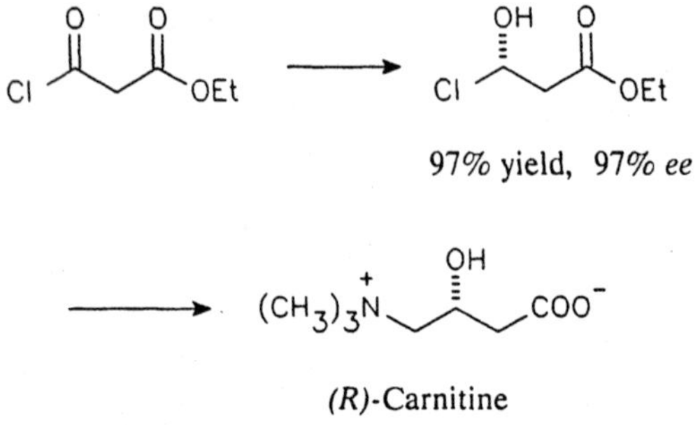

97% yield, 97% ee

(R)-Carnitine

Figure 8-18. Synthesis of *(R)*-carnitine via asymmetric hydrogenation.

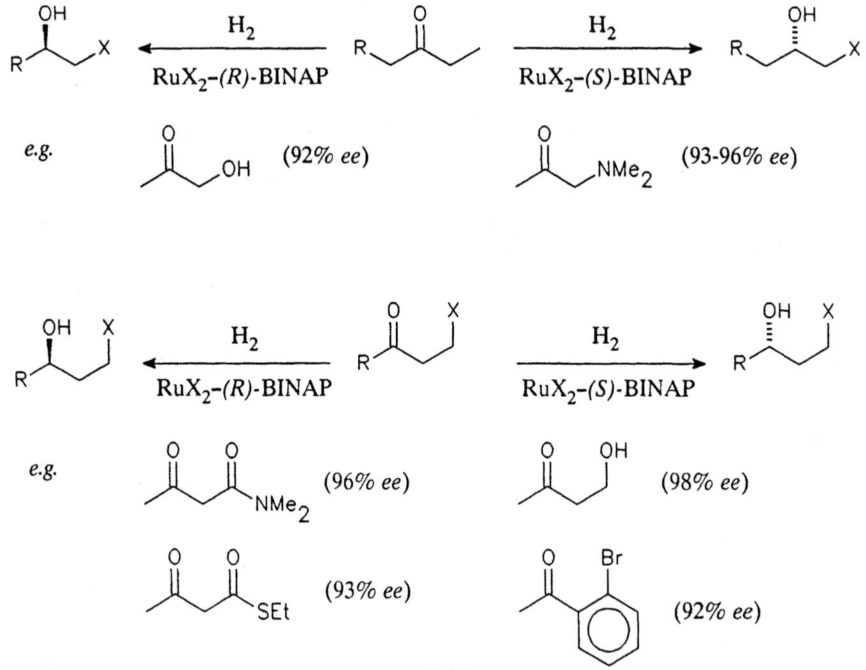

Figure 8-19. Enantioselective hydrogenation of functionalized ketones with RuX$_2$–BINAP.

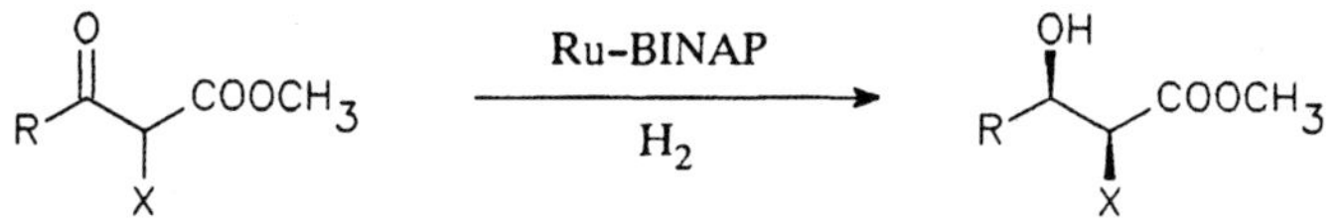

R	X	% de	% ee
CH_3	NHCOCH3	98	98
CH_3	NHCOPh	90	97
$t\text{-}C_4H_9OCH_2$	NHCOCH3	95	99
CH_3	CH_3	35	97
CH_3	Cl	80	94
CH_3	CH2NHCOPh	95	99
$PhCH_2$	Cl	95	92

Figure 8-20. Dynamic kinetic resolution of chiral β-keto esters.

This system has also been used for the remarkably selective dynamic kinetic resolution of optically labile β-keto esters [41,82,83]. The asymmetric hydrogenation of 2-substituted 3-oxocarboxylic esters (Figure 8-20) gives one of the four possible diastereomers in an enantio- and diastereoselective process. This reaction forms the basis of a commercial process developed by Takasago [41,84] for the synthesis of a key intermediate for carbapenem antibiotics (Figure 8-21).

The Ru–BINAP complexes are also effective catalysts for the enantioselective hydrogenation of 1,3-diketones [85], α-keto esters [41], and γ-keto esters [86] (Figure 8-22). The products from the latter reaction are easily converted to the corresponding chiral γ-lactones.

The spectacular success of the Ru–BINAP system tends to overshadow results obtained with other metal-chiral ligand combinations. It is worth noting, however, that in some cases good results have been obtained with Rh complexes with chiral bidentate phosphines in the asymmetric hydrogenation of carbonyl groups. For example, Rh complexes of BPPM catalyzed the enantioselective hydrogenation of ketopantolactone to (R)-pantolactone in 86% ee (Figure 8-23) [87]. The latter is a precursor of several biologically important molecules such as panthothenic acid (vitamin B_5). Subsequently, the same group showed [88] that BCPM, a close relative of BPPM, gave an ee of 92% and much higher rates.

Another close relative, Rh–MCCPM, is an effective catalyst for the enantioselective hydrogenation of β-alkylamino ketones [89,90]. The reaction forms the basis for syntheses of the pharmacologically active (S)-isomers of beta-blockers (Figure 8-24) [89] and the antiinflammatory drug, Levamisol (Figure 8-25) [90].

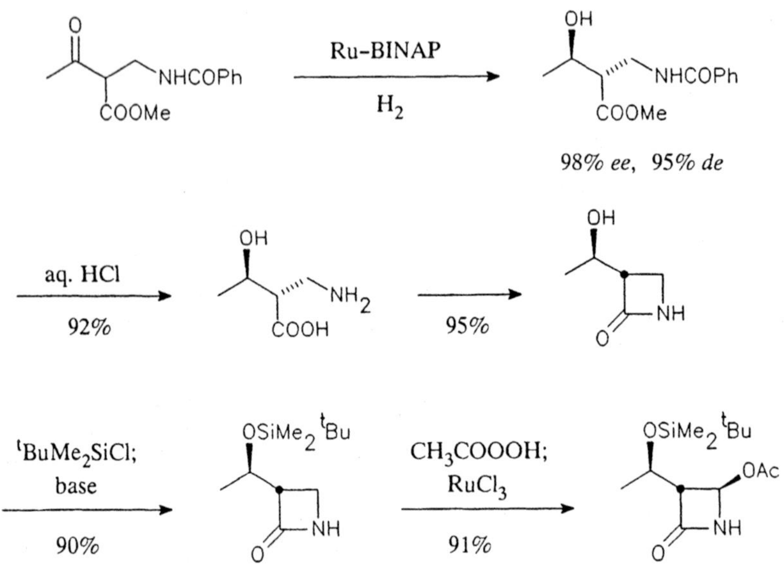

Figure 8-21. Synthesis of a carbapenem intermediate.

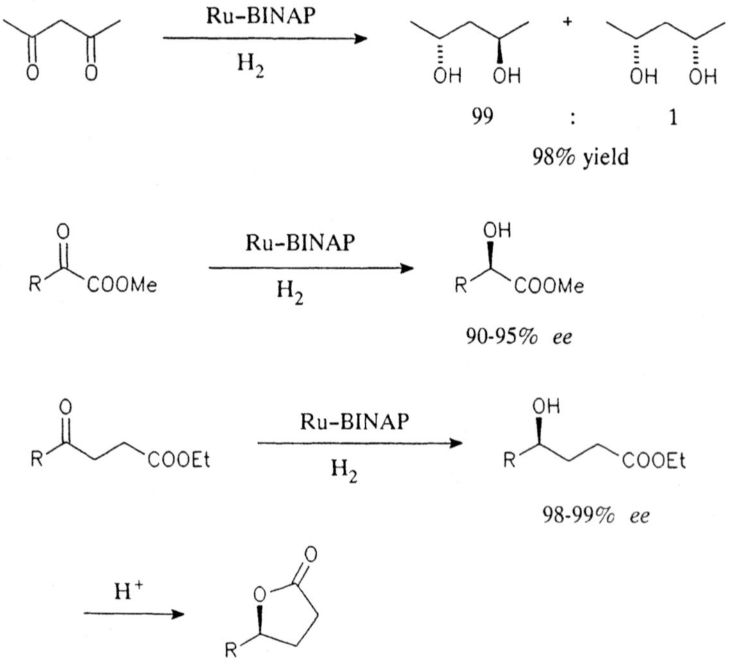

Figure 8-22. Enantioselective hydrogenations of 1,3-diketones, α-keto- and γ-keto esters.

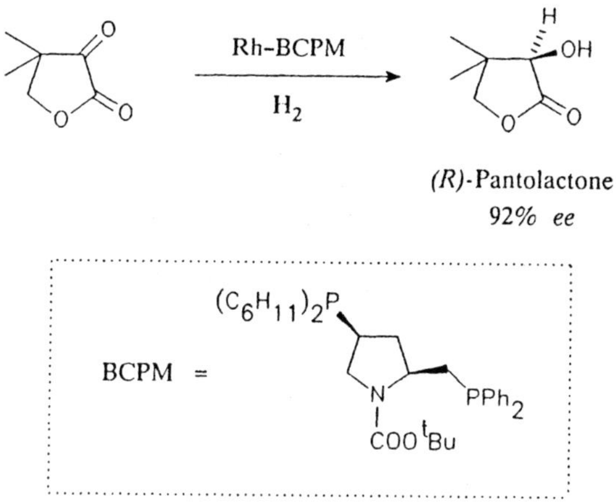

Figure 8-23. Synthesis of *(R)*-pantolactone.

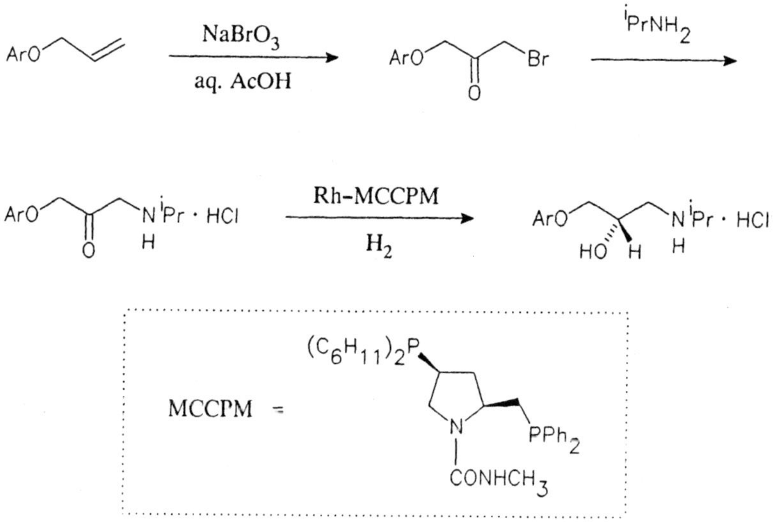

Ar	Product	*% ee*
α-naphthyl	*(S)*-Propanolol	91
p-(2-methoxyethyl)phenyl	*(S)*-Metoprolol	93

Figure 8-24. Synthesis of *(S)*-beta-blockers.

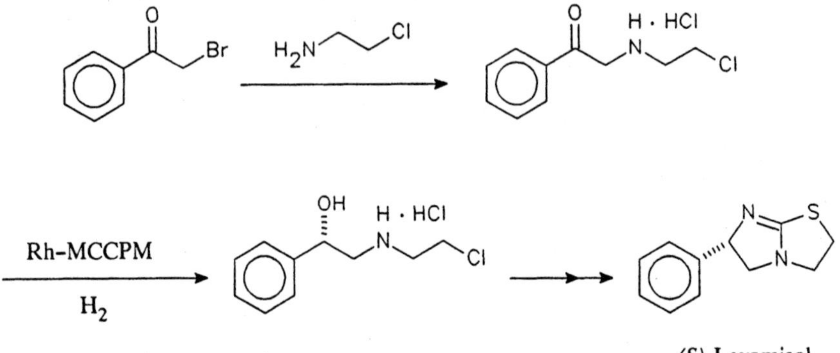

Figure 8-25. Synthesis of *(S)*-Levamisol.

In comparison with the high enantioselectivities observed with ketones containing secondary binding sites, enantioselective hydrogenations with simple, nonchelating ketones have been less successful. Bakos and coworkers [91] obtained *(S)*-2-phenylethanol in up to 76% *ee* by hydrogenation of acetophenone in the presence of a rhodium-SKEWPHOS catalyst and triethylamine as a promotor (Reaction (8-3). Chan and Landis [92] obtained similar results using a rhodium-DIOP catalyst and triethanolamine as promotor (Reaction (8-3). It was suggested that the promoting effect of bases is to remove a proton from a $Rh^{(III)}$ dihydride species, thereby forming an active anionic $Rh^{(III)}$ monohydride complex.

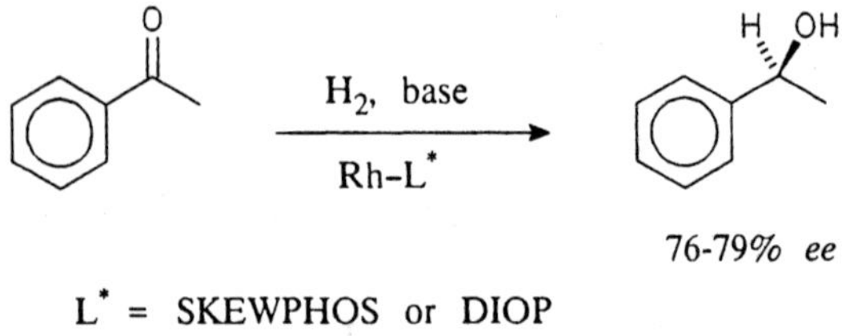

$$76\text{-}79\% \ ee \qquad (8\text{-}3)$$

$L^* = $ SKEWPHOS or DIOP

C. Heterogeneous Catalysts for C=O Hydrogenation

Two types of heterogeneous catalysts have been employed with some success in the enantioselective hydrogenation of functionalized ketones: tartaric acid-modified nickel catalysts [11,93–95] and cinchona alkaloid-modified platinum catalysts [4]. The tartrate/sodium bromide-modified Raney nickel catalyst, developed by Izumi [11] catalyzes the enantioselective hydrogenation of β-keto esters [93,94] and 1,3-diketones [95]. Brunner and coworkers [96] showed that Raney nickel can be replaced by commercially available nickel powders that are activated with H_2

prior to treatment with aqueous NaBr/tartaric acid. Hydrogenation of methyl-acetoacetate in the presence of this catalyst gave *(R)*- or *(S)*-methyl-3-hydroxy-butanoate in up to 77% *ee* (Reaction 8-4); moreover, the catalyst could be reused several times without loss of activity or enantioselectivity.

$$(8-4)$$

Optically pure *(R)*- or *(S)*-3-hydroxybutanoic acid can be easily obtained from the optically enriched ester (83% *ee*) by hydrolysis and recrystallization of the dibenzylammonium salt [97].

It is interesting to compare the three methods for the enantioselective reduction of β-keto esters from the viewpoint of industrial utility (Figure 8-26). The low price of Baker's yeast compared to Ru–BINAP is more than compensated for by the much higher productivities and enantioselectivities observed with the latter at high turnover numbers. However, although the heterogeneous catalyst cannot compete with Ru–BINAP on the basis of efficiency and scope, it is easy to prepare and is much less expensive and therefore it may be of use in some specific applications, especially when the crude product is easily purified via a simple recrystallization of a salt. The system has reportedly [4] been used in the synthesis of an intermediate for tetrahydrolipstatin, a pancreatic lipase inhibitor (Reaction 8-5).

RaNi = Raney nickel
TA = tartaric acid   (8-5)

Bläser and coworkers [4,98] have made extensive studies of cinchona alkaloid-modified Pt/Al$_2$O$_3$ catalysts in the enantioselective hydrogenation of α-keto esters. Enantioselectivities up to 95% were obtained by optimizing the catalyst, modifier, solvent, and reaction conditions. The practical utility was demonstrated [4] in the synthesis of *(R)*-2-hydroxy-4-phenylbutanoate, an ACE-inhibitor intermediate (Figure 8-27).

| | ASYMMETRIC HYDROGENATION | | YEAST REDUCTION |
	Homogeneous	**Heterogeneous**	
Substrate / g	Me ester / 40g	Me ester / 40g	Et ester / 40 g
Solvent / ml	MeOH / 40ml	none	H_2O / 2600ml
Catalyst	Ru–BINAP 0.14g	Ni / TA / NaBr *ca.* 2g	Yeast 200g
Reducing agent	H_2 / 20-100 bar	H_2 / 45 bar	Sucrose / 1 bar
Temperature	25 °C	100 °C	25 °C
Time (h)	40	20	80
Product isolation	Distillation	Distillation	Filtration/extraction/ distillation
Yield	96%	70-78%	59-76%
ee	99% *(R)* or *(S)*	77% *(R)* or *(S)*	*(S)*
Productivity g/l/h	12	*ca.* 40	*ca.* 0.1

Figure 8-26. Comparison of various techniques for asymmetric reduction.

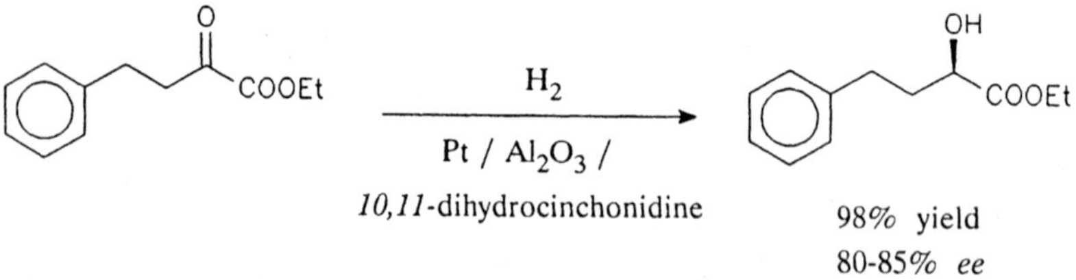

Figure 8-27. Synthesis of *(R)*-2-hydroxy-4-phenylbutanoate.

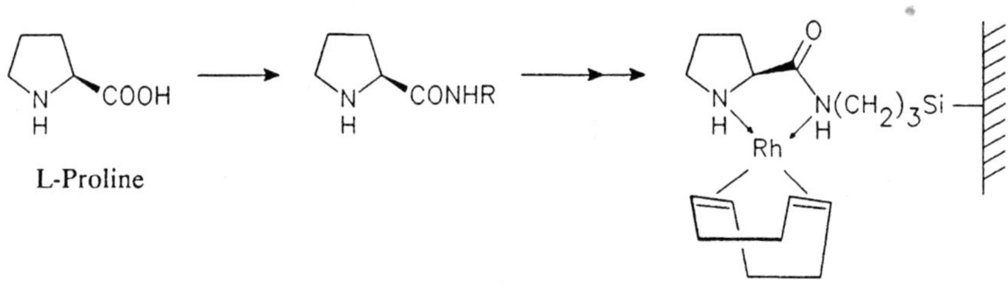

Figure 8-28. Chiral rhodium complex attached to a modified Y-zeolite.

Although some degree of success has been achieved with heterogeneous catalysts in the asymmetric reduction of functionalized ketones, the results obtained with prochiral olefins have been singularly unimpressive [4]. An interesting advance in this context is the development of a catalyst that contains Rh complexes with chiral bidentate nitrogen-based ligand attached to a modified Y-zeolite (Figure 8-28) [100]. Not only could this zeolite-bound catalyst be reused several times without loss of activity, it also gave superior (> 95%) enantioselectivities compared to the silica-supported or unsupported catalyst in the hydrogenation of N-acyl-dehydrophenylalanine derivatives. These results suggest that reaction takes place within the confined spaces of the micropores of the zeolite, where the metal complex is presumably anchored.

D.   Asymmetric Hydrogenation of C=N Bonds

Analogous to the asymmetric reduction of prochiral ketones to optically active secondary alcohols, the asymmetric reduction of prochiral ketimines constitutes, in principle, a useful route to optically active secondary amines. Moderate to good enantioselectivities were obtained in the hydrogenation of benzylimines of acetophenones in the presence of rhodium complexes of chiral phosphines [101, 102]. Two groups have reported the highly enantioselective (up to 96% ee) hydrogenation of the same substrates using rhodium complexes with water soluble, sulfonated SKEWPHOS ligands in a two-phase aqueous-organic system (Figure 8-29) [103,104].

Interestingly, it is only the monosulfonated SKEWPHOS ligand which gives such high enantioselectivities. Lensink and De Vries [104] isolated the pure monosulfonated ligand, which is a 1:1 mixture of two epimers (sulfonation creates a stereogenic phosphorus atom). The rhodium complex of this ligand is soluble only in the organic phase, presumably because the SO_3^- group acts as a counterion for Rh^I. Hydrogenation of acetophenone N-benzylimine in the presence of this complex gave the (R)-amine in 94% ee [104]. In contrast, hydrogenation in the

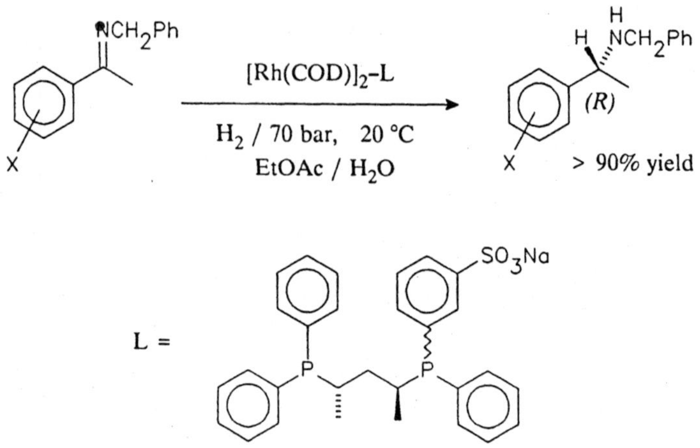

Figure 8-29. Enantioselective hydrogenation of ketimines.

| X | <------- *ee* % -------> | |
	Ref. [103]	Ref. [104]
H	96	94
4-OMe	95	92
3-OMe	89	-
2-OMe	91	-
4-Cl	-	92

presence of the rhodium complex of the disulfonated ligand, which was soluble only in water, gave a product with 2% *ee*. Obviously, the scope of this fascinating discovery in other asymmetric processes catalyzed by metal complexes of chiral phosphines is worthy of investigation.

Other avenues worth following are the catalytic asymmetric transfer hydrogenation of imines (see section III-E) and asymmetric reductive amination of ketones with in situ formation of prochiral ketimines. Finally, an alternative approach to the synthesis of optically active primary amines is via enantioselective double bond migration (Reaction 8-6) but apparently such a transformation has not been described. Its success will depend on the fact that the imine containing the double bond in conjugation with the aromatic ring is more stable than its unconjugated isomer.

$$(8-6)$$

E. Asymmetric Transfer Hydrogenation

An alternative to hydrogenation with molecular hydrogen is to use a hydrogen donor (e.g., isopropanol or formic acid) as the source of hydrogen atoms, similar to the reactions with dehydrogenases discussed in chapter 7. Brunner and co-workers [105] described the use of the triethylammonium salt of formic acid in combination with Rh-BPPM for the highly enantioselective transfer hydrogenation of dehydroamino acids and of itaconic to methylsuccinic acid (Reaction (8-7). The enantioselectivity in the latter reaction was even higher than with the corresponding hydrogenation with gaseous hydrogen [106]; this is not so surprising considering that oxidative addition of H_2 to the Rh^I-substrate complex is the step in which enantioselection occurs. Indeed, this phenomenon appears to be worthy of further attention, for example, in the enantioselective reduction of carbonyl compounds.

$$(8-7)$$

F. Asymmetric Hydrosilylation

The Si-H bond in silanes undergoes facile oxidative addition to a variety of low-valent transition metals to afford metal hydride species ($R_3Si-M-H$). The latter can subsequently add to C=C, C=N, or C=O bonds and this has been utilized to effect the enantioselective reduction of prochiral ketones. For example, hydrosilylation of acetophenone using a Rh complex of a chiral pyridyl-thiazolidine ligand (PYTHIA) gave the silyl derivative of *(R)*-2-phenyl-ethanol in 98% *ee* (Figure 8-30) [107]. The PYTHIA ligand is readily accessible in a one-step condensation of 2-acetylpyridine with the methyl ester of L-cysteine.

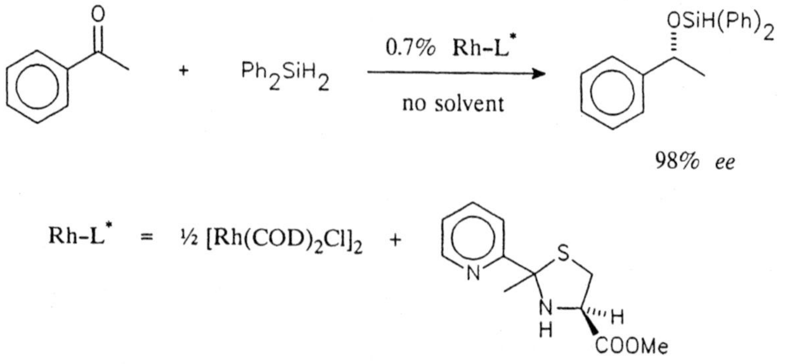

Figure 8-30. Enantioselective hydrosilylation of acetophenone.

G. Asymmetric Isomerization

The highly efficient asymmetric isomerization of prochiral allylic amines catalyzed by Rh–BINAP complexes has been extensively studied by several groups in Japan [30,37–41,108]. This led to the development and commercialization by Takasago of the well-known process for the synthesis of (–)-menthol, the most impressive achievement to date in the area of catalytic asymmetric synthesis. The process starts from myrcene and the key step is the enantioselective isomerization of diethyl-geranylamine (Figure 8-31). This step proceeds in 99% ee using Rh–((S)-BINAP)$_2$-ClO$_4$ as catalyst at 80°C and substrate-catalyst ratios of 8000:1 in THF as solvent [30,109]. Turnover numbers of 300,000 or more are obtained in practice. Moreover, the catalyst shows excellent thermal and chemical stability and can be readily recycled.

The mechanism of the enantioselective allylic hydrogen shift has been studied in detail [30,108]. Deuterium labeling experiments showed that β-hydrogen elim-ination which gives an iminium complex, is followed by exclusive 1,3-suprafacial migration of hydrogen to C(3) (Figure 8-32). Enantioselection occurs during the β-hydrogen elimination and subsequent 1,3-shift whereas the rate determining step in the process is displacement of coordinated enamine by a molecule of substrate. Indeed, it was surprising to find that enantioselection occurs at kinetically unim-portant stages of the reaction.

The citronellal intermediate in the menthol process can also be converted to (R)-(+)-citronellol or (R)-(+)-7-hydroxycitronellal, the odor of lily of the valley (Figure 8-33). Similarly, the enantiomers of these two products are available using the other enantiomer of BINAP.

The optically pure citronellal has also been used as the starting material for the synthesis of the vitamin E side chain, (3R,7R)-3,7,11-trimethyldodecanol

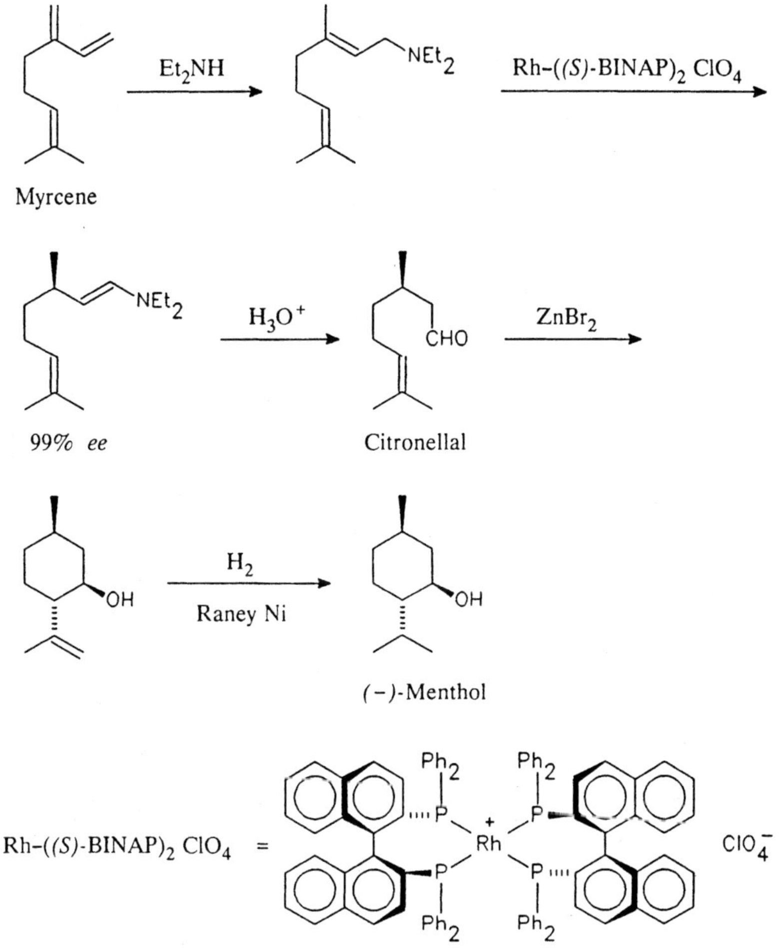

Figure 8-31. Takasago (−)-menthol process.

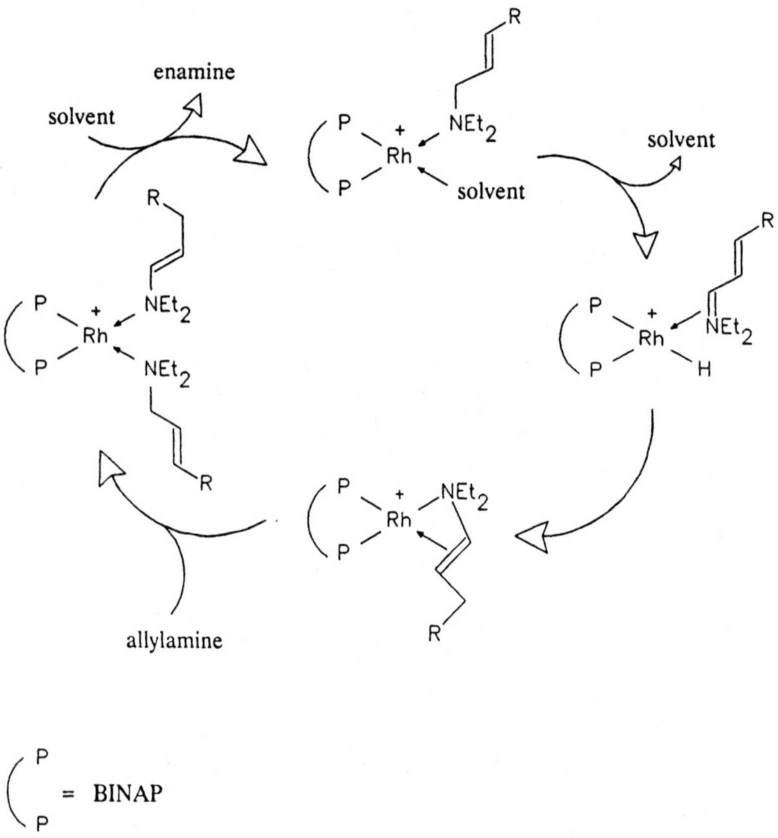

$$\left(\begin{array}{c} P \\ P \end{array}\right. = \text{BINAP}$$

Figure 8-32. Mechanism of enantioselective allylic hydrogen shift.

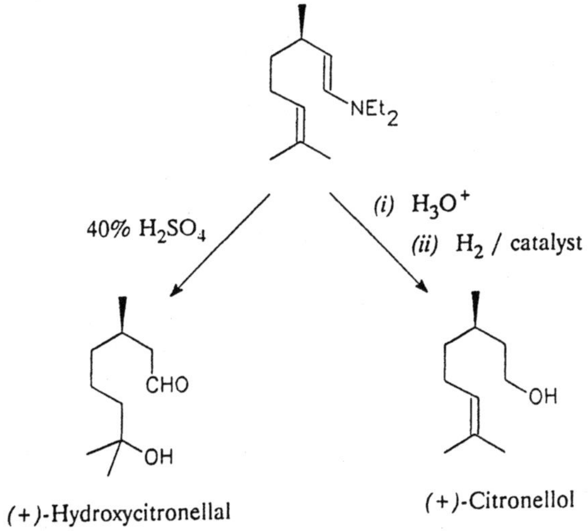

Figure 8-33. Synthesis of (+)-hydroxycitronellal and (+)-citronellol.

(Figure 8-34). The key step is the introduction of the second asymmetric center and this was achieved in high enantioselectivity by employing a Ru–BINAP-catalyzed asymmetric hydrogenation [110]. An alternative route involving a second Rh–BINAP-catalyzed allylic shift has also been described by the Takasago group [111].

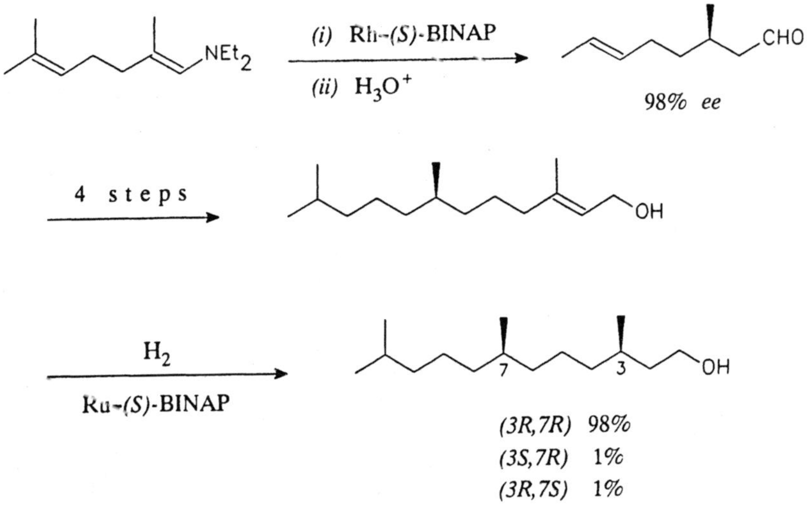

Figure 8-34. Synthesis of vitamin E side chain.

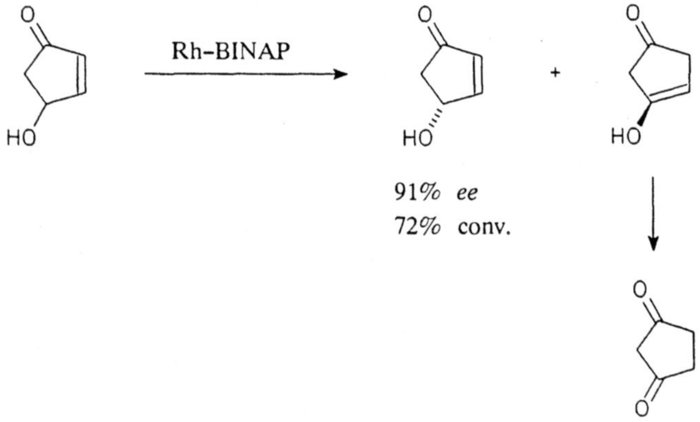

Figure 8-35. Rh–BINAP-catalyzed kinetic resolution of an allylic alcohol.

The Rh–BINAP-catalyzed asymmetric isomerization is applicable to a wide variety of allylic amine substrates [30]. This leads to the inevitable conclusion that other industrial applications will be forthcoming from this truly remarkable technology. The analogous Rh–BINAP-catalyzed isomerization of prochiral allylic alcohols to optically active aldehydes has also been studied [30] but, as yet, only moderate enantioselectivities were observed. One modest success was achieved in the Rh–BINAP-catalyzed kinetic resolution of the racemic hydroxycyclopentenone (Figure 8-35) [112]. The product is of interest as a prostaglandin intermediate but the very modest enantiomeric ratio ($E = 5$) renders the method unattractive compared to the route via enzymatic hydrolysis of the *meso*-diester described in chapter 7.

IV. ASYMMETRIC C-C BOND FORMATION

A wide variety of transition metal-catalyzed processes result in the formation of new C-C bonds, a feature that has tremendous synthetic utility. In principle, most of these reactions are amenable to asymmetric catalysis by a suitable choice of chiral ligands.

A. Asymmetric Hydroformylation and Carbonylation

The hydroformylation of olefins is an extremely important petrochemical technology [113]. The process yields a mixture of linear and branched regioisomers (Figure 8-36); only the branched isomer is chiral. Industrial research has generally been geared towards maximizing the yield of linear aldehyde [113] because it is the desired product in most cases. Commercial hydroformylation catalysts usually are

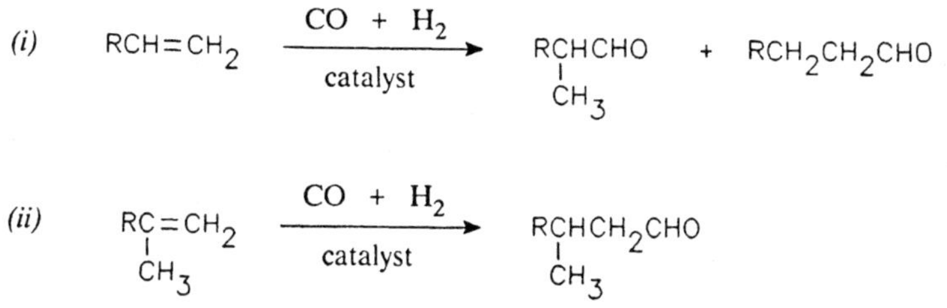

(i) $RCH = CH_2$ $\xrightarrow[\text{catalyst}]{CO + H_2}$ $\underset{\underset{CH_3}{|}}{RCHCHO}$ + RCH_2CH_2CHO

(ii) $\underset{\underset{CH_3}{|}}{RC = CH_2}$ $\xrightarrow[\text{catalyst}]{CO + H_2}$ $\underset{\underset{CH_3}{|}}{RCHCH_2CHO}$

Figure 8-36. General scheme of olefin hydroformylation.

phosphine complexes of rhodium or cobalt. In contrast, promising results in asymmetric hydroformylations have, as yet, been obtained only with the less active (compared to Rh and Co) Pt-based catalysts. SnCl$_2$ is usually employed as a cocatalyst. For example, Stille and Parinello [114] obtained good enantioselectivities with PtCl$_2$/SnCl$_2$/BPPM (Figure 8-37). It should be noted, however, that the branched aldehyde was the minor product (branched/linear ratio = 0.5–0.7). Indeed, a fundamental problem in asymmetric hydroformylation (and carbonylation) is that the use of bulky phosphine ligands that are necessary for enantiocontrol, tends to favor the formation of the linear rather than the branched aldehyde. The combination of high optical and chemical yields is, hence, a difficult proposition.

Another problem associated with asymmetric hydroformylation is the propensity of the aldehyde product for racemization under the reaction conditions. Stille and coworkers [115] circumvented this problem by carrying out the reactions in triethyl orthoformate as solvent. Although the rates were significantly lower than in benzene solvent, the more stable (towards racemization) diethylacetals were obtained in > 96% *ee*.

The diethylacetals could be readily converted to the corresponding aldehydes, without racemization, by treatment with pyridinium *p*-toluenesulfonate in acetone. The reactions are of interest as routes to the therapeutically active isomers of antiinflammatory drugs such as *(S)*-naproxen and *(S)*-ibuprofen. (See chapter 2.) Hydroformylations in triethyl orthoformate were also carried out with a polymer-supported version of the catalyst with comparable results [114].

Most attempts aimed at the asymmetric carbonylation of olefins have met with limited success [115]. One result worthy of mention is the carbonylation of *p*-isobutylstyrene and 2-vinyl-6-methoxystyrene that are catalyzed by PdCl$_2$/CuCl$_2$ in combination with a chiral biphosphate ligand [116]. *(S)*-Ibuprofen and *(S)*-naproxen were formed in 83% *ee* and 85% *ee*, respectively, albeit at rather low (< 10) substrate/catalyst ratios (Figure 8-38).

One reason why relatively little attention has been focussed on asymmetric carbonylation and hydroformylation is probably that there are simpler alternative

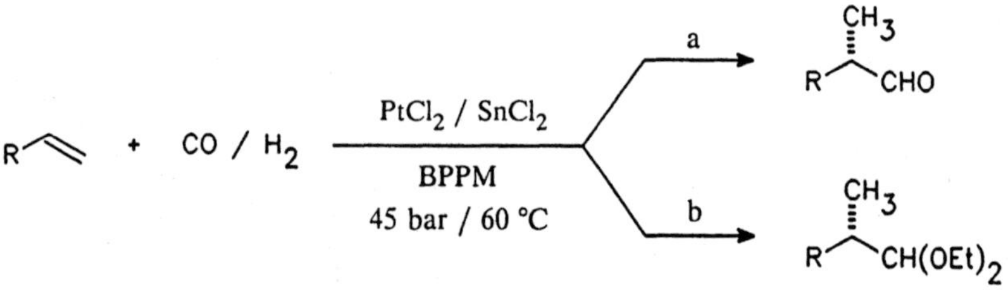

Substrate	Solvent	*ee* (%)
	a	70-78
	b	> 96
	a	81
	b	> 96
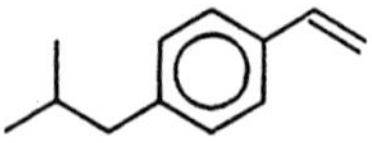	a	80
	a	73
	b	> 96
	a	82
	b	> 98

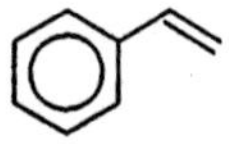

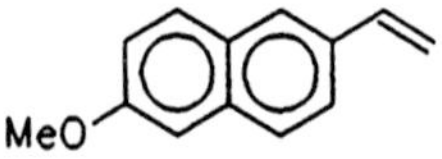

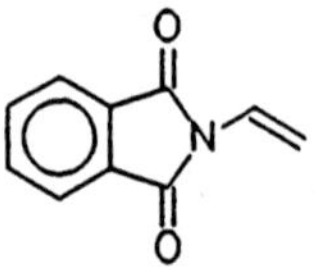

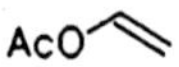

Solvent: a = PhH; b = CH(OEt)$_3$

Figure 8-37. Asymmetric hydroformylation with PtCl$_2$ / SnCl$_2$ / BPPM.

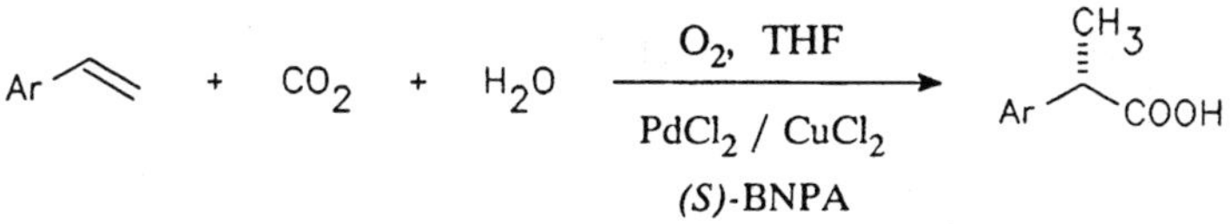

$$Ar\diagdown\diagup + CO_2 + H_2O \xrightarrow[\substack{PdCl_2 / CuCl_2 \\ (S)\text{-BNPA}}]{O_2, \ THF} \quad Ar\diagup_{COOH}^{CH_3}$$

Substrate	Yield (%)	ee (%)	Product
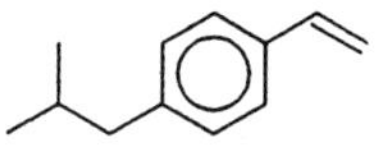	89	83	(S)-Ibuprofen
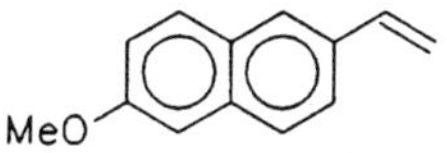	71	85	(S)-Naproxen

Figure 8-38. *(S)*-Ibuprofen and *(S)*-naproxen via asymmetric carbonylation.

routes to optically active acids, for example, asymmetric hydrogenation of unsaturated precursors or enzymatic resolutions of racemates.

B. Asymmetric Cyclopropanation

Asymmetric cyclopropanation has been extensively studied by Aratani and coworkers at Sumitomo [117] in connection with the synthesis of intermediates for optically active pyrethroids. (See chapter 2.) They found that the chiral Schiff base complex of Cu shown in Figure 8-39 is an extremely efficient catalyst for the enantioselective cyclopropanation of prochiral olefins and dienes with ethyl diazoacetate; for example, the intermediate for cypermethrin was produced in 91% *ee*. The same method was applied to the commercial production of *(1S)*-2,2-dimethylcyclopropanecarboxylate, a key intermediate in the synthesis of cilastatin (Figure 8-40). The latter is a dehydropeptidase inhibitor that is administered in combination with the beta-lactam antibiotic Imipenem (*N*-formidoylthienamycin).

Catalytic asymmetric cyclopropanation continues to attract much attention [118]. For example, copper complexes of a C_2-symmetric semicorrin ligand [119] and the structurally related *bis*-oxazoline ligands [120,121] are highly efficient catalysts (Figure 8-41); the former is readily prepared from L-pyroglutamic acid and the latter from malonic esters and chiral amino alcohols. The results obtained with these ligands in the cyclopropanation of styrene are compared with those of the Aratani system in Figure 8-41.

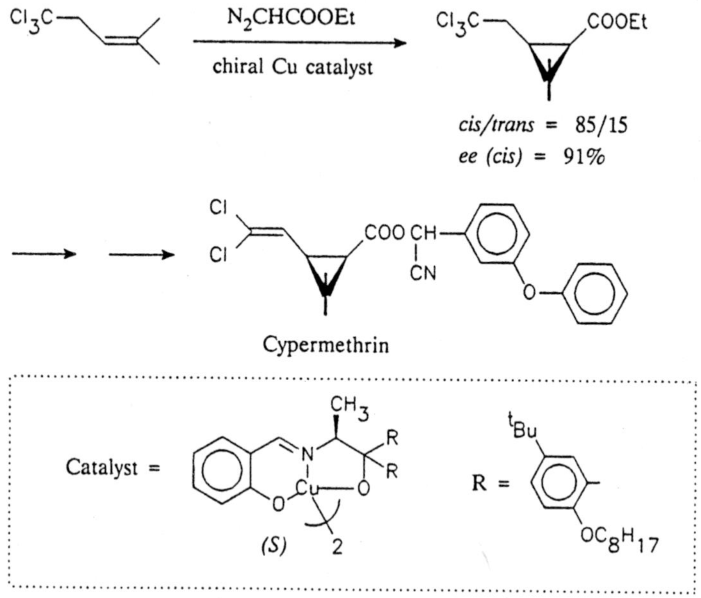

Figure 8-39. Synthesis of cypermethrin intermediate via asymmetric cyclopropanation.

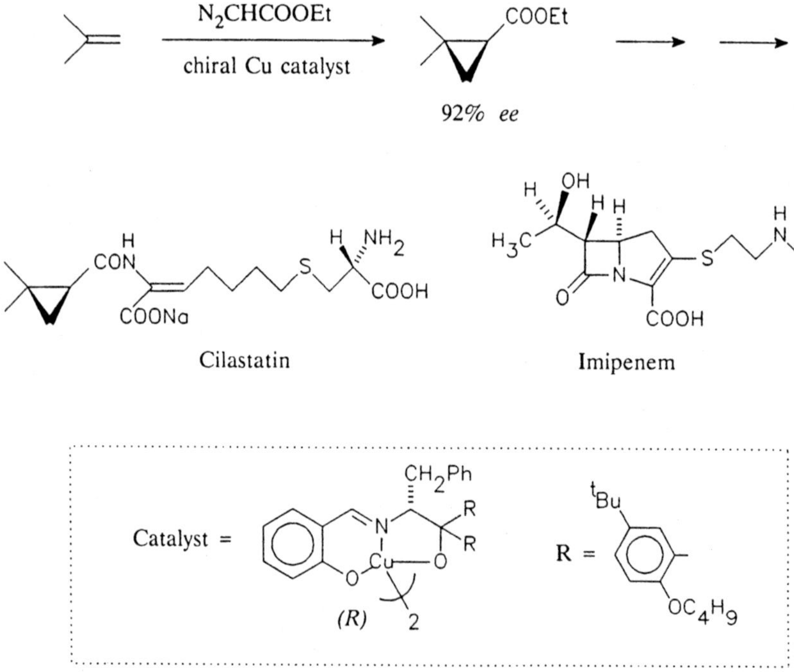

Figure 8-40. Synthesis of a cilastatin intermediate.

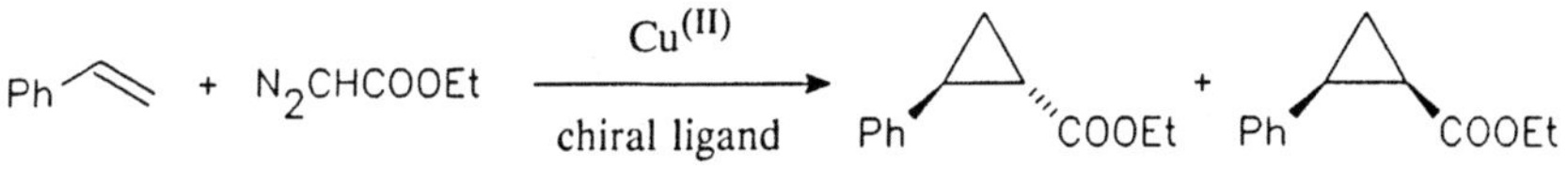

Chiral ligands :

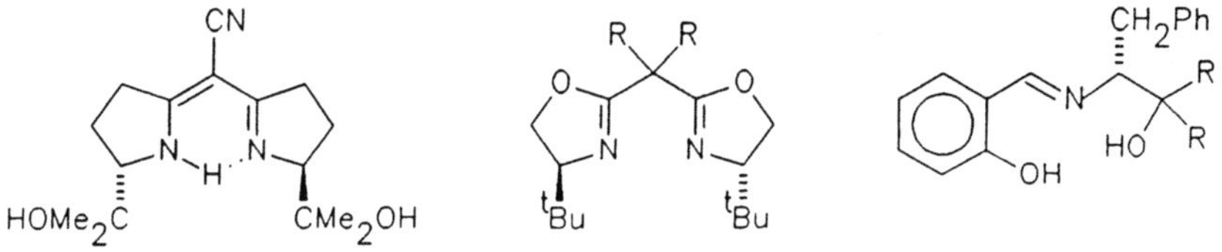

Ligand	Diastereoselectivity	Enantioselectivity		Ref.
	trans : cis	trans (% ee)	cis (% ee)	
A	75 : 25	85	68	[119]
B (R = H)	77 : 23	98	93	[120]
B (R = CH₃)	73 : 27	99	97	[120]
C (R = 2-butoxy- 5-t-butylphenyl)	86 : 14	69	54	[117]

Figure 8-41. Comparison of various chiral ligands in asymmetric cyclopropanation.

Chiral Rh^{II}-based catalysts have also been described [118] and the mechanism of the cyclopropanation, which proceeds via a cycloaddition of a metal-carbenoid species to the olefin, has been discussed at length [117,118].

C. Asymmetric Codimerization and Cyclodimerization of Olefins and Dienes

The codimerization of olefins and dienes and the cyclodimerization of dienes are examples of transition metal catalyzed C-C bond forming processes that are widely employed on an industrial scale [122]. Ni^0 complexes of triaryl phosphines or triarylphosphites are generally the catalysts of choice. It was inevitable that asymmetric variations of these reactions would be attempted and in a few cases high enantioselectivities have been reported (Figure 8-42) [123,124]. An important

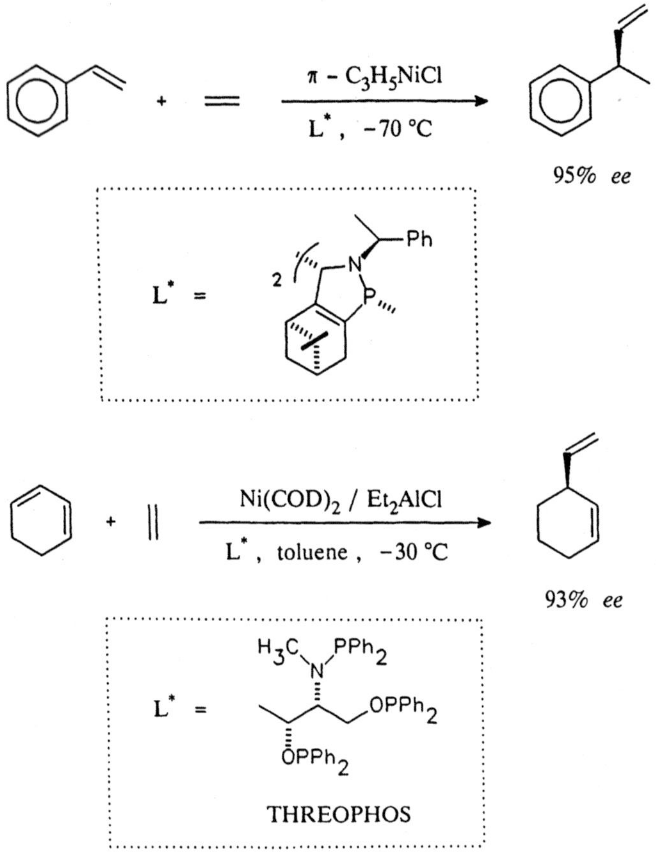
π - C3H5NiCl
L*, -70 °C
95% ee
L* =
Ph
N
P
2
Ni(COD)2 / Et2AlCl
L*, toluene, -30 °C
93% ee
H3C
PPh2
OPPh2
OPPh2
THREOPHOS

Figure 8-42. Asymmetric codimerization of olefins and dienes [123,124].

challenge in this context is the development of enantioselective catalysts for the linear dimerization of isoprene [122].

D. Asymmetric Grignard Cross-Coupling

Phosphine complexes of group 8 transition metals, especially Ni and Pd, catalyze the cross-coupling of Grignard reagents with organic halides. Since the enantiomers of a racemic Grignard reagent are in equilibrium under the reaction conditions the reaction can be carried out as a dynamic kinetic resolution by employing chiral phosphine ligands. The enantiocontrolling step is the transmetallation reaction that is assumed to proceed with inversion (Figure 8-43). High enantioselectivities were observed using Ni complexes of chiral aminophosphines derived from amino acids.

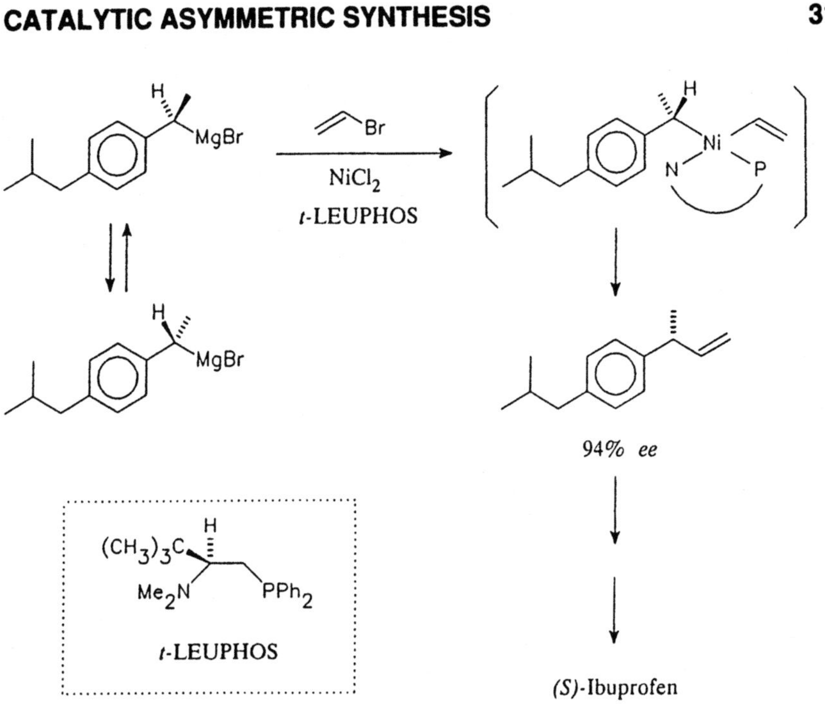

Figure 8-43. Asymmetric Grignard cross-coupling.

A precursor of *(S)*-ibuprofen, for example, was obtained in 94% *ee* [125]; however, it is unlikely that this approach can compete with other routes to *(S)*-ibuprofen. (See chapter 9.)

E. Asymmetric Addition of Dialkylzinc to Aldehydes

The nucleophilic addition of organometallics to aldehydes to give secondary alcohols is one of the basic reactions of organic synthesis. Asymmetric variations of this reaction have been designed by using dialkylzinc reagents in combination with catalytic amounts of chiral amino alcohols [40,126–133] or cinchona alkaloid bases [134]. For example, Noyori and coworkers [126] achieved an *ee* of 98% using *(−)*-3-exo(dimethylamino)borneol (*(−)*DAIB) as the catalyst (Figure 8-44).

This reaction is not only remarkable for its high enantioselectivity but also because it exhibited the phenomenon of **chirality amplification**. Thus, when *(−)*-DAIB of only 15% *ee* was used as the catalyst the *(S)*-alcohol product was formed with an *ee* (95%), which is almost the same as that observed with enantiomerically pure DAIB. This remarkable effect results from chiral and achiral catalytic systems competing in the same reaction, the turnover efficiency of the former being > 600 times that of the latter. The active catalyst is a monomeric zinc

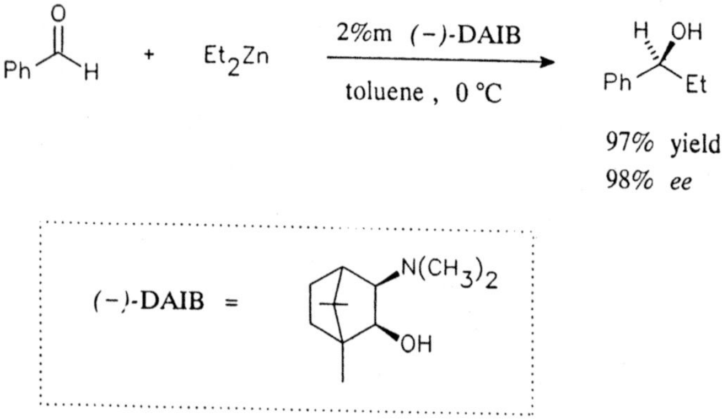

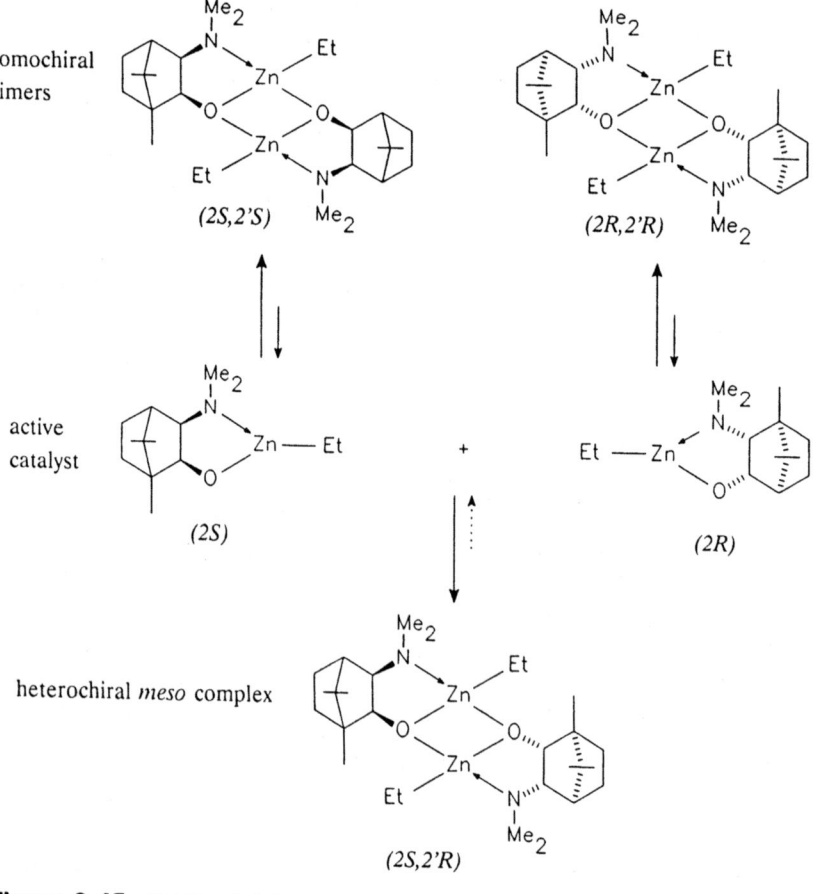

Figure 8-44. Asymmetric addition of dialkylzinc to aldehydes.

homochiral
dimers

active
catalyst

heterochiral *meso* complex

Figure 8-45. DAIB-alkylzinc complexes present in solution.

complex (Figure 8-45), although the species present in the reaction solution is predominantly the more stable dimer. When both enantiomers of DAIB are present, the solution contains a mixture of the two homochiral dimers *(2R,2'R)* and *(2S,2'S)* with C_2-chirality together with the heterochiral dimer which is a *meso* complex *(2S,2'R)*. The *meso* complex is less sterically congested than the homochiral dimers and its formation is overwhelmingly favored thermodynamically. This was demonstrated by mixing equal amounts of the two homochiral dimers which yielded the heterochiral *meso* complex. Consequently, the homochiral dimers in solution have the greater tendency to dissociate to the active monomeric catalyst. This means that when partially resolved DAIB is used the minor enantiomer will predominantly be transformed to the less reactive *meso* complex. The remaining excess of the major enantiomer gives the more reactive homochiral dimer. The active catalyst will then be predominantly derived from the latter complex, thus explaining the observed amplification of chirality. The authors noted that this observation may have implications for the origin of homochirality in nature.

In another approach, polymer-bound DAIB has recently been used as a heterogeneous recyclable catalyst for the enantioselective alkylation of aldehydes with dialkylzinc [128]. Finally, Ohno and coworkers [135] have described the use of Et_2Zn in combination with stoichiometric amounts of $Ti(O\text{-}iPr)_4$ and catalytic quantities of a chiral disulfonamide ligand (see section IV-F for a further discussion of this system). High enantioselectivities (> 90 %) were observed at reasonable (> 100) substrate-catalyst ratios (Reaction 8-8).

$$\text{PhCHO} \xrightarrow[\substack{\text{0.0005 - 0.04 equiv.} \\ \text{toluene / hexane,} \quad -20\,°C}]{\substack{Et_2Zn\,/\,Ti(O\text{-}^iPr)_2 \\ \text{NHSO}_2CF_3 \\ \text{NHHSO}_2CF_3}} \quad \substack{H\quad OH \\ Ph \quad Et}$$

97% yield
98% *ee*

(8-8)

F. Chiral Lewis Acids: Asymmetric Diels-Alder and Ene Reactions

A variety of synthetically useful C-C bond forming reactions are catalyzed by Lewis acids. Recent progress in the design of chiral Lewis acids containing B, Al, and Ti have led to the development of effective catalysts for asymmetric variations of such classics as the Diels-Alder reaction, ene reactions, and hydrocyanation [136]. For example, chiral alkoxytitanium(IV) complexes in the presence of molecular sieves catalyze asymmetric Diels-Alder [136,137] and ene reactions [138] in high enantioselectivities (Figure 8-46).

Diels-Alder reaction :

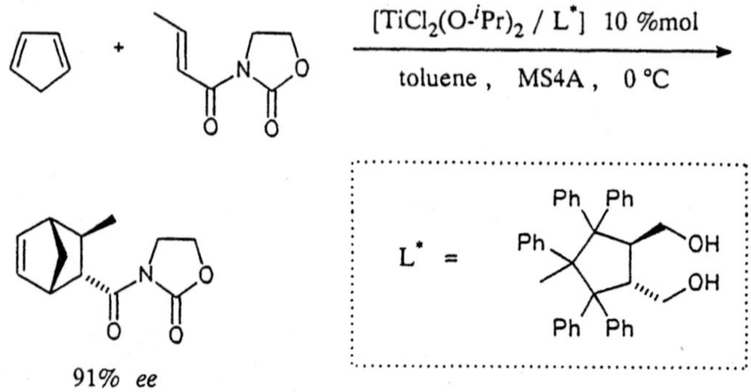

91% ee
endo/exo = 92/8

Ene reaction :

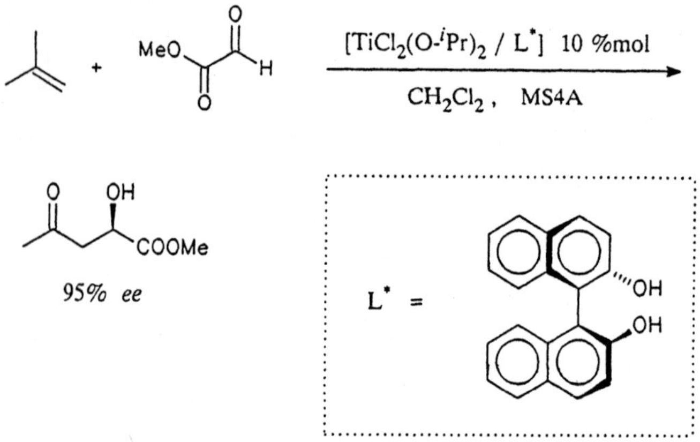

95% ee

Figure 8-46. Asymmetric Diels-Alder and ene reactions.

Although these reactions (Figure 8-46 and related processes catalyzed by chiral Lewis acids) exhibit impressive enantioselectivities, they still require relatively large quantities of catalyst and are far from being in the same league as the rhodium and ruthenium BINAP catalyzed processes discussed earlier. Indeed, as Ohno and coworkers have pointed out [135] the chiral ligands (amino alcohols and diols) that are generally used in combination with Lewis acid catalysts tend to decrease the electrophilic character (i.e., Lewis acidity) of the catalyst. In contrast, the chiral disulfonamides contain strongly electron withdrawing groups and increase the electrophilicity of the Lewis acid complex, thus creating a better scenario for ligand accelerated asymmetric catalysis. This would appear to be a step forward towards more active catalysts. Indeed, Corey and coworkers [139] reported the use of such chiral disulfonamide ligands in asymmetric Diels-Alder reactions.

G. Asymmetric Hydrocyanation

Hydrocyanation of aldehydes is catalyzed by oxynitrilases (see chapter 7) and by simple (Lewis) acids and bases. It is not surprising, therefore, that the same combinations of titanium(IV) alkoxide and chiral diols that were used for asymmetric Diels-Alder and ene reactions have also been applied to asymmetric hydrocyanation [140,141]. However, in all the reported examples, stoichiometric quantities of Ti^{IV} and chiral ligand were required, in combination with Me_3SiCN as the cyanating agent, making the commercial viability questionable.

A more attractive approach involves the use of the cyclic dipeptide (cyclo-(S)-phenylalanyl-(S)-histidyl) as a true catalyst [142]. High enantioselectivities are obtained with a broad range of aldehydes using 2 mol% of catalyst at − 20°C in toluene (Figure 8-47). The (S,S)-dipeptide gives the (R)-cyanohydrin. It was suggested [142] that the carbonyl oxygen forms a hydrogen bond with the N-H group of the histidine residue. The imidazole group of the same histidine moiety abstracts a proton from HCN to form a cyanide ion that attacks the Si face of the activated carbonyl group. The Re face is hindered by the aromatic ring of the phenylalanine residue (Figure 8-47). This system appears to be a good biomimetic

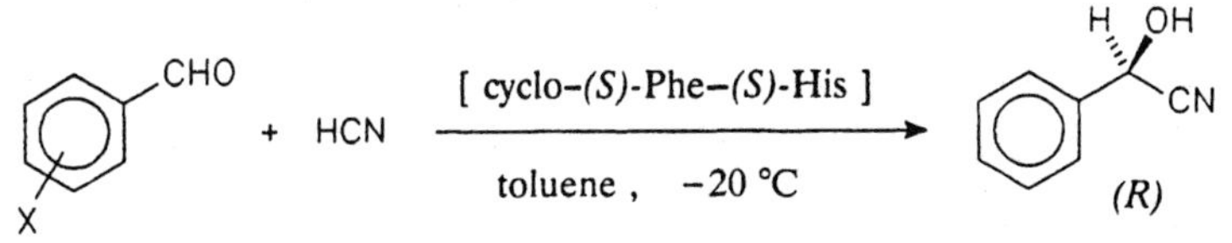

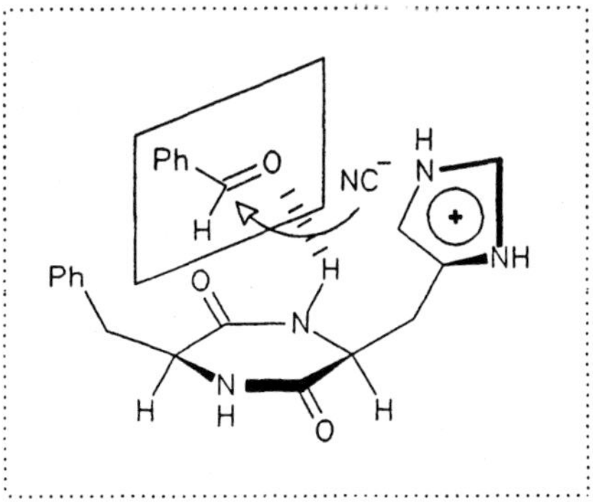

X	ee %
H	97
m-PhO	92
m-MeO	97
p-Me	96
p-MeO	78

Figure 8-47. Asymmetric hydrocyanation catalyzed by a cyclic dipeptide.

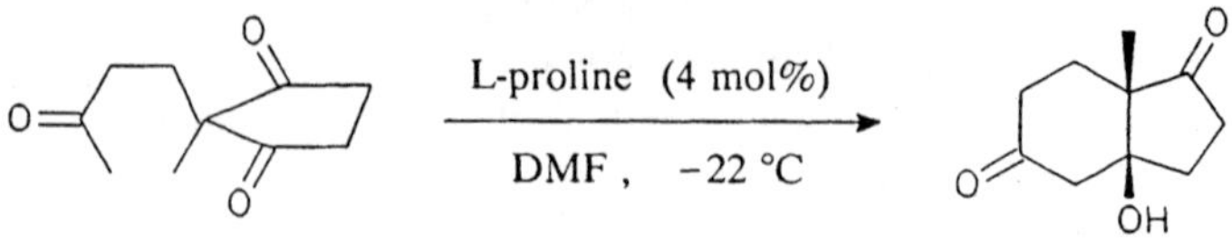

100% yield
93.4% ee

Figure 8-48. L-Proline-catalyzed asymmetric aldol condensation.

model for the oxynitrilase enzymes which catalyze the same reactions. (See chapter 7.) It has been extensively studied by industrial laboratories [143] in connection with the synthesis of the (S)-enantiomer of m-phenoxybenzaldehyde cyanohydrin, a key intermediate for optically active pyrethroids. (See chapter 2.) In this case the cyclo–(R)-Phe—(R)-His dipeptide is the catalyst [143].

H. Condensations Catalyzed by Chiral Bases

Many base-catalyzed condensation reactions are amenable to asymmetric catalysis by chiral bases. A classical example, although involving an amino acid rather than a simple amine, is the enantioselective intramolecular aldol condensation of the prochiral cyclic triketone (Figure 8-48). The reaction proceeds in 100% chemical yield and 93.4% optical yield using 4 mol% L-proline as the catalyst. The product is an intermediate in the total synthesis of 19-norsteroids.

The best catalysts for asymmetric condensation reactions are the cinchona alkaloid bases. These versatile chiral catalysts are known to mediate a variety of asymmetric condensation reactions many of which involve the formation of C-C bonds [3]. The addition of diethylzinc to aldehydes has already been mentioned. Two other examples are asymmetric Michael additions [3] and the 2,2-cycloaddition of electron-poor aldehydes and ketones, such as chloral, to ketene (Figure 8-49). When quinidine is used as the catalyst the product has the (S) configuration [145]. Acidic hydrolysis of the β-lactone product derived from chloral gives the natural (S)-malic acid via an inversion at the stereogenic carbon atom. Use of 1,1,1-trichloroacetone instead of chloral yields citramalic acid, a useful synthon for natural products.

I. Asymmetric Phase-Transfer Catalysis

Another variation on the theme of asymmetric catalysis is the use of chiral tetraalkylammonium salts as phase-transfer catalysts. In what has now become a classic example of this genre, Dolling and coworkers [146,147] obtained 98%

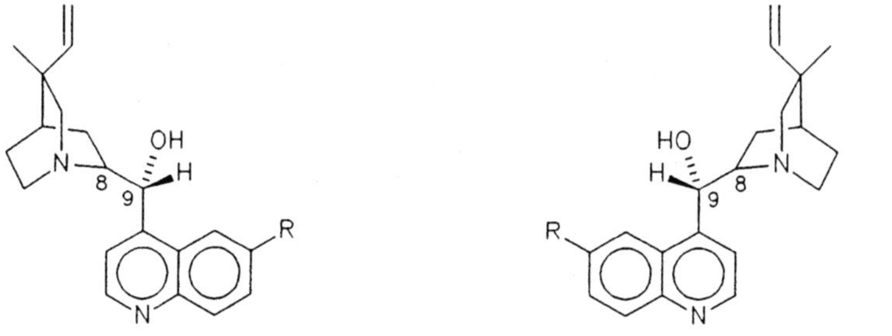

(8S,9R) – Quinine (R = OCH$_3$)
(8S,9R) – Cinchonidine (R = H)

(8R,9S) – Quinidine (R = OCH$_3$)
(8R,9S) – Cinchonine (R = H)

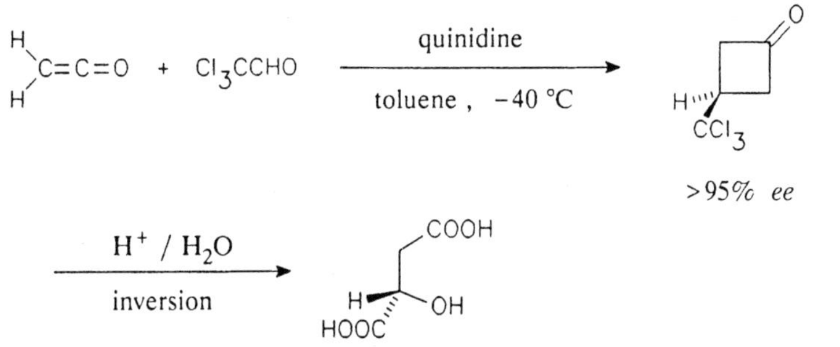

Figure 8-49. Asymmetric 2+2 cycloaddition catalyzed by cinchona alkaloids.

chemical yield and 94% optical yield in the asymmetric alkylation reaction shown in Figure 8-50. The product is a precursor of the antihypertensive drug, *(S)*-indacrinone. (See chapter 2.) Interestingly, initial results with *N*-alkyl derivatives of various cinchona alkaloids were in the modest range (20–30% *ee*). Systematic optimization of the catalyst structure and the reaction parameters (solvent, temperature, alkylating agent, concentrations, etc.) subsequently led to an *ee* of 94%. The reaction proceeds via the formation of a substrate-catalyst ion pair at the water-toluene interface, followed by reaction with the alkylating agent in the organic phase.

This important breakthrough can be expected to lead to further exploitation of the technique of asymmetric phase-transfer catalysis. Other examples of asymmetric phase-transfer catalysis [3,148–150] include alkaline epoxidation [3], Michael addition [148], amino acid synthesis [149], and base-catalyzed autoxidation of ketones under phase-transfer conditions [150].

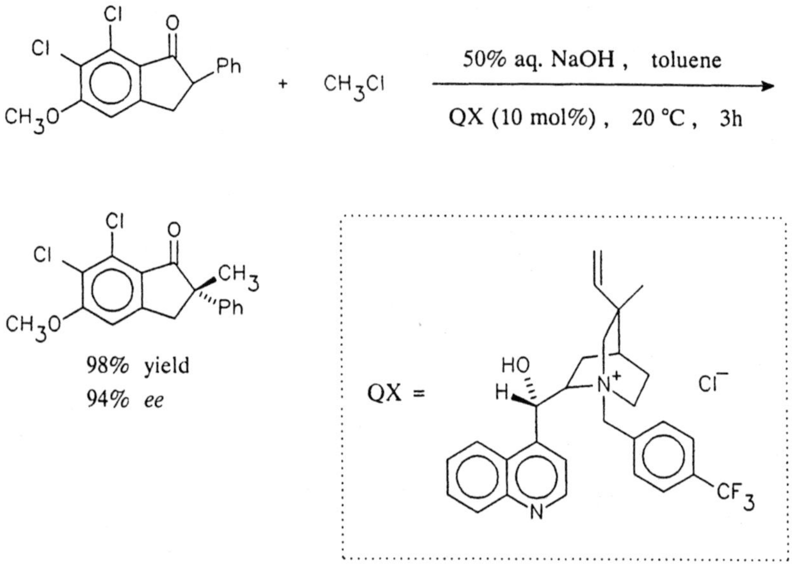

Figure 8-50. Asymmetric phase-transfer catalyzed alkylation reaction.

V. ASYMMETRIC OXIDATIONS

Considering the central role that oxidative transformations play in organic synthesis it is appropriate that much attention has been devoted to the development of catalytic asymmetric oxidations. Indeed, the design of simple catalytic systems capable of emulating redox enzymes is still a major challenge in organic synthesis. It should be noted however, that selective oxidation catalysis presents more formidable problems compared to catalytic reduction. Not the least of these is the thermodynamic instability of many (chiral) ligands under oxidizing conditions. Nevertheless, there has been considerable progress in this commercially important area.

A. Asymmetric Epoxidation

Since its discovery [34,35] in 1980 the Sharpless epoxidation of allylic alcohols has become a benchmark classic in catalytic asymmetric synthesis. A wide variety of primary allylic alcohols are epoxidized in > 90% optical yield and 70–90% chemical yield [151] using TBHP as the oxygen donor and titanium isopropoxide-diethyltartrate (DET) as the catalyst (Figure 8-51).

The original report [34] concerned the use of stoichiometric quantities of titanium and the chiral ligand. Subsequently, a catalytic procedure was developed, the key feature of which is the addition of 3-Å or 4-Å molecular sieves to remove traces of water [152]. The catalytic procedure employs 5–10 mol% of catalyst and

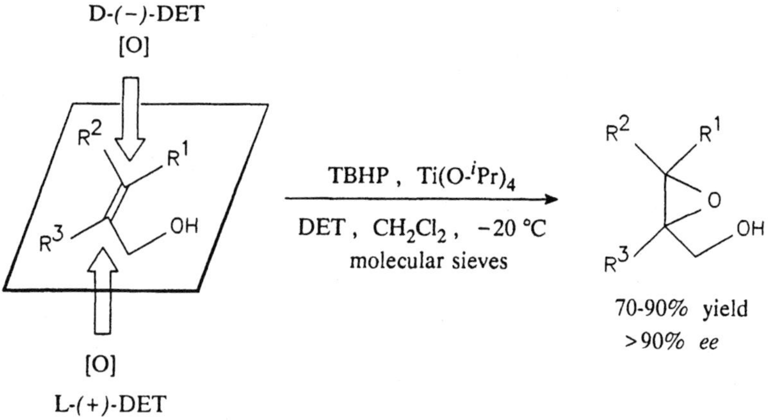

Figure 8-51. Catalytic asymmetric epoxidation of allylic alcohols.

givess roughly the same optical yields but substantially higher chemical and volume yields (i.e., productivities). The broad utility of epoxides as intermediates in organic syntheses, the ready availability of the reagents, and the ease of execution has helped the procedure to become widely accepted among organic chemists [151]. A comparison of the stoichiometric and catalytic procedures for various substrates is shown in Table 8-3.

Table 8-3 Comparison of Stoichiometric and Catalytic Procedures for Asymmetric Epoxidation [151a]

Substrate	Stoichiometric		Catalytic (5 mol%)	
	Yield (%)	*ee* (%)	Yield (%)	*ee* (%)
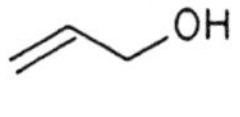 ![allyl alcohol] OH	15	73	65	90
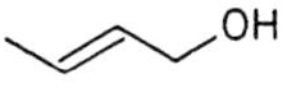 ![crotyl alcohol] OH	52	95	70	96
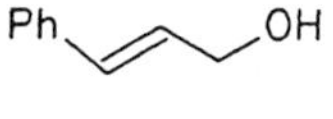 Ph OH	36	>98	89	98
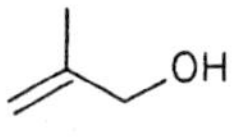 OH	0	–	50	>95
 OH	0	–	40	95

Ti(IV)-tartrate dimer =

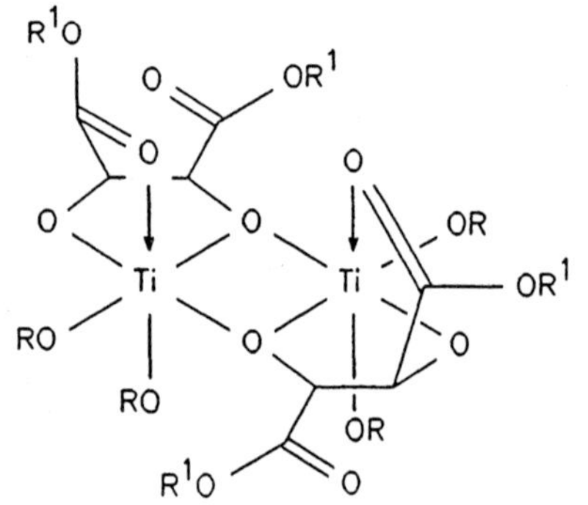

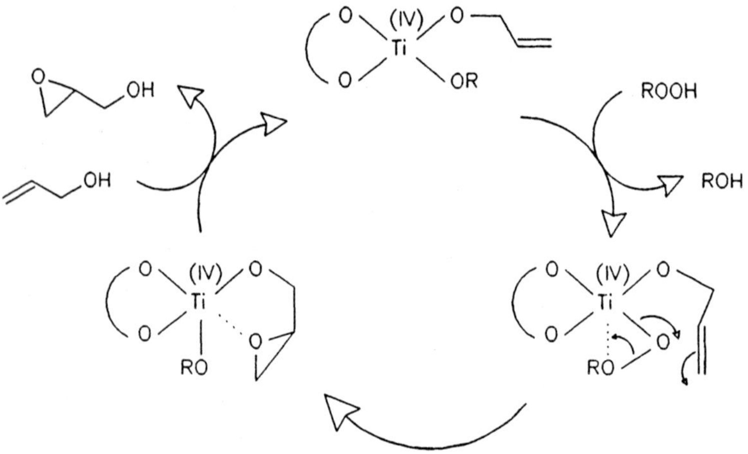

Figure 8-52. Catalytic cycle of epoxidation.

The catalytic cycle is depicted in Figure 8-52; note that the titanium remains in the tetravalent state throughout the cycle. The key step in the process is the preferential transfer of oxygen from a coordinated alkylperoxo moiety to one enantioface of a coordinated allylic alcohol. The actual catalytic species is believed to be a 2:2 titanium(IV) tartrate dimer.

Commercially relevant applications of the Sharpless epoxidation include the synthesis of *(R)*- and *(S)*-glycidol, which has been commercialized by ARCO, the epoxidation of oct-3-en-1-ol (Upjohn) and the synthesis of an intermediate for the insect pheromone, disparlure (Figure 8-53).

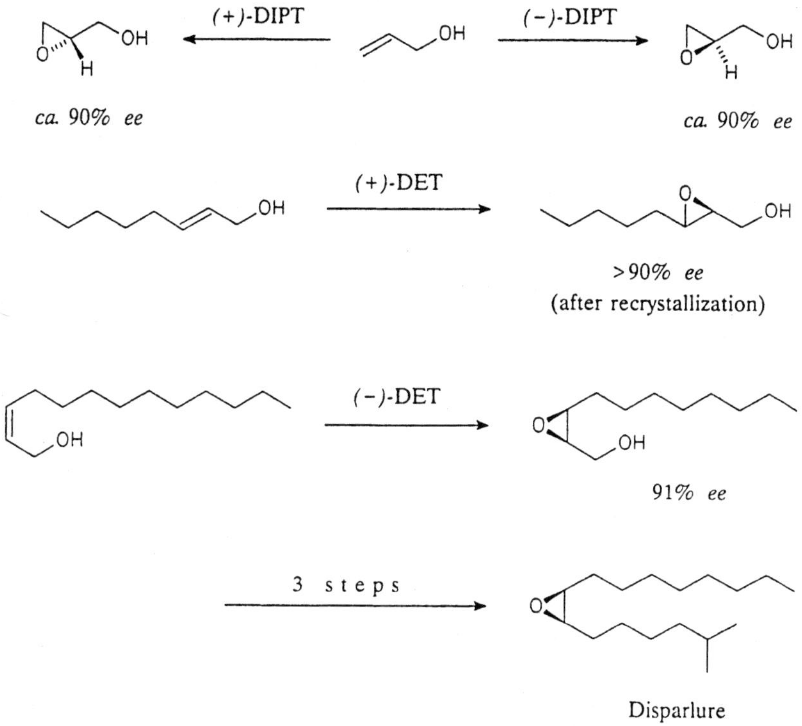

Figure 8-53. Applications of the Sharpless epoxidation.

The *(R)* and *(S)* isomers of glycidol constitute versatile C_3 chirons with potentially widespread applications in the synthesis of, for example, optically active beta-blockers and phospholipids. Unfortunately, this parent allylic alcohol gives somewhat lower enantioselectivities (90 versus $\geq 95\%$ *ee*) realized for most substituted alcohols. Sharpless and coworkers [154] have reported the preparation of crystalline arenesulfonate derivatives of the enriched glycidol that can be easily upgraded by recrystallization. For example, the tosylate and *m*-nitrobenzenesulfonate were obtained in 97% and 99% *ee*, respectively, in one recrystallization. Another problem associated with the use of glycidyl derivatives as chiral synthons is competing nucleophilic attack at the two termini ($C(1)$ and $C(3)$) of the molecule. The former leads to retention of configuration and the latter to inversion (Figure 8-54). Consequently, poor regioselectivity is accompanied by a decrease in optical purity of the starting material. Fortunately, arenesulfonates (particularly *m*-nitrobenzenesulfonate) are excellent leaving groups and exhibit high regioselectivities for attack at $C(1)$ by a variety of nucleophiles, such as phenolate anions [154]. As a result,

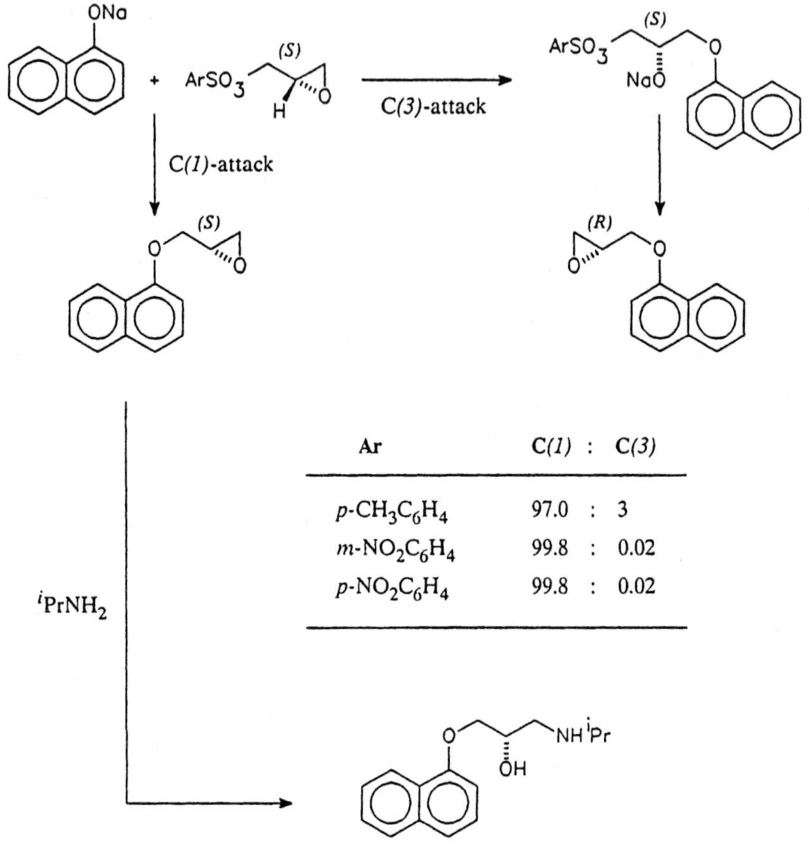

Ar	C(1) : C(3)
$p\text{-CH}_3\text{C}_6\text{H}_4$	97.0 : 3
$m\text{-NO}_2\text{C}_6\text{H}_4$	99.8 : 0.02
$p\text{-NO}_2\text{C}_6\text{H}_4$	99.8 : 0.02

(S)-Propanolol

Figure 8-54. Synthesis of *(S)*-propanolol from *(S)*-glycidyl arenesulfonates.

they are excellent synthons for the preparation of optically active beta-blockers (Figure 8-54; see also chapter 9).

The Sharpless epoxidation procedure has also been applied to the kinetic resolution of chiral secondary allylic alcohols [155] and chiral-β-hydroxyamines [156]. A heterogeneous equivalent of the Sharpless reagent, comprising a titanium-pillared clay in combination with tartrate, has also been reported [157].

Since the Sharpless epoxidation is limited to allylic alcohols as substrates, there is still a definite need for practical methods for the catalytic asymmetric epoxidation of simple nonfunctionalized prochiral olefins. An interesting development in this context is the asymmetric epoxidation of simple prochiral olefins with

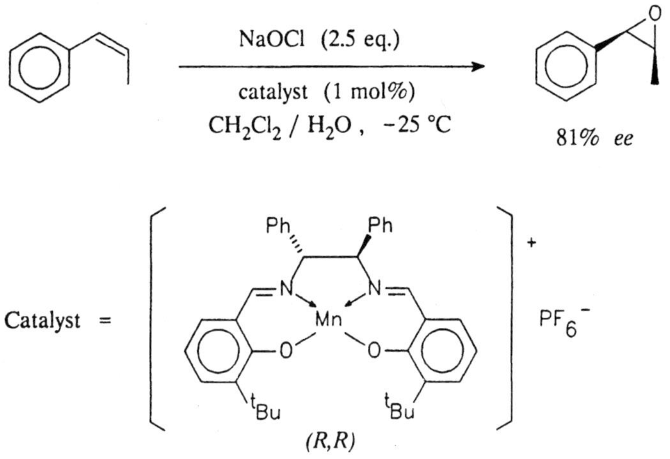

Figure 8-55. Asymmetric epoxidation with NaOCl.

sodium hypochlorite in the presence of chiral manganese(III) Schiff base complexes (Figure 8-55) reported by Jacobson and coworkers [158].

Iodosoarenes (ArIO) can also be used as the oxygen donor with these catalysts [159,160]. However, high enantioselectivities have been obtained only with a limited number of reactive cis-disubstituted olefins, such as β-methylstyrene (Figure 8-55). Nevertheless, enantioselectivities are amenable to improvement by 'electronic tuning' of the ligand via the introduction of electron donating groups into the aromatic ring.

Asymmetric epoxidation of unfunctionalized olefins was also observed with sterically crowded iron(III) porphyrins as the catalyst and iodosobenzene as the oxygen donor [161]. cis-β-Methylstyrene, for example, gave the (1S,2R)-epoxide in 72% ee. The metalloporphyrin- and metal-Schiff base complex-catalyzed epoxidations involve high-valent oxometal species as the active oxidant, in contrast to the Sharpless and related systems in which a peroxometal species is the active intermediate.

In yet another approach, Strukul and coworkers [162] used platinum complexes containing a CF_3 group and a chiral diphosphine coordinated to Pt^{II} as catalyst for asymmetric epoxidation of terminal olefins with aqueous hydrogen peroxide. The best result (41% ee) was obtained with 1-octene and CHIRAPHOS as the chiral ligand.

Based on these results, it is clear that there is still much room for improvement in methodologies for the asymmetric epoxidation of simple prochiral olefins. However, it sometimes appears that a prerequisite for high enantioselectivities coupled with broad substrate specificity is a catalyst that is of the same order of complexity as the enzymes that it is attempting to emulate. In this unceasing quest

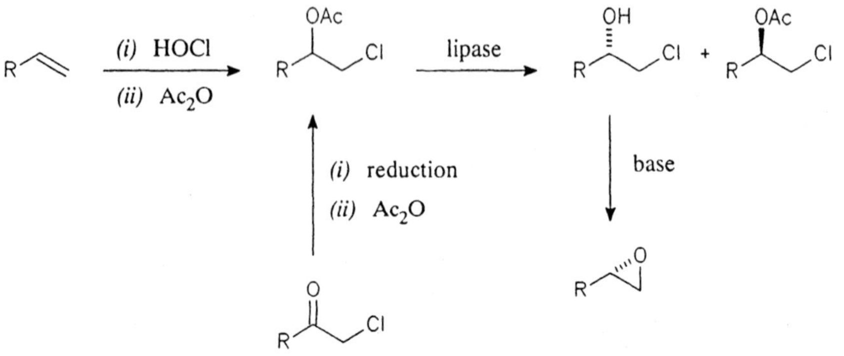

Figure 8-56. Hydrolytic route to chiral epoxides.

for 'simple' biomimetic models one should, therefore, not lose sight of the fact that there are generally alternative routes to the epoxide which may be more economically attractive. For example, a relatively straightforward approach is via lipase-catalyzed hydrolysis of a racemic ester of a chlorohydrin (Figure 8-56). The latter is generally readily available from the olefin or by reduction of an α-chloro ketone. Although this route may be less elegant than asymmetric epoxidation and has a maximum chemical yield of 50%, there is no doubt that it is economically more attractive than the above mentioned routes to simple chiral epoxides.

Another method worthy of mention is the asymmetric epoxidation of chalcones and other electron-poor olefins with H_2O_2, using poly-L-amino acids as catalysts, in alkaline media [163]. The poly-L-amino acids are commercially available or can be readily prepared by polymerization of the N-carboxyanhydride of the amino acid. SmithKline Beecham workers [164] used this method as a key step in the synthesis of a leukotriene antagonist (Figure 8-57); however, the method required 20 equivalent of H_2O_2 and 12 equivalents of NaOH (based on substrate).

The poly-L-amino acid catalysts may be considered as synthetic enzyme mimics. The mechanism presumably involves the asymmetric addition of a hydroperoxo anion (HOO) to the olefinic double bond, followed by epoxide ring closure; this is analogous to the mechanism proposed [3] for the same reactions using quaternary cinchona alkaloid bases as chiral phase transfer catalysts.

B. Asymmetric Sulfoxidation

Although the Sharpless reagent as such does not effect enantioselective oxidation of prochiral sulfides, two groups have independently developed modified systems that are effective. Kagan and coworkers [36,165] employed Ti(O-iPr)$_4$/DET/TBHP in a ratio of 1:2:2 in the presence of one equivalent of water and obtained 91% ee

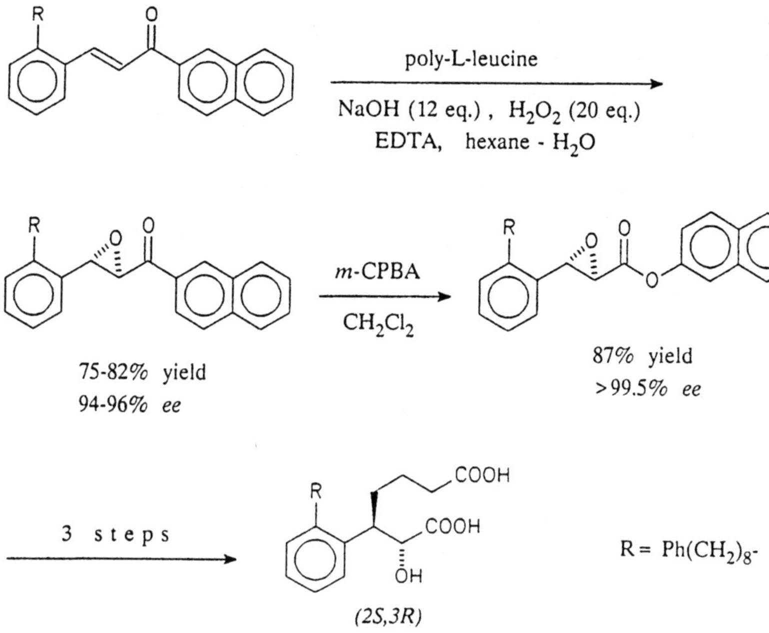

Figure 8-57. Asymmetric epoxidation of an electron-poor olefin.

in the oxidation of p-tolylmethylsulfide. The ee increased to 96%, with a chemical yield of 93%, when cumene hydroperoxide (CHP) was used instead of TBHP [166]. Di Furia and Modena [167,168] used a 1:4:2 ratio of Ti(O-iPr)$_4$/DET/ TBHP in the absence of water and obtained 88% ee with the same substrate (Figure 8-58). However, it should be noted that virtually all of the reported examples employ stoichiometric quantities of titanium alkoxide and tartrate and the use of catalytic quantities leads to substantial reductions in optical yields [166].

Sterically hindered chiral iron porphyrin complexes [169] and flavins [170], models for cytochrome P450 and flavin-dependent monooxygenases, respectively [171], have also been shown to catalyze enantioselective sulfoxidations, in moderate (up to 73%) ee's. The former utilized iodosobenzene and the latter hydrogen peroxide as the primary oxidant. In another biomimetic approach the protein, bovine serum albumin (BSA) has been used as a chiral catalyst for enantioselective sulfoxidations (ee's up to 80%) with sodium periodate or H_2O_2 as the primary oxidant [172]. However, in common with the epoxidations mentioned above, there is still much room for improvement in asymmetric sulfoxidations. From a practical viewpoint, the chloroperoxidase-catalyzed sulfoxidation described in chapter 7 appears to be the most viable process.

CH3
+ ROOH
Ti(O-iPr)4
(+)-DET
O
S
CH3
(R)

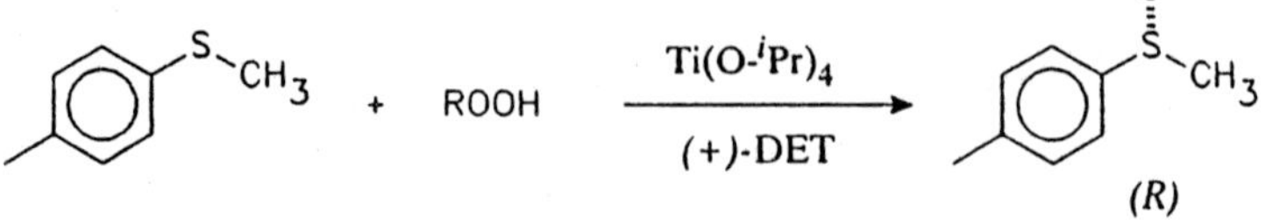

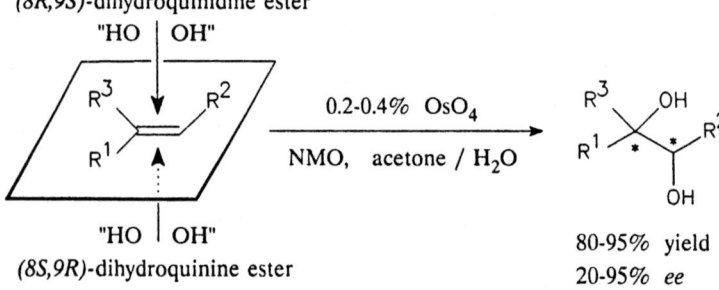

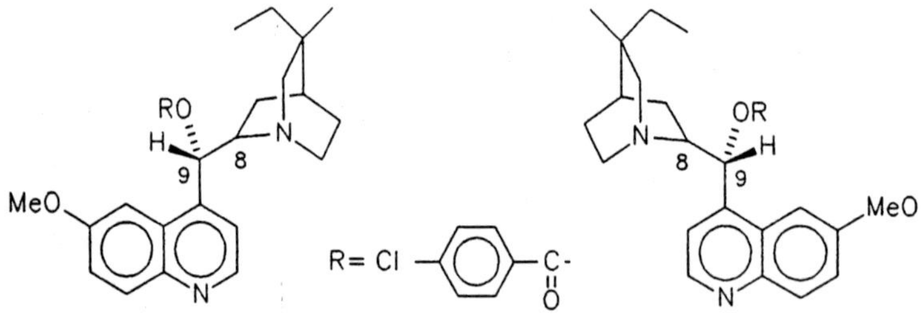

ROOH	$Ti^{(IV)}$: DET : ROOH : H_2O	*ee* (%)
TBHP	1 : 2 : 2 : 1	91
TBHP	1 : 4 : 2 : 0	88
CHP	1 : 2 : 2 : 1	96

Figure 8-58. Asymmetric sulfoxidation.

(8R,9S)-dihydroquinidine ester
"HO OH"
R3
R2
R1
0.2-0.4% OsO4
NMO, acetone / H2O
"HO OH"
(8S,9R)-dihydroquinine ester
R3
OH
R2
R1
OH
80-95% yield
20-95% ee

RO
H
9
8
N
MeO
N
R= Cl
C-
O
OR
H
N
8
9
MeO
N

(8R,9S)-dihydroquinidine (R = H) *(8S,9R)*-dihydroquinine (R = H)

Figure 8-59. Asymmetric dihydroxylation of olefins.

C. Asymmetric Dihydroxylation of Olefins

Following their success in asymmetric epoxidation, Sharpless and coworkers [173] developed an efficient method for the catalytic asymmetric dihydroxylation of olefins. The method employs catalytic amounts of OsO_4 and derivatives of cinchona alkaloids as chiral ligands together with N-methylmorpholine oxide (NMO) as the primary oxidant (Figure 8-59).

The reaction is a good example of ligand accelerated asymmetric catalysis as the alkaloid ligands enhance the rate by one to two orders of magnitude. Slow addition of the olefin is essential to obtain high *ee*'s due to a competing second catalytic cycle with low enantioselectivity (Figure 8-60). Improved enantioselectivities were observed using potassium ferricyanide as the primary oxidant under alkaline (K_2CO_3) conditions in aqueous *tert*-butanol [174]. In this case, the second cycle is precluded since hydrolysis of the monoglycolate complex precedes reoxidation of the osmium under these conditions. However, from a practical viewpoint, the improved enantioselectivities (about 5–10% *ee*) are probably offset

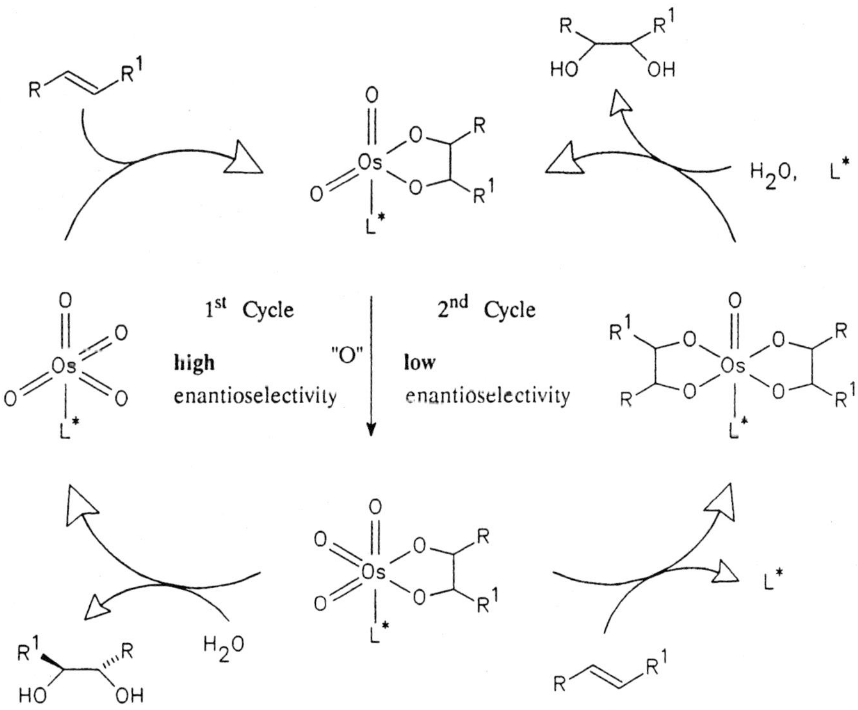

Figure 8-60. Mechanism of asymmetric dihydroxylation with NMO as primary oxidant.

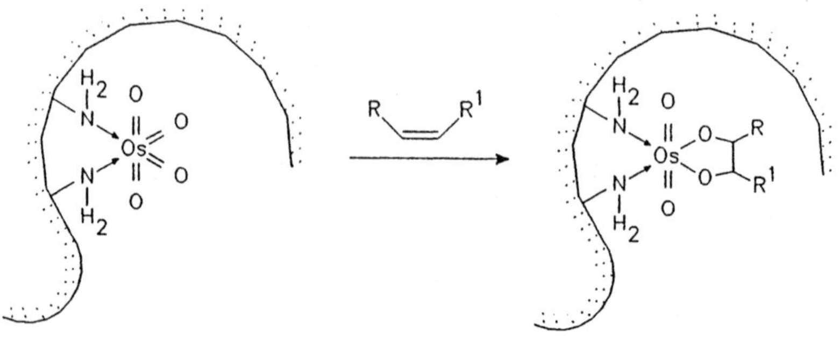

Figure 8-61. OsO₄—BSA complex.

by the use of a less attractive oxidant. When using NMO as the oxidant, the reduced form is readily recycled by oxidation with H_2O_2. Sharpless and coworkers also used a polymer-bound alkaloid ligand which facilitates recycling of the catalyst as an insoluble OsO_4 complex [175].

In contrast to the Sharpless asymmetric epoxidation, the asymmetric dihydroxylation is successful with a broad range of substituents, for example, PhCH=CHCH₃, PhCH=CHCH₂OH, PhCH=CHCH₂OAc, PhCH=CHCH₂OCH₂Ph, PhCH=CHCH-(OCH₃)₂ and PhCH=CHCOOCH₃ are all suitable substrates while only the allylic alcohol is suitable for asymmetric epoxidation.

Another interesting approach to asymmetric dihydroxylation is the use of BSA as the chiral ligand. A 1:1 complex of BSA with OsO_4 afforded *ee*'s in the range of 6–68% using TBHP as the primary oxidant [176]. The BSA presumably acts as a chiral diamine ligand (Figure 8-61) and although the enantioselectivities are modest they can probably be further improved by optimization. It is worth noting that the use of proteins, such as BSA, as relatively inexpensive chiral ligands for transition metal catalysts is an area that has been largely overlooked and probably merits further attention.

VI. THE FUTURE

The area of catalytic asymmetric synthesis is presently undergoing an explosive growth. For reasons of space, this chapter has not covered all the reactions which have been successfully subjected to asymmetric catalysis; some other examples include palladium-catalyzed allylic alkylations [176], gold-catalyzed aldol condensations [177], and rhodium-catalyzed hydroborations [178], to name but a few. Indeed, one may tentatively conclude that any catalytic process that is amenable to kinetic control can be carried out in an asymmetric mode by a suitable choice of chiral ligand and conditions. As is borne out by numerous examples in this chapter, when the necessary development effort is expended, enantioselectivities can be optimized from < 10% to > 90% *ee*. However, in most cases the choice of chiral ligand is still based largely on intuition and experience rather than rational design. In this respect the situation is no different to that with resolving agents and enzymes discussed in preceding chapters.

The ultimate in asymmetric catalysis is to dispense with the need for an (external) catalyst altogether by using the product enantiomer as a catalyst for its own formation. Such a phenomenon is referred to as **asymmetric autocatalysis** and may be the next generation of asymmetric synthesis [179]. In fact, it brings the discussion of asymmetric catalysis full circle, to the question of the origins of biomolecular chirality. Thus, if two chiral substances A and B react to form a chiral product C and C is a catalyst for its own formation this can, in principle, lead to asymmetric autocatalysis. For this to occur one enantiomer of C must catalyze its own formation and inhibit the formation of the other enantiomer. For example, the amino alcohol product of Reaction (8-9) might well be expected to be a catalyst for its own formation. Moreover, based on the observed amplification of chirality (See Section IV.E) in such systems one would expect 'seeding' with a pure enantiomer, or a mixture enriched with one enantiomer, to lead to selective formation of that enantiomer. Although there are experimental indications that asymmetric autocatalysis is possible [180], a practical example is elusive. Nevertheless, it remains a fascinating challenge for the future.

$$(8-9)$$

REFERENCES

1. Brown, J. M., *Chem. Ind.* (London), 612-617 (1988).
2. a) Bredig, G., and Fiske, P. S., *Biochem. Z.*, **46**, 7 (1912). b) See also: Bredig, G., and Minaeff, *Biochem. Z.*, **249**, 241 (1932).
3. Wynberg, H., *Topics Stereochem.*, **16**, 87-129 (1986).
4. Blaser, H. U., *Tetrahedron Asymm.*, **2**, 843-866.
5. a) Schwab, G. M., and Rudolph, L. M., *Naturwiss.*, **20**, 362 (1932). b) Schwab, G. M., Rost, F., and Rudolph, L. M., *Kolloid-Zeitschrift*, **68**, 157 (1934).
6. Akabori, S., Sakurai, S., Izumi, Y., and Fuji, Y., *Nature*, **178**, 233 (1956).
7. a) Harada, K., and Yoshida, T., *Naturwiss.*, **57**, 131 (1970). b) Harada, K., and Yoshida, T., *Naturwiss.*, **57**, 306 (1970).
8. Balandin, A. A., Klabunovskii, E. I., and Petrov, Y. I., *Dokl. Akad. Nauk. SSSR*, **127**, 557 (1959); *ibid.* (Engl.), **127**, 571 (1959).
9. Lipkin, D., and Stewart, T. D., *J. Am. Chem. Soc.*, **61**, 3295 (1939).
10. Orito, Y., Imai, S., Niwa, S., and Nguyen, G. H., *J. Synth. Org. Chem. Jpn.*, **37**, 173 (1979).

11. a) Fukawa, H., Izumi, Y., Komatsu, S., and Akabori, S., *Bull. Chem. Soc. Jpn.*, **35**, 1703 (1962). b) Izumi, Y., *Advan. Catal.*, **32**, 215 (1983).

12. Nozaki, H., Moriuti, S., Takaya, H., and Noyori, R., *Tetrahedron Lett.*, 5239 (1966).

13. Osborn, J. A., Jardine, F. H., Young, J. F., and Wilkinson, G., *J. Chem. Soc. a)*, 1711 (1966).

14. Horner, L., Siegel, H., and Buthe, H., *Angew. Chem. Int. Ed. Engl.*, **7**, 942 (1968).

15. Korpium, O., Lewis, R. A., Chickos, J., and Mislow, K., *J. Am. Chem. Soc.*, **90**, 4842 (1968).

16. Knowles, W. S., and Sabacky, M. J., *J. Chem. Soc., Chem. Commun.*, 1445 (1966).

17. a) Knowles, W. S., Sabacky, M. J., Vineyard, B. D., and Weinkauf, D. J., *J. Am. Chem. Soc.*, **97**, 2567 (1975). b) Knowles, W. S., *J. Chem. Educ.*, **63**, 222-225 (1986).

18. a) Vineyard, B. D., Knowles, W. S., and Sabacky, M. J., *J. Mol. Catal.*, **19**, 159-169 (1983). b) Knowles, W. S., *Acc. Chem. Res.*, **16**, 106-112 (1983).

19. Kagan, H. B., and Dang, T. P., *J. Chem. Soc., Chem. Commun.*, 1445 (1968).

20. Kagan, H. B., in *Asymmetric Synthesis, Vol. 5*, Morrison, J. D. (Ed.), Academic Press, New York, 1985, pp. 1-35.

21. Marko, L., and Bakos, J., in *Aspects of Homogeneous Catalysis, Vol. 4*, Ugo, R. (Ed.), Reidel, Dordrecht, 1981, pp. 145-202.

22. Fryzuk, M. D., and Bosnich, B., *J. Am. Chem. Soc.*, **100**, 5491 (1978).

23. Fryzuk, M. D., and Bosnich, B., *J. Am. Chem. Soc.*, **99**, 6262 (1977).

24. US Patent 3.949.000 (1976) to Rhône Poulenc.

25. MacNeil, P. A., Roberts, N. K., and Bosnich, B., *J. Am. Chem. Soc.*, **103**, 2273 (1981).

26. a) Brünner, H., and Pieronczyk, W., *Angew. Chem. Int. Ed. Engl.*, **18**, 620 (1979). b) Brünner, H., Pieronczyk, W., Schönhammer, B., Streng, K., Bernal, I., and Korp, J., *Chem. Ber.*, **114**, 1137 (1981).

27. a) Nagel, U., *Angew. Chem. Int. Ed. Engl.*, **23**, 435 (1984). b) Nagel, U., Kinzel, E., Andrade, J., and Prescher, G., *Chem. Ber.*, **119**, 3326 (1986).

28. Achiwa, K., *J. Am. Chem. Soc.*, **98**, 8265 (1976).

29. a) Miyashita, A., Yasuda, A., Takaya, H., Toriuni, K., Ito, K., Souchi, T., and Noyori, R., *J. Am. Chem. Soc.*, **102**, 7932 (1980). b) Mashima, K., Koyano, K., Yagi, M., Kumobayashi, H., Taketomi, T., Akutagawa, S., and Noyori, R., *J. Org. Chem.*, **51**, 629 (1986).

30. For a historical account see: Otsuka, S., and Tani, K., *Synthesis*, 665-680 (1991).

31. Selke, R., and Pracejus, H., *J. Mol. Catal.*, **37**, 213-225 (1986).

32. Vocke, W., Hänel, R., and Flöther, F. U., *Chem. Techn.*, **39**, 123 (1987).

33. a) Fiorini, M., and Giongo, G. M., *J. Mol. Catal.*, **5**, 303 (1979). b) Fiorini, M., and Giongo, G. M., *J. Mol. Catal.*, **7**, 411 (1980).

34. Katsuki, T., and Sharpless, K. B., *J. Am. Chem. Soc.*, **102**, 5974 (1980).

35. For a historical account see: a) Sharpless, K. B., *Chemtech*, 692-700 (1985). b) Sharpless, K. B., *Chem. Brit.*, 38-44 (1986).

36. Pitchen, P., Dunach, E., Deshmukh, M. N., and Kagan, H. B., *J. Am. Chem. Soc.*, **106**, 8188 (1984).

37. Tani, K., Yamagata, T., Akutagawa, S., Kumobayashi, H., Taketomi, T., Takaya, H., Miyashita, A., Noyori, R., and Otsuka, S., *J. Am. Chem. Soc.*, **106**, 5208 (1984).

38. a) Noyori, R., *Chem. Soc. Rev.*, **18**, 187-208 (1989). b) Noyori, R., and Takaya, H., *Acc. Chem. Res.*, **23**, 345 (19??).

39. Noyori, R., and Kitamura, M., in *Modern Synthetic Methods, Vol. 5*, Scheffold, R. (Ed.), Springer-Verlag, Heidelberg, 1989, pp. 115-198.

40. Noyori, R., *Science*, **248**, 1194-1199 (1990).
41. Kumobayashi, H., *Proc. 2nd Int. Conf. Pharm. Ingredients and Intermediates*, Manufacturing Chemist, London, 1992, pp. 135-138.
42. Bosnich, B. (Ed.), *Asymmetric Catalysis*, NATO ASI Series, Martinus Nijhoff, Dordrecht, 1986.
43. Brünner, H., in *Topics in Stereochemistry, Vol. 18*, Eliel, E. L., and Wilen, S. (Eds.), Wiley, New York, 1988, pp. 129-147.
44. Brünner, H., in *The Chemistry of the Metal-Carbon Bond, Vol. 5*, Hartley, F. R. (Ed.), Wiley, New York, 1989.
45. Eliel, E.L., and Otsuka, S. (Eds.), *Asymmetric Reactions and Processes in Chemistry*, American Chemical Scociety, Washington, 1982. ACS Symp. Ser., Vol. 185.
46. Kagan, H. B., in *Comprehensive Organometallic Chemistry*, Wilkinson, G. (Ed.), Pergamon Press, Oxford, 1982, pp. 463-498.
47. Scott, J. W., in *Topics in Stereochemistry, Vol. 19*, Eliel, E. L., and Wilen, S. H. (Eds.), Wiley, New York, pp. 209-226 (1989).
48. Ojima, I., Clos, N., and Bastos, C., *Tetrahedron*, **45**, 6901-6939 (1990).
49. Blystone, S. L., *Chem. Revs.*, **89**, 1663-1679 (1989).
50. Morrison, J. D., in *Asymmetric Synthesis, Vol 5*, Morrison, J. D. (Ed.), Academic Press, New York, 1985.
51. a) Pfaltz, A., *Bull. Soc. Chim. Belg.*, **99**, 729-739 (1990). b) Pfaltz, A., in *Modern Synthetic Methods*, Scheffold, R. (Ed.), Springer-Verlag, Heidelberg, 1989, pp. 199-248.
52. Brown, J. M., *Chem. Brit.*, 276-280 (1989).
53. Noyori, R., *Chemtech*, 360-367 (1992).
54. Jacobsen, E. N., Marko, I., Mungall, W. S., Schröder, G., and Sharpless, K. B., *J. Am. Chem. Soc.*, **110**, 1968-1970 (1988).
55. Pittman, C. U., in *Comprehensive Organometallic Chemistry, Vol. 8*, Wilkinson, G., Stone, F. G. A., and Abel, E. W. (Eds.), Pergamon, Oxford, 1982, p. 553.
56. Kuntz, E.G., *Chemtech*, 570 (1987).
57. Amrani, Y., Lecomte, L., Sinou, D., Bakos, J., Toth, I., and Heil, B., *Organometallics*, **8**, 542-547 (1989).
58. a) Alario, F., Amrani, Y., Colleuille, Y., Dang, T.P., Jenck, J., Morel, D., and Sinou, D., *J. Chem. Soc., Chem. Commun.*, 202 (1988). b) Benhamza, R., Amrani, Y., and Sinou, D., *J. Organometal. Chem.*, **288**, C37-C39 (1985).
59. Nagel, U., and Kinzel, E., *Chem. Ber.*, **119**, 1731 (1986).
60. Amrani, Y., and Sinou, D., *J. Mol. Catal.*, **24**, 231-233 (1984).
61. a) Toth, I., and Hanson, B. E., *Tetrahedron Asymm.*, **1**, 895-912 (1990). b) Toth, I., Hanson, B. E., and Davis, M. E., *Tetrahedron Asymm.*, **1**, 913-930 (1990).
62. Arntz, D., and Schäfer, A., in *Metal Promoted Selectivity in Organic Synthesis*, Kluwer, Amsterdam, 1991, pp. 161-189.
63. Halpern, J., in *Catalysis of Organic Reactions*, Augustine, R. L. (Ed.), Marcel Dekker, New York, 1985, p. 3.
64. Brown, J. M., *Angew. Chem. Int. Ed. Engl.*, **99**, 169-182 (1987).
65. Caplar, V., Comisso, G., and Sunjic, V., *Synthesis*, 85-116 (1981).
66. Koenig, K. E., in *Asymmetric Synthesis, Vol 5*, Morrison, J. D. (Ed.), Academic Press, New York, 1985, pp. 71-79.

67. Koenig, K. E., in *Catalysis of Organic Reactions*, Kosak, J. R. (Ed.), Marcel Dekker, New York, 1984, pp. 63-77.
68. Chan, A. S. C., Pluth, J. J., and Halpern, J., *J. Am. Chem. Soc.*, **102**, 5952-5954 (1980).
69. Halpern, J., *Pure Appl. Chem.*, **55**, 99-106 (1983).
70. Chua, P. S., Roberts, N. K., Bosnich, B., Okrasinski, S. J., and Halpern, J., *J. Chem. Soc., Chem. Commun.*, 1278-1280 (1981).
71. Ojima, I., Kogure, T., Yodo, N., Suzuki, T., Yatabe, M., and Tanaka, T., *J. Org. Chem.*, **47**, 1329-1342 (1982).
72. Ojima, I., in *Asymmetric Reactions and Processes in Chemistry*, Eliel, E.L., and Otsuka, S. (Eds.), American Chemical Society, Washington, 1982, pp. 109-138. ACS Symp. Ser., Vol. 185.
73. Meyer, D., Poulin, J. C., Kagan, H. B., Levine-Pinto, H., Morgat, J. L., and Fromageot, P., *J. Org. Chem.*, **45**, 4680-4682 (1980).
74. Takaya, H., Ohta, T., Mashima, K., and Noyori, R., *Pure Appl. Chem.*, **62**, 1135-1138 (1990).
75. Ohta, T., Takaya, H., Kitamura, M., Nagai, K., and Noyori, R., *J. Org. Chem.*, **52**, 3176-3178 (1987).
76. a) Takaya, H., Ohta, T., Sato, N., Kumobayashi, H., Akutagawa, S., Kasahara, I., and Noyori, R., *J. Am. Chem. Soc.*, **109**, 1596-1597 (1987). b) Takaya, H., Ohta, T., Sato, N., Kumobayashi, H., Akutagawa, S., Kasahara, I., and Noyori, R., *J. Am. Chem. Soc.*, **109**, 4129-4130 (1987).
77. Kitamura, M., Kasahara, K., Manabe, R., Noyori, R., and Takaya, H., *J. Org. Chem.*, **53**, 708-710 (1988).
78. a) Kitamura, M., Hsiao, Y., Noyori, R., and Takaya, H., *Tetrahedron Lett.*, **28**, 4829-4832 (1987). b) Noyori, R., Kitamura, M., Takaya, H., Kumobayashi, H., and Akutagawa, S., European Patent 0.245.960 (1987) to Takasago Perfumery Co.
79. Noyori, R., Ohkuma, T., Kitamura, M., Takaya, Sayo, N., H., Kumobayashi, H., and Akutagawa, S., *J. Am. Chem. Soc.*, **109**, 5856-5858 (1987).
80. Kitamura, M., Ohkuma, T., Inoue, S., Sayo, N., Kumobayashi, H., Akutagawa, S., Ohta, T., Takaya, H., and Noyori, R., *J. Am. Chem. Soc.*, **110**, 629-631 (1988).
81. Kitamura, M., Ohkuma, Takaya, H., and Noyori, R., *Tetrahedron Lett.*, **29**, 1555-1556 (1988).
82. Noyori, R., Ikeda, T., Ohkuma, T., Widhalm, M., Kitamura, M., Takaya, H., Akutagawa, S., Sayo, N., Saito, T., Taketomi, T., and Kumobayashi, H., *J. Am. Chem. Soc.*, **111**, 9134-9135 (1989).
83. Kitamura, M., Ohkuma, T., Tokunaga, M., and Noyori, R., *Tetrahedron Asymm.*, **1**, 1-4 (1990).
84. Sayo, N., Saito, T., Okada, Y., Nagashima, H., and Kumobayashi, H., European Patent Appl. 369.691 (1989) to Takasago Perfumery Co.
85. Kawano, H., Ishii, Y., Saburi, M., and Uchida, Y., *J. Chem. Soc., Chem. Commun.*, 87-88 (1988).
86. Ohkuma, T., Kitamura, M., and Noyori, R., *Tetrahedron Lett.*, **31**, 5509 (1990).
87. a) Achiwa, K., Kogure, T., and Ojima, I., *Tetrahedron Lett.*, 4431 (1977). b) Achiwa, K., Kogure, T., and Ojima, I., *Chem. Lett.*, 297 (1978).
88. a) Takahashi, H., Hattori, M., Chiba, M., Morimoto, T., and Achiwa, K., *Tetrahedron Lett.*, **27**, 4477 (1986). b) Morimoto, T., Takahashi, H., Fujii, K., Chiba, M., and Achiwa, K., *Chem. Lett.*, **12**, 2061 (1986).

89. Takahashi, H., Sakuraba, S., Takeda, H., and Achiwa, K., *J. Am. Chem. Soc.*, **112**, 5876-5878 (1990).
90. Takeshi, H., Tachinami, T., Aburatani, M., Takahashi, H., Morimoto, T., and Achiwa, K., *Tetrahedron Lett.*, **30**, 363 (1989).
91. Bakos, J., Toth, I., Heil, B., Szalontai, G., Parkanyi, L., and Fulop, V., *J. Organometal. Chem.*, **370**, 263-276 (1989).
92. Chan, A. S. C., and Landis, C. R., *J. Mol. Catal.*, **49**, 165-173 (1989).
93. Sachtler, W. M. H., in *Catalysis of Organic Reactions*, Augustine, R. L. (Ed.), Marcel Dekker, New York, 1985, pp. 189-206.
94. Tai, A., and Harada, T., in *Tailored Metal Catalysts*, Iwasawa, Y. (Ed.), Reidel, Dordrecht, 1986, p. 265.
95. Bakos, J., Toth, I., and Marko, L., *J. Org. Chem.*, **46**, 5427 (1981).
96. Brünner, H., Muschiol, M., Wischert, T., and Wiehl, J., *Tetrahedron Asymm.*, **1**, 159-162 (1990).
97. Kikukawa, T., Iizuka, Y., Sugimura, T., Harada, T., and Tai, A., *Chem. Lett.*, 1267-1270 (1987).
98. a) Blaser, H. U., Jalett, H. P., Monti, D. M., and Wehrli, J. T., *Applied Catalysis*, **52**, 19-32 (1989). b) Wehrli, J. T., Baiker, A., Monti, D. M., Blaser, H. U., and Jalett, H. P., *J. Mol. Catal.*, **57**, 245-257 (1989). c) Wehrli, J. T., Baiker, A., Monti, D. M., Blaser, H. U., and Jalett, H. P., *J. Mol. Catal.*, **49**, 195-203 (1989). d) Blaser, H. U., Jalett, H. P., Monti, D. M., Reber, J. F., and Wehrli, J. T., *Stud. Surf. Sci. Catal.*, **41**, 153 (1988).
99. See: Selke, R., Haupke, K., and Krause, H. W., *J. Mol. Catal.*, **56**, 315-328 (1989).
100. Corma, A., Iglesias, M., Del Pino, C., and Sanchez, F., *J. Chem. Soc., Chem. Commun.*, 1253-1255 (1991).
101. Vastag, S., Bakos, J., Toros, S., Takach, N. E., King, R. B., Heil, B., and Marko, L., *J. Mol. Catal.*, **22**, 283 (1984).
102. a) Kang, G. J., Cullen, W. R., Fryzuk, M. D., James, B. R., and Kutney, J. P., *J. Chem. Soc., Chem. Commun.*, 1466 (1988). b) Cullen, W. R., Fryzuk, M. D., James, B. R., Kutney, J. P., Kang, G. J., Herb, G., Thorburn, I. S., and Spogliarich, R., *J. Mol. Catal.*, **62**, 243 (1990).
103. Bakos, J., Orosz, A., Heil, B., Laghmari, M., Lhoste, P., and Sinou, D., *J. Chem. Soc., Chem. Commun.*, 1684-1685 (1991).
104. Lensink, C., and De Vries, J. G., *Tetrahedron Asymm.*, **3**, 235-238 (1992).
105. a) Brünner, H., and Leitner, W., *J. Organometal. Chem.*, **387**, 209 (1990). b) Brünner, H., and Leitner, W., *Angew. Chem. Int. Ed. Engl.*, **27**, 1180 (1988). c) Brünner, H., Graf, E., Leitner, W., and Wutz, K., *Synthesis*, 743 (1989).
106. Takahashi, H., and Achiwa, K., *Chem. Lett.*, 1921 (1987).
107. Brünner, H., Becker, R., and Riepl, G., *Organometallics*, **3**, 1354 (1984).
108. Tani, K., *Pure Appl. Chem.*, **57**, 1845-1854 (1985).
109. Tani, K., Yamagata, T., Tatsuno, Y., Yamagata, Y., Tomita, K., Akutagawa, S., Kumobayashi, H., and Otsuka, S., *Angew. Chem. Int. Ed. Engl.*, **24**, 217 (1985).
110. Takaya, H., Ohta, T., Sayo, N., Kumobayashi, H., Akutagawa, S., Inoue, S.I., Kasahara, I., and Noyori, R., *J. Am. Chem. Soc.*, **109**, 1596 (1987).
111. Takabe, K., Uchiyama, Y., Okisaka, K., Yamada, T., Katagiri, T., Okazaki, T., Oketa, Y., Kumobayashi, H., and Akutagawa, S., *Tetrahedron Lett.*, **26**, 5153 (1985).

112. Kitamura, M., Manabe, K., and Noyori, R., *Tetrahedron Lett.*, **28**, 4719-4720 (1987).

113. Gauthier-Lafaye, J., and Perron, R., in *Industrial Applications of Homogeneous Catalysis*, Mortreux, A., and Peltit, F. (Eds.), Reidel, Dordrecht, 1984, pp. 19-64.

114. Parinello, G., and Stille, J. K., *J. Am. Chem. Soc.*, **109**, 7122-7127 (1987).

115. See for example: Hayashi, T., Tanaka, M., and Ogata, I., *J. Mol. Catal.*, **26**, 17 (1984).

116. Alper, H., and Hamel, N., *J. Am. Chem. Soc.*, **112**, 2803-2804 (1990).

117. Aratani, T., *Pure Appl. Chem.*, **57**, 1839-1844 (1985) and references cited therein.

118. For a recent review see: Doyle, M. P., *Recl. Trav. Chim. Pays-Bas*, **110**, 305-316 (1991).

119. a) Pfaltz, A., *Bull. Soc. Chim. Belg.*, **99**, 729-739 (1990). b) Pfaltz, A., in *Modern Synthetic Methods, Vol 5*, Scheffold, R. (Ed.), Springer-Verlag, Heidelberg, 1989, pp. 199-248.

120. Evans, D. A., Woerpel, K. A., and Hinman, M., *J. Am. Chem. Soc.*, **113**, 726-728 (1991).

121. Lowenthal, R. E., Abiko, A., and Masamune, S., *Tetrahedron Lett.*, **31**, 6005 (1990).

122. For a recent review see: Mortreux, A., in *Metal Promoted Selectivity in Organic Synthesis*, Noels, A. F., Graziani, M., and Hubert, A. J. (Eds.), Kluwer, Amsterdam, 1991, pp. 47-63.

123. Wilke, G., *Angew. Chem. Int. Ed. Engl.*, **27**, 185 (1988).

124. Buono, G., Siv, C., Peiffrer, G., Triantaphylides, C., Denis, P., Mortreux, A., and Petit, F., *J. Org. Chem.*, **50**, 1781 (1985).

125. Hayashi, T., Konishi, M., Fukushima, M., Kahehira, K., Hioki, T., and Kumada, M., *J. Org. Chem.*, **48**, 2195 (1983).

126. a) Kitamura, M., Suga, S., and Noyori, R., *J. Am. Chem. Soc.*, **108**, 6071 (1986). b) Kitamura, M., Okada, S., Suga, S., and Noyori, R., *J. Am. Chem. Soc.*, **111**, 4028 (1989).

127. a) Oguni, N., Matsuda, Y., and Kaneko, T., *J. Am. Chem. Soc.*, **110**, 7877-7878 (1988). b) Oguni, N., and Omi, T., *Tetrahedron Lett.*, **25**, 2823-2824 (1984).

128. Itsuno, S., and Frechet, J. M. J., *J. Org. Chem.*, **52**, 4142-4143 (1987).

129. a) Soai, K., Ookawa, A., Kaba, T., and Ogawa, K., *J. Am. Chem. Soc.*, **109**, 7111-7115 (1987). b) Soai, K., Niwa, S., Yamada, Y., and Inoue, H., *Tetrahedron Lett.*, **28**, 4841-4842 (1987). c) Soai, K., Ookawa, A., Ogawa, K., and Kaba, T., *J. Chem. Soc., Chem. Commun.*, 467-468 (1987). d) Soai, K., Yokoyama, S., Ebihara, K., and Hayasaka, T., *J. Chem. Soc., Chem. Commun.*, 1690-1691 (1987).

130. Corey, E. J., and Hannon, F. J., *Tetrahedron Lett.*, **28**, 5237-5240 (1987).

131. Tanaka, K., Ushio, H., and Suzuki, H., *J. Chem. Soc., Chem. Commun.*, 1700-1701 (1989).

132. Mucho, G., Vannoorenberghe, Y., and Buono, G., *Tetrahedron Lett.*, **28**, 6163-6166 (1987).

133. Van Oeveren, A., Menge, W., and Feringa, B. L., *Tetrahedron Lett.*, **30**, 6427-6430 (1989).

134. Smaardijk, A. A., and Wynberg, H., *J. Org. Chem.*, **52**, 135-137 (1987).

135. a) Takahashi, H., Kawakita, T., Yoshioka, M., Kobayashi, S., and Ohno, M., *Tetrahedron Lett.*, **30**, 7095-7098 (1989). b) Yoshioka, M., Kawakita, T., and Ohno, M., *Tetrahedron Lett.*, **30**, 1657-1660 (1989).

136. For a recent review see: Narasaka, K., *Synthesis*, 1-11 (1991).

137. Narasaka, K., Iwasawa, N., Inoue, M., Yamada, T., Nakashima, M., and Sugimori, J., *J. Am. Chem. Soc.*, **111**, 5340-5345 (1989).

138. Mikami, K., Terada, M., and Nakai, T., *J. Am. Chem. Soc.*, **111**, 1940 (1989).

139. Corey, E. J., Imwinkelried, R., Pikul, S., and Xiang, Y. B., *J. Am. Chem. Soc.*, **111**, 5493 (1989).

140. a) Minamikawa, H., Hayakawa, S., Yamada, T., Iwasawa, N., and Narasaka, K., *Bull. Chem. Soc., Japan,* **61**, 4379-4383 (1988). b) Narasaka, K., Yamada, T., and Minamikawa, H., *Chem. Lett.,* 2073-2076 (1987).

141. Hayashi, M., Matsuda, T., and Oguni, N., *J. Chem. Soc., Chem. Commun.,* 1364-1365 (1990).

142. Tanaka, K., Mori, A., and Inoue, S., *J. Org. Chem.,* **55**, 181-185 (1990) and references cited therein.

143. a) Dong, W., and Petty, W. L., European Patent 0.135.691 (1986) to Du Pont. b) Jackson, W. R., Brit. Pat. Appl. 2.143.823 (1985) to ICI; *CA:* **104**, 68624c (1986).

144. Hajos, Z. G., and Parrish, D. R., *J. Org. Chem.,* **39**, 1615-1621 (1974).

145. a) Wynberg, H., and Staring, E. G. J., *J. Chem. Soc., Chem. Commun.,* 1181 (1984). b) Staring, E. G. J., Moorlag, H., and Wynberg, H., *Recl. Trav. Chim. Pays-Bas,* **105**, 374 (1986). c) Wynberg, H., and Staring, E. G. J., *J. Org. Chem.,* **50**, 1977 (1985).

146. a) Dolling, U. H., Davis, P., and Grabowski, E. J. J., *J. Am. Chem. Soc.,* **106**, 446-447 (1984). b) Hughes, D. L., Dolling, U. H., Ryan, K. M., Schoenewaldt, E. F., and Grabowski, E. J. J., *J. Org. Chem.,* **52**, 4745-4752 (1987).

147. Dolling, U. H., Hughes, D. L., Battacharya, A., Ryan, K. M., Karady, S., Weinstock, L. M., Grenda, V. J., and Grabowski, E. J. J., in *Catalysis of Organic Reactions,* Rylander, P. N., Greenfield, H., and Augustine, R. L. (Eds.), Marcel Dekker, New York, 1988, pp. 65-85.

148. Nerinckx, W., and Vandewalle, M., *Tetrahedron Asymm.,* **1**, 265-276 (1990).

149. O'Donnell, M. J., Bennett, W. D., and Wu, S., *J. Am. Chem. Soc.,* **111**, 2353-2355 (1989).

150. Masui, M., Ando, A., and Shiori, T., *Tetrahedron Lett.,* **29**, 2835-2838 (1988).

151. For reviews see: a) Sharpless, K. B., *Janssen Chimica Acta,* **6**(1), 3-6 (1988). b) Sharpless, K. B., Woodward, S. S., and Finn, M. G., *Pure Appl. Chem.,* **55**, 1823 (1983). c) Finn, M. G., and Sharpless, K. B., in *Asymmetric Synthesis, Vol 5,* Morrison, J. D. (Ed.), Academic Press, New York, 1985, pp. 247-308. d) Rossiter, B. E., in *Asymmetric Synthesis, Vol 5,* Morrison, J. D. (Ed.), Academic Press, New York, 1985, pp. 193-246. e) Rossiter, B. E., in *Catalysis of Organic Reactions,* Augustine, R.L. (Ed.), Marcel Dekker, New York, 1985, pp. 295-308. f) Masamune, S., Choy, W., Petersen, J. S., and Sita, L. R., *Angew. Chem. Int. Ed. Engl.,* **24**, 1 (1985).

152. a) Hanson, R. M., and Sharpless, K. B., *J. Org. Chem.,* **51**, 1922-1925 (1986). b) Gao, Y., Hanson, R. M., Klunder, J. M., Ko, S. Y., Masamune, H., and Sharpless, K. B., *J. Am. Chem. Soc.,* **109**, 5765-5780 (1987).

153. Jorgensen, K. A., *Tetrahedron Asymm.,* **2**, 515-532 (1991) and references cited therein.

154. Klunder, J. M., Onami, T., and Sharpless, K. B., *J. Org. Chem.,* **54**, 1295-1304 (1989).

155. Martin, V. S., Woodward, S. S., Katsuki, T., Yamada, Y., Ikeda, M., and Sharpless, K. B., *J. Am. Chem. Soc.,* **103**, 6237 (1981).

156. Miyano, S., Lu, L. D. L., Viti, M., and Sharpless, K. B., *J. Org. Chem.,* **50**, 4350-4360 (1985).

157. Choudary, B. M., Valli, V. L. K., and Durga Prasad, A., *J. Chem. Soc., Chem. Commun.,* 1186 (1990).

158. a) Zhang, W., and Jacobsen, E. N., *J. Org. Chem.,* **56**, 2296-2298 (1991). b) Jacobsen, E. N., Zhang, W., and Güler, M. L., *J. Am. Chem. Soc.,* **113**, 6703-6704 (1991).

159. Zhang, W., Loebach, J. L., Wilson, S. R., and Jacobsen, E. N., *J. Am. Chem. Soc.,* **112**, 2801-2803 (1990).

160. a) Irie, R., Noda, K., Ito, Y., and Katsuki, T., *Tetrahedron Lett.*, **32**, 1055-1058 (1991).
b) Irie, R., Noda, K., Ito, Y., Matsumotot, N., and Katsuki, T., *Tetrahedron Asymm.*, **2**, 481-491 (1991).

161. a) Groves, J. T., and Viski, P., *J. Org. Chem.*, **55**, 3628-3634 (1990). b) See also: Naruta, Y., Tani, F., and Maruyama, K., *Chem. Lett.*, 1269-1272 (1989).

162. Sinigalia, R., Michelin, R. A., Pinna, F., and Strukul, G., *Organometallics*, **6**, 728 (1987).

163. a) Julia, S., Masana, J., and Vega, J. C., *Angew. Chem. Int. Ed. Engl.*, **19**, 929-931 (1980). b) Julia, S., Guixer, J., Masana, J., Roca, J., Colonna, S., Annunziata, R., and Molinari, H., *J. Chem. Soc., Perkin Trans I*, 1314-1324 (1982).

164. Lantos, I., and Novack, V., in *Chirality in Drug Design and Synthesis*, Brown, C. (Ed.), Academic Press, New York, 1990, pp. 167-180.

165. a) Kagan, H. B., in *Stereochemistry of Organic and Bioorganic Transformations*, Bartmann, W., and Sharpless, K. B. (Eds.), VCH, Weinheim, 1987, pp. 31-48. b) Kagan, H. B., Dunach, E., Nemecek, C., and Pitchen, P., *Pure Appl. Chem.* **57**, 1911 (1985).

166. Zhao, S. H., Samuel, O., Kagan, H. B., *Tetrahedron*, **43**, 5135 (1987).

167. Di Furia, F., Modena, G., and Seraglia, R., *Synthesis*, 325 (1984).

168. Conte, V., Di Furia, F., Licini, G., and Modena, G., in *Metal Promoted Selectivity in Organic Synthesis*, Noels, A. F., Graziani, M., and Hubert, A. J. (Eds.), Kluwer, Amsterdam, 1991, pp. 91-105, and references cited therein.

169. Naruta, Y., Tani, F., and Maruyama, K., *Tetrahedron Asymm.*, **2**, 533-542 (1991).

170. Shinkai, S., Yamaguchi, T., Manabe, O., and Toda, F., *J. Chem. Soc., Chem. Commun.*, 1399-1401 (1988).

171. a) Katopodis, A. G., Smith, H. A., and May, S. W., *J. Am. Chem. Soc.*, **110**, 897-899 (1988). b) Ortiz de Montellano, P. R. (Ed.), *Cytochrome P450*, Plenum Press, New York, 1986.

172. a) Sugimoto, T., Kokubo, T., Miyazaki, J., Tanimoto, S., and Okano, M., *Bioorg. Chem.*, **10**, 311-323 (1981). b) Colonna, S., Banfi, S., Annunziata, R., and Casella, L., *J. Org. Chem.*, **51**, 891-895 (1986). c) Colonna, S., Banfi, S., Fontana, F., and Sommaruga, M.J., *J. Org. Chem.*, **50**, 769-771 (1985).

173. a) Jacobsen, E. N., Marko, I., Mungall, W. S., Schröder, G., and Sharpless, K. B., *J. Am. Chem. Soc.*, **110**, 1968-1970 (1988). b) Jacobsen, E. N., Marko, I., France, M. B., Svendsen, J. S., and Sharpless, K. B., *J. Am. Chem. Soc.*, **111**, 737-739 (1989). c) Wai, J. S. M., Marko, I., Svendsen, J. S., Finn, M. G., Jacobsen, E. N., and Sharpless, K. B., *J. Am. Chem. Soc.*, **111**, 1123-1125 (1989). d) Marko, I. E., *Proc. Chiral 89 Symp.*, Spring Innovations, Stockport, UK, 1989, pp 13-20.

174. a) Kwong, H. L., Sorato, C., Ogino, Y., Chen, H., and Sharpless, K. B., *Tetrahedron Lett.*, **31**, 2999-3002 (1990). b) Ogino, Y., Chen, H., Kwong, H. L., and Sharpless, K. B., *Tetrahedron Lett.*, **32**, 3965-3968 (1991). c) Sharpless, K. B., Amberg, W., Beller, M., Chen, H., Hartung, J., Kawanami, Y., Lübben, D., Manoury, E., Ogino, Y., Shibata, T., and Ukita, T., *J. Org. Chem.*, **56**, 4585-4588 (1991).

175. Kim, B. M., and Sharpless, K. B., *Tetrahedron Lett.*, **31**, 3003-3006 (1990).

176. Togni, A., *Tetrahedron Asymm.*, **2**, 683-690 (1991) and references cited therein.

177. Ito, Y., Sawamura, M., and Hayashi, T., *J. Am. Chem. Soc.*, **108**, 6405 (1986).

178. Hayashi, T., Matsumoto, Y., and Ito, Y., *Tetrahedron Asymm.*, **2**, 601-612 (1991) and references cited therein.

179. Wynberg, H. W., *Chimia*, **43**, 150-152 (1989).
180. a) Alberts, A. H., and Wynberg, A. H., *J. Am. Chem. Soc.*, **111**, 7265-7266 (1989). b) Alberts, A. H., and Wynberg, A. H., *J. Chem. Soc., Chem. Commun.*, 453-454 (1989).

ADDITIONAL READING

1. Whitesell, J. K., Symmetry and Assymmetric Induction, *Chem. Rev.*, **89**, 1581-1590 (1989).
2. Rosini, C., Franzini, L., Raffaelli, A., and Salvadori, P., Synthesis and Applications of Binaphthylic C_2-Symmetry Derivatives as Chiral Auxiliaries in Enantioselective Reactions, *Synthesis*, 503-517 (1992).
3. Crosby, J., Synthesis of Optically Active Compounds. A Large Scale Perspective. *Tetrahedron*, **47**, 4789-4846 (1991).

9

Industrial Processes: Comparing
Different Approaches

All generalizations are false, including this one.

G. K. Chesterton

The preceding chapters have outlined the different methodologies that can be considered by a researcher when embarking on the industrial synthesis of a pure enantiomer. Chirality introduces an extra dimension into the task of designing an industrial synthesis, one that is often the overriding factor in choosing a synthesis route. One general conclusion is that it is dangerous to draw general conclusions with regard to the superiority of one type of technology compared to another. The most economically viable route varies from one product to another, even within groups of closely related products such as α-amino acids. Hence, each case has to be judged on its own merit.

Certain points of consideration are common to all syntheses of chiral compounds, such as when to introduce the optical activity. A tentative general conclusion was that this should be as early as possible in the overall synthesis, to avoid carrying expensive ballast through the process, but this may be countered by other factors such as the ease of racemization of the unwanted isomer. Another point worth considering is the beneficial effect of having a recrystallization subsequent to the 'chiral' step, thus providing the opportunity for upgrading the optical purity.

The choice of examples in this chapter is largely a personal one but an attempt has been made to choose examples that represent a cross-section of the market for pure enantiomers and also that illustrate general principles. Since many of the

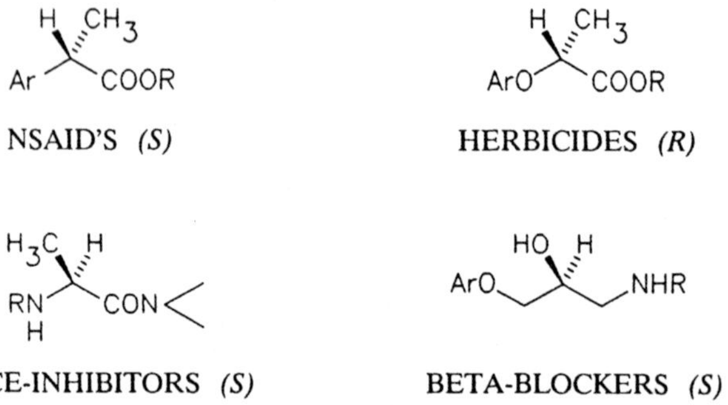

H CH₃
\|
Ar   COOR

NSAID'S *(S)*

H CH₃
\|
ArO   COOR

HERBICIDES *(R)*

H₃C H
\|
RN   CON
\|
H

ACE-INHIBITORS *(S)*

HO H
\|
ArO   NHR

BETA-BLOCKERS *(S)*

Figure 9-1. Structures of the eutomers of various product groups.

commercially important chiral products show striking similarities in their basic structures (Figure 9-1) it is not surprising that the competing synthesis routes are often markedly similar.

I. *l*-MENTHOL

Worldwide production of *l*-menthol amounts to about 4000 tons per annum. Traditionally it is produced from Japanese mint oil (*Mentha arvensis*); however, a desire to be independent of the capricious harvesting of a natural product stimulated the development of many alternative syntheses of *l*-menthol [1]. In the Haarman and Reimer synthesis, *m*-cresol is converted via thymol to a mixture of all eight isomers of menthol. The *dl*-menthol is separated by distillation and the residue is isomerized and recycled. The racemic menthol is subsequently resolved by preferential crystallization of an ester (e.g., benzoate). Alternatively, an enzymatic method that employs an esterase has been developed [2] for the resolution of racemic menthol but it does not appear to offer any substantial economic advantage. Other routes have also been developed [3] based on raw materials from the chirality pool (e.g., (−)-α-phellandrene or (+)-3-carene (Figure 9-2); the latter process has been commercialized by Malti-Chem in India. The optical activity does not have to be introduced into the molecule in these routes.

The elegant catalytic asymmetric synthesis route of Takasago has emerged as an economically viable technology and now accounts for a substantial share of *l*-menthol production. (See chapter 8.) Although the Takasago route starts from a chiral terpene, β-pinene, the first step involves conversion to an achiral molecule, myrcene. Hence, in contrast to the other terpene-based routes, the chirality has to be introduced into the molecule (by asymmetric synthesis) in the Takasago route.

The evolution of *l*-menthol manufacturing processes parallels the development of synthetic methodologies for optically active products in general. It began with

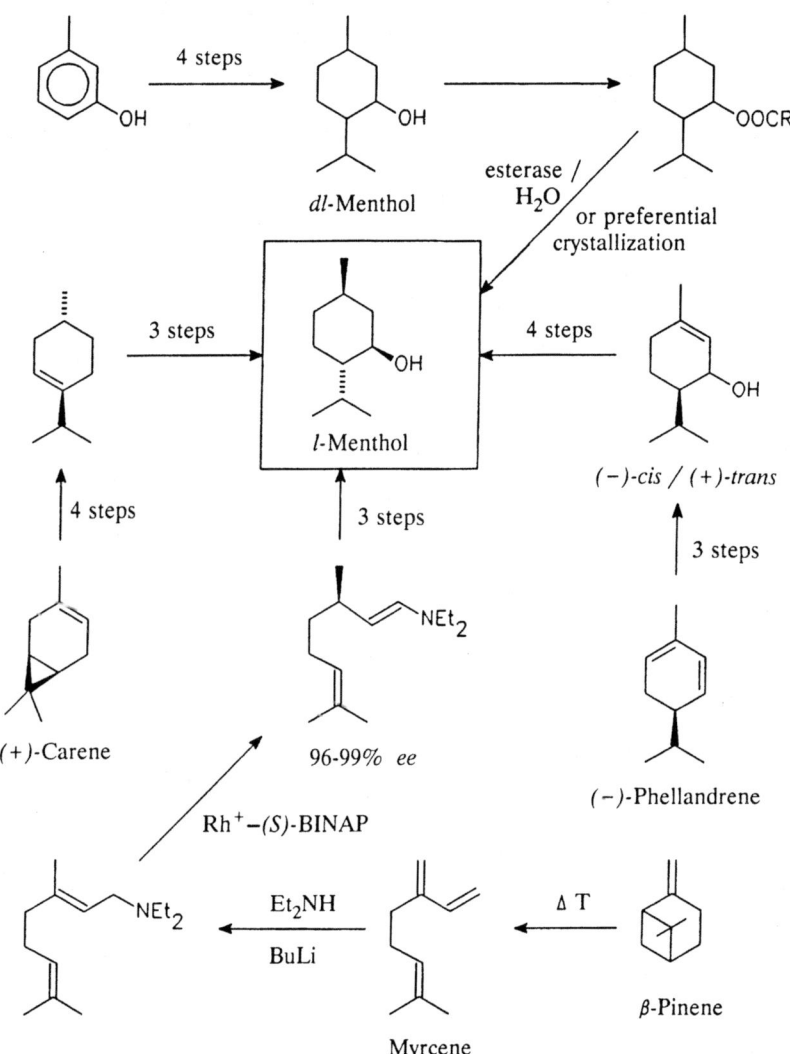

Figure 9-2. Different routes to *l*-menthol.

isolation from natural sources and evolved through methods involving the resolution of synthetic racemate or the use of raw materials from the chirality pool to the use of catalytic asymmetric synthesis.

II. α-PHENOXYPROPIONIC ACID HERBICIDES

The various approaches to the synthesis of the biologically active *(R)*-enantiomers of α-phenoxypropionic acid herbicides are outlined in Figure 9-3. One can make

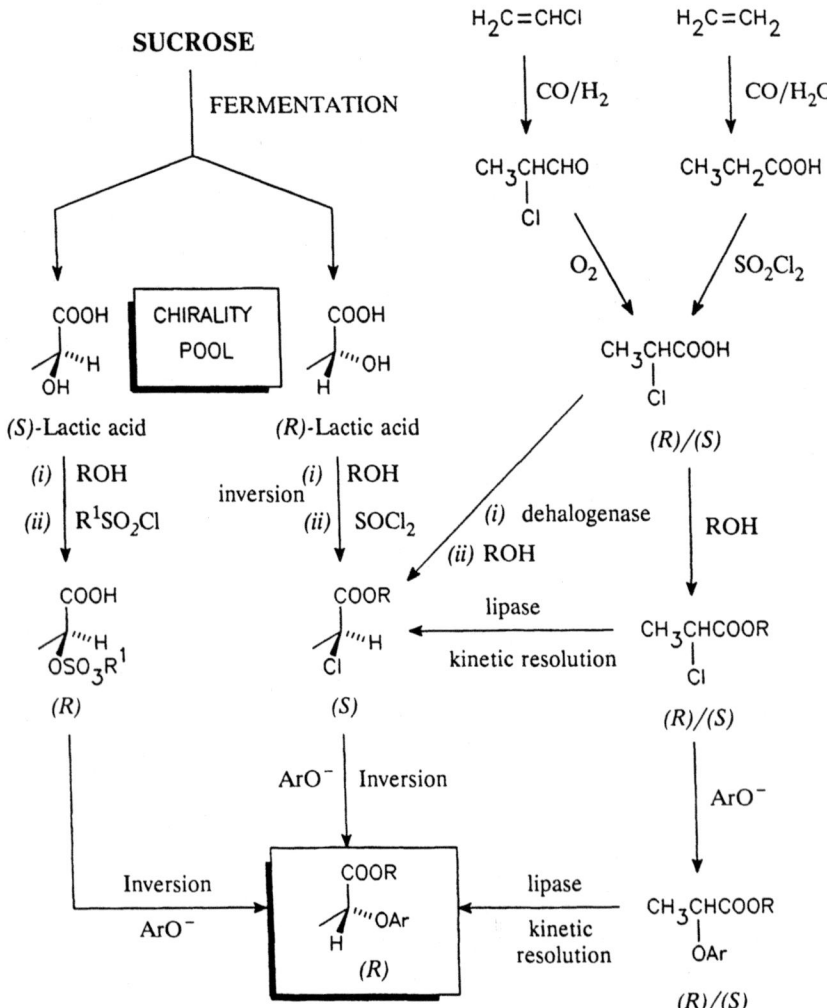

Figure 9-3. Alternative routes to *(R)*-α-phenoxypropionic acids (esters).

use of the chiral pool, represented by D- or L-lactic acid which are both produced commercially by fermentation (see chapter 4), or carry out a (kinetic) resolution at a particular stage in the synthesis.

Routes involving *(S)=L-α-*chloropropionic acid, or an ester thereof, as the key chiral synthon appear to be the most attractive commercially. According to one report [4], AKZO has started up production of *(R)*-mecoprop with a plant capacity of 14000 tons per annum. This followed a ban on the use of the racemic forms of these herbicides in The Netherlands. The key step in the AKZO process is reaction of an L-α-chloropropionic acid ester (probably a isobutyl ester) with 4-chloro-2-methyl phenolate anion (Reaction 9-1).

$$(R)\text{-Mecoprop}\ (R = H) \qquad (9\text{-}1)$$

The L = *(S)*-α-chloropropionic acid (ester) is produced by at least two routes: reaction of an *(R)*-lactic acid ester with $SOCl_2$ [5], and kinetic resolution of *(R,S)*-α-chloropropionic acid using an *(R)*-specific dehalogenase. (See chapter 7.) The former is probably practiced by BASF and/or Rhône-Poulenc, while the latter has reportedly been commercialized by ICI. Which route is more attractive depends on, among other factors, whether the required intermediate is the free acid [6] or an ester thereof, and on the value assigned to the *(S)*-lactic acid coproduct in the ICI process. Moreover, as noted in Chapter 4, the economics of *(R)*-lactic acid from fermentation of sucrose are very sensitive to the scale of operation.

Alternatively, *(S)*-chloropropionic acid (or an ester thereof) can be produced by lipase-catalyzed enantioselective hydrolysis of the racemic ester [7] or esterification [8] of the racemic acid. The wrong isomer can be readily racemized under acidic conditions [9]. However, this process does not appear to compete effectively with the two processes mentioned above, probably because the enantioselectivities are not high enough.

Finally, the most elegant synthesis of *(R)*-α-phenoxypropionic acid herbicides is the direct conversion of the racemate into the *(R)*-isomer via microorganism-mediated inversion of the *(S)*-enantiomer. For example, *(R,S)*-fluazifop is converted in the presence of growing cells of a *Rhodococcus* sp. to the pure *(R)* enantiomer in essentially quantitative yield [10]. However, this method does not appear to be competitive, probably because of low productivities, lack of availability of the microorganism, and there are not the economies of scale that accrue from processes which produce a common chiral intermediate for various phenoxypropionic acid herbicides.

III. α-ARYLPROPIONIC ACID ANTIINFLAMMATORIES

The therapeutically active *(S)*-enantiomers of the arylpropionic acid group of nonsteroidal antiinflammatory drugs (NSAIDs) are structurally analogous to the *(R)*-isomers of the α-phenoxypropionic acids (Figure 9-1). The commercial importance of this group of drugs (Figure 9-4 for examples) is underlined by the large diversity of routes that have been considered for their production. The two most important members of the group are naproxen and ibuprofen, which are marketed as the single *(S)*-enantiomer and as the racemate, respectively. (See chapter 2.)

The major producer of naproxen employs a route involving classical resolution of the racemate via diastereomeric salt crystallization. The racemate can be produced by a variety of routes starting from β-naphthol. The most important ones are outlined in Figure 9-5 and several alternative routes to *(S)*-naproxen are outlined in Figure 9-6 [11].

The only processes that are operated commercially (to our knowledge)are the classical Syntex route and Zambon's elegant stereoconvergent route, which involves the use of L-tartaric acid as a chiral auxiliary. (See chapter 5.) Lipase-mediated enantioselective hydrolysis of naproxen esters has also been described (see chapter 7, Figure 7-28) but does not appear to offer any distinct advantage as compared to the classical resolution. An interesting alternative is suggested by the enantioselective hydrolysis of the racemic nitrile precursor of ibuprofen to give *(S)*-ibuprofen in 95% *ee* [12]. The reaction is mediated by an *(S)*-specific nitrilase

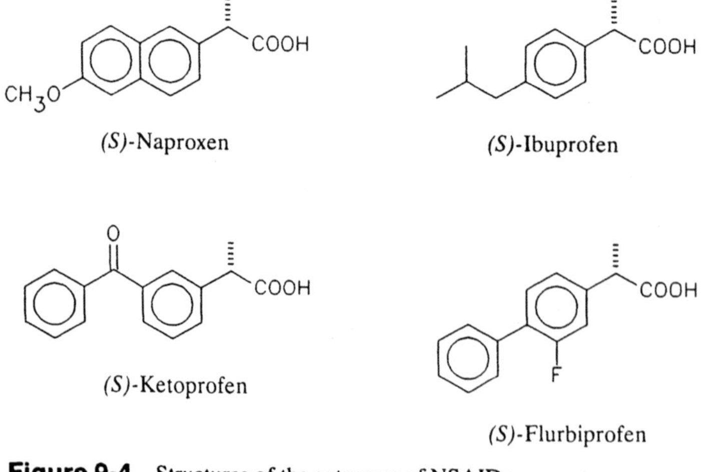

(S)-Naproxen *(S)*-Ibuprofen

(S)-Ketoprofen

(S)-Flurbiprofen

Figure 9-4. Structures of the eutomers of NSAIDs.

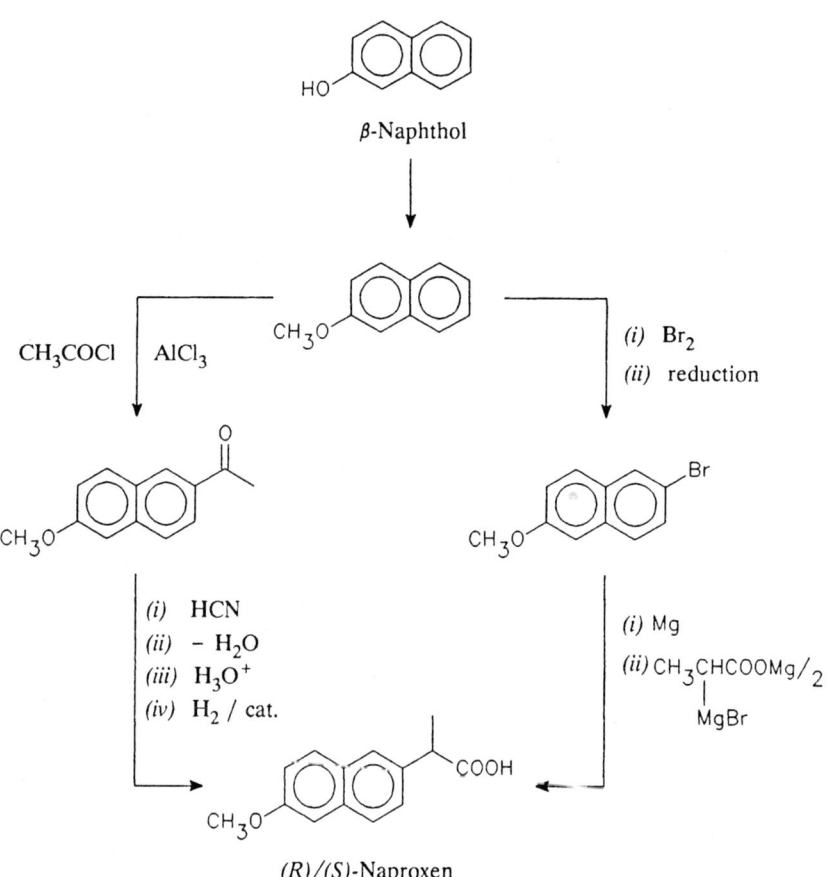

Figure 9-5. Synthesis of racemic naproxen.

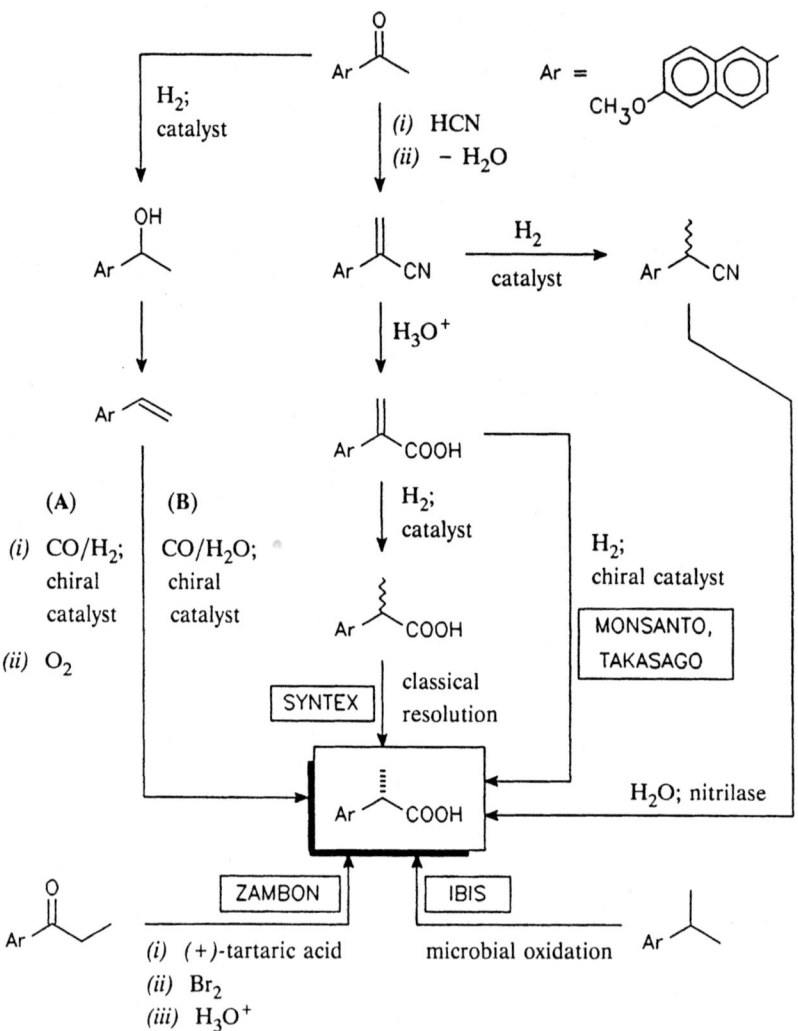

Figure 9-6. Alternative routes to *(S)*-naproxen.

contained in resting cells of an *Acinetobacter sp.* and is one of the few examples of an enantioselective enzymatic hydrolysis of a nitrile (Reaction 9-2).

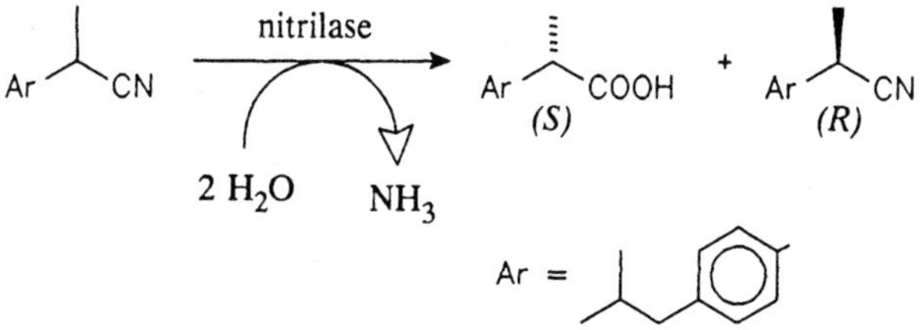

$$Ar = \text{(9-2)}$$

This route has the advantage that it is a precursor process. (See chapter 3.) Moreover, the unwanted nitrile isomer is readily racemized at alkaline pH under mild conditions (30°C, dilute ammonia). Another interesting route has been developed by Merck scientists [13] (Figure 9-7); however, since it starts from racemic product, it does not appear to offer any advantages over the classical route.

Although some catalytic asymmetric syntheses are potentially attractive (see chapter 8) to our knowledge they have not been commercialized. One interesting possibility is the one-step synthesis of *(S)*-naproxen by microbial oxidation of 6-methoxy-2-isopropylnaphthalene (Figure 9-6), developed by IBIS [14], but the route has not been commercialized, probably because of low productivities, lack of availability of microorganism, etc. The same probably applies to the production of *(S)*-naproxen from the racemate by 'biokinetic isomerization' (Reaction 9-3)[15].

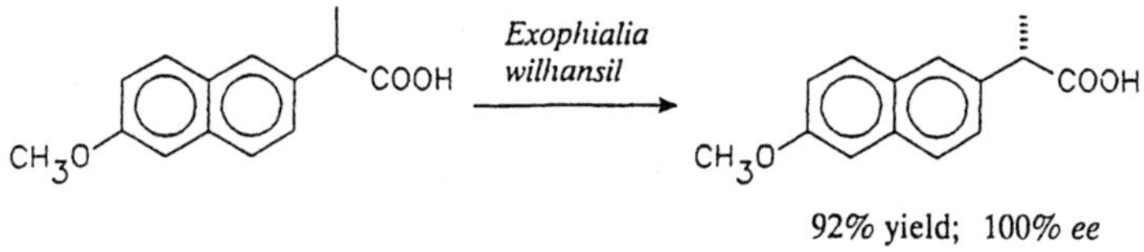

92% yield; 100% *ee*   (9-3)

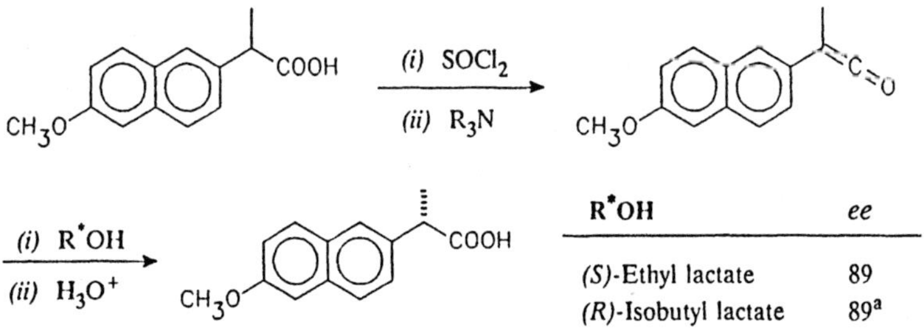

R*OH	ee
(S)-Ethyl lactate	89
(R)-Isobutyl lactate	89[a]
(R)-Pantanoyllactone	99[a]

[a] The product was *(R)*-naproxen

Figure 9-7. Merck route to *(S)*-naproxen.

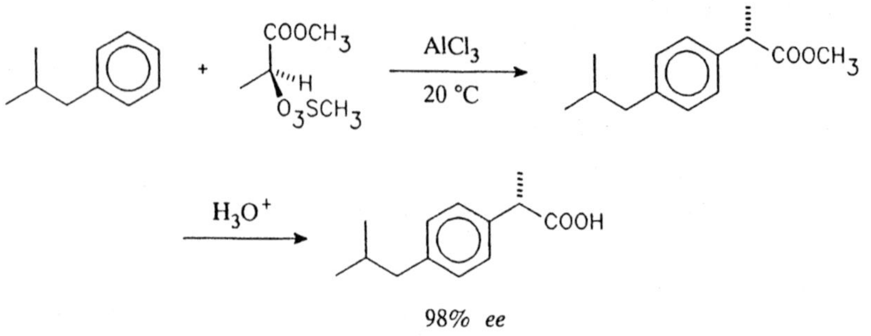

Figure 9-8. Chirality pool route to *(S)*-ibuprofen.

Considering the structural resemblance to the α-phenoxypropionic acid herbicides one might expect a 'chirality pool' route based on optically active lactic acid (derivatives) to be an attractive option. Indeed, the selective preparation of *(S)*-ibuprofen via Friedel-Crafts alkylation of *p*-isobutylbenzene by the mesylate of methyl *(S)*-lactate has been described [16] (Figure 9-8).

IV. BETA-BLOCKERS

No review of the industrial synthesis of optically active compounds would be complete without a discussion of optically active beta-blockers: the chemist's dream and the marketing manager's nightmare. There is no group of products that has attracted so much attention from manufacturers of chiral intermediates but with so little financial reward.

Although they are in competition with ACE-inhibitors and calcium antagonists, the beta-blockers have been a very successful group of antihypertensive drugs for more than two decades. They are all chiral molecules but the commercially most important ones—exemplified by propranolol, atenolol and metoprolol—are all sold as racemates (see chapter 2) and are synthesized via the general route shown in Figure 9-9.

An enormous amount of both industrial and academic research was devoted to the development of chiral intermediates for optically active beta-blockers. This frenetic effort was stimulated by the expectation that the racemates would eventually be replaced by the pure *(S)*-eutomers. This has not happened. *(S)*-Timolol is the only example of a beta-blocker of commercial importance that is sold as a pure enantiomer. Although this product has been a financial success, the total world demand for the bulk product amounts to only about 2 tons. This is not much of a bone for manufacturers of chiral intermediates to fight over. Fortunately, the chiral C_3 synthons, developed with optically active beta-blockers in mind, can also be applied to the synthesis of other commercially interesting products.

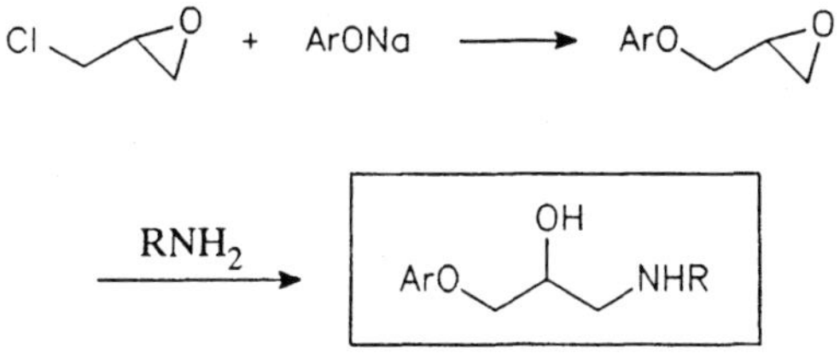

Name	Ar	R
Propanolol	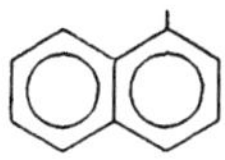	$(CH_3)_2CH$
Atenolol	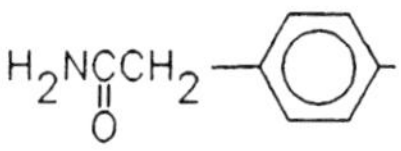	$(CH_3)_2CH$
Metoprolol	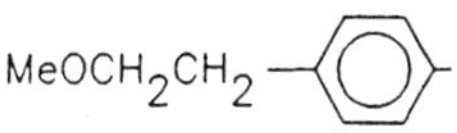	$(CH_3)_2CH$

Figure 9-9. Synthesis of racemic beta-blockers.

The probable commercial synthesis of *(S)*-timolol [17] is illustrated in Figure 9-10 and bears little resemblance to the general synthesis of racemic beta-blockers (Figure 9-9). The key chiral intermediate is produced by a classical resolution by diastereomer crystallization.

An obvious strategy for the synthesis of optically active beta-blockers is to adapt the route used for racemates (Figure 9-9) by replacing racemic epichlorohydrin with the pure enantiomer [18]. However, it was soon recognized [19] that nucleophilic attack on optically pure epichlorohydrin and related molecules proceeds via two competing pathways, leading to retention or inversion of configuration. The relative contributions of the two pathways are very dependent on the nature of the leaving group (Figure 9-11). Obviously, simultaneous operation of both pathways leads to product of lower optical purity. Examination of the observed results leads to the conclusion that chloride is not the ideal leaving group. Sharpless and coworkers have shown [20] that *m*- and *p*-nitrobenzene sulfonates are ideal leaving groups, giving > 99.9% retention.

Examples of C_3 chirons, developed with optically active beta-blockers in mind, are shown in Figure 9-12. Some of them **(I-VII)** can be used for any *(S)*-beta-blockers, while others (**VIII** and **IX**) are more advanced intermediates that already contain the alkylamino or aryloxy moiety, respectively.

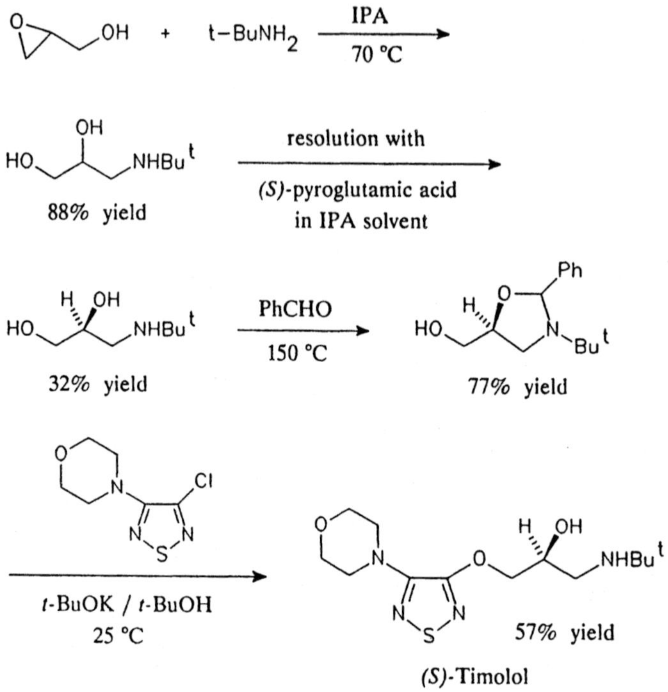

Figure 9-10. Commercial synthesis of (S)-timolol [17].

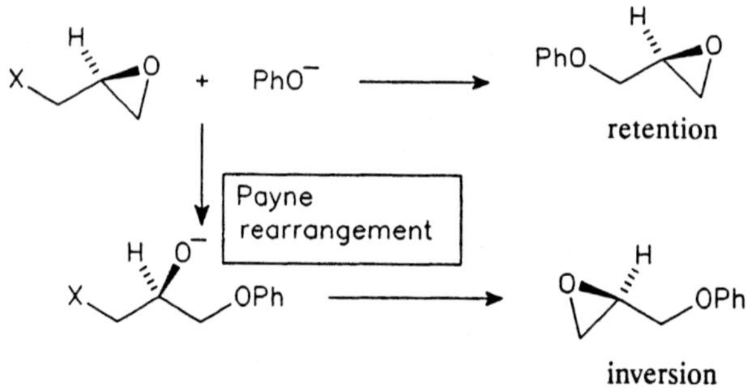

X	Base	% retention
Cl	K_2CO_3	5
CH_3SO_3	K_2CO_3	20
CH_3SO_3	NaH/DMF	85
CF_3SO_3	K_2CO_3	98

Figure 9-11. Competing pathways in nucleophilic substitution of epichlorohydrin.

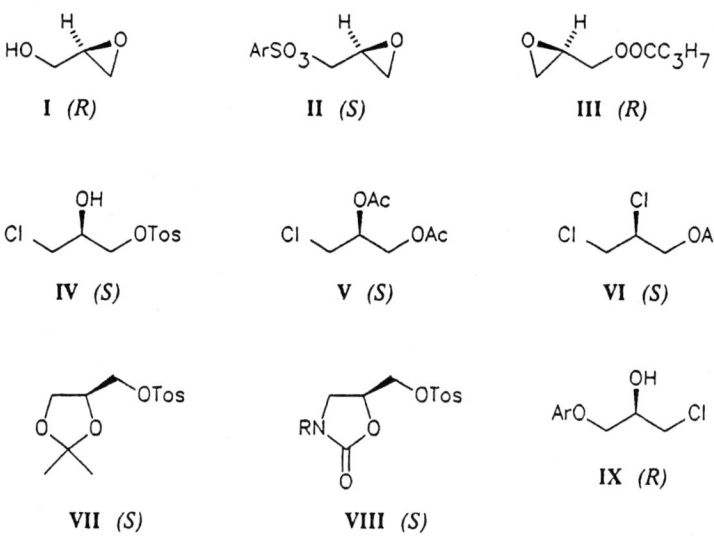

Figure 9-12. Examples of C₃ chirons.

Prior to 1980, such chiral intermediates were synthesized via circuitous routes from carbohydrate raw materials, such as D-mannitol [21]. Since then many enzymatic [22] and other [23] routes have been developed that start from simple raw materials such as allyl alcohol, glycidol, or epichlorohydrin (Figure 9-13).

Certain key intermediates are produced by more than one route, for example, Andeno produces optically active glycidyl derivatives (**I**) and (**III**) via enzymatic hydrolysis of racemic glycidyl butyrate [22] while ARCO produces (**I**) and (**II**) using the Sharpless catalytic epoxidation procedure. Although the latter has the advantage of a theoretical yield of 100%, the enzymatic kinetic resolution route can be carried out at very high substrate concentrations, even dispensing with a solvent altogether, which gives much higher productivities.

Kanegafuchi scientists have devoted much effort to developing intermediates (**IV**) [24] and (**VIII**) [25]. For example, (**IV**) is produced in 78–90% yield and greater than 99% *ee* by lipoprotein lipase-catalyzed hydrolysis of the racemate of the corresponding acetate or butyrate (Reaction 9-4) [24]. The same synthon has been synthesized by lipase-catalyzed transesterification [26] as shown in Reactions 9-5 and 9-6. Similarly, (**VIII**) is produced via lipase-catalyzed hydrolysis of a racemic ester, which in turn is derived from glycidol (Figure 9-13). Interestingly, the unwanted *(R)*-enantiomer can be inverted to the required *(S)*-enantiomer [27]; however, this requires five additional chemical steps which makes the economics questionable (Figure 9-14).

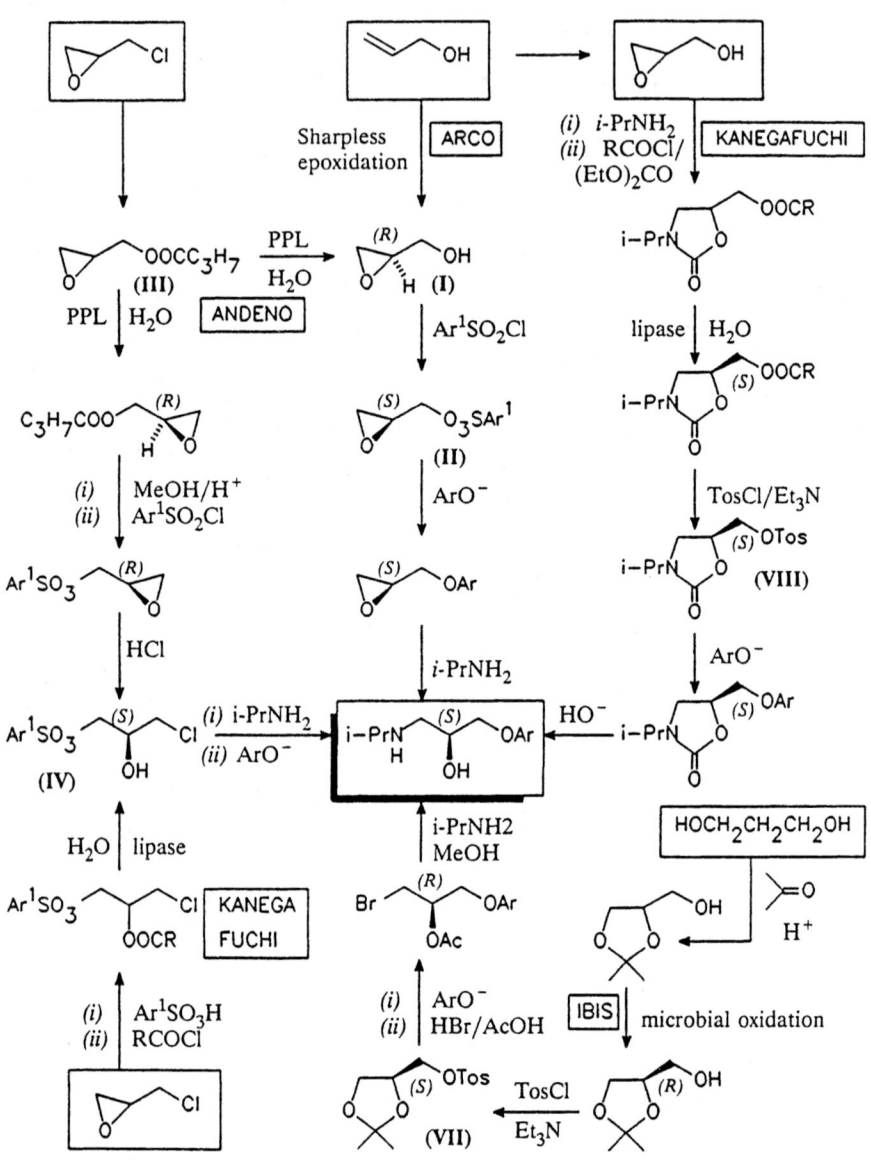

Figure 9-13. Alternative routes to *(S)*-beta-blockers.

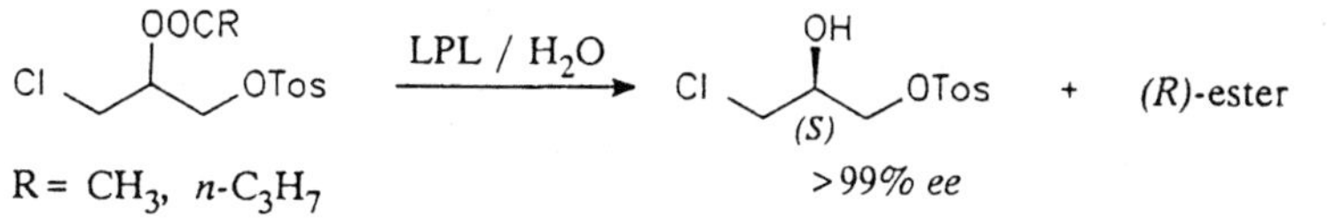

$$\text{(9-4)}$$

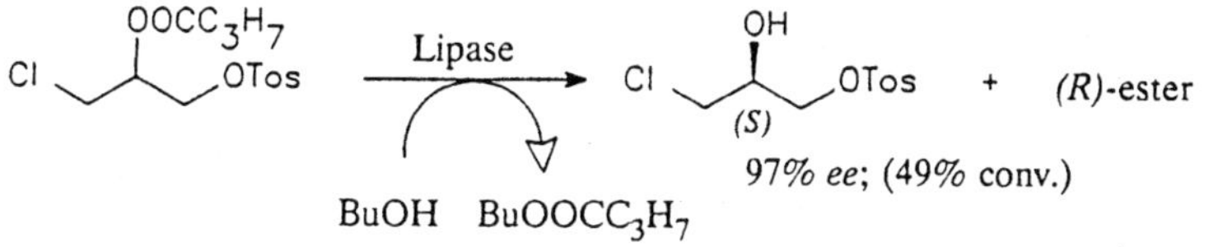

$$\text{(9-5)}$$

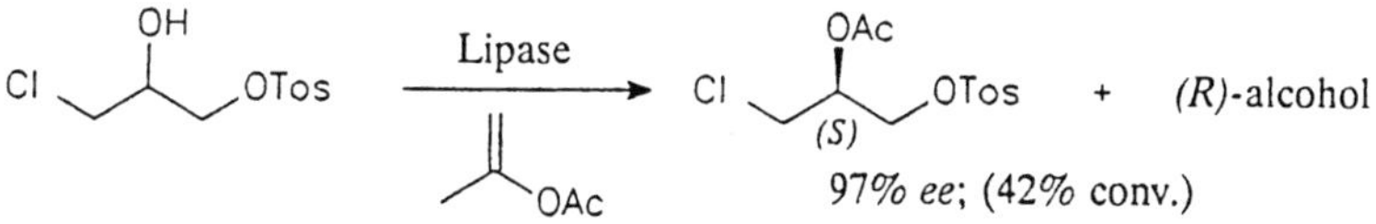

$$\text{(9-6)}$$

In yet another variation on this theme, scientists at the Sagami Research Institute [28,29] synthesized the chiral synthons (V) and (VI) via lipase-catalyzed hydrolysis of the corresponding racemic acetates (Figure 9-15). These acetates are prepared from the readily available starting materials, 3-chloro-1,2-propanediol and 2,3-dichloro-1-propanol, respectively. Enantioselectivities were, however, lower than those observed for (IV) and (VIII) (see above). Schneider and coworkers [30] have described the synthesis of the advanced intermediate (IX) by lipase-catalyzed hydrolysis of the racemic acetate or transesterification of the alcohol (Figure 9-16). Furthermore, as noted in Chapter 8, the lipase-mediated hydrolysis of chlorohydrin

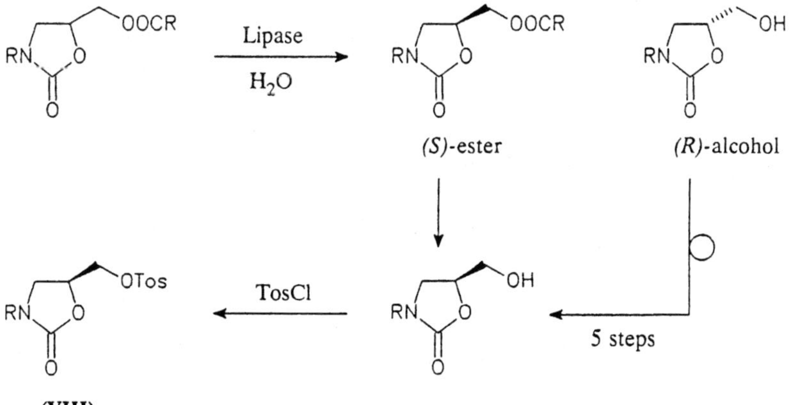

Figure 9-14. Kanegafuchi synthesis of the chiral oxazolidone (VIII).

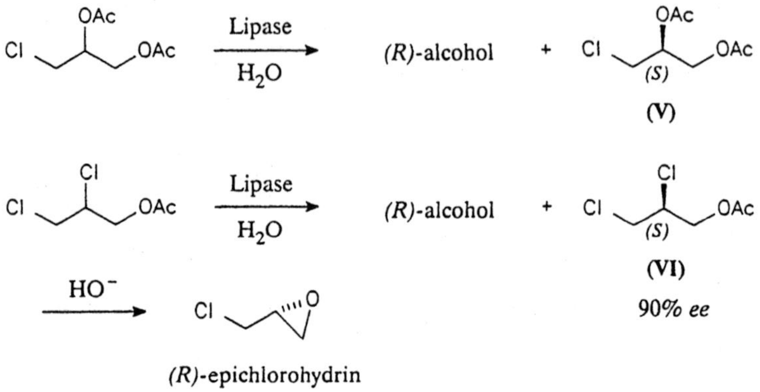

Figure 9-15. Sagami synthesis of C$_3$ chirons.

esters constitutes a general method for synthesizing precursors of optically active epoxides which, in many cases, is more attractive than asymmetric epoxidation.

Finally, yet another approach to optically active beta-blockers involves microbial (ep)oxidation of suitable precursors. For example IBIS [31] has developed a commercial synthesis of *(R)*-isopropylidene glycerol *((R)*-IPG) by a *Rhodococcus* sp.–mediated oxidation of the racemate which is readily prepared from glycerol. Treatment of *(R)*-IPG with *p*-toluenesulfonyl chloride and triethylamine gives the C$_3$ chiron, **(VII)**. Interestingly, the use of *Pseudomonas* sp. leads to products of the

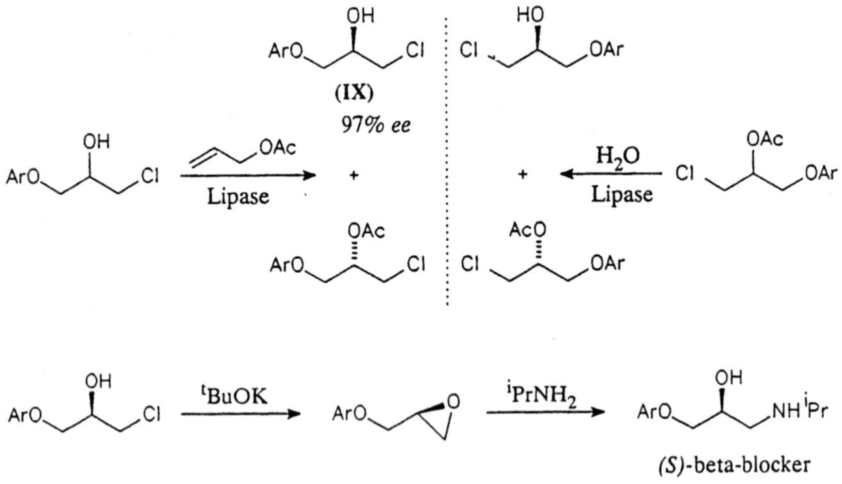

Figure 9-16. Enzymatic synthesis of an advanced *(S)*-beta-blocker intermediate.

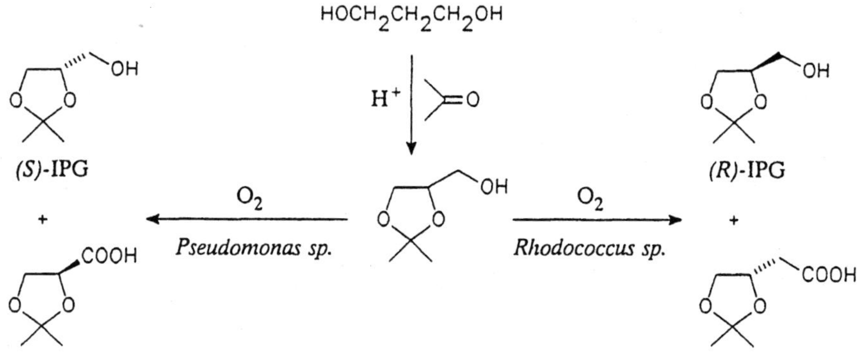

Figure 9-17. Synthesis of *(R)*- and *(S)*-isopropylidene glycerol by microbial oxidation.

opposite configuration (Figure 9-17) [32]. Similarly, microbial epoxidation has been used to prepare *(S)*-epichlorohydrin [33] and a precursor of *(S)*-metoprolol [34] from the corresponding olefins (Figure 9-18).

Although these constitute short routes to key chiral intermediates and offer high enantioselectivities, they tend to suffer from very low productivities and lack of commercial availability of the catalyst, as compared to the lipase-based processes. Moreover, the second process in Figure 9-18 is limited in that it does not produce a common intermediate.

In short, C_3 chirons for optically active beta-blockers have proven to be extremely interesting target molecules that have tested the ingenuity of organic chemists involved in the industrial synthesis of optically active compounds. Methods have evolved from carbohydrate-based routes to enzymatic transformations, catalytic asymmetric epoxidation, or catalytic asymmetric hydrogenation

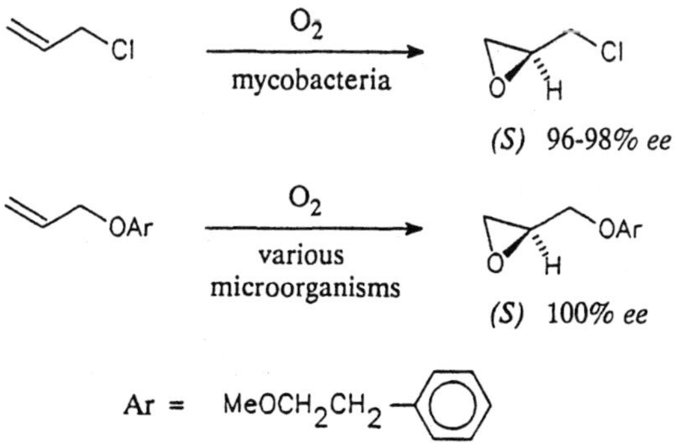

Figure 9-18. Synthesis of optically active beta-blocker precursors by microbial epoxidation.

for preparing advanced intermediates. (See chapter 8.) Although asymmetric synthesis or enzymatic kinetic resolution appear to have the economic edge, note that C_3 chirons via carbohydrate conversions [21] are undergoing a period of revival. (See also chapter 5.)

V. ASPARTAME

The methods developed for the industrial synthesis of the artificial sweetener, aspartame, represent a microcosm of the industrial production of optically active amino acids. The key intermediates are L-aspartic acid and L-phenylalanine (Reaction 9-7):

Aspartame

L-Phenylalanine   L-Aspartic acid   (9-7)

Because of the commercial importance of aspartame the synthesis of these key intermediates has attracted much attention. The commercial production of L-aspartic acid is dominated by one process: enzymatic addition of ammonia to fumaric acid mediated by immobilized whole cells of *E. coli*. (See chapter 7.) In contrast, a variety of methods were developed for the production of L-phenylalanine [35], the most important of which are outlined in Figure 9-19.

Virtually all the synthetic routes start from benzaldehyde, which is readily available and relatively inexpensive. By analogy with the other amino acids, obvious approaches are enzymatic hydrolysis of the benzylhydantoin and catalytic asymmetric hydrogenation of α-acetamidocinnamic acid. The latter is (or has been) used commercially in Italy by ISIS. (See chapter 8.) Alternatively, hydrolysis of benzalhydantoin, the condensation product of benzaldehyde and hydantoin, gives phenylpyruvic acid that can be converted to L-phenylalanine by enzymatic amination or transamination. (See chapter 7.) Another approach that was commercialized by Genex corporation is the phenylalanine lyase-catalyzed addition of ammonia to cinnamic acid; however, this production process, which is completely analogous to the production of L-aspartic acid from fumaric acid, has reportedly been abandoned.

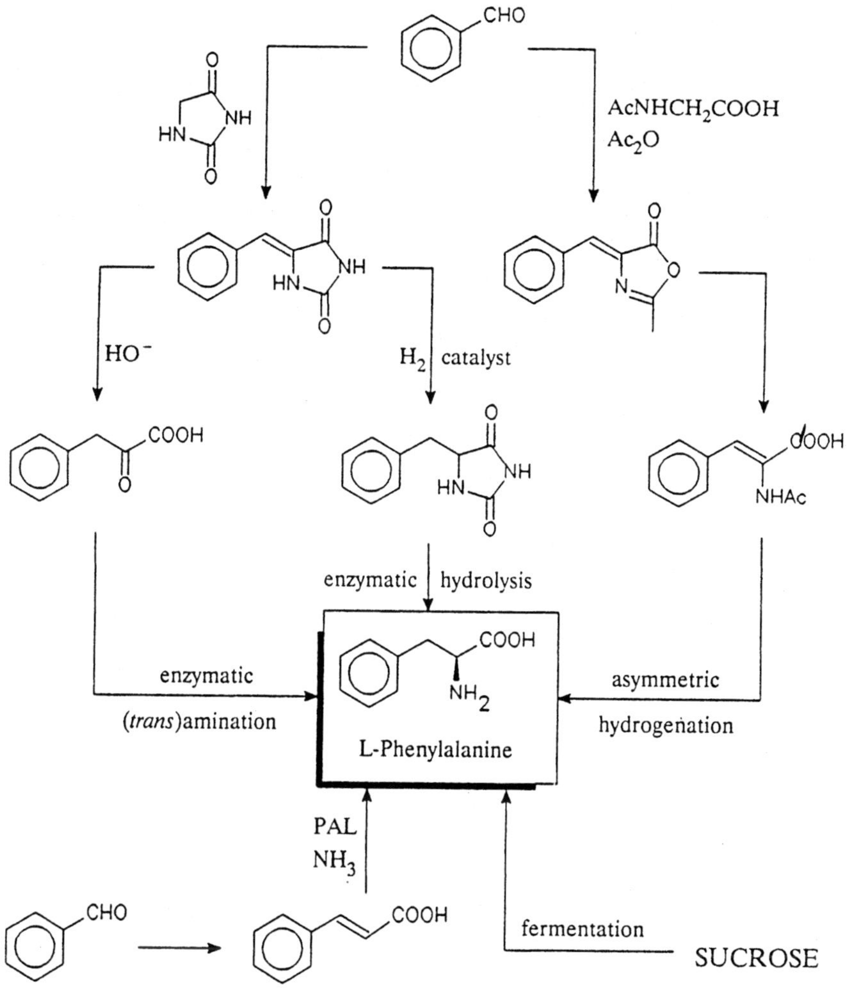

PAL= phenylalanine ammonia lyase

Figure 9-19. Alternative routes to L-phenylalanine.

Indeed, the process which has emerged victoriously is de novo fermentation, largely because of the substantial increases in productivity resulting from genetic engineering of the microorganisms involved. (See chapter 4.) It is a further illustration of the fact that de novo fermentation is often the method of choice for the production of microbial metabolites.

For the manufacture of aspartame itself, there is yet another possibility, one that bypasses the synthesis of L-phenylalanine altogether. This is accomplished in the DSM-Toyo Soda process for the production of aspartame by regio- and enantiospecific enzymatic coupling of DL-phenylalanine methyl ester with an aspartic acid derivative. This process was discussed in chapter 7. However, the substantial decrease in the price of L-phenylalanine, resulting from optimization of the fermentation process, has led to a situation where the price of the pure enantiomer is probably not much higher than that of the racemate. The latter is produced by hydrolysis of the corresponding hydantoin which, in turn, is produced from benzaldehyde as shown in Figure 9-19.

VI. ACE-INHIBITORS

As noted in chapter 2, the ACE-inhibitors comprise a group of highly successful antihypertensive agents, all of which are chiral molecules and are sold as pure optical isomers. The two protagonists, captopril and enalapril, are currently two of the top ten selling drugs worldwide.

Captopril contains two stereogenic carbon atoms. In the commercial synthesis, one of these is provided by L-proline which is readily available from fermentation. The synthetic challenge is provided by the other chiral building block, β-mercapto-isobutyric acid. Indeed, the evolution of processes for captopril synthesis is a perfect illustration of the golden rule of 'early resolution' discussed in chapter 3. In the original captopril synthesis, the resolution step involved separation, by crystallization, of the mixture of diastereomers formed by coupling of L-proline to racemic β-mercaptoisobutyroyl chloride (Figure 9-20). This involves carrying 50% isomeric ballast through this step and cumbersome recovery of the relatively expensive L-proline from the wrong diastereomer. This process, used for the production of initial development quantities, was subsequently replaced by the more economical route involving resolution of the β-mercaptoisobutyric acid, the first chiral intermediate in the synthesis. Thus, the required (R)-enantiomer is manufactured by Andeno by classical resolution of the racemate. Subsequent coupling of the acid chloride with L-proline affords (S)-acetyl captopril in higher volume yield with less reagents and solvent (as compared to the 'racemic route') and without the need for L-proline recovery.

Kanegafuchi, on the other hand, produces (R)-β-hydroxyisobutyric acid by (precursor) fermentation of isobutyric (or methacrylic) acid. (See chapters 4 and 5.) This is subsequently converted to (R)-β-acetylmercaptoisobutyric acid via the

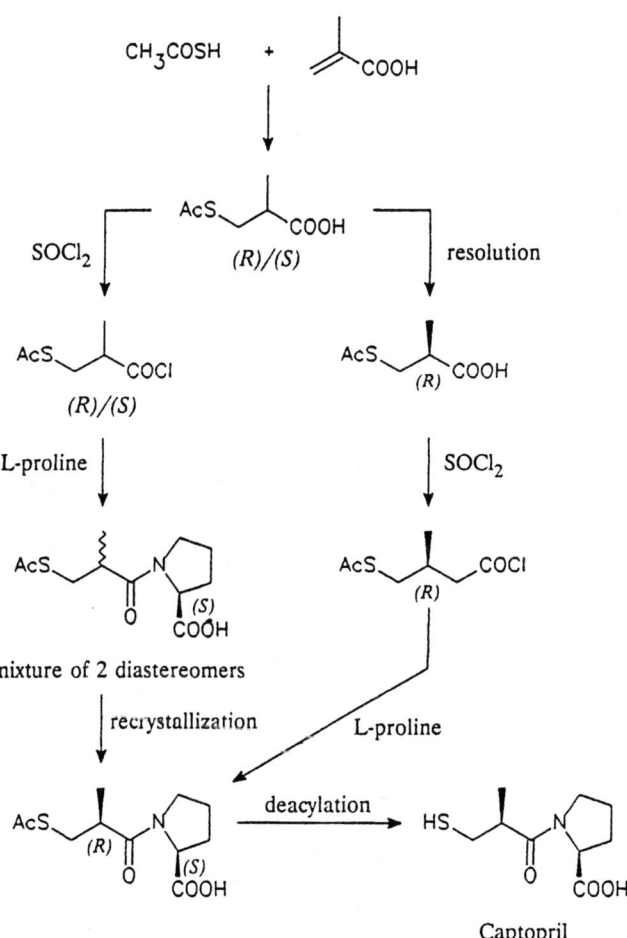

Figure 9-20. Early versus late resolution in captopril synthesis.

corresponding chloro compound. The latter can also be converted to captopril by a shorter route (Figure 9-21) but, to our knowledge, the producer of captopril (Bristol-Myers Squibb) has not elected to do so. Furthermore, note that racemic β-chloroisobutyric acid should, in principle, be amenable to resolution by enzymatic hydrolysis. Apparently, an efficient enzymatic resolution has not (yet) been forthcoming. Similarly, attempts to resolve the racemic β-acetylmercaptoisobutyric acid enzymatically (see chapter 7) have not (yet) produced a method that is competitive with classical resolution. Most of the other ACE-inhibitors contain the L-homophenylalanine moiety as a common structural feature (Figure 9-22).

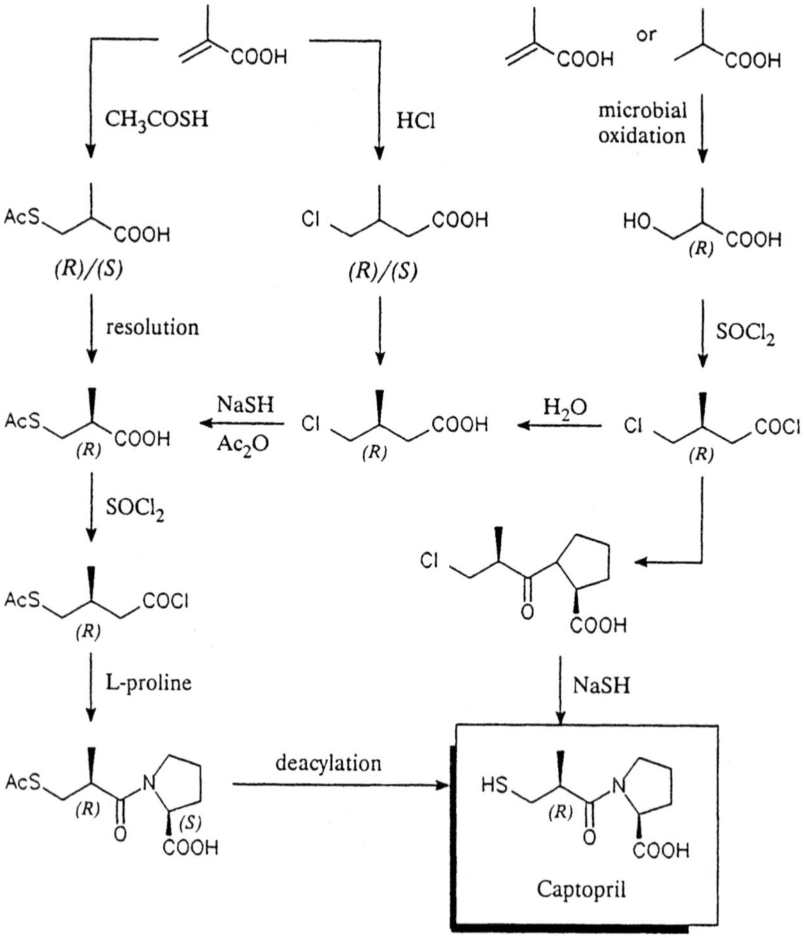

Figure 9-21. Alternative routes to captopril.

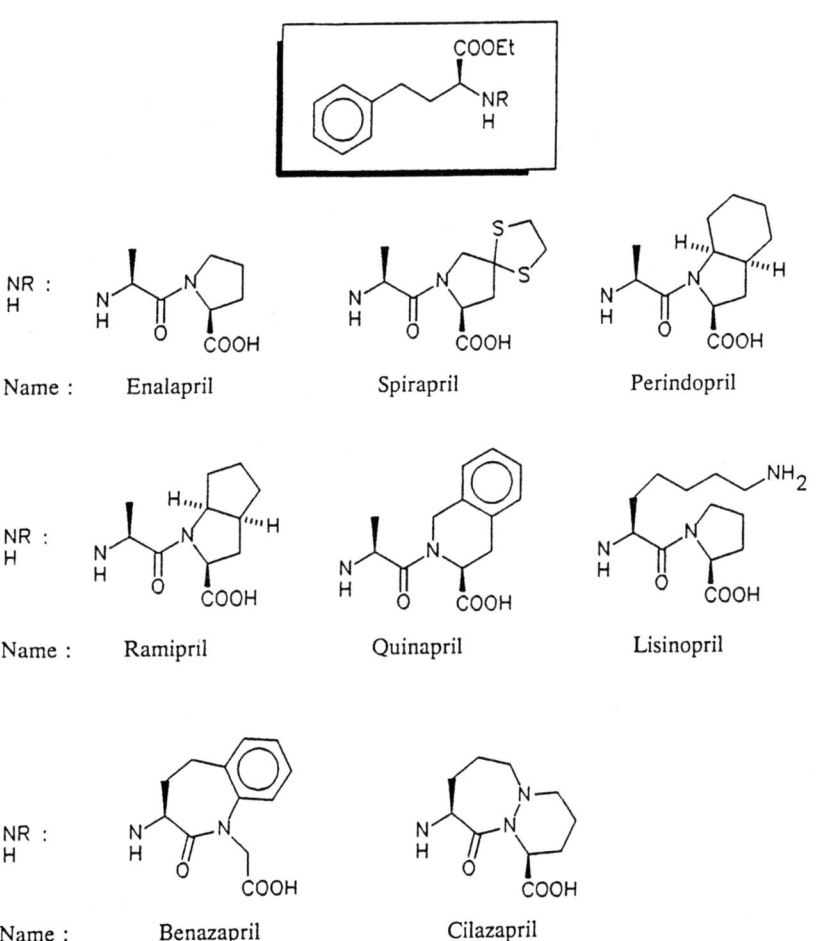

Figure 9-22. Structures of ACE-inhibitors containing the L-homophenylalanine moiety.

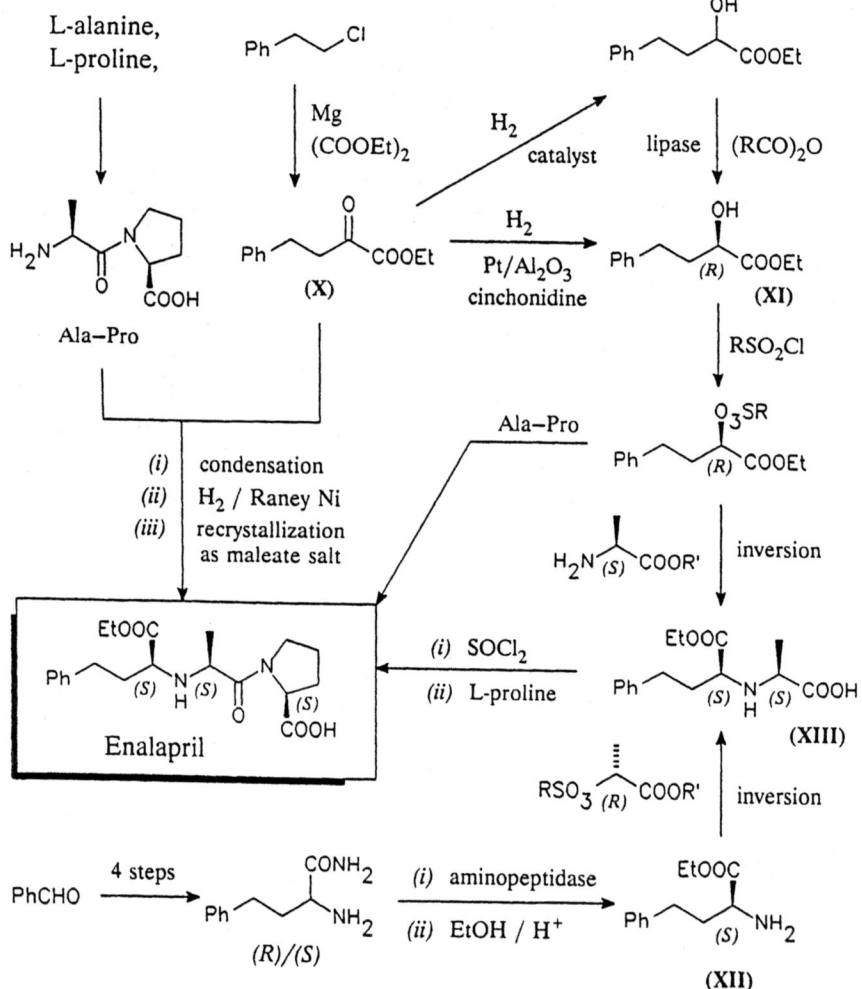

Figure 9-23. Alternative routes to enalapril.

The most important strategies that have been followed [36] for the synthesis of these products are illustrated in Figure 9-23 for enalapril. The Merck process involves reductive amination of the α-keto ester (**X**) by L-alanyl-L-proline (Ala-Pro). Asymmetric induction is observed, giving a 87:13 mixture of the *(S,S,S)*-and *(R,S,S)*-isomers. Crystallization of this mixture as the maleate salt gives the *(S,S,S)*-isomer (i.e., enalapril) in 78% yield.

By analogy with the captopril synthesis, the expensive purification of a mixture of diastereomers could be circumvented by employing a chiral intermediate instead of (**X**). The obvious candidates are the ethyl ester of L = *(S)*-homophenylalanine (**XII**) or the *(R)*-α-hydroxyester (**XI**). Coupling of these chirons with a derivative of *(R)*-lactic acid or L = *(S)*-alanine, respectively, yields the key intermediate (**XIII**); coupling of the latter (e.g., as the acid chloride) with L-proline gives enalapril. Similarly, coupling of (**XIII**) with the appropriate amino acid gives spirapril, perindopril, ramipril, or quinapril. The precursor (**XI**), on the other hand, can be converted to all of the ACE-inhibitors depicted in Figure 9-22 by coupling with the appropriate dipeptide (e.g., Ala-Pro and Lys-Pro yield enalapril and lisinopril, respectively).

Whether or not these processes can compete with the established route depends very much on the relative prices of the α-keto ester and (**XI**) or (**XII**). Two methods have been described for the synthesis of the α-hydroxy ester (**XI**): asymmetric hydrogenation of the α-keto ester over a Pt/Al$_2$O$_3$-cinchonidine catalyst (see chapter 8) or regular hydrogenation followed by enzymatic resolution (see chapter 7). It would seen unlikely, however, that either route can compete with the Merck route, considering that 2–3 extra steps are needed to produce one diastereomer instead of an 87:13 mixture. The L-homophenylalanine ester (**XII**), on the other hand, can be prepared by aminopeptidase-catalyzed hydrolysis of the corresponding amide (see chapter 7), which is readily produced from benzaldehyde in four steps. In principle, this route could be competitive with the Merck route. In short, the industrially relevant syntheses of enalapril and structurally related drugs is dominated by the use of raw materials that are available from the chirality pool, (e.g., L-alanine, L-proline, *(R)*-lactic acid), possibly in combination with an enzymatic hydrolysis or an asymmetric hydrogenation.

VII.   *(R)*-CARNITINE

L = *(R)*-Carnitine is a vitamin-like nutrient, sometimes called vitamin B$_T$, that plays an essential role in fat metabolism. It acts as a carrier, a sort of biological phase transfer catalyst, for transporting fatty acids into the mitochondria where they undergo beta oxidation thereby generating energy. Skeletal and cardiac muscle cells rely on this mechanism as a source of metabolic energy. Only *(R)*-carnitine occurs naturally in meat and other foods of animal origin, and it is only the *(R)* enantiomer that is physiologically effective; interestingly, *(S)*-carnitine is a com-

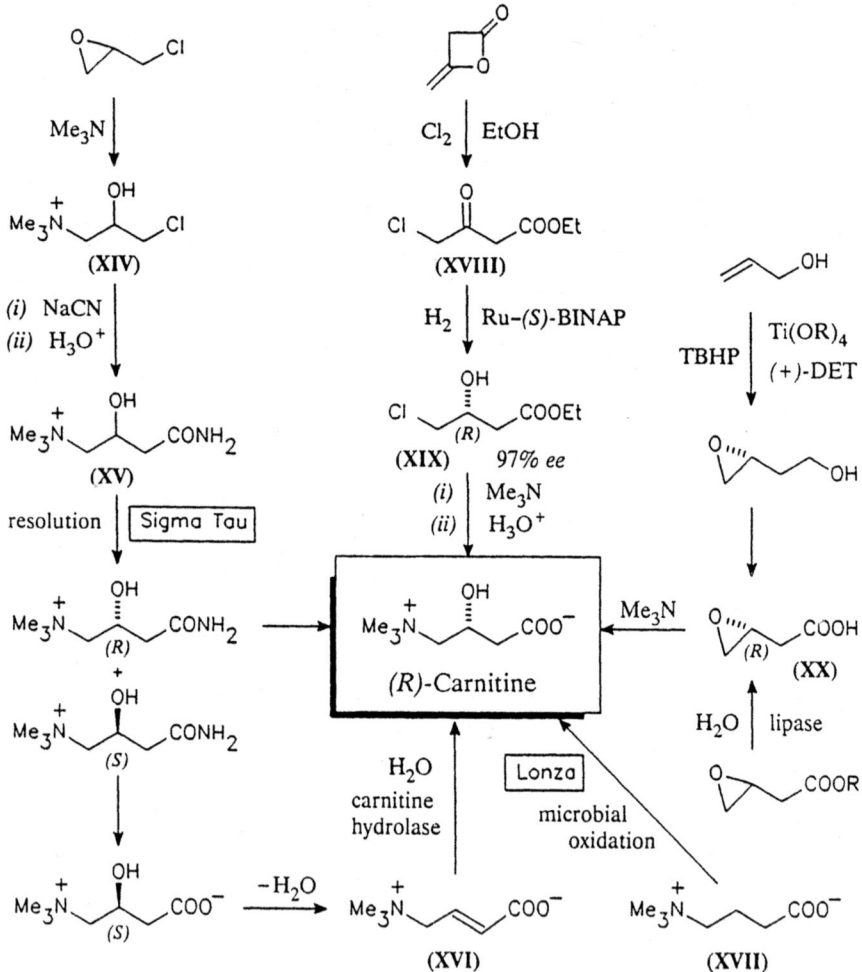

Figure 9-24. Alternative routes to *(R)*-carnitine.

petitive inhibitor. Animal foods, particularly meat, are rich dietary sources of *(R)*-carnitine. Alternatively, it is synthesized in the human liver, from L-lysine. Hence, vegetarians consuming lysine-deficient diets are likely to be deficient in carnitine. Ascorbic acid (vitamin C) is also needed (as a cofactor) for *(R)*-carnitine biosynthesis and a deficiency of vitamin C can lead to carnitine deficiency, which manifests itself in muscle weakness, fatigue, and elevated triglyceride levels in the blood. Consequently, *(R)*-carnitine finds widespread application as a dietary supplement in infant, sport and geriatric nutrition, and as an antiobesity agent.

Notwithstanding the very simple structure of carnitine, a wide variety of methods have been used for its synthesis, and therein lies the charm of this molecule. Bearing in mind the structural resemblance to beta-blockers, it is not surprising that many routes to *(R)*-carnitine involve the same C_3 chirons as key building blocks. Indeed, as with beta-blocker intermediates, many reported syntheses of *(R)*-carnitine begin with carbohydrates. A synthesis starting from vitamin C [38] has already been described in chapter 5 (Figure 5-11). Similarly, routes have also been described based on D-arabinose [38], D-malic acid [39], and other chirality pool materials such as β-pinene [40]. (See Figure 5-24.) However, all of these routes are circuitous, with inherent low overall yields that make their commercial viability doubtful to say the least.

The major producer of *(R)*-carnitine, Sigma Tau (Italy), uses a route that starts from epichlorohydrin and involves a classical resolution of racemic carnitinamide as the key step (Figure 9-24). The key step is apparently very efficient, giving a high yield of material with high optical purity in one crystallization; the resolving agent is probably camphoric acid.

The *(S)*-enantiomer of carnitinamide (or of carnitine itself) cannot be racemized and, normally speaking, it would be processed as waste. However, Sigma Tau has developed a process [41] for converting *(S)*- to *(R)*-carnitine by dehydration to the crotonobetaine (**XVI**) followed by carnitine hydrolase-mediated enantioselective hydration. The carnitine hydrolase is produced by several microorganisms, such as *E. coli*, *P. vulgaris*, and *P. mirabilis* [42]. *(R)*-Carnitine is produced in 100% *ee* in this step. Lonza, on the other hand, has commercialized a process [43] for *(R)*-carnitine production by microbial oxidation of butyrobetaine (**XVII**). This reaction is one of the steps in the biosynthesis of *(R)*-carnitine from L-lysine and gives a product of 100% optical purity. Lonza has devoted much effort to optimizing the productivity of the microorganisms involved [43]. This process has been discussed in detail in chapter 4.

Obviously, many other routes are possible in addition to the two that have been commercialized. Any of the precursors of the racemic carnitinamide could, in principle, be resolved. For example, Lonza scientists have reported the resolution of (**XIV**) as its salt with L-tartaric acid [44]. This is, however, apparently not competitive with the precursor fermentation route.

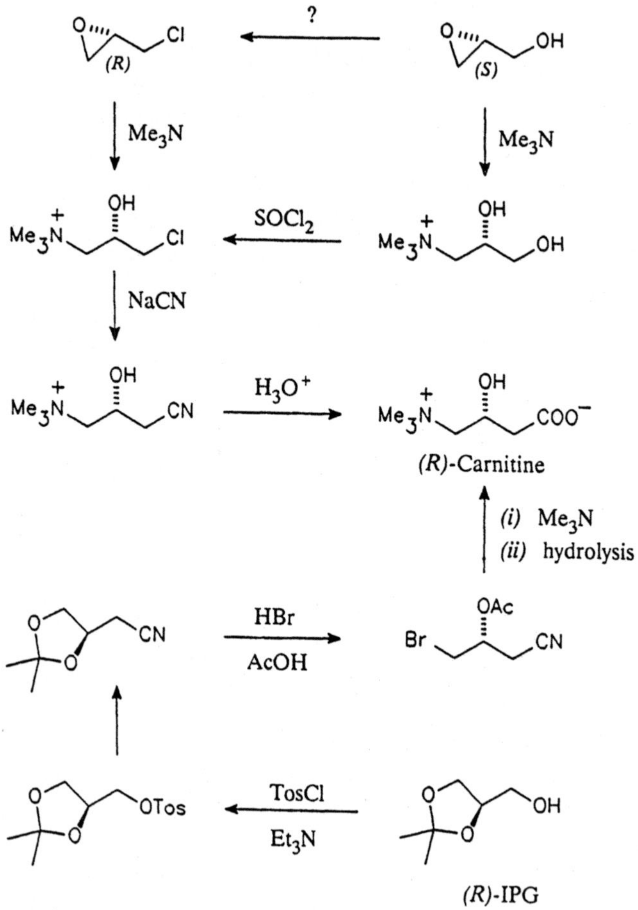

Figure 9-25. *(R)*-Carnitine syntheses based on C$_3$-chirons.

A potentially attractive route to *(R)*-carnitine would appear to be the Ru–BINAP-catalyzed asymmetric hydrogenation (see chapter 7) of 4-chloroacetoacetate ester (**XVIII**) which yields the *(R)*-hydroxy compound (**XIX**) in 97% *ee* [45]. However, to our knowledge, this route has not (yet) been commercialized. 4-Chloroacetoacetate esters can also be reduced to the corresponding *(R)*-hydroxy compound using baker's yeast [46] but enantioselectivities and productivities are substantially lower. Yet another approach is through the *(R)*-epoxy acid (**XX**) which has been prepared by Sharpless epoxidation of homoallyl alcohol [47] or via enzymatic resolution of the racemic acid [48]. However, both routes suffer from being too long and/or relatively low optical yields and productivities.

As mentioned earlier, many C_3 chirons, such as *(R)*-epichlorohydrin, *(S)*-glycidol, and *(R)*-isopropylidene glycerol (*(R)*-IPG), are suitable precursors for *(R)*-carnitine (Figure 9-25). However, none of these routes can compete with the Sigma Tau or Lonza routes due to the high price of the chiral synthons and/or lengthy procedure. In conclusion, *(R)*-carnitine production is dominated by routes involving classical resolution or precursor fermentation with catalytic asymmetric synthesis as a possible future contender.

VIII. CALCIUM ANTAGONIST: DILTIAZEM

The calcium antagonists are another group of highly successful antihypertensive agents having largely unrelated structures. One of these drugs, diltiazem, is included in the 'chiral top ten.' (See chapter 2, Table 2-2.) The original synthetic route, developed by Tanabe, involved late resolution by diastereomeric salt crystallization. (See chapter 7, Figure 7-30.) This route is gradually being displaced by one developed by Andeno that involves an early resolution of a common intermediate, the glycidate ester, (**XXI**) (Figure 9-26). The early resolution route is more economically attractive despite the fact that the maximum theoretical yield of the key step is 50%. This is more than compensated for by the overall route being 2–3 steps shorter and by avoiding the transport of isomeric ballast through the process.

ICI scientists [49] have reported an asymmetric synthesis of diltiazem in which the key step employs the osmium-catalyzed enantioselective olefin dihydroxylation pioneered by Sharpless and coworkers. (See chapter 8.) However, this route is unlikely to be competitive as it involves two additional steps starting from anisaldehyde, compared to the enzymatic route and gives material of lower optical purity (88% versus 100%) (Figure 9-27). It also suffers from the fact that it does not 'connect' very well with the established route.

A potentially attractive asymmetric synthesis route is one comprising enantioselective catalysis of Darzen's condensation. If successful, this would give the key optically active glycidate ester in 100% theoretical yield and would connect well with the established routes. Unfortunately, reported examples [50] employ stoichiometric quantities of expensive chiral auxiliaries or chiral bases and give mediocre

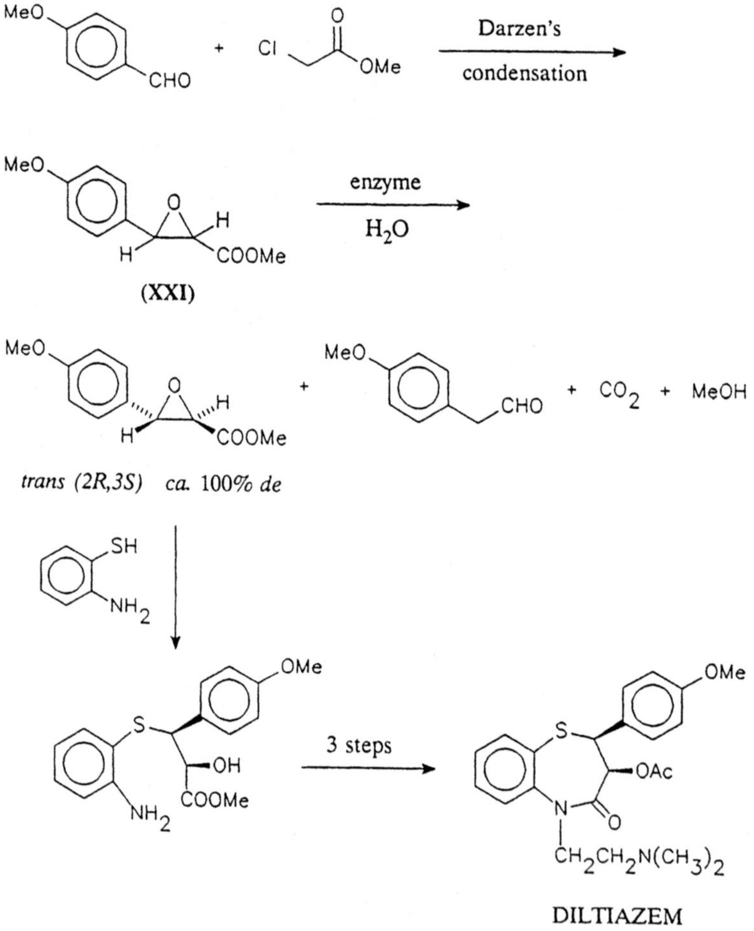

Figure 9-26. Enzymatic resolution route to diltiazem.

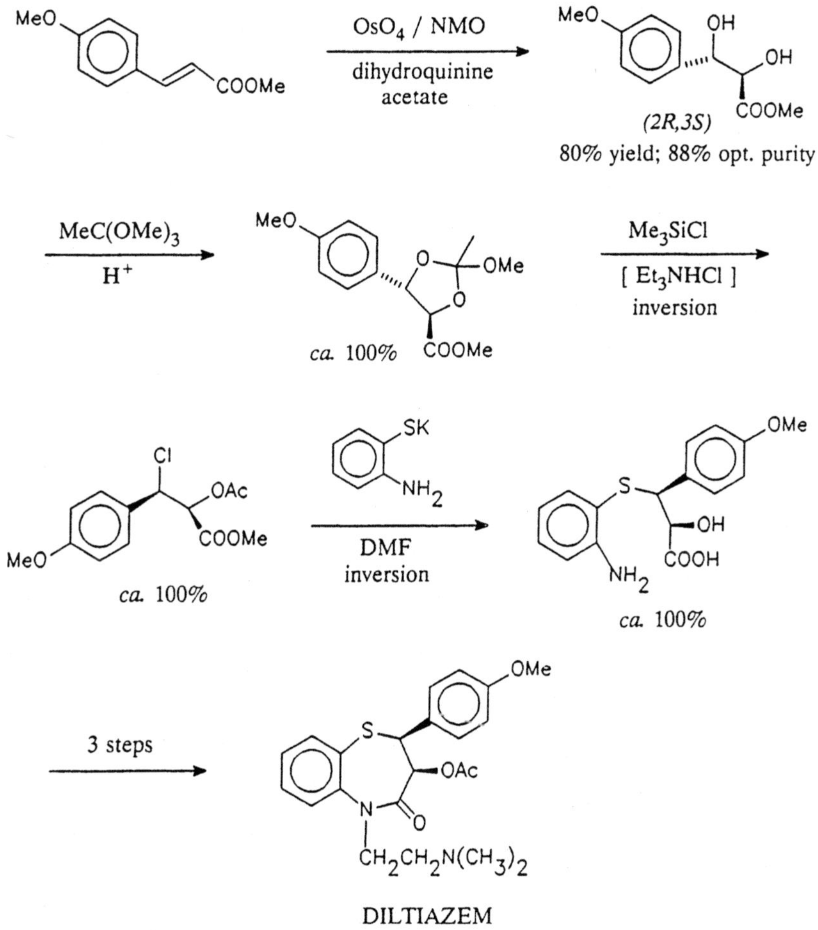

Figure 9-27. Asymmetric synthesis of diltiazem.

enantioselectivities. Nevertheless, a catalytic asymmetric Darzen's condensation remains an important synthetic challenge.

IX.   BETA-LACTAM ANTIBIOTICS

The beta-lactam antibiotics account for five of the top ten drugs and are probably the single most important group of drugs. They comprise the penicillins, cephalosporins and, more recently, the penems, carbapenems, and monobactams (Figure 9-28). Penicillin G and cephalosporin C, the key raw materials for the synthesis of penicillins and cephalosporins, are produced by de novo fermentation. (See chapter 4.) They are converted by enzymatic and/or chemical transformations to 6-aminopenicillanic acid (6-APA), 7-aminocephalosporanic acid (7-ACA), and 7-aminodeacetoxycephalosporanic acid (7-ADCA). (See chapter 4).

Several of the commercially more important penicillins and cephalosporins contain D-phenylglycine or D-*p*-hydroxyphenylglycine as the acyl side-chain. (See chapter 4, Figure 4-14.) Since they are unnatural amino acids, they are not available from de novo fermentation. Interestingly, two processes compete for the manufacture of these commercially important chiral intermediates: classical resolution and enzymatic kinetic resolution. As noted in chapter 7, the hydantoinase process—a subtractive, precursor process with in situ racemization—is more economical than other enzymatic resolution methods.

In the production of D-phenylglycine (Figure 9-29) most manufacturers, such as DSM-Andeno, employ a classical resolution of the racemate which is readily available from a Strecker reaction on benzaldehyde. Kanegafuchi probably uses enzymatic resolution of the DL-hydantoin that is also prepared from benzaldehyde.

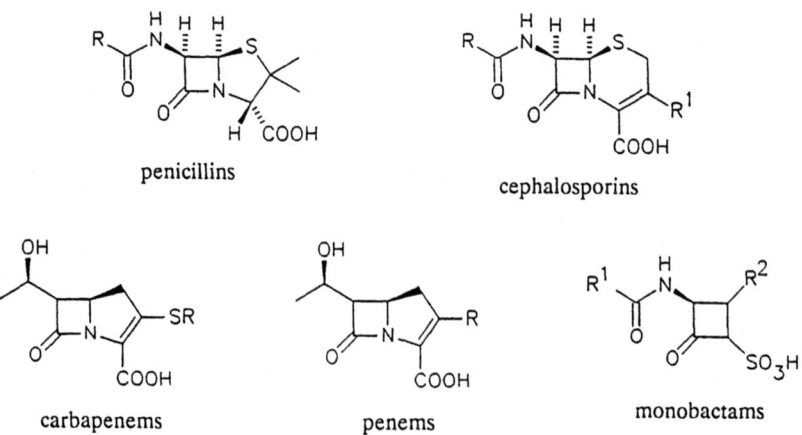

Figure 9-28. Structures of beta-lactam antibiotics.

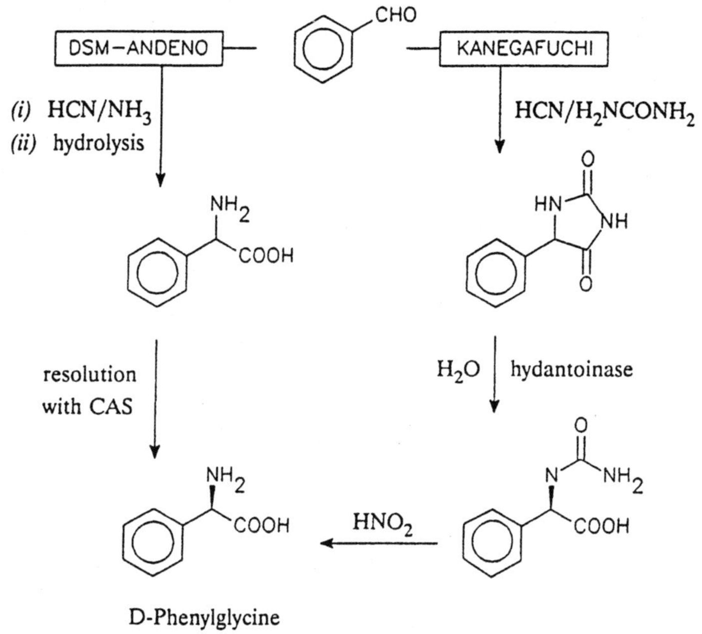

Figure 9-29. Two routes to D-phenylglycine.

However, this process has difficulty in competing with the more established classical resolution.

In the manufacture of D-p-hydroxyphenylglycine (Figure 9-30), however, the hydantoinase route is better able to compete with the classical route. The classical route, which employs bromocamphorsulphonic acid as the resolving agent, is not so firmly entrenched as in the case of phenylglycine. The raw material in both routes is phenol since p-hydroxybenzaldehyde is too expensive.

Since both routes employ essentially the same raw materials, the relative cost prices will be very much dependent on the actual yields and productivities attained in practice and the cost of resolving agent or biocatalyst consumed. The hydantoinase process has the advantage of not requiring a racemization step but, in the Kanegafuchi variant at least, it requires an extra chemical step. Moreover, as noted in chapter 6, crystallization-induced asymmetric transformations of readily prepared derivatives of p-hydroxyphenylglycine appear to be feasible [51].

In the case of the carbapenems, penems, and monobactams, it is the beta-lactam nucleus which is the synthetic target. A key intermediate that can be converted to both the carbapenem and penem nucleus is the azetidinone (**XXII**). Two commercial processes for the synthesis of (**XXII**) are depicted in Figure 9-31. One route, developed by Kanegafuchi [52], employs *(R)*-β-hydroxybutyric acid as

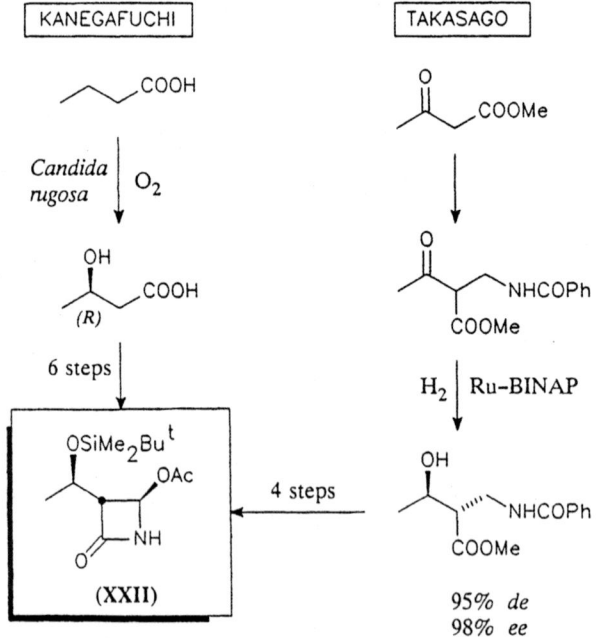

Figure 9-30. Alternative routes to D-*p*-hydroxyphenylglycine.

Figure 9-31. Two routes to a key carbapenem intermediate.

the key raw material; it is available from microbial oxidation of *n*-butyric acid. (See chapter 5.). The other process, operated by Takasago [53], employs a Ru–BINAP-catalyzed asymmetric hydrogenation as the key step. (See chapter 7.) The monobactam nucleus is produced commercially [54] from the appropriate amino acid, L-threonine or L-serine. (See Figure 5-20.) To our knowledge, other methods cannot compete with this chirality pool approach.

X. SYNTHETIC PYRETHROID INSECTICIDES

The synthetic pyrethroids are a commercially important group of environmentally friendly insecticides. (See chapter 2.) They are all chiral molecules, many of them having three stereogenic centers, and there has been a marked trend towards marketing these products as the optically pure eutomers. Most of the commercially important pyrethroids are esters with the general structure (**XXIII**) or (**XXIV**) in which the alcohol moiety is the cyanohydrin of *m*-phenoxybenzaldehyde (Figure 9-32) [55]. With the exception of deltamethrin, all of these products are marketed as esters of the racemic cyanohydrin.

Various approaches have been used for the synthesis of the required *(2R,3R)* (= *2R,cis*) isomer of the cyclopropanecarboxylic acid [55]. The most straightforward approach is to resolve the racemic *cis*-acid by diastereomeric salt formation [56] or enzymatic hydrolysis of an appropriate ester [57], as has been described for the structurally related chrysanthemic esters.

Two alternative approaches to the key dichlorovinylcyclopropane carboxylic acid intermediate (**XXV**) are illustrated in Figure 9-33. One starts with simple raw materials and builds up the cyclopropane ring of the right stereochemistry using catalytic asymmetric cyclopropanation. (See chapter 8.) The other involves selec-

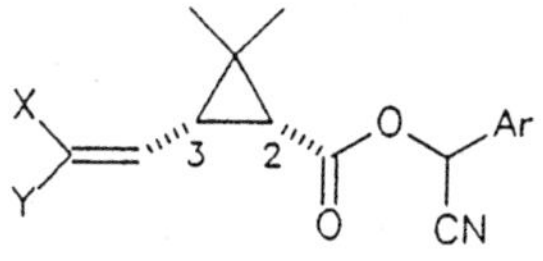

Alphametrin: X = Y = Cl
Deltamethrin: X = Y = Br
Curare: X = Cl; Y = CF$_3$

(2R,3R)-(**XXIII**)

Ar = *m*-phenoxyphenyl

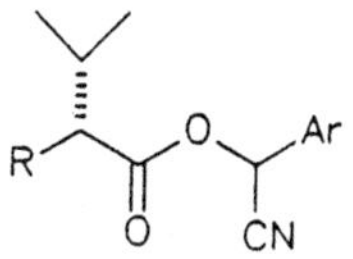

Esfenvalerate: R = *p*-chlorophenyl
Fluvalinate: R = *(2-chloro-4-trifluoro-methyl)phenylamino*

(**XXIV**)

Figure 9-32. Structures of synthetic pyrethroids.

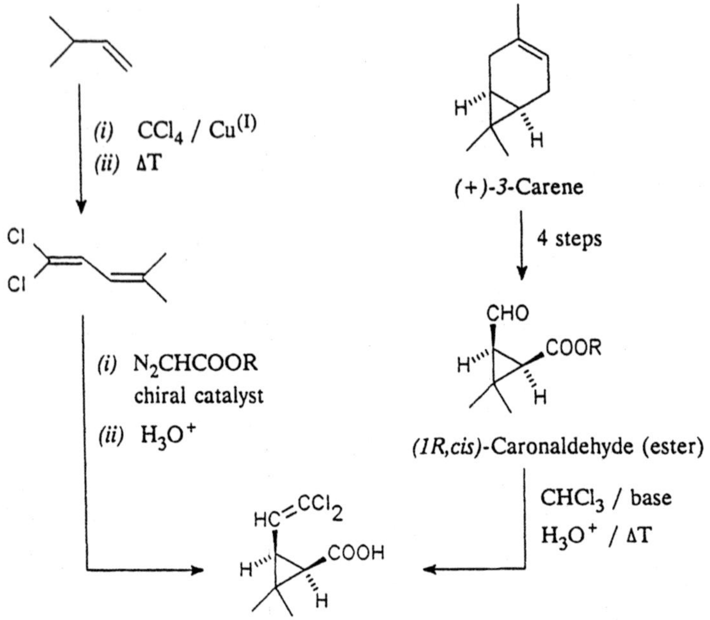

(1R,2R)-(**XXV**) = *(1R,cis)*

Figure 9-33. Two alternative routes to a key pyrethroid intermediate.

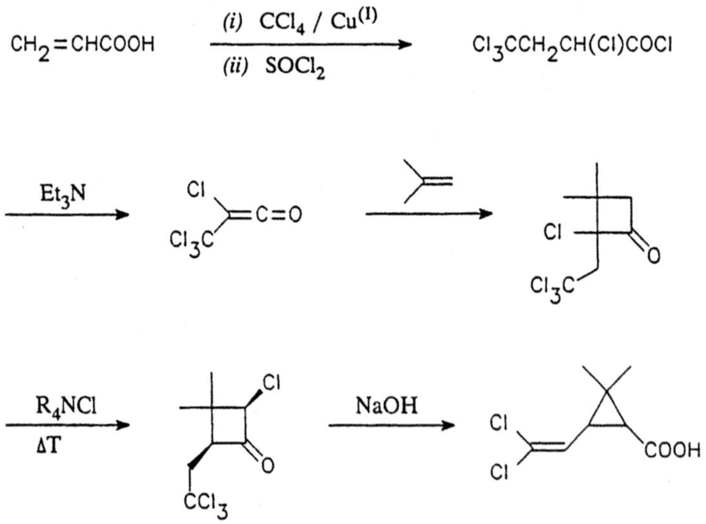

(**XXVI**)

Figure 9-34. The cyclobutanone route to a pyrethroid intermediate.

tive dismantling of *(+)-3-carene*, a raw material from the chirality pool, which already contains a cyclopropane moiety with the required stereochemistry. Although the carene route scores high on elegance the asymmetric cyclopropanation route is shorter and undoubtedly more economical.

Another approach, developed by Ciba-Geigy [58], for the synthesis of racemic **(XXV)** employs a Favorski-type rearrangement of the α-chlorocyclobutanone **(XXVI)**. The latter is prepared by cycloaddition of the appropriate ketene to isobutene (Figure 9-34). An elegant adaption of this route (Figure 9-35) was subsequently developed by Ciba-Geigy scientists [59] for the synthesis of the *(1R,cis)* isomer **(XXV)**. The bisulfite adduct of the *cis*-cyclobutanone **(XXVI)** is treated with

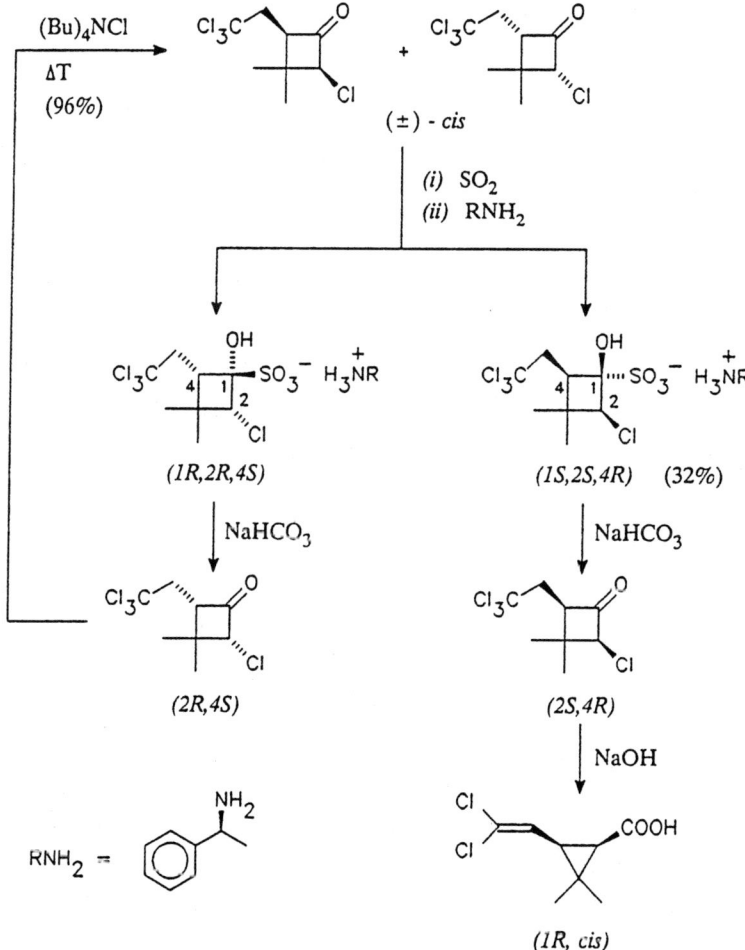

Figure 9-35. Resolution of the key cyclobutanone.

(S)-α-methylbenzylamine, thereby introducing a new stereogenic center. Only two of the four possible diastereomeric salts are formed and the salt of the required *(1S,2S,4R)* bisulfite adduct selectively crystallizes out. Treatment with NaHCO₃ gives the *(2S,4R)* isomer of (**XXVI**) that is subsequently converted to (**XXV**). The *(1R,2R,4S)* bisulfite adduct is converted to the enantiomeric cyclobutanone that is epimerized by heating with a catalytic amount of tetrabutylammonium chloride.

The key intermediate for esfenvalerate, *(S)*-2-(4-chlorophenyl)-3-methyl-butanoic acid (**XXVII**), has been prepared by preferential crystallization of its diethylamine salt [60], or diastereomeric salt formation with *(S)*-α-methylbenzyl-amine [61] or *(+)*-α-phenyl-β-*p*-tolylethylamine [62]. The *(R)*-enantiomer can be racemized by heating a mixture with the corresponding racemic nitrile (precursor of the acid) and NaOH [63].

Alternatively, (**XXVII**) can be prepared by reaction of the corresponding ketene with formic acid in the presence of a catalytic amount of cyclo((*R*)-phenyl-alanyl–(*R*)-histidyl) [64]. It is, of course, no coincidence that the very same catalyst is used for the synthesis of another key intermediate by asymmetric hydrocyanation. It is a perfect example of a chiral catalyst available from one project being used with great success in another. The reaction is also reminiscent of the ketene route to *(S)*-α-arylpropionic acids discussed earlier. It would be interesting, therefore, to apply the method outlined in Figure 9-36 to the synthesis of *(S)*-α-arylpropionic acids.

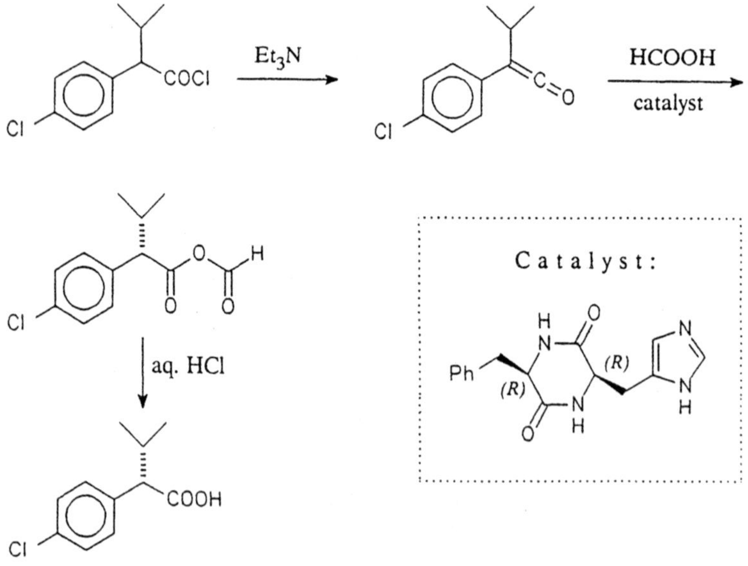

Figure 9-36. Asymmetric addition of formic acid to a ketene.

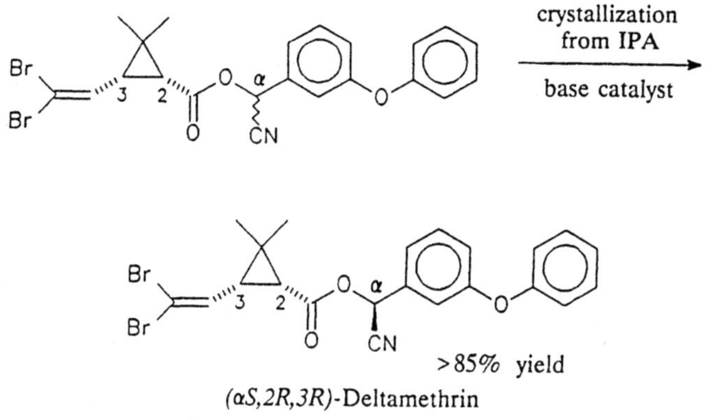

Figure 9-37. Deltamethrin synthesis by asymmetric transformation.

A third key intermediate for optically active pyrethroids is the *(S)*-isomer of *m*-phenoxybenzaldehyde cyanohydrin. As discussed in chapters 7 and 8, there are several good methods for asymmetric hydrocyanation of aldehydes, for example, using an oxynitrilase or the cyclic dipeptide cyclo(*(R)*-phenylalanyl-*(R)*-histidyl) as catalyst or lipase-catalyzed acylation with in situ racemization. (See chapter 7.) There is, however, yet another, very elegant solution to this synthetic problem and that is to dispense with the need to synthesize the optically active cyanohydrin intermediate altogether. When a saturated solution of the *(α-RS,2R,3R,)*-deltamethrin in isopropanol (with a catalytic amount of an organic or inorganic base) is allowed to stand, the required *(α-S,2R,3R,)*-isomer crystallizes out in greater than 85% yield [65]. This method, a perfect illustration of a crystallization-induced asymmetric transformation, is the one actually used by Roussel-Uclaf for the production of deltamethrin (Figure 9-37). The method also works with other pyrethroids, such as cypermethrin [66] and fenvalerate [67], the *(α-R,S)*-mixtures being readily prepared by one-step coupling of *m*-phenoxybenzaldehyde with the acid chloride and NaCN [68].

The synthesis of optically active pyrethroids appears to have stimulated organic chemists to commendable feats of ingenuity as is illustrated in Figure 9-38. The alcohol moiety of the synthetic pyrethroid, prallethrin, is synthesized by lipase-mediated hydrolysis of the racemic acetate to a mixture of *(S)*-acetate and *(R)*-alcohol [69]. This mixture is treated with methanesulfonyl chloride to afford a mixture of *(S)*-acetate and *(R)*-mesylate. Subsequent hydrolysis of this mixture gives the desired *(S)*-alcohol as the sole product, the acetate by retention and the mesylate by inversion of configuration.

Obviously, the theme of this chapter could be illustrated with many more examples, for instance, by extolling the virtues of enzymatic [70] versus chemo-

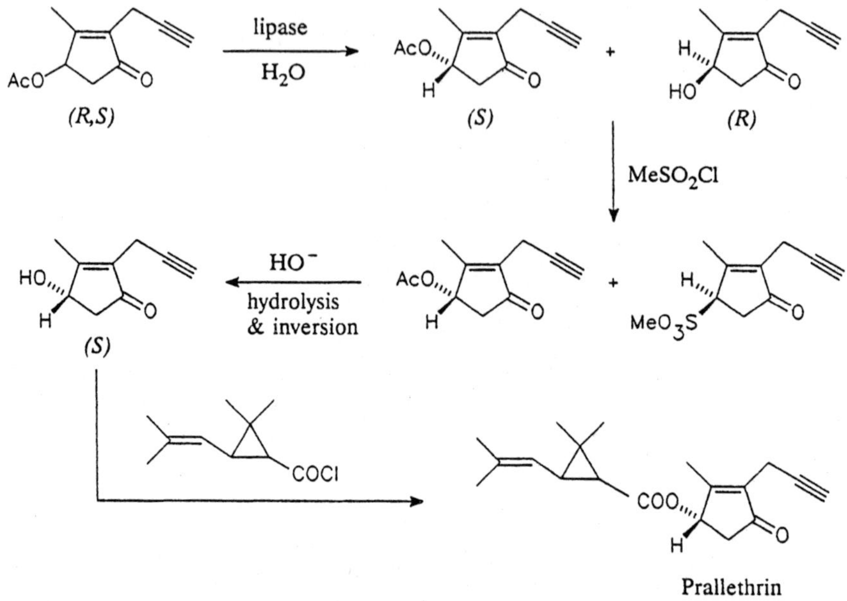

Figure 9-38. Chemoenzymatic synthesis of prallethrin intermediate.

catalytic [71] approaches to the synthesis of the pure enantiomers of the anti-depressants, tomoxetine, fluoxetine, and nisoxetene. However, the elegant integration of biocatalysis and organic chemistry illustrated in Figure 9-38 would seem to be an appropriate point to end our discussion on industrially relevant syntheses.

REFERENCES

1. Kirk-Othmer, *Encyclopedia of Chemical Technology, 3rd Ed., Vol. 22*, Wiley, 1983, pp. 742-745.
2. Crosby, J., *Tetrahedron*, **47**, 4789-4846 (1991).
3. Bauer, K., and Garbe, D., *Common Fragance and Flavor Materials*, VCH, Weinheim, 1985, p. 40.
4. Van Zuylen, A., *Chem. Weekblad*, June, 1992.
5. Metivier, P., and Rajoharisson, European Patent Appl., 401.104 (1989) to Rhône-Poulenc.
6. Gras, G., German Patent 3.024.265 (1979) to Rhone-Poulenc.
7. Dahod, S. K., Siuta-Mangano, P., *Biotechnol. Bioeng.*, **30**, 995-999 (1987).
8. Bodnar, J., Gubicza, L., and Szabo, L. P., *J. Mol. Catal.*, **61**, 353-361 (1990).
9. Siuta-Mangano, P., and Dahod, S. K., United States Patent, 4.613.689 (1986) to Stauffer.
10. Bewick, D. W., European Patent Appl., 133.034 (1985) to ICI.

11. For a recent review see: Sonawane, H. R., Bellur, N. S., Ahuja, J. R., and Kulkarni, D. G., *Tetrahedron Asymm.*, **3**, 163-192 (1992).

12. Yamamoto, K., Ueno, Y., Otsubo, K., Kawakami, K., and Komatsu, K., *Appl. Environ Microbiol.*, **56**, 3125-3129 (1990).

13. Larsen, R. D., Corley, E. G., Davis, P., Reider, P. J., and Grabowski, E. J. J., *J. Am. Chem. Soc.*, **111**, 7650-7651 (1989).

14. Phillips. G. T., Robertson, B. W., Watts, P. D., Matcham, G. W. J., Bertola, M. A., Marx, A. F., and Koger, H. S., European Patent Appl., 205.215 (1986) to IBIS.

15. Reid, A. J., Phillips, G. T., Marx, A. F., and Desmet, M. J., European Patent, 0.338.645-A (1989) to Shell.

16. a) Piccolo, O., Spreafico, F., and Visentin, G., *J. Org. Chem.*, **50**, 1946 (1985). b) Piccolo, O., Azzena, U., Melloni, G., Delogu, G., and Valoti, E., *J. Org. Chem.*, **56**, 183 (1991).

17. a) Reinhold D. F., German Patent, 2.556.040 (1976) to Merck. b) Lotti, V. J., European Patent Appl., 105.996 (1984) to Merck; *CA*, **101**, 60142 (1984).

18. Westfelt, L., *Proc Chiral 90 Symp.*, Spring Innovations, Stockport, UK, (1990) pp. 29-32.

19. McClure, D. E., Arison, B. H., and Baldwin, J. J., *J. Am. Chem. Soc.*, **101**, 3666-3668 (1979).

20. Klunder, J. M., Onami, T., and Sharpless, K. B., *J. Org. Chem.*, **54**, 1295-1304 (1989).

21. For a recent review see: Emons, C. H. H., Kuster, B. F. M., Vekemans, J. A. J. M., and Sheldon, R. A., *Chimica Oggi (Chemistry Today)*, 10 (Nov./Dec.), 59-65, 1992.

22. Kloosterman, M., Elferink, V. H. M., Van Iersel, J., Roskam, J. H., Meijer, E. M., Hulshof, L. A., and Sheldon, R. A., *Tibtech*, **6**, 251 (1988).

23. For reviews see: a) Hanson, R. M., *Chem. Rev.*, **91**, 437-475 (1991). b) Jurczak, J., Pikul, S., and Bauer, T., *Tetrahedron*, **42**, 447 (1986). c) Altenbach, H. J., *Nachr. Chem. Tech. Lab.*, **36**, 33 (1988).

24. a) Hamaguchi, S., Ohashi, T., and Watanabe, K., *Agric. Biol. Chem.*, **50**, 375-380 (1986). b) Hamaguchi, S., Ohashi, T., and Watanabe, K., *Agric. Biol. Chem.*, **50**, 1629-1632 (1986). c) Hamaguchi, S., Ohashi, T., and Watanabe, K., European Patent Appl., 189.878 (1986) to Kanegafuchi.

25. a) Hamaguchi, S., Hasegawa, J., Kawaharada, H., and Watanabe, K., *Agric. Biol. Chem.*, **48**, 2055-2059 (1984). b) Hamaguchi, S., Asada, M., Hasegawa, J., and Watanabe, K., *Agric. Biol. Chem.*, **48**, 2331-2337 (1984). c) Kan, K., Miyama, A., Hamaguchi, S., Ohashi, T., and Watanabe, K., *Agric. Biol. Chem.*, **49**, 207-210 (1985). d) Hamaguchi, S., Yamamura, H., Hasegawa, J., and Watanabe, K., *Agric. Biol. Chem.*, **49**, 1509-1511 (1985). e) Hamaguchi, S., Asada, M., Hasegawa, J., and Watanabe, K., *Agric. Biol. Chem.*, **49**, 1661-1667 (1985).

26. Chen, C. S., Liu, Y. C., and Marsella, M., *J. Chem. Soc., Perkin Trans. I*, 2559-2561 (1990).

27. Kan, K., Miyama, A., Hamaguchi, S., Ohashi, T., and Watanabe, K., *Agric. Biol. Chem.*, **49**, 1669-1674 (1985).

28. Iriuchijima, S., and Kojima, N., *Agric. Biol. Chem.*, **46**, 1153-1158 (1982).

29. Iriuchijima, S., Keiyu, A., and Kojima, N., *Agric. Biol. Chem.*, **46**, 1593-1597 (1982).

30. a) Ader, U., and Schneider, M. P., *Tetrahedron Asymm*, **3**, 201-204 (1992). b) Ader, U., and Schneider, M. P., *Tetrahedron Asymm*, **3**, 205-208 (1992). c) Ader, U., and

Schneider, M. P., *Tetrahedron Asymm*, **3**, 521-524 (1992). d) See also: Bevinakalti, H. S., and Banerji, A. A., *J. Org. Chem.*, **56**, 5372-5375 (1991).

31. Bertola, M. A., Koger, H. S., Phillips, G. T., Marx, A. F., and Claasen, V. P., European Patent Appl., 244.912 (1987) to Gist-Brocades and Shell.

32. Elferink, V. H. M., Breitgoff, D., Kloosterman, M., Kamphuis, J., Van den Tweel, W. J. J., and Meijer, E. M., *Recl. Trav. Chim. Pays-Bas*, **110**, 63-74 (1991) and references cited therein.

33. Haberts-Crützen, A. Q. H., Carlier, S. J. N., De Bont, J. A. M., Wistuba, D., Schurig, V., Hartmans, S., and Tramper, J., *Enzym. Microb. Technol.*, **7**, 17-21 (1985).

34. Robinson, B. W., Bertola, M. A., Koger, H. S., Marx, A. F., and Watts, P. D., European Patent Appl., 193.228 (1986) to Gist-Brocades.

35. For a review see: Mirviss, S. B., Dahod, S. K., and Empie, M. W., *Ind. Eng. Chem. Res.*, **29**, 651-659 (1990).

36. Sheldon, R. A., Zeegers, H. J. M., Houbiers, J. P. M., and Hulshof, L. A., *Chimica Oggi (Chemistry Today)*, May, 1991, pp. 35-47 and references cited therein.

37. Colucci, W. J., and Gandour, R. D., *Bioorg. Chem.*, **16**, 307-334 (1988).

38. a) Bock, K., Lundt, I., and Pederson, C., *Acta Chem. Scand.*, **B37**, 341 (1983). (b) See also: Jung, M. E., and Shaw, T. J., *J. Am. Chem. Soc.*, **102**, 6304 (1980).

39. Bellamy, F. D., Bondoux, M., and Dodey, P., *Tetrahedron Lett.*, **31**, 7323-7326 (1990).

40. Pellegata, R., Dosi, I., Villa, M., Lesma, G., and Palmisano, G., *Tetrahedron*, **41**, 5607 (1985).

41. Jung, H., Jung, K., and Kieber, H. P., European Patent, 320.460 (1989) to Sigma Tau.

42. Yokozeki, K., Takahashi, S., Hirose, Y., and Kubota, K., *Agric. Biol. Chem.*, **52**, 2415-2421 (1988).

43. Kulla, H. G., *Chimia*, **45**, 81-85 (1991).

44. Tenud, L., and Gosteli, J., European Patent., 157.315 (1984) to Lonza; *CA*, **104**, 167989 (1986).

45. Kitamura, M., Ohkuma, T., Takaya, H., and Noyori, R., *Tetrahedron Lett.*, **29**, 1555-1556 (1988).

46. a) Zhou, B., Gopalan, A. S., Van Middlesworth, F., Shieh, W., and Sih, C. J., *J. Am. Chem. Soc.*, **105**, 5925 (1983). b) Fuganti, C., Grasselli, P., Casati, P., and Carmeno, M., *Tetrahedron Lett.*, **26**, 101-104 (1985).

47. Rossiter, B. E., and Sharpless, K. B., *J. Org. Chem.*, **49**, 3707 (1984).

48. Bianchi, D., Gabri, W., Cesti, P., Francalanci, F., and Ricci, M., *J. Org. Chem.*, **53**, 104-107 (1988).

49. Watson, K. G., Fung, Y. M., Gredley, M., Bird, G. J., Jackson, W. R., Gountzous, H., and Matthews, B. R., *J. Chem. Soc., Chem. Commun.*, 1018-1019 (1990).

50. a) Palmer, J. T., European Patent Appl., 0.342.904 (1989) to Marion Laboratories. (b) Coffen, D.L., Madan, P. B., and Schwartz, A., European Patent Appl., 0.343.474 (1988) to Hoffmann La Roche.

51. Boesten, W. H. J., Netherlands Patent Appl., 90.00386 and 90.00387 (1990) to Stamicarbon.

52. a) Ohashi, T., and Hasegawa, J., *J. Synth. Org. Chem. Japan*, **45**, 331-345 (1987). b) See also: Iimori, T., and Shibasaki, M., *Tetrahedron Lett.*, **26**, 1523 (1985).

53. Noyori, R., Ikeda, T., Ohkuma, T., Widhalm, M., Kitamura, M., Takaya, H., Akutagawa, S., Sayo, N., Saito, T., Taketomi, T., and Kumobayashi, H., *J. Am. Chem. Soc.*, **111**, 9134 (1989).
54. Floyd, D. M., Fritz, A. W., and Cimarusti, C. M., *J. Org. Chem.*, **47**, 176-178 (1982).
55. For reviews see: a) Arlt, D., Jautelat, M., and Lantzsch, R., *Angew. Chem. Int. Ed. Engl.*, **93**, 719-738 (1981). b) Elliot, M., *Pestic. Sci.*, **27**, 337 (1989).
56. Naumann, K., German Pat., 2.826.952 (1980) to Bayer.
57. Schneider, M., Engel, N., and Boensmann, H., *Angew. Chem. Int. Engl.*, **96**, 52-54 (1984).
58. Martin, P., Greuter, H., and Bellus, D., *J. Am. Chem. Soc.*, **101**, 5853 (1979).
59. a) Tombo, G. M. R., and Bellus, D., *Angew. Chem. Int. Engl.*, **103**, 1219-1241 (1991). b) Greuter, H. Dingwall, J., Martin, P., and Bellus, P., *Helv. Chim. Acta*, **64**, 2812 (1981).
60. Nohira, H., Terunuma, D., and Koube, S., European Patent Appl., 60.466 (1982) to Sumitomo; *CA*, **98**, 71689 (1983).
61. Martel, J., Tessier, J., Teche, A., and Demoute, J. P., German Patent, 2.902.478 (1979) to Roussel Uclaf; *CA*, **91**, 175030 (1979).
62. Takuma, K., and Morino, H., European Patent. Appl., 107.972 (1984) to Sumitomo; *CA*, **101**, 130401 (1984).
63. Friend, P. S., European Patent Appl., 163.094 (1985) to Shell Oil; *CA*, **104**, 168147 (1986).
64. Stoutamire, D. W., United States Patent, 4,570,017 (1986) to Shell Oil.
65. French Patent, 2.382.422 (1947) to Roussel Uclaf.
66. Kral, V., Dvorak, D., Zavada, J., Stibor, I., Mostecky, J., and Votava, V., Czechoslovakian Patent, 257.094 (1989); *CA*, **112**, 35450 (1990).
67. Petty, W. L., European Patent Appl., 60.580 (1982) to Shell; *CA*, **98**, 71697 (1983).
68. Sheldon, R. A., Been, P., Wood, D. A., and Mason, R. F., German Patent, 2.708.590 (1977) to Shell; *CA*, **89**, 42758 (1978).
69. Hirohara, H., Mitsuda, S., Ando, E., and Komaki, R., in *Biocatalysts in Organic Synthesis*, Tramper, J., Van der Plas, H. C., and Linko, P. (Eds.), Elsevier, Amsterdam, 1985, pp. 119-134.
70. Schneider, M. P., and Goergens, U., *Tetrahedron Asymm.*, **3**, 525-528 (1992).
71. Gao, Y., and Sharpless, K. B., *J. Org. Chem.*, **53**, 4051 (1988).

10

Epilogue: Future Prospects

You see things that are and say "why", I dream of things that never were and say "why not".

George Bernard Shaw

The synthesis of optically pure compounds continues to attract much attention. Indeed, it is difficult to keep abreast of new developments in synthetic methods or with the proliferation of international symposia devoted to the topic.

Catalytic asymmetric synthesis and enzymatic transformations appear to occupy the fast lanes. In the former, the development of new chiral ligands leads to steadily increasing enantioselectivities in many reactions (Figure 10-1), for example, the phthalazine ligands such as (I) in osmium-catalyzed asymmetric dihydroxylation of olefins [1] and the Ti(IV) complex (II) of diacetone glucose as a catalyst for asymmetric hetero Diels-Alder reactions [2].

New axially dissymmetric ligands analogous to Binap continue to appear. The BICHEP ligand (III), for example, gives high enantioselectivities in rhodium-catalyzed asymmetric hydrogenations of prochiral olefins [3] as illustrated in Figure 10-2.

Chiral cyclopentadienyl metal complexes of early transition elements and lanthanides have emerged as catalysts for asymmetric synthesis [4]. A chiral titanocene has been used, for example, as a catalyst for the enatioselective hydrogenation of imines (Figure 10-3) with *ee*'s up to 98% [5].

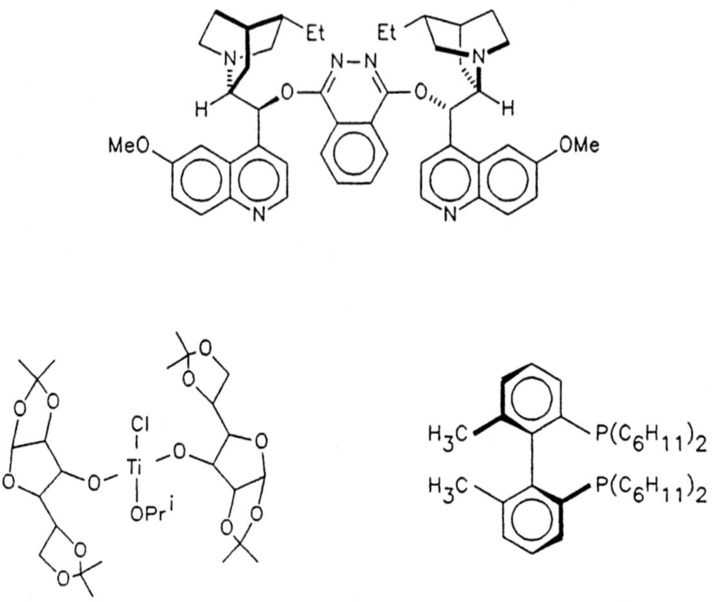

Figure 10-1 New chiral ligands.

Considering that Nature has taken millions of years to optimize the chiral ligands involved in biocatalytic processes, the potential for improvement in abiological asymmetric catalysis by tailoring chiral ligands is obviously enormous.

In the area of biocatalysis, on the other hand, availability and stability are often the key limiting factors rather than enantioselectivity. Many oxidoreductases, for example, would find wide application in industrial organic synthesis if their availability and stability were improved. As more and more oxidoreductases

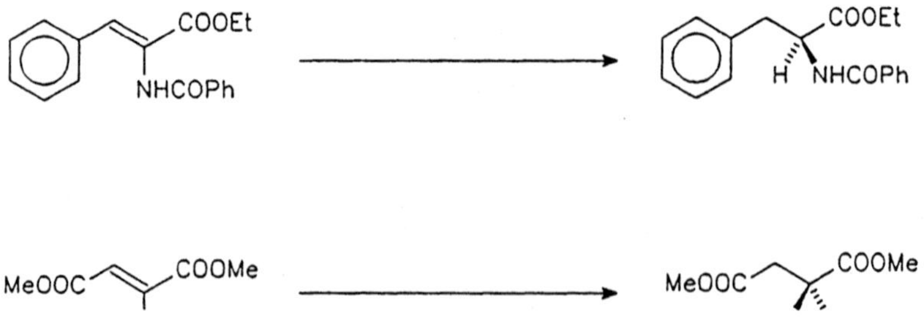

Figure 10-2. BICHEP as a chiral ligand in asymmetric hydrogenations.

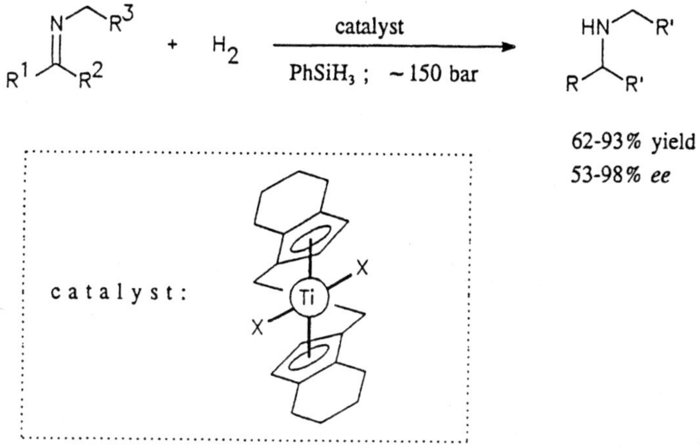

Figure 10-3. Asymmetric titanocene-catalyzed hydrogenation of imines.

become available from thermophilic microorganisms [6], or by cloning enzymes of plant origin in microorganisms, the scope of redox enzymes is rapidly being extended. Consequently, we expect that in the near future oxidoreductases will find wide application in catalytic redox processes.

New methodologies for enzyme stabilization continue to emerge. For example, a recently described approach [7] involves the multisite attachment of proteases through the ε-lysine residues to an aminoglucose-based macromolecule (Figure 10-4). This provides both structural stability and a water-like microenvironment to the enzyme. These so-called carbohydrate-protein conjugates (CPCs) are remarkably stable, retaining their activity at elevated temperatures in aqueous media while displaying comparable catalytic behavior to the native proteases.

Significant advances are being made in "mapping" the active sites of enzymes. For example, substantial information has been accumulated with regard to the structure of the active site of lipases [8] and related enzymes [9]. Such information has provided a structural rationale [8] for the phenomenon of interfacial activation (enhanced activity at lipid-water interfaces) of lipases. Apparently the active site is covered by a hydrophobic "lid" which opens when the enzyme binds the substrate. Not only are we learning more and more about the structure of the active sites of enzymes, advances in protein engineering allow us to modify this structure at will to give tailor-made biocatalysts [10]. There appears to be no limit to the possibilities. Indeed, when one considers that the broad application of enzymes in organic synthesis began in earnest only about five years ago, the future must hold great expectations. Moreover, if catalytic antibodies ever achieve the potential

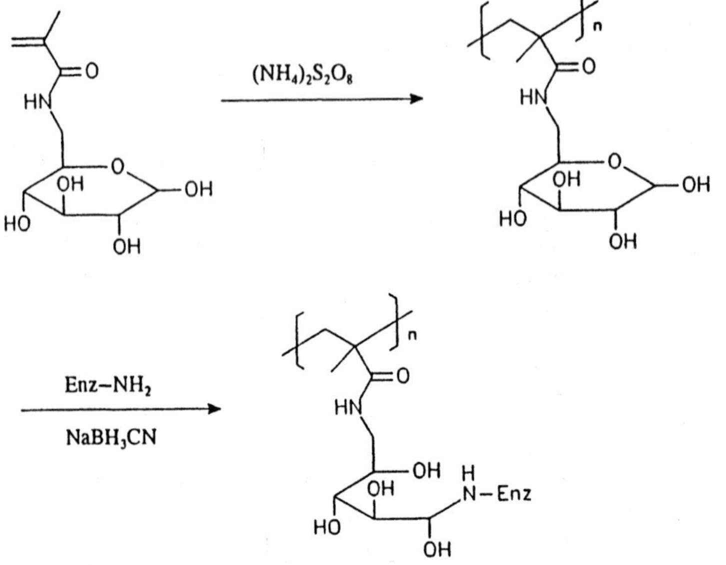

Figure 10-4. Synthesis of carbohydrate protein conjugates (CPCs).

promulgated by many enthusiasts [11], the sky's the limit. And one hardly dares to mention the implications of antibody engineering.

Notwithstanding the remarkable advances being made in catalytic asymmetric synthesis and biocatalysis one should not forget the more mundane classical approaches. Incredible as it may seem, new concepts continue to appear even in the area of classical resolution via diastereomeric salt formation. For example, Acs and coworkers have reported [12] a new resolution method based on the fact that dibasic acid resolving agents (such as tartaric acid) can form both acidic and neutral salts of different thermal stability. Thus, a liquid racemic base was allowed to react at room temperature with half an equivalent of L-tartaric acid. The mixture solidified within a few minutes. When this solid was heated in vacuo it released a portion of the base. Both the distillate and the residue contained optically enriched base, of opposite configuration. The efficiency of the separation varied from a few to 65% depending on the structure of the amine. Finally, the concept of "molecular imprinting"—designing synthetic polymers containing cavities with specific molecular dimensions capable of chiral discrimination—appears to be coming of age [13]. Although practical applications are limited as yet to chromatographic separations this approach could also conceivably lead to industrial scale methods.

In conclusion, enormous progress has been made in the last decade in the industrial synthesis of optically active compounds. The synthetic organic chemist now has a veritable arsenal of practical methods at his disposal. Hence, "lack of a viable technology" is no longer a tenable excuse for marketing bioactive substances in racemic form. Chirotechnology [14] has come a long way since Pasteur first tackled racemic tartrate with a pair of tweezers.

REFERENCES

1. Sharpless, K. B., Amberg, W., Bennani, Y. L., Crispino, G. A., Hartung, J., Jeong, K. S., Kwong, H. L., Morikawa, K., Wang, Z. M., Xu, D., and Zhang, X. L., *J. Org. Chem.*, **57**, 2768-2771 (1992).
2. Tietze, L. F., and Saling, P., *Syn. Lett.*, 281-282 (1992).
3. Myashita, Karino, H., Shinamura, J., Chiba, T., Nagano, K., Nohira, H., and Takaya, H., *Chem. Lett.*, 1849-1852 (1989).
4. For a review see: Halterman, R. L., *Chem. Rev.*, **92**, 965-994 (1992).
5. Willoughby, C. A., and Buchwald, S. L., *J. Am. Chem. Soc.*, **114**, 7562-7564 (1992).
6. Bradshaw, C. W., Hummel, W., and Wong, C. H., *J. Org. Chem.*, **57**, 1532-1536 (1992).
7. Hill, T. G., Wang, P., Huston, M. E., Wanchow, C. A., Oehler, L. M., Smith, M. B., Bednarski, M. D., and Callstrom, M. R., *Tetrahedron Lett.*, **32**, 6823-6826 (1991).
8. a) Brady, L., Brzozowski, A. M., Derewenda, Z. S., Dodson, E., Dodson, G. G., Tolley, S., Turkenburg, J. P., Christiansen, L., Huge-Jensen, B., Norskov, L., Thim, L., and Menge, U., *Nature*, **343**, 767-770 (1990). b) Brzozowski, A. M., Derewenda, U., Derewenda, Z. S., Dodson, G. G., Lawson, D. M., Turkenburg, J. P., Bjorkling, F., Huge-Jensen, B., Patkar, S. A., and Thim, L., *Nature*, **351**, 491-494 (1991).
9. Martinez, C., De Geus, P., Lauwereys, M., Matthijssens, G., and Cambillau, *Nature*, **356** 615-618 (1992).
10. Zhong, Z., Liu, J. L. C., Dinterman, L. M., Finkelman, M. A. J., Mueller, W. T., Rollence, M. L., Whitlow, M., and Wong, C. H., *J. Am. Chem. Soc.*, **113**, 683-684 (1991).
11. Hilvert, D., *Pure Appl. Chem.*, **64**, 1103-1108 (1992).
12. Acs, M., Szili, T., and Fogassy, E., *Tetrahedron Lett.*, **32**, 7325-7328 (1991).
13. Wulff, G., in *Bioorganic Chemistry in Healthcare and Technology*, Pandit, U. K. and Alderweireld, F. C. (Eds.), Plenum Press, New York, 1991.
14. Stinson, S. C., *Chemical & Engineering News*, 28 Sept, 1992, pp. 46-79.

Glossary of Terms

Absolute rotation. The specific rotation of an optically pure substance.

Achiral. A molecule is achiral when it is superimposable on its mirror image (i.e., when it exhibits reflectional symmetry).

Antibody. A protein produced by the immune system of an organism in response to a foreign substrate (antigen) and characterized by its specific binding to that substance.

Asymmetric. Lacking all elements of symmetry.

Asymmetric autocatalysis. An asymmetric synthesis in which the enantiomeric product catalyzes its own formation.

Asymmetric center. See **Stereogenic center**.

Asymmetric induction. (a). Preferential formation of one enantiomer from a prochiral substrate induced by, for example, a chiral catalyst. (b). Preferential formation of one diastereomer by the creation of a new stereogenic center in a chiral molecule.

Asymmetric synthesis. There are two types of asymmetric synthesis. (a). **Enantioselective syntheses** involve the preferential formation of one enantiomer from a prochiral substrate. (b). **Diastereoselective syntheses** involve the selective formation of one diastereomer by creating a new stereogenic center in a chiral molecule (see **Asymmetric induction**).

Asymmetric transformation. Selective conversion of a racemate to one enantiomer or a diastereomeric mixture to one diastereomer induced by crystallization

393

of one isomer in conjunction with in situ racemization or epimerization of the other isomer.

Atropisomerism. Chirality resulting from restricted rotation around a single bond (i.e., axial dissymmetry).

Auxotrophy. The inability of an organism to synthesize a particular organic compound required for its growth.

Biocatalyst. An enzyme in cell-free or whole-cell form that catalyzes metabolic reactions in living organisms and/or substrate conversions in various chemical reactions.

Chiral. An object or molecule is chiral when it is nonsuperimposable on its mirror image, that is, when it lacks reflectional symmetry. Chiral is not synonymous with optically active or enantiomerically pure.

Chiral amplification. An asymmetric synthesis in which the product has a higher enantiomeric purity than the chiral catalyst mediating its formation. It may involve **asymmetric autocatalysis.**

Chiral auxiliary. An optically active molecule that is used in stoichiometric amounts to orchestrate asymmetric induction at a newly formed stereogenic center without being incorporated in the product.

Chiral symmetry breaking. This occurs when one enantiomer catalyzes its own production while inhibiting the formation of its mirror image (see **Asymmetric autocatalysis**).

Chromosome. A selfreplicating structure consisting of DNA complexed with various proteins and involved in the storage and transmission of genetic information.

Clone. A population of genetically identical cells produced from a common ancestor. Sometimes used to denote a number of recombinant DNA molecules carrying the same inserted sequence.

Coenzymes. See **Cofactors.**

Cofactors. Low-molecular weight organic compounds or metal ions that are required by an enzyme for its activity. They may act as cosubstrates (e.g., NADH, FADH, ascorbic acid, and ATP) or as cocatalyst (e.g., biotin, pyridoxal phosphate); in the latter case they are sometimes referred to as **coenzymes.** The coenzyme may be loosely or tightly bound to the enzyme. When it is tightly bound it is generally referred to as a **prosthetic group.** A cofactor (coenzyme) binds with its associated protein (**apoenzyme**), which is functionally inactive, to form the active enzyme (**holoenzyme**).

Configuration. The actual three dimensional spatial arrangement of the atoms in a molecule (i.e., of the substituents around a stereogenic center).

Conformation. One of numerous spatial arrangements of a molecule achieved by rotation about single bonds.

d or *l*. Dextrorotatory or levorotatory according to the experimentally determined rotation of the plane of monochromatic polarized light to the right or left. Equivalent to *(+)* and *(−)*, respectively.

D or L. Absolute configuration of a molecule according to the Fischer convention (i.e., according to experimental correlation with that of D- or L-glyceraldehyde as a reference).

Diastereomeric excess *(de)*. Mole fraction denoting the ratio of two diastereomers in a mixture; analogous to enantiomeric excess for enantiomers.

Diastereomers. Stereoisomers that are not mirror-images of each other (i.e., that are not enantiomers).

Diastereoselective synthesis. Preferential formation of one diastereomer as a result of the creation of a new stereogenic center in a chiral molecule. It may involve differentiation of **diastereotopic faces** or **diastereotopic groups** by a reagent.

Diastereospecific synthesis. Preferential formation of one diastereomer by selective conversion of a mixture of diastereomers. The equivalent of kinetic resolution in racemates.

Diastereotopic faces. If the approach of a reagent from either one or the other face of a molecule leads to a pair of diastereomers then the faces are said to be diastereotopic.

Diastereotopic groups. If replacement of one of two equivalent groups at a particular atom (e.g., $C_{XYZZ} \rightarrow C_{XYZA}$) leads to a pair of diastereomers the groups are said to be diastereotopic.

Dissymmetric. Lacking an alternating axis of symmetry and hence usually existing as enantiomers.

Distomer. The optical isomer(s) of a chiral biologically active substance not (predominantly) responsible for the desired biological effect.

Dynamic kinetic resolution. Kinetic resolution of a racemate simultaneous with in situ equilibration of the two enantiomers.

Enantiomeric excess *(ee)*. Mole fraction denoting the ratio of enantiomers in a mixture:

$$\% \, ee = \frac{[R] - [S]}{[R] + [S]} \times 100$$

where [R] and [S] are the concentrations of the *(R)*- and *(S)*-enantiomers.

Enantiomeric ratio (E). (a). The ratio of the rate constants for reaction of a pair of enantiomers with an optically active reagent or catalyst (e.g., $E_R = k_R/k_S$). E is a measure of the efficiency of a kinetic resolution and varies with the extent of conversion. (b). The ratio of the two enantiomers formed in an enantioselective synthesis (e.g., *S/R*). In principle, it is independent of conversion.

Enantiomers. Stereoisomers that are not superimposable on their mirror images.

Enantioselective synthesis. Preferential formation of one enantiomer from a prochiral substrate via differentiation of **enantiotopic faces** or **enantiotopic groups** (i.e., an asymmetric synthesis).

Enantiospecific synthesis. Preferential formation of one enantiomer by reaction of an optically active reagent or catalyst with a mixture of enantiomers (i.e., a kinetic resolution of a racemate). For example, an L-specific enzyme reacts preferentially with the L-enantiomer.

Enantiotopic faces. If the approach of an optically active reagent or catalyst from either one or the other face of a molecule leads to a pair of enantiomers than the faces are said to be enantiotopic. A prochiral olefin, for example, contains enantiotopic faces.

Enantiotopic groups. If replacement of one of two equivalent groups at a particular atom (e.g., C_{XYZZ} -> C_{XYZA}) leads to a pair of enantiomers the groups are said to be enantiotopic.

Enzyme. A protein that functions as a catalyst for one or more reactions.

Eudismic ratio (ER). The ratio of biological activities of the eutomer and the distomer of a particular bioactive substance.

Eutomer. That optical isomer of a biologically active substance that is predominantly responsible for the desired biological effect.

Extremophiles. Organisms which require extreme physiochemical conditions for their optimum growth. For example, **thermophiles** or **psychrophiles, halophiles, acidophiles** or **alkalophiles,** and **barophiles** function at extremes of temperature, salt concentration, pH, and pressure, respectively.

Fermentation. Growth of microorganisms (e.g., bacteria, yeasts) in the presence of a carbon (energy) source and various nutrients. **De novo fermentation** refers to the formation of organic substances (e.g., L-amino acids) by multistep conversion of a carbohydrate feedstock such as glucose. **Precursor fermentation** denotes a single chemical step effected by subjecting a substrate to growing microorganisms (e.g., a fermenting yeast reduction of a ketone to an alcohol).

Gene. An ordered sequence of nucleotide bases that encodes for a particular protein.

Genetic manipulation or engineering. The use of in vitro techniques to produce DNA molecules containing novel combinations of **genes** or altered sequences and the insertion of these into **vectors** that are subsequently incorporated into **host organisms** or **cells** in which they are capable of continued propagation of the modified genes.

Homochiral. Refers to objects or molecules of the same "handedness" or configuration (e.g., left hands or a series of L-amino acids or D-sugars).

Hydrolases. Enzymes that catalyze hydrolytic processes.

Isoenzymes. A group of related enzymes within an organism that catalyze the same reaction but possess different amino acid sequences.

Isomerases. Enzymes that catalyze intramolecular rearrangements (e.g., racemases and epimerases).

Kinetic resolution. A process whereby one of a pair of enantiomers reacts preferentially with an optically active reagent or catalyst (e.g., an enzyme).

Ligases. Enzymes that catalyze the ligation (joining together) of two molecules, with concomitant hydrolysis of the pyrophosphate bond in ATP, to form C-C, C-O, C-S, P-O, or C-N bonds.

Lyases. Enzymes that catalyze the cleavage of C-C, C-O, C-N, and other bonds by reactions other than hydrolysis or oxidation. Lyases catalyze the addition of H_2O, NH_3, etc., to double bonds and the reverse reaction.

***Meso* compounds.** Compounds possessing two or more stereogenic centers but also exhibiting a plane of symmetry. *Meso* compounds are achiral and do not exist as enantiomers (e.g., *meso*-tartaric acid).

Metabolism. The entire physical and chemical processes involved in the maintenance and reproduction of life in which nutrients are broken down to generate energy and simpler molecules (**catabolism**) which may be subsequently converted to more complex molecules (**anabolism**).

Metabolite. Any intermediate or product resulting from metabolism. **Primary metabolites** are the primary building blocks (e.g., amino acids) formed from metabolism of carbohydrates and nutrients. **Secondary metabolites** (e.g., vitamins and hormones) are formed via further biosynthetic reactions of the primary metabolites.

Microorganisms (microbes). Microscopic living entities. Four classes of microorganism are of importance for fermentation: actinomycetes (filamentous bacteria), bacteria, molds, and yeasts. **Prokaryotes** are unicellular microorganisms lacking a membrane-bound cell nucleus and **eukaryotes** have cells containing their genetic material in a membrane-enveloped nucleus.

Mutagenesis. The introduction of permanent inheritable changes (i.e., mutations) into the DNA of an organism. When this occurs specifically at one site encoding for a single amino acid in a protein this is referred to as **site-directed mutagenesis.**

Mutation. An inheritable change in the nucleotide sequence of genomic DNA which may be induced by chemical mutagens or radiation or by the use of in vitro recombinant DNA techniques.

Optical activity. The unique property exhibited by enantiomers of rotating the plane of polarized light.

Optical purity. The ratio of the measured specific optical rotation of a substance as compared with that of the pure enantiomer under the same conditions. Not a reliable measure of enantiomeric purity. **Enantiomeric excess** is preferred to denote the enantiomeric purity of a substance.

Optical isomer. Synonym for enantiomer but the latter is now favored.

(specific) Optical rotation. The experimentally measured optical rotation of a substance for specified wavelength, temperature, solvent and concentration (e.g., $[\alpha]_D^{25}$ refers to the sodium D line and 25°C).

Oxidoreductases. Enzymes that catalyze oxidations and reductions, classified into several groups according to their respective electron or oxygen donors and acceptors.

Plasmids. An extra chromosomal genetic element consisting generally of a circular duplex of DNA which can replicate independently of chromosomal DNA. Plasmids are used as **vectors** for cloning DNA in bacteria or yeast host cells.

Prochiral. A molecule is said to be prochiral when addition to a double bond or replacement of one of two equivalent groups at a particular atom leads to the creation of a new stereogenic center in the molecule (e.g., a prochiral olefin or diol).

Prosthetic group. The nonamino acid portion of a protein (enzyme) (e.g., the heme group in heme-containing oxidoreductases). See **Cofactor**.

Proteases. Enzymes that catalyze the hydrolysis of proteins.

Protein engineering. A technique used to produce proteins (enzymes) with altered and/or novel amino acid sequences. This could involve genetic engineering of DNA sequences using in vitro recombinant DNA techniques or posttranslational chemical modifications of proteins or de novo polypeptide synthesis.

Racemate or racemic mixture. A mixture composed of equal amounts of a pair of enantiomers; also denoted as *dl*, *(±)* or *R/S*.

Racemization. The process whereby an enantiomer is transformed into a racemic mixture.

***R* or *S*.** Absolute configuration of a molecule about a particular stereogenic center. Assigned by application of the sequence rule otherwise known as the Cahn-Ingold-Prelog rules. The *R* and *S* refer to whether a circle passes through certain atoms surrounding the stereogenic center clockwise to the right (*R*, *rectus*) or counterclockwise to the left (*S*, *sinister*).

Receptor. A molecule, generally a protein, in or on a cell which specifically recognizes and binds to a substance acting as a molecular messenger (e.g., a hormone, neurotransmitter, drug, fragrance).

Stereodifferentiation. The differentiation by reagents of: *(i)* stereoheterotopic features in a molecule (e.g., **enantiotopic** and **diastereotopic groups** and **faces**; asymmetric synthesis), or *(ii)* different enantiomers or diastereomers (kinetic resolution). In both cases one speaks of enantiodifferentiation or diastereodifferentiation. Stereodifferentiation is synonymous with chiral recognition, chiral discrimination, etc.

Stereogenic center. The structural feature(s) in a molecule responsible for its chirality (e.g., a stereogenic carbon atom is connected to four nonequivalent groups). It is synonymous with (and preferred to) asymmetric center. Chiral center is not correct in that chirality is a property of the molecule as a whole.

Stereoheterotopic groups. Groups that occupy stereochemically nonequivalent positions in space. Groups that produce enantiomers by substitution are said to be **enantiotopic** and those that produce diastereomers are said to be **diastereotopic**.

Stereoisomers. Isomers that differ **only** in the way that the atoms are oriented in three dimensional space.

Sticky ends. The staggered ends of complementary sequences of DNA resulting from cleavage by restriction enzymes.

Strain. A genetically homogeneous population of organisms of common origin at a subspecies level that can be differentiated from other populations by morphological, physiological, biochemical or taxonomic features.

Topicity. The name given to spatial relationships within molecules. The hydrogen atoms in CH_3X are indistinguishable in a chiral environment and are said to be **homotopic**. In CH_2XY, on the other hand, the hydrogen atoms occupy stereochemically nonequivalent positions in space and are **stereoheterotopic** (in fact, **enantiotopic**).

Transferases. Enzymes that catalyze reactions in which a group is transferred from one compound to another. Examples of groups that are transferred include chloride, acyl, glycosyl, alkyl, etc.

Index

401